SOIL EROSION
AND
CROP PRODUCTIVITY

SOIL EROSION
AND CROP PRODUCTIVITY

Editors

R. F. Follett and B. A. Stewart

Consulting Editor

Iris Y. Ballew

Managing Editor
Domenic A. Fuccillo

Editor-in-Chief ASA Publications
Dwayne R. Buxton

**American Society of Agronomy, Inc., Crop Science Society of America, Inc.
Soil Science Society of America, Inc., Publishers
Madison, Wisconsin, USA
1985**

American Society of Agronomy, Inc.
Crop Science Society of America, Inc.
Soil Science Society of America, Inc.
677 South Segoe Road, Madison, Wisconsin 53711 USA

Library of Congress Cataloging in Publication Data

Soil Erosion and Crop Productivity

Bibliography
Includes index.
Contents: Soil erosion, Crop production, Conservation, Natural resources

ISBN 0-89118-087-7

Printed in the United States of America

CONTENTS

v

5 Assessments of Soil Erosion and Crop Productivity with Process Models (EPIC)

J. R. WILLIAMS AND K. G. RENARD

6 Assessment of Soil Erosion and Crop Productivity with Economic Models

PAUL T. DYKE AND EARL O. HEADY

7 Economic and Social Perspectives on T Values Relative to Soil Erosion and Crop Productivity

PETER J. NOWAK, JOHN TIMMONS, JOHN CARLSON, AND RANDY MILES

8 Assessment: A Farmer's Perspective

9 Processes of Soil Erosion by Water

G. R. FOSTER, R. A. YOUNG, M. J. M. RÖMKENS,
AND C. A. ONSTAD

10 Wind Erosion: Processes and Prediction

LEON LYLES, G. W. COLE, AND L. J. HAGEN

11 Criteria for Determining Tolerable Erosion Rates

G. F. HALL, T. J. LOGAN, AND K. K. YOUNG

12 Effects of Soil Erosion on Soil Properties as Related to Crop Productivity and Classification

W. E. LARSON, T. E. FENTON, E. L. SKIDMORE, AND
C. M. BENBROOK

13 Experimental Approaches for Quantifying the Effect of Soil Erosion on Productivity

L. D. MEYER, A. BAUER, AND R. D. HEIL

14 Regional Effects of Soil Erosion on Crop Productivity—Northeast

W. SHAW REID

15 Effects of Soil Erosion on Crop Productivity of Southern Soils

G. W. LANGDALE, H. P. DENTON, A. W. WHITE, JR.,
J. W. GILLIAM, AND W. W. FRYE

22 Simulation of Tillage Residue and Nitrogen Management

J. A. E. MOLINA, M. J. SHAFFER, R. H. DOWDY, AND
J. F. POWER

23 Structures and Methods for Controlling Water Erosion

J. M. LAFLEN, R. E. HIGHFILL, M. AMEMIYA, AND
C. K. MUTCHLER

24 Methods for Controlling Wind Erosion

D. W. FRYREAR AND E. L. SKIDMORE

28 Concerns and Policy Directions of the U.S. Department of Agriculture

FOREWORD

The value of soil is rarely appreciated because of its seemingly universal abundance. Except where covered with buildings and roads, or in rocky places in spectacular parks, the entire land surface appears to be covered with soil. Only a small fraction of soils, however, are suited for cultivation. It is this small fraction upon which an ever-expanding civilization must depend most for food and fiber. Preservation and wise stewardship of our soil resouces must be among our highest priorities.

Soil is a product of physical and chemical weathering: rocks are ground fine by glaciation or stream action, decompose chemically, degrade biologically, and form soil. Some of the soil may be carried by winds and deposited as loess, or carried by streams and deposited along their course or in lakes and seas to form deltas. The deltas, themselves a product of erosion, later become rich farmland. These same erosional processes carry away good soil from farmland that is producing crops. However, the time frames of soil formation and soil destruction differ. The first takes place over centuries and millennia, while the second takes place in days, months, and years.

Erosion caused by violent flooding, inundating rains, or high winds can devastate a productive farm in a short time. Erosion, at a seemingly low but continuous rate, can destroy a productive field in a few generations. However with modern agronomic science crop production per unit area has increased in the USA, even in the face of continued erosion. As the authors discuss in this volume, modern agronomic practices increase while erosion decreases production. These relationships must be clearly understood if we are to manage our soil resources optimally.

Topics discussed result from extensive research by qualified agricultural scientists, economists, sociologists, and historians. The three societies invited them to present their results at a symposium on the impacts of soil erosion on crop productivity. The book is an outgrowth of that symposium and recommended to those who must advise on and develop private practices and public policy dealing with soil conservation, land use, and crop production.

Officers of the three associated societies, over a period of several years, have been particularly interested and active in the creation of this symposium and its publication. The current officers of these societies, on behalf of the membership, extend thanks to these people and, particularly to the authors, reviewers, and the organizing and editorial committees.

William E. Larson	Robert F Barnes	Edward C. A. Runge
American Society of Agronomy	*Crop Science Society of America*	*Soil Science Society of America*

PREFACE

Soil erosion is a relentless, universal geologic process, usually difficult to control within acceptable limits and easily accelerated by man. Virtually every country in the world has experienced a period of severe man-made erosion. The increased demand for food due to population increases is causing marked acceleration of erosion in many Third World countries. Accelerated erosion is usually the result of improper management of productive soils or exploitation of marginal lands. The cost of erosion, in terms of yield reductions, is difficult to determine. Dregne (1978) has estimated, however, that the loss of 2.5 cm (1 inch) of topsoil is enough to reduce U.S. wheat yields by 4 032 000 t/year (60 000 000 bushels/year). Many wheat lands in the USA have, no doubt, lost more than 2.5 cm of topsoil. Soil erosion reduces crop production principally by decreasing nutrient supplies, water infiltration, and soil water-holding capacity. Progressive soil erosion may result in poor soil tilth, poor soil aeration, and restricted plant rooting depth (Langdale and Schrader, 1982; Larson et al., 1983).

Serious technical and socioeconomic questions concerning the impacts of soil erosion on crop productivity must be addressed. Giltmier (1982) recently gave the example that the U.S. Congress was told during the 1940s and 1950s that Iowa would become less productive agriculturally because its soil was flowing down the Mississippi River. But Iowa is now producing more corn (*Zea mays* L.) than ever before. Arguments can and are being made that soil erosion is only a minor problem, limited to a few soils.

On the other side of the issue, a recent conference of more than 100 scientists and leaders identified "sustaining soil productivity" as one of the six soil and water resources research priorities; soil erosion was identified as a principal cause for loss of soil productivity (Larson et al., 1981). The above types of controversy prevent the public from judging the issues based upon a solid understanding of the facts. This problem, in turn, creates difficulty in searching for the right political solution at the very time when a much broader political commitment, based on solid public support for the conservation of this nation's soil resources, is most needed.

Exposure of barren rock is an extreme, but not uncommon, example of the impact of soil erosion on crop productivity; the results are so obvious as to require no documentation. On deep, medium-textured soils, crop productivity losses from soil erosion may be corrected simply by adding N and P fertilizers. Most adverse effects lie between these extremes, however, and despite the difficulties involved, an accurate appraisal must be made of the long-term effects of soil erosion on crop productivity for major soil and crop conditions in the USA. Additionally, the declining productivity of existing eroded cropland must be evaluated in terms of a decreasing agricultural land base as more and more agricultural lands are converted to other uses.

Sampson (1982), in analyzing the politics of conservation, concluded that the main challenge is not the politics of partisanship, but the politics of professionalism. He further states that "the questions that are keeping a greater public commitment from emerging are the questions created by professional infighting . . . masked as professional analysis." The time is here for professionals across all appropriate disciplines to bring factual data together to provide answers allowing and promoting an understanding on the part of the public. Professionals also must identify gaps in knowledge and begin research to answer the most relevant questions. Therefore, the objectives of this conference on "Soil Erosion and Crop Productivity" were to begin to define:

1) The physical extent of the problem whereby soil erosion is decreasing crop productivity on agricultural lands of the USA;

2) Philosophical, socioeconomic, and institutional causes of excessive or accelerated erosion in the USA;

3) "State-of-the-art" regarding methodology of measuring the impact of soil erosion on crop productivity;

4) Current technological and institutional advances for erosion control and productivity maintenance by physiographic regions of the USA; and

5) A strategy for influencing policy and institutional decisions regarding the impacts of soil erosion on crop production to protect the long-range interests of the public in land productivity.

REFERENCES

Dregne, H. E. 1978. The effect of desertification on crop production in semi-arid regions. p. 113–127. *In* Glen H. Cannel (ed.) Proceedings of an international symposium on rainfed agriculture in semi-arid regions. University of California, Riverside.

Giltmier, J. W. 1982. What priority conservation? J. Soil Water Conserv. 37:250–251.

Langdale, G. W., and W. D. Shrader. 1982. Soil erosion effects on soil productivity of cultivated croplands. p. 41–51. *In* B. L. Schmidt et al. (ed.) Determinants of soil loss tolerance. Spec. Pub. 45. American Society of Agronomy, Madison, WI.

Larson, W. E., F. J. Pierce, and R. H. Dowdy. 1983. The threat of soil erosion to long-term crop production. Science 219:458–465.

----, L. M. Walsh, B. A. Stewart, and D. H. Boelter (eds.). 1981. Soil and water resources: research priorities for the nation. Soil Science Society of America, Madison, WI.

Sampson, R. N. 1982. Building a political commitment to conservation. J. Soil Water Conserv. 37:252–254.

R. F. Follett
Fort Collins, Colorado

B. A. Stewart
Bushland, Texas

CONTRIBUTORS

R. R. Allmaras
Soil Scientist, Agricultural Research Service, U.S. Department of Agriculture, University of Minnesota, St. Paul, Minnesota

M. Amemiya
Professor of Agronomy and Extension Agronomist, Department of Agronomy, Iowa State University, Ames, Iowa

Richard W. Arnold
Director, Soil Survey Division, Soil Conservation Service, U.S. Department of Agriculture, Washington, D.C.

Iris Y. Ballew
Head of Publications Unit, Soil Conservation Service, U.S. Department of Agriculture, Washington, D.C.

Sandra S. Batie
Professor of Agricultural Economics, Department of Agricultural Economics, Virginia Polytechnic Institute and State University, Blacksburg, Virginia

Armand Bauer
Soil Scientist, Northern Great Plains Research Center, Agricultural Research Service, U.S. Department of Agriculture, Mandan, North Dakota

Charles M. Benbrook
Executive Director, Board on Agriculture, National Research Council, National Academy of Sciences, Washington, D.C.

O. L. Bennett
Laboratory Director, Appalachian Soil and Water Conservation Research Laboratory, Agricultural Research Service, U.S. Department of Agriculture, Beckley, West Virginia

Orville G. Bentley
Assistant Secretary for Science and Education, U.S. Department of Agriculture, Washington, D.C.

A. L. Black
Center Director, Northern Great Plains Research Center, Agricultural Research Service, U.S. Department of Agriculture, Mandan, North Dakota

George J. Buntley
Professor of Plant and Soil Science-Extension, Plant and Soil Science Department, University of Tennessee, Knoxville, Tennessee

Earl Burnett
Laboratory Director, Grassland, Soil and Water Research Laboratory, Agricultural Research Service, U.S. Department of Agriculture, Temple, Texas

John E. Carlson
Professor of Rural Sociology, Department of Agricultural Economics, University of Idaho, Moscow, Idaho

G. W. Cole
Agricultural Engineer, Agricultural Research Service, U.S. Department of Agriculture, Kansas State University, Manhattan, Kansas

Jerry R. Cox
Range Scientist, Arid Land Ecosystems Research Unit, Agricultural Research Service, U.S. Department of Agriculture, Tucson, Arizona

xix

Pierre Crosson	Senior Fellow, Resources for the Future, Washington, D.C.
H. P. Denton	Visiting Assistant Professor, Soil Science Department, North Carolina State University, Raleigh, North Carolina
R. H. Dowdy	Supervisory Soil Scientist, Agricultural Research Service, U.S. Department of Agriculture, University of Minnesota, St. Paul, Minnesota
Paul T. Dyke	Agricultural Economist, Grassland, Soil and Water Research Laboratory, Economic Research Service, U.S. Department of Agriculture, Temple, Texas
Bart Eleveld	Associate Professor, Department of Agricultural and Resource Economics, Oregon State University, Corvallis, Oregon
T. E. Fenton	Professor of Agronomy, Department of Agronomy, Iowa State University, Ames, Iowa
R. F. Follett	National Program Leader, Agricultural Research Service, U.S. Department of Agriculture, Fort Collins, Colorado
G. R. Foster	Hydraulic Engineer, National Soil Erosion Laboratory, Agricultural Research Service, U.S. Department of Agriculture, Purdue University, West Lafayette, Indiana
D. P. Franzmeier	Professor, Agronomy Department, Purdue University, West Lafayette, Indiana
Wilbur W. Frye	Professor of Agronomy, Department of Agronomy, University of Kentucky, Lexington, Kentucky
Donald W. Fryrear	Supervisory Agricultural Engineer, Cropping Systems Research Laboratory, Agricultural Research Service, U.S. Department of Agriculture, Big Springs, Texas
J. W. Gilliam	Professor and Head, Soil Science Department, North Carolina State University, Raleigh, North Carolina
L. J. Hagen	Agricultural Engineer, Agricultural Research Service, U.S. Department of Agriculture, Kansas State University, Manhattan, Kansas
G. F. Hall	Professor, Agronomy Department, Ohio State University, Columbus, Ohio
Earl O. Heady	Distinguished Professor, Center for Agriculture and Rural Development, Iowa State University, Ames, Iowa
R. D. Heil	Director, Colorado Agricultural Experiment Station, Colorado State University, Fort Collins, Colorado
R. E. Highfill	Agricultural Engineer, Soil Conservation Service, U.S. Department of Agriculture, retired. Cape Coral, Florida

Robert W. Jolly	Associate Professor, Department of Agricultural Economics, Iowa State University, Ames, Iowa
H. A. Krauss	Conservation Agronomist, Soil Conservation Service, U.S. Department of Agriculture, Spokane, Washington
J. M. Laflen	Supervisory Agricultural Engineer, Agricultural Research Service, U.S. Department of Agriculture, Iowa State University, Ames, Iowa
G. W. Langdale	Soil Scientist, Southern Piedmont Conservation Research Center, Agricultural Research Service, U.S. Department of Agriculture, Watkinsville, Georgia
W. E. Larson	Professor and Head, Department of Soil Science, University of Minnesota, St. Paul, Minnesota
Jerry S. Lee	Director, Resource Inventory Division, Soil Conservation Service, U.S. Department of Agriculture, Washington, D.C. Currently, Director, South National Technical Center, Fort Worth, Texas
Terry J. Logan	Professor of Agronomy, Agronomy Department, Ohio State University, Columbus, Ohio
Leon Lyles	Supervisory Agricultural Engineer, Agricultural Research Service, U.S. Department of Agriculture, Kansas State University, Manhattan, Kansas
J. V. Mannering	Professor of Agronomy, Agronomy Department, Purdue University, West Lafayette, Indiana
D. K. McCool	Agricultural Engineer, Agricultural Research Service, U.S. Department of Agriculture, Washington State University, Pullman, Washington
Donald E. McCormack	National Leader, Soil Technology, Soil Conservation Service, U.S. Department of Agriculture, Washington, D.C.
Ralph J. McCracken	Deputy Chief for Natural Resource Assessments, Soil Conservation Service, U.S. Department of Agriculture, Washington, D.C.
James M. McGrann	Associate Professor, Department of Agricultural Economics, Texas A&M University, College Station, Texas
Charles T. McLaughlin	Farmer and Past President, Iowa Association of Soil Conservation District Commissioners, Britt, Iowa
L. D. Meyer	Supervisory Agricultural Engineer, Sedimentation Laboratory, Agricultural Research Service, U.S. Department of Agriculture, Oxford, Mississippi
R. J. Miles	Assistant Professor, University of Tennessee, Currently, Assistant Professor, Department of Agronomy, University of Missouri, Columbia, Missouri
Fred P. Miller	Professor and Head, Department of Agronomy, University of Arkansas, Fayetteville, Arkansas

John A. Miranowski — Director, Natural Resource Economics Division, Economic Research Service, U.S. Department of Agriculture, Washington, D.C.

W. C. Moldenhauer — Supervisory Soil Scientist, National Soil Erosion Laboratory, Agricultural Research Service, U.S. Department of Agriculture, Purdue University, West Lafayette, Indiana

J. A. E. Molina — Professor of Soil Microbiology, Department of Soil Science, University of Minnesota, St. Paul, Minnesota

C. K. Mutchler — Hydraulic Engineer, Sedimentation Laboratory, Agricultural Research Service, U.S. Department of Agriculture, Oxford, Mississippi

Peter C. Myers — Chief, Soil Conservation Service and Assistant Secretary Designate for Natural Resources and Environment, U.S. Department of Agriculture, Washington, D.C.

L. Darrell Norton — Soil Scientist, National Soil Erosion Laboratory, Agricultural Research Service, U.S. Department of Agriculture, Purdue University, West Lafayette, Indiana

Peter J. Nowak — Associate Professor, Department of Sociology, Iowa State University, Ames, Iowa

C. A. Onstad — Laboratory Director, North Central Soil Conservation Research Laboratory Agricultural Research Service, U.S. Department of Agriculture, Morris, Minnesota

R. I. Papendick — Supervisory Soil Scientist, Agricultural Research Service, U.S. Department of Agriculture, Washington State University, Pullman, Washington

Donald F. Post — Professor of Soil Science, Department of Soils, Water, and Engineering, University of Arizona, Tucson, Arizona

J. F. Power — Supervisory Soil Scientist, Agricultural Research Service, U.S. Department of Agriculture, University of Nebraska, Lincoln, Nebraska

Daryll Raitt — Agricultural Economist, Economic Research Service, U.S. Department of Agriculture, University of Missouri, Department of Agricultural Economics, Columbia, Missouri

Wayne D. Rasmussen — Chief, Agricultural History Branch, Economic Research Service, U.S. Department of Agriculture, Washington, D.C.

W. Shaw Reid — Professor of Soil Science, Department of Agronomy, Cornell University, Ithaca, New York

Kenneth G. Renard — Center Director, Southwest Rangeland Watershed Research Center, Agricultural Research Service, U.S. Department of Agriculture, Tucson, Arizona

M. J. M. Römkens
Soil Scientist, Sedimentation Laboratory, Agricultural Research Service, U.S. Department of Agriculture, Oxford, Mississippi

David L. Schertz
National Conservation Tillage Specialist, Soil Conservation Service, U.S. Department of Agriculture, Washington, D.C.

M. J. Shaffer
Soil Scientist, Agricultural Research Service, U.S. Department of Agriculture, University of Minnesota, St. Paul, Minnesota

E. L. Skidmore
Soil Scientist, Agricultural Research Service, U.S. Department of Agriculture, Kansas State University, Manhattan, Kansas

B. A. Stewart
Laboratory Director, Conservation and Production Research Laboratory, Agricultural Research Service, U.S. Department of Agriculture, Bushland, Texas

John F. Timmons
Professor Emeritus of Economics and Charles F. Curtiss Distinguished Professor of Agriculture, Department of Economics, Iowa State University, Ames, Iowa

P. W. Unger
Supervisory Soil Scientist, Conservation and Production Research Laboratory, Agricultural Research Service, U.S. Department of Agriculture, Bushland, Texas

David J. Walker
Associate Professor of Agricultural Economics, Department of Agricultural Economics, University of Idaho, Moscow, Idaho

A. W. White, Jr.
Soil Scientist, Southern Piedmont Conservation Research Center, Agricultural Research Service, U.S. Department of Agriculture, Watkinsville, Georgia

D. E. Wilkins
Agricultural Engineer, Columbia Plateau Conservation Research Center, Agricultural Research Service, U.S. Department of Agriculture, Pendleton, Oregon

J. R. Williams
Hydraulic Engineer, Grassland Soil and Water Research Laboratory, Agricultural Research Service, U.S. Department of Agriculture, Temple, Texas

M. Gordon Wolman
Professor of Geography, Department of Geography and Environmental Engineering, Johns Hopkins University, Baltimore, Maryland

D. L. Young
Agricultural Economist, Department of Agricultural Economics, Washington State University, Pullman, Washington

Keith K. Young
Soil Scientist, Soil Conservation Service, U.S. Department of Agriculture, Washington, D.C.

R. A. Young
Agricultural Engineer, North Central Soil Conservation Research Laboratory, Agricultural Research Service, U.S. Department of Agriculture, Morris, Minnesota

ORGANIZING COMMITTEE

Soil Erosion and Crop Productivity Symposium

Ronald F. Follett, Chair, National Program Leader, Agricultural Research Service, U.S. Department of Agriculture, Fort Collins, Colorado

R. R. Allmaras, Soil Scientist, Agricultural Research Service, U.S. Department of Agriculture, University of Minnesota, St. Paul, Minnesota

Min Amemiya, Professor, Department of Agronomy, Iowa State University, Ames, Iowa

Frank Bell, Professor, University of Tennessee, retired. Knoxville, Tennessee

D. C. Hanway, Professor, Department of Agronomy, University of Nebraska, Lincoln, Nebraska

Arnold King, Conservation Agronomist, Soil Conservation Service, U.S. Department of Agriculture, South National Technical Center, Fort Worth, Texas

E. L. Skidmore, Soil Scientist, Agricultural Research Service, U.S. Department of Agriculture, Kansas State University, Manhattan, Kansas

EDITORIAL COMMITTEE

Soil Erosion and Crop Productivity

Ronald F. Follett, Co-Chair, National Program Leader, Agricultural Research Service, U.S. Department of Agriculture, Fort Collins, Colorado

Bobby A. Stewart, Co-Chair, Laboratory Director, Conservation and Production Research Laboratory, Agricultural Research Service, U.S. Department of Agriculture, Bushland, Texas

Garren O. Benson, Professor, Department of Agronomy, Iowa State University, Ames, Iowa

A. R. Hidlebaugh, Soil Survey Research, Soil Conservation Service, U.S. Department of Agriculture, retired. Basye, Virginia

Brian L. McNeal, Professor and Head, Department of Soil Science, University of Florida, Gainesville, Florida

1 Soil Erosion and Crop Productivity: A Call for Action

Orville G. Bentley

U.S. Department of Agriculture
Washington, DC

This symposium addresses the topic of soil erosion and the array of complex issues involved in soil conservation and management. No challenge is more important than protecting our natural resources while maintaining our food and fiber production system. All sectors of U.S. agriculture, including those with special interest in research and education, have important roles in responding to this challenge.

Rexford Resler (1979) of the American Forestry Association, placed soil and water issues in perspective:

> It is easy to forget that our future and the stability of the free world depend largely on how respectful we are of the most common physical resources—soil and water—on the surface of the earth for the sustenance of all life.
> The time is right for significant changes—changes that can most readily occur when the attitudes of the people, the awareness of our leadership, and the laws of the land are aligned.

1-1 SOIL EROSION: NOT A NEW PROBLEM

Soils of the earth have been changing since the beginning of time. Natural changes in the earth's surface have brought devastation as well as the bounty of fertile topsoil on which our civilization depends for its food and fiber.

For centuries, farmers and naturalists have stressed the need to save the land. Each generation has decried the loss of precious topsoil and the ensuing decline in productivity. Nevertheless, despite this awareness, people have continued to abuse the soil which has led to poverty for some and has

Published in R. F. Follett and B. A. Stewart, ed. 1985. *Soil Erosion and Crop Productivity.*
© ASA-CSSA-SSSA, 677 South Segoe Road, Madison, WI 53711, USA.

forced migration by others. The ensuing competition for good land and water rights often destabilized relations between tribes and countries and created tensions and even wars.

In our own country, the westward migration in the 19th century was in part motivated by depletion of the fragile soils in the eastern USA after about 150 years of farming.

1-2 SOIL CONSERVATION EFFORTS SINCE THE 1930s

During this century, the concern over depleting our soils has received considerable attention. Through research, the problem has been defined and quantified. Agricultural leaders and conservationists have alerted the nation to the crisis. Measures must be taken now to stem soil losses by applying conservation practices to better manage our land and water resources. The problem continues to haunt the outlook for agriculture, despite our unparalleled scientific and technological advances over the last 50 years.

The evidence supporting this conclusion is well-documented. Repeated studies show that, while sheet and rill erosion varies from region to region, the national average annual soil loss is about 11.2 t/ha (5 tons/acre), measured against a goal of 4.5 t/ha (2 tons/acre). In the Corn Belt where row crop production is extensive, the estimated average annual loss is more than 17.9 t/ha (8 tons/acre). In the Great Plains, the cyclical losses from wind erosion are substantial. For example, in parts of Texas and New Mexico, losses to wind erosion of 22.4 t/ha (10 tons/acre) have been reported.

Additional cropland is lost each year to salinity and depletion of irrigation water. Groundwater "mining" is particularly serious in areas of the southern Great Plains, where more water is being withdrawn than is being replaced.

Another serious threat to cropland protection is suburban sprawl and other nonagricultural uses of farmland.

Each year some 1.21 million ha (3 million acres) of rural land are shifted to nonagricultural uses. One-third of this land—0.40 million ha (one million acres)—is prime agricultural land. Some of this land is returned to agriculture when prices are high, as in the 1970s.

Our rangelands also lose soil through over-grazing, poor management, and erosion—conditions that are exacerbated by drought, abnormally heavy rainfall, or lack of snow cover.

Older civilizations have had similar problems. Devastating erosion has occurred in India, East Africa, China, and the Middle East; salinity is not a new phenomenon, and polluted rivers are as old as civilization.

Likewise, many marvelously preserved lands have been under cultivation for thousands of years. People can find a way to live in harmony with the environment and thus, to a degree, can control their destiny. Against the litany of concerns about soil losses in the 1980s and beyond, we should recall that much has been done since the USA was shocked into action 50 years ago by the dust storms in the Great Plains.

During the height of the drought in the 1930s, we began to focus national attention upon the devastating effect of water erosion on the fragile soils of the eastern and southern USA and upon the plight of families trying to make a living on those soils.

Hugh Bennett and others like him woke us to the importance of our soil resources. And, fortunately for us all, the 19th century laissez-faire attitude on land policy, based on moving to new frontiers for a new start, gave way to the conservation ethic and the realization that the supply of land was exhaustible.

In the past 30 to 40 years, we have also realized that the issues involved are enormously complex. These issues touch nearly every facet of our lives, both in a public sense and individually. Also involved are off-farm issues of water quality, sedimentation of waterways and reservoirs, and the quality of the environment. Not surprisingly, we have steadily raised issues of soil and water policy to a national level through laws and programs.

Numerous structural developments have had and are still having a significant impact on the management of our soil and water. The creation of the Soil Conservation Service in 1935 was one such benchmark.

Federal conservation programs over the past 45 years have cost $21.3 billion, adjusted for inflation. Farmers, states, and local groups have contributed $21.6 billion, also adjusted for inflation. Yet, after a commitment of $43 billion, the problem remains, and the need for additional effort grows each year.

A series of land retirements and extensive rehabilitation of forests and rangeland, funded in part by the Federal government, expressed the national sense of urgency.

The 1977 National Resources Inventory (NRI) established a data base to support a more systematic assessment of erosion in the USA and an analysis of its effects on productivity.

1-3 THE RCA APPRAISAL

The Soil and Water Resources Conservation Act of 1977 (RCA) calls for a national program to guide the USDA's soil and water conservation activities on private and nonfederal lands through 1987. Prepared in response to this legislation and with input acquired through public hearings, the RCA report was recently submitted to Congress. Based on an appraisal of existing resources, trends, and projected resource needs, it provides guidelines for current USDA conservation programs, while redirecting some funds and personnel. The report encourages state and local governments to assume additional responsibility for developing and implementing conservation programs, giving priority to protecting soil resources, conserving water used in agriculture, and reducing upstream flood damage.

In his memorandum accompanying the 1982 RCA report to Congress, President Reagan said:

> The natural resources on our rural lands are vital to the present and future welfare of the American people. The soil and water on these lands are basic to the production of food and fiber for domestic and world needs. Maintaining the productivity of these resources is essential to American agriculture and to the health of the nation's economy.

The RCA appraisal makes it clear that some changes in Federal soil and water conservation programs will be necessary. Simply increasing Federal contributions to soil and water conservation is not the answer to our resource problems, since the role of the Federal government is subject to limitations imposed by economic conditions and to the individual landowners' willingness to cooperate.

The recommended program has six key features:

1. *National Conservation Priorities.* For the first time, the program sets clear national priorities to guide Federal conservation efforts. The top priorities are reducing soil erosion, conserving water, and reducing upstream flood damages.
2. *Development and Promotion of Cost-effective Conservation Measures.* The program encourages the development and adoption of conservation measures that are most cost-effective in reducing erosion and solving other resource problems. Examples include conservation tillage and range-management systems.
3. *Targeting.* The program calls for targeting an increased share of USDA resources—people and dollars—to critical problem areas where the need for conservation is greatest. It also directs USDA research and education efforts toward solving those soil and water problems that impair agricultural productivity and cause permanent damage to basic resources. The effort will take no more than 25% of total conservation funds and will be phased in at a rate of 5% a year over 5 years.
4. *Matching Grants.* The program proposes matching grants to encourage local and state governments to participate more fully in planning and implementing conservation programs.
5. *Conservation Pilot Projects.* The program calls for undertaking pilot projects to test new conservation methods and incentives to help farmers and ranchers practice conservation at reasonable cost.
6. *Intergovernmental Cooperation.* The program aims for improved coordination among the various Federal, state, and local agencies charged with conservation responsibilities. It fosters closer cooperation and coordination within the USDA among the eight agencies with responsibilities for conservation programs.

When he commended Secretary Block for the Department's work in preparing the RCA appraisal and program, President Reagan noted that 83 000 people had commented on the draft, including members of Congress and governors of 37 states, Puerto Rico, and Guam. More than half the respondents were farmers or ranchers.

Implementing the RCA program will require expanded knowledge, coupled with well-conceived, long-term education. These efforts must involve both the public and private sectors, educational foundations, associations such as the Soil Conservation Districts, scientific societies, Soil Conservation Service, Forest Service, Cooperative Extension Service, State Agricultural Experiment Stations, Agricultural Research Service, and, above all, the people who live and work on the land.

1-4 NEEDS AND PRIORITIES FOR RESEARCH AND EDUCATION

We in the public sector, government and universities, are continually challenged to use our research and education resources more effectively. A major goal of this symposium is to evaluate current research needs as a means of improving our effectiveness.

For fiscal year 1984, the Agricultural Research Service (ARS) has committed $64.5 million, or 13.7% of its budget, for soil and water research. The state Agricultural Experiment Stations have committed $23.3 million, or 10.1% of their budgets, for similar research. The new ARS 6-year-plan projects increased emphasis on soil and water research. This research will be funded by redirecting programs and can be expanded if new money becomes available.

Considering the effort to make our country more "conservation minded," one wonders why, after 50 years, we aren't making more progress. And why must we rediscover so much that we already learned through painful and trying experiences? Many reasons are given for the slow progress in applying conservation practices. These reasons include lack of incentives, economic conditions, and low farm income. Since we must assume that the dilemma is the result of rational actions by well-informed people, what are the barriers to understanding and alleviating the problem?

Fortunately, current research is addressing some of these issues. Pierre Crosson (1983), for example, compared results of three research efforts that estimated the effect of erosion on crop yields. He found that the results were strikingly similar and concluded that:

> The Yield Soil Loss Simulation indicates that continued erosion at the 1977 level would reduce national average yields 50 years later by 8% from what they otherwise would be.
>
> The work by Larson and associates suggests a loss of 5 to 10% over 100 years.
>
> And my work at the Resources for the Future indicates that because of erosion, 1980 yields of corn and soybeans in major producing areas for those crops were about 2.5% less than they would have been, given the growth yields from 1950 to 1980.

Thus, our research and education work, including extension programs and the RCA process, suggests that technology and the improvement in agricultural productivity are not keeping pace with the demands now placed on the nation's soil and water resources.

On the basis of findings of the Economic Research Service, Mel Cotner concludes that:

> Alternative actions to reduce erosion and maintain long-term soil productivity can be grouped in three major categories:
> 1. Accelerate farmer-landowner adoption of conservation practices and soil conserving crop rotations;
> 2. Accelerate development of technology to increase crop yields to lessen the pressure to use fragile lands for intensive crop production; and
> 3. Reduce emphasis on agricultural exports that are being used to achieve favorable balance of trade with other nations.

The sharp contrast among these three strategies—cultural, technical, and market—point up the complexities that research and educational activities must consider.

Perhaps more than for any other issue facing agriculture, multidisciplinary cooperation focused on soil erosion and related water issues is required in research and education.

Cotner's (1982) following points further emphasize the challenge to research and education:

> If current erosion rates continue in the future, production costs would be about 5% higher than if erosion were controlled through greater use of conservation-tillage practices.
>
> If technology grows at the rate of 1.6% a year, in contrast to the 1% growth rate on these analyses, erosion could be reduced by one-third. Simultaneously, the cost of producing the same level of agricultural output would be reduced by about one-fourth.
>
> These changes result primarily from less land being needed for production and converting fragile, erosive lands to pasture or other extensive uses.

Let us also contemplate Vernon Ruttan's comments about technology in general:

> Over the last 50 years, U.S. agriculture has been transformed from a resource-based industry to a science-oriented industry. It has been transformed from a traditional to a high technology sector. There are relatively few sectors in the U.S. economy that have been able to maintain their technological leadership. Agriculture is one of those sectors.
>
> The future growth of the U.S. economy will depend very heavily on those sectors that are able to maintain their technological leadership. We are part of a world in which scientific and technological leadership in agriculture can no longer be ours by default.

The biological, technical, and economic research important to soil and water resources and related policies can make important contributions to this leadership.

Furthermore, the maintenance of technological leadership will place a special burden on each of the three major elements—Federal, state, and private—to recognize the responsibility to support policies that will maintain the strength of the other two.

The USA's agricultural system must answer some serious questions about the direction that agricultural sciences should take for the remainder of this century. Some of the questions concern planning, coordination, and priorities within the Federal/state system.

Agriculture's research and education system is structurally sound. The system links public and private initiatives in a common effort to provide a scientific and technological base for our food and fiber industries.

These linkages don't just happen. They need constant reinforcement. Similarly, we need to encourage cooperation among many scientific disciplines to achieve our goals in research and education. Through research, we can protect our natural resources and our environment for future generations.

As we consider questions about our nation's soil and water resources and crop productivity, we must develop a common vision and a unity of direction and understanding.

REFERENCES

Cotner, M. R. 1982. Summary, interpretations RCA model runs. Staff memo, 18 June. National Resource Economics Division, USDA Economics Research Service, Washington, DC.

----, and W. K. Easter. Soil Conservation policies, institutions, and incentives. p. 283–301. Soil Conserv. Soc. of Am., Ankeny, Iowa.

Crosson, Pierre. 1983. Impact of erosion on land productivity and water quality in the U.S. Paper presented on 16 June 1983 at Second Int. Conf. on Soil Erosion and Conser. Honolulu, HI.

Larson, W. E., F. J. Pierce, and R. H. Dowdy. 1983. The threat of soil erosion to long-term crop production. Science 219:458–465.

Resler, R. A. 1979. In paper prepared for the Soil Conserv. Soc. of Am.

Ruttan, V. W. 1983. B. Y. Morrison Memorial Lecture. HortScience 18:809.

Soil and Water Resources Conservation Act of 1977 (Public Law 95-192), "A National Program for Soil and Water Conservation," 1982, Final Program Report and Environmental Impact Statement.

2 Soil Erosion and Crop Productivity: A Worldwide Perspective

M. Gordon Wolman
The Johns Hopkins University
Baltimore, Maryland

A worldwide perspective on crop productivity and soil erosion is easily achieved if one keeps clearly in mind two different scales of space and time. On a world scale, virtually all estimates (Buringh et al., 1975; Dudal, 1982) suggest that the lands of the earth potentially can produce perhaps 10-fold or more food than they currently do, and hence theoretically can support much larger human populations. In contrast, large numbers of people, perhaps millions, go hungry in the world today. Many of these people live in areas of poor and often rapidly eroding soils. In a fully integrated free world, the potential food-producing capacity of the world as a whole would ensure that everyone would have ample food. Thus, on the large scale, world food prospects are encouraging. At smaller spatial scales, the reverse is true. Since the physically finite world is an essential element in analysis of resource issues, treating a large world as if it were a single place can be misleading and counterproductive.

Time introduces different factors and issues. Despite significant erosion on poor lands and on many good lands, current and recent effects of erosion on productivity have been masked by progressively increasing productivity on many of the world's lands as a result of improved technology and management. However, it is not self-evident that this compensatory process can be maintained indefinitely. Here, time and space are related. Soils, perhaps better called the rooting zone of plants, are rapidly disappearing in some places. Their future is short, barring the heroic use of technology. Good soils in good climates are the most important from the standpoint of production and productivity. In these regions the significance of erosion over a very long time is not self-evident. Questions arise even if

Published in R. F. Follett and B. A. Stewart, ed. 1985. *Soil Erosion and Crop Productivity.*
© ASA-CSSA-SSSA, 677 South Segoe Road, Madison, WI 53711, USA.

one acknowledges that erosion does reduce potential productivity over the long run. For example, are such potential losses important? If so, how important? And over what time scale? Who should pay the costs now to provide insurance for the future? Who should be paid? How much insurance is needed?

This paper provides some evidence to support these statements, as well as a perspective from which to evaluate elements of uncertainty related to "guesstimates" about the present and future significance of erosion. This evidence is, of necessity, brief and illustrative, not exhaustive. The paper deals successively with erosion, crop production and productivity, the relationship between soil erosion and productivity, and problems associated with developing estimates of the significance of erosion to potential world productivity. Such a broad frame appears essential to a world perspective on the relationship of soil erosion to crop productivity.

2-1 NATURAL AND ACCELERATED EROSION

Rates of natural erosion are controlled by geology, topography, vegetation, and climate. Climate acts in two opposite ways: as the potential erosive force through the temporal distribution and intensity of rainfall and runoff, and, through the vegetative cover, as an element resisting erosion. A number of equations have been developed to characterize both the "erosivity" of climate and the resistivity of the landscape to erosion (Wischmeier and Smith, 1978; Fournier, 1960). However, because erosion per se is difficult to measure over large areas, for large drainage basins and continents the sediment yield of rivers provides a surrogate measurement, albeit a very uncertain one. Sediment yield of rivers represents erosion minus deposition on land and in river channels and valleys. Soil formation is not uniform even over individual agricultural fields, and material eroded at one point is deposited at another. Neither the source nor the timing of erosion is discernible from sediment yield in large river systems. Nevertheless, some sense of the scale of denudation may be discernible from observations of yield. Assuming that the sediment is derived uniformly from the entire surface area, estimated rates of denudation of continental masses range from about 1.12 mm/yr for Australia (including Tasmania, New Zealand, and New Guinea) to 0.009 mm/yr for Africa. Spatial variations, however, are very large. For example, in the loess plateau of the Yellow River drainage, annual erosion is two to three times the mean of 1 mm for the basin as a whole. (The annual sediment discharge of 2×10^9 t represents about 10% of the world sediment yield.) Similar variations are seen in the Amazon basin, where a small fraction of the area in the Andes accounts for the bulk of the total sediment load (Gibbs, 1967). Highest erosion rates over large areas occur in the southern and southeastern subcontinents of Asia, and in Central Asia where depth of erosion exceeds 0.15 mm/yr. At the opposite extreme, large Siberian rivers indicate depths of erosion of about 0.002 to 0.006 m (L'vovich, 1979, p. 222-237).

Modern denudation measurements, of course, include the effects of both man and nature. Not shown by the current data is the fact that variations in climatic conditions, or changes in base level, may produce natural erosion rates greater than those currently experienced under modern agriculture. Thus, stratigraphic studies in the Midwest plains of the USA indicate that during some periods before settlement denudation was higher than after agricultural settlement (Ruhe and Daniels, 1965). (Similarly, man-induced erosion may initiate gullying in systems repetitively gullied earlier under natural conditions.)

Information on sediment yields around the world is relevant to the present discussion in several ways. First, the rates of erosion vary enormously throughout the globe under natural climatic and geologic conditions. Second, as a result, the relative significance of human agricultural activities is also highly variable and not always readily distinguished. Where erosion rates in semiarid regions are naturally high, variations due to human disturbances may fall within the range induced by natural climatic and other variations, but at the same time, the human impact, by destroying vegetation, may lower the threshold at which erosion will occur under a given climatic regime. Deforestation for cropping may induce erosion rates 5 to 100 times the natural background on small areas (Wolman, 1967).

By definition, accelerated erosion is a rate of erosion that exceeds the presumed rate under natural, sometimes called geologic, conditions. From very rough computations and assumptions, Golubev (1980) concluded that at the continental scale agriculture had increased erosion, actually measured as sediment yield, about fivefold. For the USA, Schumm and Harvey (1982) concluded that, because current rates of erosion are probably twice background values, soils over large areas should be considered a nonrenewable resource, inasmuch as soil formation rates are likely to be lower than erosion rates.

Much more vivid than these average figures are the well-known descriptions of devastating erosion observed in various parts of the world. Since Marsh's classic, *Man and Nature or Physical Geography as Modified by Human Action* (1864), which documented the relationship between land use and accelerated erosion in many parts of the world, successive observers from Jacks and Whyte (1939) to Eckholm (1976), have provided dramatic examples of land erosion from the Himalayas to the Amazon basin. Indeed, Marsh (1864, p. 44) stated that,

> The earth is fast becoming an unfit home for its noblest inhabitant, and another era of equal human crime and human improvidence, and of like duration with that which traces of that crime and that improvidence extend, would reduce it to such a condition of impoverished productiveness, of shattered surface, of climatic excess, as to threaten the depravation, barbarism and perhaps even extinction of the species.

These reports and Marsh's statement are relevant today since they reflect observations on selected regions and generalizations about the future in those regions and elsewhere.

Observations of erosion in many steep mountain lands clearly demonstrate that sheet and gully erosion can quickly remove thin soils to bedrock when natural cover is removed and no management is practiced to impede the loss. Similarly, overgrazing in the Great Plains or in the Sahel has been clearly linked to erosion by wind and water. Since the impetus given to concepts of soil conservation by H. H. Bennett in the USA in the 1930s, broad field observations of erosion have been complemented by thousands of plot studies demonstrating the effect of different vegetative covers, and their absence, on rates of erosion at specific sites (University of Hawaii, 1983). Erosion rates on plots are many times higher on bare soil or under row crops than under more continuous cover or on controlled slopes.

Worldwide, then, evidence demonstrates that man can, and has, accelerated erosion over geologic norms or background levels. One can conclude, as noted earlier, that in some places the soil is a nonrenewable resource. From the standpoint of crop productivity and erosion, however, a number of cautions are in order.

First, soils under agriculture cannot be considered solely as products of nature. Rather, in many instances the soil is an artifact of man (Henin et al., 1969). Second, because natural rates of erosion are so variable from region to region, man's impact is not always easy to evaluate. Third, within the usual context of erosion and soil formation, relatively little attention is given to recovery or renewal. Last, because much of the interest in soil erosion today relates to prospects for tomorrow, the variation of soils at both regional and smaller scales becomes extremely important. Thus, because of the heterogeneity of soil materials and the variability of rates of soil formation and erosion on hillslopes even at the agricultural field scale, large-scale extrapolation of erosion rates to determine the impact on soil productivity in the future is uncertain (Daniels, 1982). Moreover, much eroded soil is not removed but redistributed, and the effect of such changes on productivity is not self-evident.

The well-established fact that soil erosion has been accelerated by human use of the land in many areas does not in itself provide a measure of the impact of such erosion on agriculture or society. Additional information and associations are needed.

2-2 CROP PRODUCTIVITY

Productivity is used here in the physical sense as yield per unit area. For the world, food production has increased rapidly in the last three or four decades (Wortman and Cummings, 1978). At the country level, although some show marked declines, many poor and developing countries also showed increases in annual yields of cereals on the order of 1 to 4 or 5% between 1969–1971 and 1975–1977 (International Agriculture Development Service, 1980). Annual variations as well as declines in yields in the Sahel in Africa between 1963 and 1974 were clearly related to variations and progressively diminishing precipitation in much of this region (National Oceanic and Atmospheric Administration and University of Missouri-Columbia, 1979b). For the continents as a whole, from a base of 100 for 1961–65, only Africa showed a decline in food production per capita, to 88 in 1979, although gains were very limited in South Asia (103) and West Asia (108) (Holdgate et al., 1982, p. 259).

These changes in production and productivity do not represent changes at a given place. Instead, increased productivity in one region, derived from such management practices as the application of fertilizers and irrigation water, compensates for declines elsewhere. In other places, potential losses in productivity from erosion may be compensated by added inputs at a given site. Thus, in many ways, recognition of increases in overall productivity is roughly comparable to the evidence of accelerated erosion losses over large areas, coupled with documentation of evidence of severe erosion and striking increases in productivity in specific regions. Both kinds of evidence are relevant to an evaluation of the significance of worldwide soil erosion in relation to crop productivity, but neither provides a direct link between the two.

2-3 SOIL EROSION AND CROP PRODUCTIVITY

Despite considerable evidence relating soil loss from erosion to losses in productivity, the results are still fragmentary and not yet adequate for management decisions, particularly if these decisions relate to the economics of agriculture and to farmers' decisions about soil conservation practices. Moreover, most of the data are from experimental plots, and most are from temperate regions rather than the dry and wet tropics, which are areas of great global importance (Lal, 1983). The selected data used here indicate the range of the results and the approaches used to relate reduction in yields to erosion. These approaches include historical observations, crop-yield models, plot experiments, and extensions of soil erosion equations.

Perhaps the boldest claim to significance in the relationship between soil erosion and crop productivity is the frequently stated view that a number of civilizations, particularly in the Middle East and Mediterranean area, declined as a result of soil erosion (Hymans, 1952). Certainly, the loss of soil is well documented in many of these areas, but war, migration, social upheaval, and politics are inseparable from the temporal association of soil loss and mismanagement of the land (Adams, 1965; Wittfogel, 1957). Field evidence, as well as models relating losses in soil properties and plant growth, does suggest that erosion of shallow soils on limestone common in areas of the Middle East may have produced nearly irreparable losses in potential productivity, reversible only with enormous inputs of human energy. Pressures on the land, which required sophisticated management to alleviate, may well have compounded problems of government and society, but excessive human-induced and the inability to produce food, then as now, appear to have arisen from persistent neglect and failure to use knowledge and tools at hand.

Some estimates place the loss of fertility over large areas from erosion and other causes associated with continuous cultivation as high as 40% for portions of the USSR and 25% in the USA (Golubev, 1980, quoting several sources). "Fertility" implies nutrient losses, but it cannot be readily defined in an historic context without reference to many factors of production and productivity. Erosion results in loss of nutrients and changes in soil structure, which controls infiltration and retention of moisture. Both in turn influence productivity. Progressive loss of the rooting zone of the plants (100 to 150 cm) may be viewed as irreversible at some level of technology, whereas nutrients are replaceable. Clearly, the degree of reversibility is partly a function of technology and hence of available resources. Nevertheless, the distinction between reversible, or compensable, and irreversible losses to potential productivity from soil erosion is important.

The significance of irreversible losses is suggested by a recent study in Haiti. Reporting the results of the use of an analog corn-yield model in a study of repeated food shortages in Haiti, not matched by comparable shortages in adjacent Caribbean areas experiencing the same climatic conditions, Steyaert (1980, p. 11) noted that yield in the simulated period (1921 to 1960) for the "eroded" soil was 30% lower than for the "uneroded" soil. Moreover, "the variance of corn yield for the 'eroded' soil was more than *four* times as large as for the 'uneroded' soil." Higher yields from the eroded soil in some years may be an artifact. Even so, deep declines in poor years are striking, and the effect is to increase both uncertainty and risk, the precise concerns of the poorest farmers. The results are characterized as "pseudo-drought," a consequence of the loss of moisture-holding capacity associated with soil erosion (National Oceanic and Atmospheric Administration and University of Missouri-Columbia, 1979a), and an illustration of irreversibility as defined here.

Data from the USA suggest that potential wheat productivity is reduced from 2 to 10% with removal of 2.5 cm of topsoil by wind. For corn on deep, medium-textured soils, from which all topsoil was removed, the potential losses are 8 to 30% (Lyles, 1975; Langdale and Shrader, 1982). Attempting to summarize some of the experimental data from the USA, Sanders (1981) noted that erosion of 25 cm of some Iowa soils resulted in reductions of about 50% in maize yield, while a thin soil in Nigeria experienced similar yield reductions with loss of only 5 cm of soil. On Alfisols in western Nigeria, Lal (1976) reports declines of maize under plowing on experimental plots of 20 to 25% on varying slopes, and progressive declines of maize and cowpeas of as much as 50% as soil was removed artificially to depths of 12 or 13 cm. Declining yields were attributed to losses in organic matter, nitrogen, moisture retention, and infiltration. Similar experiments in western Australia showed possible yield reduction of 2 to 7.5% for wind erosion of 1 mm, and 10 to 25% for losses of 8 mm. Natural erosion of 5 mm and 14 mm of soil in the USSR is reported to have reduced winter wheat yields by 50% and 70%, respectively (Marsh, 1980), yield reductions which appear to be inordinately high. Because natural rains compact the soil surface, potential reductions in yields under natural, sequential erosion from rain may well exceed those where equivalent soil depths have been removed artificially.

As soil properties important to plant growth diminish with erosion, so too may the effectiveness of fertilizer applications. Thus, eroded sediments are not only enriched as the residual portion is impoverished, but the effectiveness, for example, of nitrogen applications on the residual may be one-third to one-fifth that on uneroded soils (Lal, 1983, Fig. 2).

Despite meager data, one can conclude that, on many soils in many regions of the world, erosion without conservation management is likely to reduce productivity. Rates of decline are obviously related to the characteristics of the soil profile in a given climatic zone. Yields cannot decline much where they are already low and the soil already degraded, although the percentage reduction may be great. Because of the enormous variety of soil profile characteristics, one cannot generalize a curve relating reduction in potential productivity at one site to progressive loss of soil over large areas. This variability, coupled with limited data, poses problems in extending site-specific information to quantify the impact of erosion on crop production worldwide.

2-4 EXTRAPOLATION IN SPACE AND TIME

Extrapolating information about soil erosion and loss in crop productivity to the world involves three parts: (i) knowledge of rates of erosion, (ii) knowledge of the relationship between erosion and crop reduction, and (iii) measures of the area covered by different rates of erosion and the accompanying reductions in productivity. Because rates of erosion are not well known for many areas, various equations have been developed to estimate soil erosion from hillslopes. The universal soil loss equation (USLE), developed in the USA, has been extensively applied, or modified, for use elsewhere (Wischmeier and Smith, 1978). Based on field plot experiments relating soil loss to climatic parameters, soil erodibility, gradient, hillslope length, vegetation cover, and management practices, the USLE and similar correlative equations provide a basis for estimating erosivity of climate and erosion under varying assumed plant covers. Extrapolation of erosion rates alone requires moving from plot, to field, to ever-larger spatial units, an uncertain process given the heterogeneity of the soil and landscape, the spatial variability of transport and deposition of sediment, and the temporal variability of climatic factors. Despite these pitfalls, the world soils map (FAO/ UNESCO, 1974) and maps of degradation, expressed in tons per hectare, begin to provide some rough guides to the spatial distribution of erosion over large areas. Riquier (1982), summarizing extensive work by the United Nation's Food and Agriculture Organization, describes a parametric method by which classes of degradation related to a variety of processes, including erosion, can be derived from estimates of the environmental factors controlling each process. Risks of degradation are calculated on the basis of relatively stable environmental factors, such as climate, topography, and soil. Then, the application of coefficients corresponding to present land use or natural vegetation gives the present soil degradation. Maps of soil degradation risks in Africa north of the equator and in the Middle East were published in 1979 (Food and Agriculture Organization, 1979). In general, degradation risks roughly coincide with degradation rates in tons per hectare over large areas on these small-scale maps. At the same time, processes responsible for risks of degradation, such as water erosion and leaching in tropical regions, or wind and salinization in arid zones, go together. Maps of North Africa and the Middle East show that large areas are at risk, although degradation rates mapped over large areas are calculated, not observed, values.

Using a wide variety of sources, Buringh (1981, also cited in Holdgate, 1982, p. 257) provides perhaps the most comprehensive, current estimate of expected changes in world land uses directly related to the relationship between cropland (cultivated land) and erosion. Average annual losses of agricultural land, consisting of cropland, grassland, and forest land, totaling in million ha (megahectares, Mha) in 1975, were estimated to be 8 MHa by conversion to nonagricultural use, 3 MHa by erosion, 2 MHa by desertification, and 2 Mha by toxification. Estimated cropland in 1975 of 1500 Mha included 400 Mha in the highest land class, 500 Mha in the medium classes, and 600 Mha in the lowest class. Assuming present rates of land-use change remain unaltered, by the year 2000 cropland amounts will be 345 Mha (highest class), 745 Mha (medium class), and 710 Mha (lowest class)—a total of 1800 Mha. These figures represent estimates of net change associated with losses from various causes and reclamation of other lands, hence the increase in cropland. Of particular interest here is an estimated loss of cropland by erosion of 40 Mha (8%) in the medium class and 10 Mha (1.7%) in the lowest class. Because highest-class cropland is often on flatter lands and less subject to erosion, it is assumed that no land in this class is lost to erosion. Total estimated land lost by erosion is 50 Mha for cropland, 15 Mha for low-quality grassland, and 10 Mha for medium-quality forest land, for a total of 75 Mha. Land conversion to nonagricultural use, land lost to desertification, and land lost by toxification account for larger areas of change. In some areas land lost to nonagricultural use, land surrounding Cairo for example, may be of high agricultural potential. Desertification poses more complex problems of definition in that the capacity for recovery is less well known, thereby making the definition of loss more uncertain.

A loss of 50 Mha represents 3.3% of the total estimated cropland in 1975, and a somewhat smaller percentage of the cropland in 2000 because of assumed reclamation. Roughly similar values are suggested for the equivalent of productive cropland in the USA in the same period (Larson et al., 1983, p. 464). Perceptions of the significance of a loss of over 3% in 20 years will vary. More dramatically, a loss of 50 Mha over 20 years represents a loss of about 5 ha per minute (Buringh, 1977, p. 483), or the equivalent of about one farm per minute in some parts of the developing world.

In a model estimating world food production based on labor-oriented agriculture, Buringh and van Heemst (1977) estimate the hypothetical potential production of food grains on each continent. Regions of the continents are classified into four production classes from the highest (an estimated grain yield of 2325 kg/ha, representing the mean yield plus one-half the standard deviation) through Class 4 lands that require irrigation (1700 kg/ha). Intermediate Class 2 (1575 kg/ha) and Class 3 (1025 kg/ha) represent regions in which soil conditions or water deficiencies influence crop yields. About 30 of the 151 regions in the world generate roughly 60% of estimated world production, because the bulk of estimated potential production throughout the world is from Class 1 (Buringh, 1977, p. 25). The two intermediate production classes account for quite different percentages of total estimated potential grain production on each continent: North America 6%, Asia 15%, Europe 20%, Australia 31%, South America 40%, Africa 68%, and the world 26%. Both Africa and South America derive a large part of total estimated grain production from Class 2 croplands. Only in Africa does Class 2 land contribute a greater percentage than Class 1. To the extent that losses of productivity from erosion are likely to be higher on poorer quality lands than on better lands, erosion may increase global differences in productivity potential. Because tropical lands subject to erosive rainfalls represent major reserves of agricultural land, there is likely to be a significant threat to future productivity, as much from land instability as from losses in nutrients.

Increasing food production by labor-intensive agriculture implies large increases in the land area under cultivation and hence greater potential for erosion and other degradation (Buringh and van Heemst, 1977). By definition, the higher the productivity per unit of land of a given quality, the less significant to total world food production is the contribution from poor quality lands. While modern agriculture on presently cultivated land produces somewhat over "three times the quantity of food grains as labor-oriented agriculture" (Buringh and van Heemst, 1977, p. 31), the potential of the former is many times higher. Buringh (1977) and Dudal (1982) make a convincing case for the necessity of intensifying productivity on high quality and potentially high quality lands, from the standpoint of both efficiency and reduction of potential degradation through erosion and other processes.

2-5 BETWEEN CALAMITY AND COMPLACENCY

Competing claims regarding the condition and prospect for any resource or part of the environment, whether soils, metals, air, or water, are often couched in terms of doomsday or cornucopia. At any given time, degradative trends in land use being experienced in one or more places, if extrapolated under plausible assumptions, could be shown to lead to both irreparable losses and to severely degraded states. The present is no exception.

In many places throughout the world, just such degradative changes are taking place. A loss of 5 ha/min, or half that, emphasizes the point. Such losses are stimulated primarily by population pressures, aided and abetted by poverty, ignorance, and indifference (Dudal, 1982). Assuming nothing is done to stem these processes, those who depend on such lands can be expected to continue to suffer severe deprivation. Locally, increasing population pressure in steep terrain, such as the foothills of the Himalayas or portions of Central America, indirectly induces rates of erosion for which no reasonable controls are possible. Those who depend for a living on these areas do not appear in the computations of total food production. They will starve, leave, or perhaps import food. Effective terrace systems have been constructed by simple means on the most inhospitable terrain in the Andes, Himalayas, and Mediterranean regions. Certainly local salvation at a small scale may result from these desirable efforts. Somewhere between this worst condition and the best condition is presumed the desired focus for both research on techniques and the application of what is known.

In the current debate it is important to recall that not all the land described by Marsh (1864) in the 19th century was permanently lost to human use, nor was each spot in Africa as described by Jacks and Whyte (1939). Similarly, much land designated irreparable in the USA in the 1930s has not proven to be so (Malin, 1947). Potential losses of productivity of croplands can be reduced and reversed, even in the tropics where current research rebuts the contention that degradation is inevitable under continuous cropping (Greenland and Lal, 1977; Sanchez et al., 1982). This is particularly important, since the tropics constitute a potential major source of food in an area of rapidly expanding agriculture.

Unfortunately, roughly 50 years of conservation science and demonstration have not convinced farmers throughout the world of the wisdom of adopting conservation practices, even where appropriate practices are known. (Parenthetically, more plot experiments which demonstrate that bare soil erodes more rapidly than soil covered by vegetation are unlikely to add much to the store of useful knowledge.) Increasingly, the evidence suggests that conservation farming does not pay, in the farmer's reasonable but limited time horizon. In this respect, Schultz's blunt admonition regarding agricultural development in general, made nearly 20 years ago, is still relevant: "[Agricultural] programs are unsuccessful primarily because no profitable rewarding new agricultural inputs have been available to farmers which they could adopt and use" (Schultz, 1965). Thus Hudson (1980) makes a persuasive case that, along with acquisition of scientific information needed to adapt conservation techniques appropriate to the diverse conditions of agriculture throughout the world, much more attention needs to be given to the economic and social incentives and disincentives which influence farmers' decisions to use or, more often, not to use known techniques of conservation management. Better documentation of loss and potential recovery is essential in agricultural planning and development, but action clearly does not await the arrival of new findings.

The evidence for cornucopia is as impressive as the evidence for calamity. Both extreme expressions may be needed to get the attention of people and policymakers. Global assessment is important in planning for agricultural development, in assessing the adequacy of food resources for the world's population, and in evaluating agricultural policies. The global context is clearly misleading if used to mask the importance of human survival and simple human dignity. In a portion of the Philippines, over 50% of the farms are on steep or very steep lands, and population pressure forces more settlement on such land (Gwyer, 1978).

In his keynote address at the annual meeting of the Soil Conservation Society, Dudal (1982) put the matter eloquently and simply: "It is not enough for the world as a whole to have the capability of feeding itself; it is necessary to produce more food where it is needed." One may argue that money in the hands of these same people might obviate the need to eke a living from the most marginal lands, but presently the prospects for such money in many areas are less promising than staving off starvation by working poor land.

Undoubtedly, the development of more elaborate, and verifiable, models of the potential impact of erosion on productivity for different soils in different regions throughout the world will improve, or refine, estimates of the aggregate potential impact of soil erosion on the world scale. Better models may also provide improved techniques for evaluating alternatives and allocating scarce resources where they can be used most effectively and efficiently at the local, country, or regional level. At the same time, better global assessments are unlikely to alter the recurring findings that the impact of soil erosion on potential crop productivity is highly variable from place to place, presently and potentially devastating to human beings in some areas of the world, and amenable to control where it is most important to do so to assure adequate food for the people of the world.

REFERENCES

Adams, R. McC. 1965. Land behind Baghdad. Univ. of Chicago Press, Chicago, IL

Buringh, P. 1977. Food production potential of the world. World Dev. 5:477–485.

----. 1981. An assessment of losses and degradation of productive agricultural land in the world. Working Group on Soils Policy, second mtg., FAO, Rome.

----, and H. D. J. van Heemst. 1977. An estimation of world food production based on labour-oriented agriculture. Centre for World Food Market Res., Wageningen, The Netherlands.

----, ----, and G. J. Staring. 1975. Computation of the absolute maximum food production of the world. Agricultural University of Wageningen, Department of Tropical Soil Sci., Wageningen, The Netherlands.

Daniels, R. B. 1982. Soil depletion study, Southern Iowa Basins. U.S. Committee SCOPE Land Transformation Study. Natl. Acad. Sci., Washington, DC.

Dudal, R. 1982. Land degradation in a world perspective. J. Soil Water Conserv. 37:245–249.

Eckholm, E. P. 1976. Losing ground. W. W. Norton, New York.

FAO. 1979. A provisional methodology for soil degradation assessment. FAO, Rome.

FAO/UNESCO. 1974. Soil map of the world. FAO, Rome.

Fournier, F. 1960. Climat et erosion. Presses Universitaires de France, Paris.

Gibbs, R. J. 1967. The geochemistry of the Amazon River system. Part I: The factors that control the salinity, the composition and the concentration of suspended solids. Geol. Soc. of Am. Bull. 78:1203–1232.

Golubev, G. N. 1980. Agriculture and water erosion of soils: A global outlook. Working paper 80-129. Int. Inst. for Applied Systems Anal. Laxenburg, Vienna, Austria.

Greenland, D. J., and R. Lal. 1977. Soil conservation and management in the humid tropics. John Wiley and Sons, New York.

Gwyer, G. 1978. Developing hillside farming systems for the humid tropics: The case of the Philippines. Oxford Agrarian Studies. 7:1-37.

Henin, S., R. Gras, and G. Monnier. 1969. Le profil cultural: L'etat physique du sol et ses consequences agronomique. Masson, Paris.

Holdgate, M. W., M. Kassas, and G. F. White. 1982. United Nations Environment Program. The World Environment 1972-1982. Tycooly, Dublin.

Hudson, N. W. 1980. Soil degradation and civilization. Occasional paper no. 9. National College of Agricultural Engineering, Silso Bedford, England.

Hymans, E. 1952. Soil and civilization. Thames and Hudson, London.

International Agriculture Development Service. 1980. Agricultural development indicators: A statistical handbook. New York.

Jacks, J. V., and R. O. Whyte. 1939. Vanishing lands: A world survey of soil erosion. Doubleday Publ. Co., New York.

Lal, R. 1976. Soil erosion problems on an alfisol in western Nigeria and their control. Mono. No. 1. Int. Institute of Tropical Agriculture. Ibadan, Nigeria.

----. 1983. Soil erosion and its relation to productivity in tropical soils. Int. Conf. on Soil Erosion and Conservation (Hawaii).

Langdale, G. W., and W. D. Shrader. 1982. Soil erosion effects on soil productivity of cultivated cropland. In B. L. Schmidt et al. (ed.) Determinants of soil loss tolerance. Spec. Pub. 45. American Society of Agronomy and Soil Science Society of America, Madison, WI.

Larson, W. E., F. J. Pierce, and R. H. Dowdy. 1983. The threat of soil erosion to long-term crop production. Science 219:458-465.

L'vovich, M. I. 1979. World water resources and their future. Engl. trans. American Geophysical Union, Washington, DC.

Lyles, L. 1975. Possible effects of wind erosion on soil productivity. J. Soil Water Conserv. 30:279-283.

Malin, J. C. 1947. The grassland of North America: Prolegomena to its history. J. C. Malin, Lawrence, KS.

Marsh, B. a.'B. 1980. Economics of soil loss: a top priority research need. Int. Soil Conserv. Conf. (England).

Marsh, G. P. 1864. Man and nature or physical geography as modified by human action. Sampson Low, London.

National Oceanic and Atmospheric Administration and University of Missouri, Columbia. 1979a. A study of the Caribbean Basin drought/food production problem: Final report. U.S. Department of State, University of Missouri, Columbia.

----. 1979b. Weather-crop yield relationships in drought prone countries of Sub Sahel. University of Missouri, Columbia.

Riquier, J. 1982. A world assessment of soil degradation. Nat. Resour. 18:18-21.

Ruhe, R. V., and R. B. Daniels. 1965. Landscape erosion—geologic and historic. J. Soil Water Conserv. 20:52-57.

Sanchez, P. A., D. E. Bandy, J. H. Villachica, and J. J. Nicholaides. 1982. Amazon basin soils: Management for continuous crop production. Science 216:821-827.

Sanders, D. W. 1981. Relating potential productivity to soil loss. FAO Land and Water Tech. Newsl. p. 21-25.

Schultz, T. W. 1965. Economic crises in world agriculture. University of Michigan Press, Ann Arbor.

Schumm, S. A., and M. D. Harvey. 1982. Natural erosion in the U.S.A. p. 15-22. In B. L. Schmidt et al. (ed.) Determinants of soil loss tolerance. Spec. Pub. 45. American Society of Agronomy and Soil Science Society of America, Madison, WI.

Steyaert, L. T. 1980. Early-warning of drought related food shortages in developing countries. EDIS (U.S. Department of Commerce) 11:3–11.

University of Hawaii. 1983. Int. Conf. on Soil Erosion and Conserv. East-West Institute, Hawaii.

Wischmeier, W. H., and D. D. Smith. 1978. Predicting rainfall erosion losses: A guide to conservation planning. Handb. 537. USDA, Washington, DC.

Wittfogel, K. A. 1957. Oriental despotism: A comparative study of total power. Yale University Press, New Haven, CT.

Wolman, M. G. 1967. A cycle of sedimentation and erosion in urban river channels. Geogr. Ann. 49:385–295.

Wortman, S., and R. W. Cummings, Jr. 1978. To feed this world. Johns Hopkins University Press, Baltimore, MD.

3 Historical Perspective of Soil Erosion in the United States

Fred P. Miller
University of Arkansas
Fayetteville, Arkansas

Wayne D. Rasmussen
Economic Research Service
U.S. Department of Agriculture
Washington, DC

L. Donald Meyer
Agricultural Research Service
U.S. Department of Agriculture
Oxford, Mississippi

3-1 SOIL: ITS IMPORTANCE TO THE SUSTENANCE OF HUMANKIND

All life depends on stocks and flows of nutrients. The development of civilization was largely paced by the corresponding development of agriculture. Soils provided the medium through which these nutrient flows and energy conversions took place. Before agricultural development, these nutrient and energy flows were attained by a nomadic hunting and gathering existence. The transition from a nomadic culture to a conscious soil-dependent plant and animal husbandry predates recorded history. It probably began some 9000 years ago and was surely one of the great forward steps of humanity.

Soil is nature's earthly carpet—a crustal fabric reflecting mineralogical parentage and formative environments, both past and present. It is a fragile, thin, organically-enriched mineralogical membrane covering much of the earth's crust. Jacks and Whyte (1939) emphasized this point by stating that "below that thin layer comprising the delicate organism known as the soil is a planet as lifeless as the moon" (p. 4). Soil is the foothold for much of the life on earth. Soil provides the medium from which most of the

Published in R. F. Follett and B. A. Stewart, ed. 1985. *Soil Erosion and Crop Productivity.* © ASA-CSSA-SSSA, 677 South Segoe Road, Madison, WI 53711, USA.

sustenance for humankind is derived. This thin, complex, crustal carpet uniquely integrates many attributes of the lithosphere, atmosphere, hydrosphere, and biosphere. Coffey (1912) stated early in this century that soil "is the one great formation in which the organic and inorganic kingdoms meet and derives its distinctive character from this union" (p. 8).

The historical match of civilization and soils has not been random. The Neolithic people recognized distinctions among soils as suggested by the pattern of prehistoric occupation of soils in western Europe. Early civilizations, commonly located in proximity to transportation routes, were also conscious of agricultural soil quality. Written records of soil classification systems based on productivity have been found for the early Chinese, Egyptian, Mesopotamian, Indus, and Roman civilizations, some more than 5000 years old.

The parade of civilizations through time has often marched to the cadence of the natural resource trinity—soil, water, and climate. History is replete with attempts to march to different drummers. These often tragic parades have names such as the early civilizations of the Middle East, now collapsed from a variety of frailties, both social and earthly. The skeletal landscapes of these early civilizations bear witness to their earthly frailty and abuse. The U.S. Dust Bowl, the Sahel, and other ecological tragedies reveal a flaw in humankind's ability to assess the carrying capacity of the resource trinity and to manage its components.

Bradley (1935) wrote that "the fabric of human life has been woven on earthen looms. It everywhere smells of the clay" (p. 331). The soil carpet in many areas of the world is badly worn and is threadbare in places. Neglect and exploitation have ravaged the carpet through erosion. The fabric of human life in these areas is weak, and human suffering stalks the land. In other areas, the soil carpet, although used for centuries, has endured treading and cultivation by humankind without significant scars and still provides sustenance. Herein lies one of the basic lessons for human survival: Unlike most other resources, soil, if properly managed, can be used indefinitely for agricultural production without being consumed.

Soil is worthy of an ethic dedicated to its conservation, not to be preserved in an outdoor museum but to be used wisely. Dubos (1976) perceived the earth as "neither an ecosystem to be preserved unchanged nor as a quarry to be exploited for selfish and short-range economic reasons, but as a garden to be cultivated for the development of the human adventure" (p. 462).

3–2 HISTORICAL PERSPECTIVE

Three periods of accelerated soil erosion have had particularly acute impacts on the earth's land resources over the last few millenia (Dregne, 1982). Few areas of the settled world have escaped man-induced erosion. The first period—centered in the Mediterranean, Middle East, and China—occurred between 1000 and 3000 years ago as a result of deforestation, overgrazing, extensive cultivation, and the ravages of army invasions (Dregne, 1982). Lowdermilk (1948) has documented this period in his classic report on these erosion-ravaged regions.

The second period of accelerated erosion took place during the 19th and 20th centuries in those areas impacted by large European migrations and by the introduction of export-oriented agricultural economies (Dregne, 1982). This period was exemplified by the settlement and expansion of the USA.

The third and latest period of excessive accelerated erosion began in the early 20th century, continues today, and is most severe in developing countries where population numbers and land pressures are increasing rapidly (Dregne, 1982). Latin America, Africa, and southern Asia are the most impacted and endangered areas.

The Europeans who settled the USA beginning more than four centuries ago, unlike their American Indian predecessors, exploited the natural resources they found. These settlers, as individuals, acted rationally in trying to wrest a living from a wilderness, whose land, covered mostly by heavy forests, seemed limitless. Yet, the settlers' needs for food and a place to grow it were immediate. Furthermore, ordinary farmers had little or no idea of the hazards of soil erosion or how it might be controlled. Their experience was forged from regions of low-intensity rainfall and more temperate climates conducive to the maintenance of soil organic matter.

Neither the exploitative behavior of these early farmers nor their environmental impacts are unique in history. When compared to previous civilizations or to other frontier developments, such ecological irreverence seems more the norm than the exception (Carter and Dale, 1974; Lowdermilk, 1948; Marsh, 1864; Thomas, 1956). As Carter and Dale (1974) pointed out, these early Americans were following a pattern as old as civilization itself. Nevertheless, these settlers probably caused more waste and ruin in a shorter time than any people before them, because they had more land to exploit and better equipment with which to exploit it (Beasley, 1972; Carter and Dale, 1974).

The historical perspective of soil erosion in the USA involves more than assessments of natural resource exploitation and soil degradation and their consequence as measured and studied through the earth sciences. The causes of soil erosion extend beyond climatic abnormalities and edaphic attributes. Soil degradation does not come about solely as a result of ignorance by its perpetrators or simply as a result of environmental accidents. Soil erosion follows from human demands and impacts on the earth's resources. These demands and impacts are shaped by human beliefs, attitudes, policies, economic necessities, and practices (Nicholson, 1970). Therefore, it is as appropriate to understand religion, history, economics, philosophy, and psychology as it is to study soil science and climatology in a comprehensive assessment of soil erosion. Human desires, motives, and emotions, including greed and the drive for self-preservation, are as crucial to the cause of soil erosion as rainfall erosivity and soil erodibility.

Even if landowners and farmers do recognize erosion as a problem, their short economic time frame often works against the establishment of self-implemented methods to deter erosion. Any benefits from soil conservation accrue on a time scale that is frequently beyond the short-term remunerative needs of farmers. This large time gap between soil conservation costs and conservation benefits is a major deterrent to the implementation of conservation practices even among those with the ethic to try. This situation is as real today as it was 200 years ago. Thus, the short-term human planning horizon produces a perception of erosion that overlooks its insidious nature and cumulative effects.

Any strategy to control soil erosion, therefore, must address these human perceptions and attributes as well as the LS, C, and P factors in the universal soil loss equation (USLE). It is little wonder that our voluntary strategy to control soil erosion in the USA over the last half century still results in billions of tons of soil washed from segments of our farmlands, despite the many conservation programs of the past and the individual efforts of millions of landowners to abate erosion.

White (1967) and others (Bouillenne, 1962; Sears, 1980; Tuan, 1966) have described humankind's historical arrogance toward nature. White argued that this attitude is dominant in Western traditions stemming from Westerners' religious beliefs. But Tuan (1966) argued that the tendency toward environmental degradation and the desire to maximize one's well-being characterize all human existence.

Sears (1980) made the point that the first English settlers were bent upon moving their British world into the New World. According to Sears, one of the most significant traits of Anglo-Saxon psychology is the need for lofty motives to rationalize getting whatever may be wanted. More a capacity for self-deception than a matter of perfidy or hypocrisy, this trait nevertheless lies at the root of much of the mischief done to the resources of North America (Sears, 1980).

White (1967) stated that "the emergence in widespread practice of the Baconian creed that scientific knowledge means technological power over nature can scarcely be dated before about 1850. . . . Its acceptance as a normal pattern of action may mark the greatest event in human history since the invention of agriculture, and perhaps in nonhuman terrestrial history as well" (p. 1203). This ingrained human perception is applicable to the soil erosion problem. With bigger and more powerful machinery today, rills and small gullies are smoothed over and "healed" in one pass, not only wiping out the physical evidence of erosion, but also erasing its existence from one's psyche.

3–3 SETTLEMENT OF THE NORTH AMERICAN CONTINENT

Soil erosion is a paradigm of environmental stress induced by man. To some extent, it is a measure of humankind's wear on the environment as well as a reflection of the intensity of man's stewardship ethic. Bidwell and

Falconer (1925) maintained that soil exploitation was the keynote of western agricultural expansion across the USA.

The U.S. portion of the North American continent (excluding Alaska) was originally dominated by forest and grass-rangeland. This land area of 770 million ha consisted of approximately 40% forest, 40% grass-rangeland and 20% desert and mountains (Osborn, 1948). The eastern third of this land was dominated by magnificant deciduous forests and lesser areas of coniferous forests with some grassy glades occurring more frequently from Ohio westward.

The sparsely populated land was inhabited by American Indians who had been farming, at least to some extent, for several centuries. Partly because of their small numbers and partly because they were content to produce only what they needed to live, the continent's natural resources were virtually intact at the time of the first European settlements.

These early settlements became the catalyst for mass migrations of Europeans to this new land, migrations which lasted for three centuries. The vast and resource-rich continent that lay before these settlers and the speed with which it was populated are unprecedented in history (Carter and Dale, 1974). As a testimonial to the resources these settlers inherited, the dense forests were considered an obstacle to agriculture even though they were utilized for fuel and timber. The volume of this timber resource was so large that the center of the commercial lumbering industry did not move beyond western New York until after 1850 (Sears, 1980).

These forest and grassland resources covering hundreds of millions of hectares of the area now within U.S. borders, coupled with temperate climates and productive soils, were in themselves a fabulous prize, to say nothing of the minerals, wildlife, and other riches that awaited their taking. It took nearly a century after the early American settlements for the European nations to become convinced of the importance of this new land and the resource bounty its political control would insure.

By the time of the American Revolution, a 160-km strip of land from southern Maine to Georgia had been settled, and nearly half to three-quarters of this area had been cleared. The population was less than four million within the entire 13 colonies (USDA, 1981). But even before the Revolution, erosion had been recognized as a problem (McDonald, 1941)—a disease of the land that was to be carried westward with the European migrations.

Trimble (1974) studied the settlement and exploitation of the Southern Piedmont portion of the area extending from Virginia to Alabama. Beginning about 1700, the land was progressively cleared, farmed to exhaustion, and then abandoned. The Southern Piedmont was settled completely by 1830. The erosive land use spread from north to south. By the Civil War, the entire region was subject to extremely destructive erosion rates, which continued at high to medium intensity levels from 1860 to 1920 (Fig. 3–1). Since 1920 much of the land of this region has reverted to forest and pasture (Trimble, 1974).

While this eastern strip of land took more than a century to settle, the westward migration had already begun to spill over the Alleghenies. By

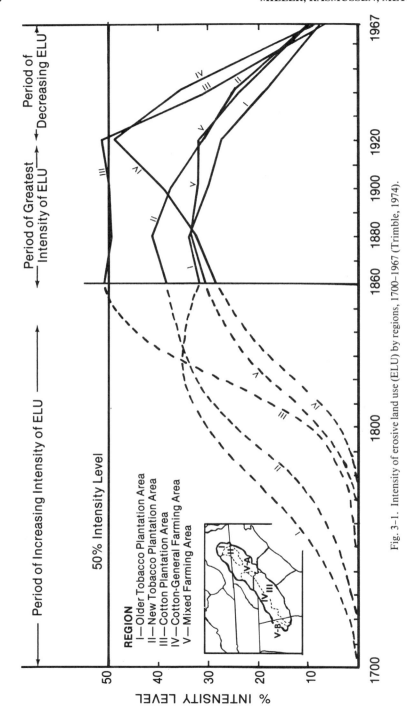

Fig. 3-1. Intensity of erosive land use (ELU) by regions, 1700-1967 (Trimble, 1974).

1830, most of the better, well-drained lands east of the Mississippi River were occupied (Bennett, 1939). The push westward, however, was not uniform. Sears (1980) outlined two settlement patterns: The movement of European farmers through the northern states into the Northern Great Plains and prairies, and the southern migration and settlement, which differed because of the culture of its settlers and the character of its land.

The northern farmer moved westward with unrestrained destruction of native Indian culture. Individual property rights were emphasized without regard for public policy. Until the grain-belt states were reached and self-interest dictated better husbandry practices, farming areas were systematically impoverished and the lands exhausted (Sears, 1980). Except for the Teutonic groups among them, most of these settlers were not first-class farmers and for the most part, lacked an agricultural tradition honed through the experience of previous generations. But their considerable practical and mechanical sense, combined with their work ethic, resulted in a brilliant record of commercial and industrial growth.

The farm itself often came to be regarded as a mere stepping stone into the professional or commercial life of the rapidly growing towns (Sears, 1980). This attitude and the poor land-stewardship ethic derived from it resulted in a vast deterioration of soil resources in the wake of settlement across the northern USA. The emphasis on commerce and industrialization became decisive in the outcome of the young nation's Civil War.

The southern migration of Europeans west of the Alleghenies resulted in a culture quite different from that of their northern counterparts (Sears, 1980). The gentlefolk of old England were granted large estates, which were worked by the impoverished and the dregs of society who flooded this new land. Also, the indentured, including many Scotch-Irish, were numerous among those who worked these landholdings and eventually settled in the Southeast. Later, the black slaves, who proved unprofitable in the North, were found to be well suited to the tobacco, cotton, indigo, and sugarcane plantations of the warmer South. The small, personally owned and operated farm could not compete in the staple markets with the larger slave-operated plantations (Sears, 1980).

Sears (1980) also points out another circumstance that was destined to exact a staggering toll on the ensuing generations. The wealthy landed gentry viewed education as a luxury to be dispensed only to those with the means to pay for it. Not only blacks, because of their status as slaves, but also poorer whites were denied educational opportunities. These poor whites were driven from their small holdings by plantation expansion and competition, and kept from employment on the plantations by the abundance of black labor. Because the better lands in the South lay along the coastal plain and the broad valleys running inland from it, these lands were occupied and controlled by the plantation landlords. The poorer whites were relegated to the lower quality lands, resulting in their literally taking to the hills, eventually developed their own culture in the backwoods and hollows of the Appalachians and the Ozarks (Sears, 1980).

On the southern plantations there was some sense of land stewardship, as exemplified by George Washington and Thomas Jefferson (McDonald, 1941; Sears, 1980). But the continued export of each harvest eventually took its toll on both the soil resource and its fertility. And the poorer whites, relegated to the hills and lower quality lands, burned, cleared, pastured, and cultivated their way westward, leaving behind a skinned and impoverished land. They eventually reached the eastern portion of the great grasslands via the southern route (Sears, 1980).

During these westward migrations, the vastness of the resource base and the open and cheap land areas still available to the west became the settler's talisman and worked against a psychology of permanence (Sears, 1980). The knowledge that these western lands were still available tended to salve the anxiety of failure and was not conducive to fostering a conservation ethic or to promoting a sense of stability.

To illustrate this impermanence syndrome, Trimble (1974) cited one wit of the late 1830s who summarized the situation in the southern Piedmont by noting that "the scratching farmer's cares and anxieties are only relieved by his land soon washing away. As that goes down the rivers he goes over the mountains" (p. 53).

Gray (1933) also pointed out the tendency to deplete land and then migrate west by stating that:

Over the upland soils from Virginia to Texas the wave of migration passed like a devastating scourge. Especially in the rolling piedmont lands the planting of corn and cotton in hill and drill hastened erosion, leaving the hillsides gullied and bare (p. 446).

As the northern and southern migrations into the semi-arid portion of the new nation continued, the exploitation of the soil resources behind these migrations was supplemented by yet another chapter of soil erosion—the great Dust Bowl era. The vast expanse of prairie encountered by the western migration was unparalleled in the history of human settlements. The European background of the encroaching settlers and any experience that may have been acquired in the humid East left unprepared those who first entered the tallgrass prairie and then the semi-arid shortgrass region. The latter ecosystem proved the undoing of those who broke the sod. Tempted by their tradition, lured by the bait of immediate profits, and encouraged by financial and industrial interests (Sears, 1980), the settlers' breaking of the sod in this region coupled with the vagaries of its climate triggered one of the great ecological disasters of human history. On 12 May 1934 the droughts common to this marginal cropping region resulted in the first of many dust storms. It was a tragic event in American history which scarred both the land and its people. The impacts of this Dust Bowl were felt from Texas to the Dakotas, damaging some 60 to 80 million ha of land (Bennett, 1939).

The westward trek of humanity across the USA is a story of human courage, tragedies, successes, failures, and a willingness to reach for new horizons—regardless of the motives. In the process, a great nation was built, but not without significant deterioration and scarring of its land resources. By 1890, nearly all of the better lands in the USA had been settled. And by the 1920s and early 1930s, American agricultural land use reached its zenith of areal extent.

3-4 SOIL EROSION: A PARADIGM OF EXPEDIENCY

From the beginning of the country's feeble settlements, soil erosion was a voluminous by-product of U.S. agriculture. It was an ecological price extracted for an exploitative land settlement philosophy developed from basic human instincts as modified by cultural inheritance. The lessons of such behavior had been etched in the earth by civilizations past, but like most historical precedents, humankind proved itself a slow learner.

The wealth of the young nation was generated from the harvest and export of portions of its vast natural resource base. Those resources that could not be converted immediately to commodities for export or domestic use were sidestepped, overrun, or destroyed in the interest of expediency. Forests were pushed aside, cleared, and burned. Animal resources were harvested and their habitats destroyed or damaged. Some animal species were simply killed for pastime, a testimonial to their abundance as well as the ignorance and perhaps arrogance of the harvesters. And, of course, the soil resource associated with these lands was often left exposed, allowing the soil to be ravished by rain and wind. The unlimited development of the nation's land and the harvesting of its natural resources continued to fuel an unfettered economic growth of the nation into the 20th century.

But, as Madden (1974) pointed out, this development philosophy had two serious flaws. First, most Americans treated land as if it were indestructible and inexhaustible. Second, land was treated as a commodity, and its biological role as a crucial link in the web of life was ignored. For 100 years after the USA gained its independence, the dominant political philosophy was to settle the nation's interior by encouraging the disposition of land to settlers. The process of land disposal enabled most Americans to gain wealth and live better. It dominated early American history. Heedless and headlong, land transfer to private ownership was a major social and political force throughout the 19th century. The policies derived from these forces moved two-thirds of the land in the 48 conterminous states out of the public domain.

The Land Ordinance of 1785 and The Northwest Ordinance of 1787 (White, 1983) encouraged the western movement by providing for the survey and sale of western lands and the admittance of these regions as new, equal states when a certain population was reached. The laws were modified from time to time until, in 1862, the Homestead Act offered 65 ha (160 acres) of public land to anyone 21 years of age or the head of a family who would settle on it and farm it for 5 years. This act was a part of an agrarian reform package that also established the U.S. Department of Agriculture and funded the land grant colleges and a transcontinental railroad.

From the Civil War to World War I, land use policy emphasized making more land available for farming. Several federal and state programs encouraged the drainage of swampy land. Even more important were efforts to encourage irrigation in the West, culminating in the Newlands Act of 1902.

Land became the capital transferred to ordinary people by their government, and its availability attracted millions of immigrants from all parts of the world. But the land-stewardship ethic of these people was neither strong nor cultivatable in the face of immediate resource needs and an exploitative philosophy that was fostered by the hope that failure could result in the acquisition of even better lands elsewhere.

3-5 RECOGNITION OF SOIL EROSION

A few people, even in colonial times, saw the dangers of this exploitative use of land and unchecked soil erosion. One such man was Jared Eliot, minister, physician, and farmer in Connecticut. Eliot saw that water washing soil from hills might enrich a valley but would leave the hills sterile. He advocated planting clover and grass, spreading manure, and plowing in a way to reduce erosion. He published his observations (Eliot, 1760), the first of a large body of literature on soil erosion. A generation later, Samuel Deane of Maine urged many of the same practices, particularly various crop rotations (McDonald, 1941).

After the Revolution, a number of southern farmers recognized the dangers of soil erosion and worked to control it. Outstanding among these was John Taylor of Virginia. In newspapers, books, and a little magazine of his own, Taylor pointed out the damage caused by erosion. He urged plowing on the contour, controlling gullies by filling them with green bushes, and manuring the soil (Rasmussen, 1975). John Lorain, a contemporary of Taylor's who farmed first in Maryland and then in Pennsylvania, spoke out against unwise plowing and urged crop rotations that included grass. Lorain pointed out that growing wheat and corn could lead to erosion in Pennsylvania but that the South seemed to be hardest hit, partly because of the continuous growing of tobacco and cotton on clean-tilled fields subject to heavy rains. The scarcity of capital and the cheapness of land drove farmers toward an extensive agricultural production system, regardless of the consequences to the land.

Under these conditions Edmund Ruffin started farming in Virginia in 1813. Before long, he became convinced that the poverty of Virginia was created in large part by man-made soil erosion, the major cause of which was the continuous growing of clean-cultivated row crops. Ruffin devoted much of his life to encouraging farmers to lime the soil, practice erosion-preventing methods of cultivation, plow under green crops, and adopt crop rotations (Craven, 1932; Ruffin, 1855).

In spite of the efforts of men from Eliot to Ruffin, little impression was made on the average farmer in these early days. Southern agriculture underwent a major change after 1793, when Eli Whitney invented the cotton gin. Cotton soon became dominant in southern agriculture. In the 1830s, Cyrus H. McCormick and Obed Hussey patented practical grain reapers, and other inventors came up with horse-drawn grain drills, cultivators, threshers, and other machines.

The Civil War saw the first U.S. agricultural revolution when a seemingly unlimited demand for farm products, high prices, and labor shortages encouraged farmers to turn to the new horse-powered machinery to increase production. They farmed more land than ever but gave little thought to soil erosion.

After the Civil War, the settlement of the prairies and plains continued, with the flow of migrants from the East being almost overshadowed by immigrants from Europe and other parts of the world. The new land grant colleges and the USDA were engaged in research that would help farmers increase production—a thoroughly worthwhile goal, but one that too often neglected the impact of production technology on the environment.

3-6 THE SOIL CONSERVATION MOVEMENT

While attempts were made to call the abuse of the land to the attention of farmers and officials, little headway resulted until after the first conservation movement from 1890 to 1920. The chief reformer during this period was Gifford Pinchot. Although Pinchot was aware of the soil erosion problem, his primary interest as a forester was to establish the national forests (Held and Clawson, 1965). During this period, soil conservation was neglected and soil conservation programs were left out of USDA programs.

Despite the low status that soil conservation had during this time, some agriculturists, particularly in the land grant colleges and the USDA, urged care of the soil. The editors of some eastern and southern farm journals, as well as political leaders from those regions, pleaded with farmers to make better use of the soil they had rather than abandon it for the unknown West.

One of the first farmers' bulletins published by the USDA pointed out that thousands of hectares of valuable but eroded cropland were abandoned each year. This bulletin, *Washed Soils: How to Prevent and Reclaim Them,* published in 1894, urged farmers to reclaim and use the land they had. In 1910, the USDA published Farmers Bulletin 406, *Soil Conservation,* and Farmers Bulletin 414, *Corn Cultivation,* in which farmers were told that erosion must be stopped if production was to be maintained. Probably some farmers heeded these admonitions, but apparently most did not.

Then, in 1928, an increasingly prominent crusader for soil conservation, Hugh Hammond Bennett, pointed out in a USDA circular entitled *Soil Erosion a National Menace* (Bennett and Chapline, 1928) that the problem affected everyone, not just farmers. Bennett's continued emphasis on this theme made him an effective proponent for conservation. Bennett was to soil conservation what Pinchot and Muir were to the national forests and parks, respectively. It is difficult to overestimate his influence in bringing about a national soil-conservation consciousness and program.

The year after Bennett's bulletin appeared, Congressman James P. Buchanan of Texas amended the Agricultural Appropriations Act for 1930 to provide $160 000 to study the causes of erosion and methods for its control. Bennett was put in charge of the program, working through regional soil-erosion experiment stations, which was the beginning of a major program of research on soil and water conservation.

The Great Depression and the resulting New Deal gave additional impetus to conservation. Soil erosion had become a serious national resource problem. Relief legislation authorized soil-erosion control work to aid the unemployed, while the Civilian Conservation Corps carried out reforestation and soil conservation projects.

Another effort to carry out conservation programs was financed with $5 million of Public Works Administration funds. After some maneuvering, the work was assigned to the Department of the Interior to be carried out by a temporary Soil Erosion Service under the direction of Bennett. A research program was developed, but after further political feints, the work was transferred to the USDA. Shortly thereafter, on 27 April 1935, Congress passed the Soil Conservation Act. This act declared soil erosion to be a national menace, established the soil conservation program on a permanent basis, and changed the name of the responsible agency to the Soil Conservation Service. After nearly 200 years of effort, from Eliot to Bennett, the people of the USA, through their Congress, took responsibility for addressing the issue of protecting the nation's soil.

3-7 SOIL EROSION IN THE USA:
A SUBJECTIVE ASSESSMENT

The magnitude of environmental deterioration, including soil erosion, on the U.S. portion of the North American continent prior to the 1930s is not well documented. Evidence of this environmental impact is largely subjective, given in historical accounts, often detailed and graphically eloquent, but seldom quantitatively comprehensive in geographic extent, time frame, or degree of injury. Despite the lack of quantitative information on environmental damage and soil erosion prior to the early 1900s, the literature on this subject and early U.S. agriculture is immense (Bennett, 1939; Bidwell and Falconer, 1925; Edwards, 1940; Gray, 1933; Lord, 1938; Marsh, 1864; Petulla, 1977; Rasmussen, 1975 and 1982; Schlebecker, 1975; Sears, 1980; Smith, 1971; and Thomas, 1956).

In the 1800's and early 1900's, numerous state and federal bulletins, reports, and other publications, along with many private accounts, detailed the impacts of erosion on various segments of land. The commentary of various witnesses to the erosive demise of many landscapes provides a unique description of an ecological problem as seen through the eyes of those who experienced it personally. Such accounts echo a common concern, if not disgust, among those who wrote about it. Recorded in these accounts is an intensity of feeling and despair reserved only to those who are stunned by the immediacy of a tragedy and where such shock has not been tempered by either time healing the landscape or by the more muffled accounts of historians.

An 1853 appraisal of Laurens County, SC (Trimble, 1974), was written in apocalyptic prose:

> The destroying angel has visited these once fair forests and limpid streams. . . .The farms, the fields. . .are washed and worn into unsightly gullies and barren slopes— everything everywhere betrays improvident and reckless management. . . (p. 54).

And, still prior to the Civil War, C. C. Clay in a speech before Congress (Gray, 1933) described soil exhaustion in Madison County, AL:

> In traversing that county one will . . . observe fields, once fertile, now unfenced, abandoned and covered with those evil harbingers, foxtail and broomsedge. . . . Indeed, a country in its infancy, where fifty years ago, scarce a forest tree had been felled by the axe of the pioneer, is already exhibiting the painful signs of senility and decay, apparent in Virginia and the Carolinas (p. 446).

Even U.S. government documents of the day waxed eloquent on the erosion situation, providing the reader with awesome analogies but few data. An example is the 1852 agricultural volume of the Report of the U.S. Commissioner of Patents[1] which reads:

> Twice the quantity of rain falls in the Southern States in the course of a year than falls in England, and it falls in one-third the time. . . (p. 12). Cotton has destroyed more land than earthquakes, eruptions of burning volcanic mountains or anything else. Witness the red hills of Georgia and South Carolina that have produced cotton till the last dying gasp of the soil forbids any further attempt at cultivation and the land turned out to nature reminding the traveler, as he views the dilapidated country, of the ruins of Greece. . . (p. 72). And these evils to the community and to posterity, greater than could be effected by the most powerful and malignant foreign enemies of any country, are the regular and deliberate work of benevolent and intelligent men, of worthy citizens and true lovers of the country (p. 386).

Hilgard, in his 1860 *Report on the Geology and Agriculture of Mississippi,* expressed shock and dismay at the severe erosion damage to his state's soil resources (Hilgard, 1860). His concern was expressed at both the damage and attitude of farmers. A European-educated geologist and chemist, Hilgard spent most of his career as a soil scientist. He perceived the severity of the damage and reduced productivity of agricultural soils by erosion. His classic report is both a scientific assessment and a philosophical

---

[1] The following quotations were taken from three separate sources within the same reference (Report of the Commissioner of Patents, 1853). The USDA had its origin in the U.S. Patent Office in 1837 through Henry Ellsworth's program of distributing seeds from abroad to farmers. In 1839, Congress appropriated $1000 to Ellsworth for gathering agricultural statistics, conducting experiments, and distributing seeds. He eventually established an Agricultural Division within the Patent Office which expanded its operations until 1862, when Congress established an independent USDA (Petulla, 1977).

treatise on the erosion problem and the human attitudes which brought it about. He was troubled by the attitude of Mississippi farmers who, when told that they were mining their land and should use better management by returning manures and marl, turned up their noses in contempt of such old-fashioned advice. They commonly remarked, according to Hilgard, that whenever their land gave out, there was plenty more to be had. Hilgard wrote:

> While I am far from attributing sentiments like these to the majority, or even to any large part of the planters of Mississippi, I have not so rarely met the opinions like the above. . . .And be it remembered, that the burden thus imposed upon posterity. . .is quite out of proportion with the temporary advantage the present generation may derive from it; . . .even the present generation is rife with complaints about the exhaustion of soils—in a region which, thirty years ago, had but just received the first scratch of the plow-share! In some parts of the State, the deserted homesteads and fields. . .might well remind the traveler of the descriptions given of the aspect of Europe after the Thirty Years War. . . .
>
> I do not mean to say, that the early settlers could or should, under the circumstances which surrounded them, have pursued a different course, and commenced, at once, a regular system of agriculture. . . .But what was justifiable in them, is no longer so with their successors. As members of a Christian commonwealth, it is their right to use, but not to abuse, the inheritance which is theirs, and to hand it down to their children as a blessing, not as a barren, inert incubus, wherewith to drudge through life, as a penalty for their fathers' wastefulness (p. 239).

Hilgard's perception of the problem could well have been applied to just about every cleared and cultivated region of the young nation during this time. In 1875, Hilgard went to the University of California at Berkeley. He was appointed Professor of Agriculture and Botany, and he immediately set up an Experiment Station on the Berkeley campus, the first in the country to combine theoretical and field research in agriculture (Petulla, 1977).

Those, like Hilgard, who subjectively but clearly perceived the land abuses in the humid states failed to establish an enlightened caretaking ethic for the land or to deter the westward momentum propelled by economic, social, political, or simply adventurous motivations. As the semi-arid shortgrass lands were overrun, grazed, and then the sod broken for cultivation, the ravages of erosion took a different form. Wind was the energy source that lifted the soil and transported it eastward. The impact on people from this ecological mismanagement, however, was much the same as in previous examples of soil exploitation. Sears (1980) quotes the owner of a large tract in the shortgrass country that had been plowed for wheat in the 1930s:

> We're through. It's worse than the papers say. Our fences are buried, the house is hidden to the eaves, and our pasture which was kept from blowing by the grass, has been buried and is worthless now. We see what a mistake it was to plow up all that land, but it's too late to do anything about it (p. 158).

Others were not so articulate about the problem. They simply packed their few belongings and headed west, much as they and their predecessors had done when the water-induced erosion to the east had rendered the land scarred and unyielding. These hardy but tragic people became the human pulp for Steinbeck's *The Grapes of Wrath.* They left behind a damaged ecosystem totaling many tens of millions of hectares.

3-8 SOIL EROSION IN THE USA: A QUANTITATIVE ASSESSMENT

In his preface, Bennett (1939) examined the background of events which led to the national conservation program. Cited among the prominent mileposts along the way are the findings of the 1911 Fairfield County, SC, soil survey which disclosed that:

> 90 000 acres [36 450 ha] of formerly cultivated land had been so cut to pieces by gullies that it had to be classed as *rough gullied land* and that an additional 46 000 acres [18 630 ha] of formerly rich bottomland had been converted into swampy *meadow* land because the streams, gorged with the products of erosion, had lost their original channel capacities (p. vii).

Bennett (1939) noted that this was probably the first survey of a large area in the USA which pointed specifically and quantitatively to the wholesale ravages of unrestrained soil erosion.

In an erosion report of Lafayette County, MS, Lentz et al. (1929) described a practice of land abandonment and subsequent erosion that was common in the Southeast. They pointed out that most of the land available for agriculture in this area was cleared and devoted to cotton production before the Civil War:

> . . .As fields under cultivation were depleted of fertility, they were 'turned out' to pasture, and attention directed to newly-cleared areas. Most of the badly eroded lands may be traced back to these old fields that were turned out to pasture and repeatedly burned, thus leaving the bare soil in a condition to be severely eroded by the heavy rains (p. 4).

Lentz et al. commented that the amount of land under cultivation in the 1920s was far less than that cultivated before the Civil War, partly due to changes in labor and economic conditions, but probably more to soil depletion and erosion. Land abandonment was not common in the South until the period of reconstruction began (Lentz et al., 1929). Much of this land, however, had become gullied even while under cultivation. But Lentz et al. observed that "rapid soil removal as a rule did not occur until after the land had been abandoned" (p. 12).

Because the Piedmont region of the Southeast is one of the oldest cultivated regions in the country and is subject to high rainfall erosivity, it contains some of the most severely eroded land in the country (Bennett, 1939; Trimble, 1974). One of the first comprehensive erosion surveys of a large area was published in 1935 (Fuller, 1935) and consisted of more than 2.7 million ha (6.7 million acres) covering 35 counties in the old plantation belt of the Southern Piedmont. Bennett worked with Fuller in developing the basic principles of this survey (Bennett, 1939). The survey showed that past erosion had damaged 95% of the upland portion of the survey area to some degree. About 44% of the upland area had reached the stage of gullying, and 22% of the formerly cultivated land, abandoned as a result of erosion, had grown up to voluntary stands of "oldfield" pine. More than 400 000 ha (1 million acres) had been ruined by erosion in this region alone (Bennett, 1939).

Bennett (1939) cited the case of Steward County, GA, where uncontrolled, accelerated erosion had ruined approximately 28 000 ha (70 000 acres) through gullying and deep sheet erosion prior to World War I. Bennett wrote that some of these gullies, being more than 30 m (100 ft) deep, were unprecedented in North America.

Bennett (1939) also described the erosion damage suffered by Oklahoma, then one of the newer agricultural states. McDonald (1938) chronicled the settlement of this region during the latter part of the 19th century and the early 20th century. Bennett (1939) referred to the 1930 general erosion survey of Oklahoma, which showed that some soil had been washed or blown from 5.3 million ha (13 million acres) of the 6.5 million ha (16 million acres) in cultivation and that more than 2.4 million ha (6 million acres) had reached the stage of gullying. Bennett (1939) states that in all probability, the Oklahoma situation "is the most outstanding example of rapid land decline on a wide scale to be found in the annals of human history" (p. 68).

Bennett (1939) further cites an erosion survey of 20 Southern Great Plains counties in southeastern Colorado, southwestern Kansas, and the panhandle areas of Oklahoma and Texas, where nearly 98% of the 6.6 million ha (16.3 million acres) surveyed had been affected by accelerated erosion. This damage resulted mostly from wind erosion, with more than half the area affected seriously.

The historical assessment of erosion in the USA is probably best documented in Bennett's (1939) accounts. He was a soil scientist by training and traveled widely throughout the USA. Until the national erosion survey was conducted under his direction in the mid 1930s, most of the erosion damage assessments had been based on subjective accounts, regional erosion surveys, and soil surveys.

Despite an educational and erosion investigation program begun in the late 1920s, by Bennett and others (Bennett and Chapline, 1928; Bennett, 1931), Bennett stated in his 1939 book that the soil erosion impact on the land area of the USA may never be known. Subsequent erosion assessments support Bennett's observation (Larson et al., 1983).

Nevertheless, the most comprehensive assessment of soil erosion in the USA prior to the 1958 Conservation Needs Inventory was the 1934 national reconnaisance erosion survey. This erosion survey was conducted by the Soil Erosion Service of the Department of the Interior. The survey covered every county in the USA and was carried out by Soil Erosion Service soil technologists over a two-month period in the late summer and early fall of 1934 (Bennett, 1935a; Lowdermilk, 1935). The maps from this reconnaisance survey vary in both scale and detail, but they indicate the predominant erosion conditions. The maps are preserved in the National Archives. From this survey and previous survey information and data, it was possible to estimate the real extent and seriousness of erosion damage to the U.S. soil resource base. The information derived from this 1934 survey was used to develop the nation's erosion control policy at that time (National Resources Board, 1935). Table 3-1 provides a summary of these data.

These data show that more than half of the U.S. land area had been affected by erosion before and during the 1930s. Excluding the 40 million ha (100 million acres) of essentially ruined cropland, the process of soil erosion had already damaged or was threatening more than one-third of the arable land (Bennett, 1939). Approximately three-fourths of the U.S. cropland in the 1930s was susceptible to some degree of soil erosion.

Figure 3-2 shows the general distribution of erosion in the USA as surveyed in 1935 (Bennett, 1939). Because the map is generalized, each map unit represents an area within which at least 25% of the land has been categorized as indicated. Bennett (1939) used these and other data to generate the data in Table 1, which are based on land use.

Within less than a decade, from 1928 to 1935, the estimate of erosion damage to U.S. soil had increased dramatically. Bennett (1928), writing in *The Geographical Review,* stated that "probably not less than ten million acres [4 million ha] of land formerly cultivated in the U.S. have been permanently ruined by soil erosion" (p. 583). Just seven years later, Bennett (1935b) revealed that the figure for ruined and severely damaged cultivated land had reached 40 million ha (100 million acres).

Not only was some cropland ruined or severely damaged for future production, but the production potential of eroded land remaining in cultivation was reduced. Bennett (1939) reported that the average per hectare U.S. yields of corn and cotton stayed about the same from 1871 to 1930,

Table 3-1. Estimated extent of erosion damage in the conterminous USA (Bennett, 1939).

Erosion condition	Area
	million ha
Total U.S. land area (48 states)	771
Damaged land of all types (includes cropland)	
Essentially ruined	24
Severely eroded	90
Moderately eroded or beginning	314
Damaged cropland (includes harvested, idle, failure, fallow)	
Essentially ruined for cultivation	20
Severely damaged	20
One half to all topsoil gone	40
Moderately eroded, erosion beginning	40
Land not now damaged	
Land in forest, swamp, marsh, etc.	284
Land on which damage not defined	
Desert, badlands, western mountains	59

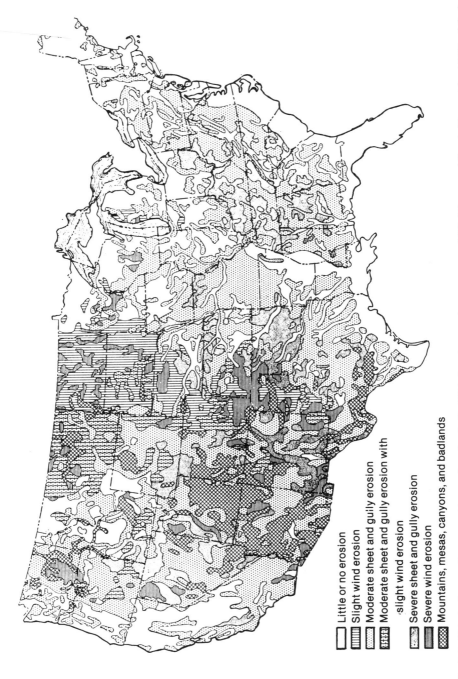

Little or no erosion

Slight wind erosion

Moderate sheet and gully erosion

Moderate sheet and gully erosion with
slight wind erosion

Severe sheet and gully erosion

Severe wind erosion

Mountains, mesas, canyons, and badlands

Fig. 3–2. Principal areas, degrees, and types of erosion in the USA, based on erosion surveys, ca. 1935. The areas delineated have at least 25% of the land affected by the type and degree of erosion severity indicated. (Based on Bennett, 1939.)

even though agricultural technology improved during those decades. He attributed this lack of yield increase to loss in productive potential due to erosion. He further pointed out that these effects of erosion were much more serious in some localities, because many additional thousands of fields had to be abandoned due to lost productivity and others were so gullied that they were impassable with farm machinery. Stallings (1950) reported similar data and conclusions.

Current technological advances and the energy subsidies undergirding much of this technology account for much of today's agricultural production. Therefore, contemporary crop production requires proportionately less from the soil (for example, nutrients) compared to a century ago or even a few decades ago. This technological differential in time is reflected in historical estimates of the impact of soil erosion on the nation's forfeited productivity and requires another look at Bennett's (1939) concept of "ruined" land. In the 1930s, Bennett (1939) claimed that soil erosion had reduced the nation's agricultural productivity potential by 35%, whereas present erosion-induced productivity losses are unknown but have been estimated to range from 15% to less than 5% when projected into the future (Pimentel et al., 1976; Benbrook and Miller, 1981; Larson et al., 1983).

Although Bennett and others attempted to quantify the impact of erosion on soil productivity and physical deterioration, their terminology for the soil damage classes implies a degree of destruction and permanency of injury that is inapplicable to many of these lands today. Much of the "ruined" land described by Bennett produces timber and forage today, and some of this land is even being cultivated. Thus the term "ruined" land, as so frequently quoted from Bennett and used as an erosion classification unit (Bennett, 1939), must be understood as a designation that, while appropriate in Bennett's day, has lost its harshness. Many of those lands that were severely eroded and classed as "ruined" in the mid-1930s remain irreparably harmed. But some, if not much or most, of the land that Bennett (1939) described as "ruined" is or can be productive under various levels of management. Of course, the intensity of these management levels and their resource requirements are higher than would have been the case without erosion, and future productivity of much of this land must be technologically augmented or remain far below the land's original potential. The possibility for partial restoration may dilute Bennett's word "ruined" but in no way lessens the severity of the erosion he depicted.

Subsequent surveys (USDA, 1981) have refined the information base on the erosion damage to U.S. lands. It is difficult, however, to compare these surveys with the erosion estimates of the 1930s because different survey techniques were used. In comparing these various surveys, Larson et al. (1983) pointed out that the bases of these erosion estimates are obscure, inadequate, and variable, thereby rendering unanswerable the question of whether soil erosion on U.S. cropland has increased or decreased since the 1930s. To complicate the issue further, land use had changed from one survey period to the next, coupled with a decrease in the amount of hay and forage land as the number of draft animals has declined. Furthermore, the severity of erosion and its effect on production are not uniform from one soil region to another.

Losing 25 cm (10 in.) of soil from the Udolls of Iowa or Xerolls of the Palouse region would not have the same effect on productivity as similar amounts of soil loss from the Udults of the Southern Piedmont. In each area, the same tonnage of soil may be swept seaward, but the damage to the original soil profile and the productivity of the remaining profile are quite different if crop production or rooting depth are criteria by which erosion damage is based. The relatively thin A and E horizons of the Ultisols of the Southeast are commonly underlain by clay-enriched argillic horizons as well as fragipan horizons which commonly provide a hostile physical and chemical rooting environment. Once the topsoil and the eluvial horizon are removed from these soils, roots can develop only in the subsoil. Root growth is often restricted because of high physical resistance to root penetration, inadequate moisture availability, low fertility and pH, and aluminum toxicity. Erosion damage to Corn Belt Mollisols is not as severe as similar erosion losses from southeastern Udults because the mollic epipedons are thicker than Ultisol topsoil and therefore the remaining soil profile provides a physical and chemical environment more conducive to root development. Erosion of Corn Belt Mollisols does, however, remove a more valuable epipedon segment compared to similar losses of the less productive epipedons of the southeastern Ultisols.

Erosion, therefore, has affected U.S. agricultural production most severely in those areas of thin topsoils, erosive rainfall, and infertile and dense subsoils where moisture-holding capacity is low—namely, the Ultisol region. Much of the land that has been withdrawn from crop production is in the Southeast where either gullying, topsoil loss, or rolling and dissected topography have rendered cultivation impossible and unprofitable. Similarly, large areas of severely eroded and erosion-dissected land in east-central Oklahoma and northern Texas have been withdrawn from cultivation. Much of this land has reverted to forest, pasture, or rangeland. Therefore, many of the hectares cropped today are not the same hectares cropped during the 1930s. Innovations in fertilizer technology, drainage, irrigation, and grading have resulted in U.S. agriculture concentrating on better quality and correspondingly less erodible lands today compared to 50 years ago (USDA, SCS, 1980). The consequence of soil erosion on crop productivity, therefore, is not easily generalized and must be assessed on the basis of a soil taxonomy in which the subsoil and substrata are characterized as a medium for root development.

Nevertheless, soil erosion is still a serious problem in many areas of the country. Current estimates of annual gross losses due to sheet, rill, and wind erosion from the nation's nonfederal cropland, rangeland, pastureland, and forest land total about 5 billion t (USDA, 1981). In addition, about 1 billion t of soil erode annually from streambanks, gullies, roadsides, and construction sites located on nonfederal lands. Despite changes in land use and in erosion sources, these contemporary soil losses from the nonfederal landscape are not greatly different from those of the 1930s.

Annual soil erosion losses in excess of 5 billion t from the nation's nonfederal lands are reported from estimates made nearly 50 years ago (Bennett, 1939). Annual cropland soil losses were reported by Bennett (1939) in the neighborhood of 3 billion t during the 1930s compared to about 2 billion t eroded from cropland in the form of sheet and rill erosion during the 1970s (USDA, 1981). While gullying is still a problem in some areas, it is not as prevalent as in the 1930s. Natural or geologic erosion may account for as much as 15 to 30% of the erosion from these lands, but the impact of man on the erosion process is obvious.

Judson (1968) used sediment delivery to calculate the impact of humankind on the North American continent. Sediment delivery to the oceans has increased threefold over geologic sediment discharge rates according to this assessment. Since sediment delivery represents only a small fraction of the total amount of soil eroded, the accelerated erosion rate for many landscapes often exceeds the geologic erosion rate by one or more orders of magnitude. Figure 3-3, developed from data on Mid-Atlantic area landscapes (Wolman, 1967), shows the schematic sequence of sediment yield from various land uses. As the forests were cleared and cultivated and as the grasslands were plowed, similar erosion and sediment yield responses occurred for each landscape, although the amplitude of the curve would be modified by the erodibility of the soil. This schematic diagram shows clearly that the landscape's response to change in use has had a tremendous impact on erosion rates.

The map in Fig. 3-4 indicates the areas of the USA currently affected by erosion. Comparing this map with that of the 1935 map (Fig. 3-2) shows the result of the shift in land use over the last half century. Much of the previously scarred land is today eroding very little because of its withdrawal from cultivation and conversion to permanent cover. The sediment load and turbidity of streams draining these early cultivated lands have abated significantly since the early part of this century (Trimble, 1974). The major erosion areas in the USA have shifted, therefore, as the population moved

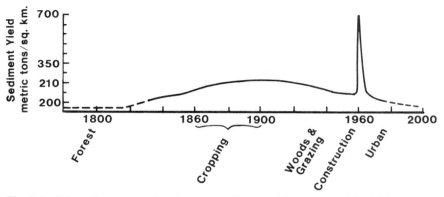

Fig. 3-3. Schematic sequence: Land use and sediment yield from a fixed Mid-Atlantic area landscape (Wolman, 1967).

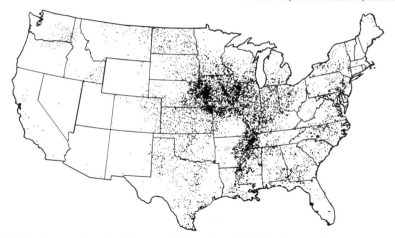

Fig. 3–4. Total tons of cropland sheet and rill erosion: 1977. One dot equals 250 000 t of soil eroded annually (USDA, 1981).

westward and the more erodible, less productive and gullied lands gave way to more productive but still erodible soil regions.

3–9 SUMMARY

The sweep of humanity across North America since the 1700s has altered the face of the land. Much of the land and its native vegetation have been changed by clearing, burning, cultivation, abandonment, introduction and invasion of non-native species, drainage, irrigation, grading, addition of soil amendments, and other manipulations. Some of these modifications of the native ecology resulted in ecological reactions that manifest themselves in the form of hydrologic changes, shifts in fauna and flora, macro- and micro-climatic alterations, and accelerated erosion.

Evidence of the erosional impact remains on the land today in the form of gullied and dissected landscapes. Extensive areas exist where much or all of the topsoil has been removed. Early surveys indicate that 20 to 40 million ha (50 to 100 million acres) have been severely damaged or ruined in the USA. Many of these severely eroded and gullied lands have been withdrawn from cultivation, although they still support some level of agronomic or forest production. Despite these impacts, the soil resources of the nation in most areas still resemble the magnificent soil dowry inherited by the early settlers.

The greatest impact of erosion on the nation's soils occurred from 1830 to 1930. But the problem of erosion persists. It seldom takes the form of dramatic "black rollers" and awesome gullies. But the insidious, incremental losses of soil through sheet, rill, and wind erosion continue to remove nearly 3 billion t of soil annually from the nation's cropland and more than 2 billion t each year from U.S. pastureland, rangeland, and forest land (USDA, 1981).

While the talisman of the early settlers was the knowledge of more and better land to the west, the talisman of modern farmers is advancing production technology coupled with a calculus of what constitutes serious soil loss that differs greatly from that of soil scientists and conservationists.

3-10 EPILOGUE

The settlement of the USA was swift by historical clocks and exploitative of the land's resources. And yet the expeditious manner in which the land and its resources were harvested and developed released many people from agrarian ties, thereby permitting commerce to flourish and various institutions to be nourished. As a result, the developing nation grew to a position of economic dominance and military superiority among the hundreds of nations sharing this planet. Such a circumstance begs several questions.

1. Would this preeminence among nations have been possible under a more restrained settlement policy and stronger conservation ethic?
2. Would such a policy and stewardship ethic have sacrificed the rapid generation of exploitation-induced wealth from which our industry and commerce were nourished and grew?
3. Would we as a nation then have developed into a more pastoral society, perhaps jeopardizing the stabilizing influence of our country's contribution to today's free world?
4. Was soil erosion a necessary or accidental external cost of nation-building?
5. What would the nation's soil condition be now if a soil-conservation effort and land-stewardship consciousness had not occurred?
6. Would our agricultural productivity have suffered enough to have impacted significantly upon our capacity for fulfilling domestic needs and sustaining our current export trade?

While the erosion impact on U.S. agriculture has been real, the erosion-induced productivity reduction has not seriously hampered our nation's capacity for agricultural production within the framework of today's needs and resource availability. But this lack of any catastrophic impact should not allow our concern over erosion to wane.

The long-term harm of both historical and contemporary erosion in the USA may manifest itself in at least two ways. First, the production forfeited due to erosion may hasten the time when our unutilized potential cropland will have to be brought into production. Our rich endowment of arable land has allowed us the luxury of consuming portions of our soil resources through erosion and nonagricultural uses while still retaining a large and productive cropland resource. Second, erosion-induced productivity reduction of our soil resources will increase the consumption of energy-intensive technology. Some erosion-damaged soils will eventually reach the threshold of their economic productivity even with technological inputs, just as the severely damaged soils have already been withdrawn from production.

The land still available in the USA for crop production coupled with advances in agricultural technology have allowed the USA to weather its erosion losses without serious national impacts. Similar, or perhaps even lesser, amounts of erosion on the smaller soil-resource endowments of other nations have proved catastrophic (Lowdermilk, 1948). If the past is prologue, then the erosion experience of the USA must be viewed in the context of history where extenuating circumstances have allowed the USA to suffer significant erosion without jeopardizing its agricultural productivity as measured by today's agricultural demands.

Although erosion has made an impact on both current and future agricultural production in the USA, there is still disagreement over its extent, its effects on soil productivity and the environment, and its socioeconomic impacts (Larson et al., 1983). This holds for both contemporary and historical erosion in the USA. That many landscapes are devoid of their original soil resource and that soil erosion continues to gnaw away at other landscapes provide a testimonial that soil erosion remains a universal paradigm of humankind's environmental impact. If soil erosion had been lessened throughout history, so too would history have been rewritten.

REFERENCES

Beasley, R. P. 1972. Erosion and sediment pollution control. The Iowa State University Press, Ames.

Benbrook, C., and A. Miller. 1981. Soil and water conservation: Production pressure, conventional wisdom, and research needs. Background paper for Natl. Workshop on Soil and Water Resources: Research Priorities for the Nation, Madison, WI. 23–27 Feb. Am. Soc. Agron.

Bennett, H. H. 1928. The geographical relation of soil erosion to land productivity. Geogr. Rev. 18(4):583.

----. 1931. National program of soil and water conservation. J. Am. Soc. Agron. 23:357–371.

----. 1935a. Report of the chief of the Soil Conservation Service (to the Secretary of Agriculture). Mimeograph, USDA–SCS, Washington, DC.

----. 1935b. Facing the erosion problem. Science 81:321–326.

----. 1939. Soil conservation. McGraw-Hill Book Co., New York.

----, and W. R. Chapline. 1928. Soil erosion a national menace. USDA Agric. Circ. 33. U.S. Government Printing Office, Washington, DC.

Bidwell, P. W., and J. I. Falconer. 1925. History of agriculture in the northern United States, 1620–1860. Publication 358. Carnegie Institution of Washington. The Lord Baltimore Press, Baltimore, MD.

Bouillenne, R. 1962. Man, the destroying biotype. Science 135:706–712.

Bradley, J. H. 1935. Autobiography of earth. Coward-McCann, Inc., New York.

Carter, V. G., and T. Dale. 1974. Topsoil and civilization. University of Oklahoma Press, Norman.

Coffey, G. N. 1912. A study of the soils of the United States. p. 1–114. In USDA Bureau of Soils Bull. 85. U.S. Government Printing Office, Washington, DC.

Craven, A. O. 1932. Edmund Ruffin, southerner. Appleton-Century-Crofts, New York.

Dregne, H. E. 1982. Historical perspective of accelerated erosion and effect on world civilization. p. 1–14. In B. L. Schmidt et al. (ed.) Determinants of soil loss tolerance. Spec. Publ. 45, American Society of Agronomy, Madison, WI.

Dubos, R. 1976. Symbiosis between the earth and humankind. Science 193:459–462.

Edwards, E. E. 1940. American agriculture: The first 300 years, farmers in a changing world. p. 171-276. *In* USDA Yearbook. U.S. Government Printing Office, Washington, DC.

Eliot, J. 1760. Essays upon field husbandry in New England. Edes and Gill. Boston, MA.

Fuller, G. L. 1935. Georgia land use problems. p. 99-107. *In* Georgia Exp. Stn. Bull. 191.

Gray, L. C. 1933. History of agriculture in the southern United States to 1860. Publication 430. Carnegie Institution of Washington, DC. Waverly Press, Inc., Baltimore, MD.

Held, R. B., and M. Clawson. 1965. Soil conservation in perspective. The Johns Hopkins University Press, Baltimore, MD.

Hilgard, E. W. 1860. Report on the geology and agriculture of the state of Mississippi. E. Barksdale, state printer, Jackson, MS.

Jacks, G. V., and R. O. Whyte. 1939. Vanishing lands: A world survey of soil erosion. Doubleday, Doran, NY.

Judson, S. 1968. Erosion of the land. Am. Sci. 56(4):356-374.

Larson, W. E., F. J. Pierce, and R. H. Dowdy. 1983. The threat of soil erosion to long-term crop production. Science 219:458-465.

Lentz, G. H., J. D. Sinclair, and H. G. Meginnis. 1929. Erosion Report—Lafayette County, Mississippi. USDA Forest Service Report. U.S. Government Printing Office, Washington, DC.

Lord, R. 1938. To hold this soil. USDA-SCS Misc. Publ. No. 321. U.S. Government Printing Office, Washington, DC.

Lowdermilk, W. C. 1935. Soil erosion and its control in the United States. SCS-MP-3, USDA–SCS. U.S. Government Printing Office, Washington, DC.

----. 1948. Conquest of the land through 7000 years. SCS-MP-32. USDA–SCS. U.S. Government Printing Office, Washington, DC.

McDonald, A. 1938. Erosion and its control in Oklahoma Territory. MP-301. USDA. U.S. Government Printing Office, Washington, DC.

----. 1941. Early American soil conservationists. Misc. Publ. No. 449. USDA–SCS. U.S. Government Printing Office, Washington, DC.

Madden, C. H. 1974. Land as a national resource. p. 6-30. *In* C. L. Harriss (ed.) The good earth of America, planning our land use. Prentice-Hall, Englewood Cliffs, NJ.

Marsh, G. P. 1864. Man and nature. Harvard University Press, Cambridge, MA; 1965 reprint.

National Resources Board. 1935. Soil erosion: A critical problem in American agriculture. Prepared for the Land Planning Committee of the National Resources Board. U.S. Government Printing Office, Washington, DC. p. 1-112.

Nicholson, M. 1970. The environmental revolution: A guide for the new masters of the world. McGraw-Hill Book Co., New York.

Osborn, F. 1948. Our plundered planet. Little Brown and Co., Boston, MA.

Petulla, J. M. 1977. American environmental history: The exploitation and conservation of natural resources. Boyd & Fraser Publishing Co., San Francisco, CA.

Pimentel, D., E. C. Terhune, R. Dyson-Hudson, S. Rochereau, R. Samis, E. A. Smith, D. Denman, D. Reifschneider, and M. Shepard. 1976. Land degradation: Effects on food and energy resources. Science 194:149-155.

Rasmussen, W. D. (ed.) 1975. Agriculture in the United States: A documentary history. Random House, New York.

----. 1982. History of soil conservation, institutions and incentives. p. 3-18. *In* H. G. Halcrow et al. (ed.) Soil conservation policies, institutions and incentives. Soil Conserv. Soc. of Amer., Ankeny, IA.

Report of the Commissioner of Patents. 1853. Part II Agriculture for the year 1852. 32nd Congress, 2nd Session. House of Representatives Exec. Doc. No. 65. Washington, DC.

Ruffin, E. 1855. Essays and notes on agriculture. J. W. Randolph, Richmond, VA.

Schlebecker, J. T. 1975. Whereby we thrive: A history of American farming, 1607-1972. Iowa State University Press, Ames.

Sears, P. B. 1980. Deserts on the march. 4th ed. Oklahoma University Press, Norman.

Smith, F. E. 1971. Conservation in the United States. A documentary history: Land and water 1900-1970. Chelsea House Publishers, New York.

Stallings, J. H. 1950. Erosion of topsoil reduces productivity. Tech. Publ. No. 98. USDA–SCS. U.S. Government Printing Office, Washington, DC.

Thomas, W. L., Jr. (ed.) 1956. Man's role in changing the face of the earth. University of Chicago Press, Chicago, IL.

Trimble, S. W. 1974. Man-induced soil erosion on the Southern Piedmont, 1700–1970. Soil Conserv. Soc. of Am., Ankeny, IA.

Tuan, Y. 1966. Man and nature. Landscape 15(3):30–36.

USDA. 1981. Soil and Water Resources Conservation Act. 1980 RCA Appraisal, Parts I, II, and Summary. U.S. Government Printing Office, Washington, DC.

USDA–SCS. 1980. America's soil and water: Condition and trends. U.S. Government Printing Office, Washington, DC.

White, C. A. 1983. A history of the rectangular survey system. U.S. Department of Interior. U.S. Government Printing Office, Washington, DC.

White, L., Jr. 1967. The historical roots of our ecological crisis. Science 155:1203–1207.

Wolman, M. G. 1967. A cycle of sedimentation and erosion in urban river channels. Geogr. Ann. 49A:385–395.

4   An Appraisal of Soil Resources in the USA

R. J. McCracken, J. S. Lee, R. W. Arnold, and D. E. McCormack
Soil Conservation Service
U.S. Department of Agriculture
Washington, DC

Can the USA's soil resources continue to meet the nation's long-term food needs and sustain our export capability? Or will we someday face declining soil productivity because of soil erosion? These questions are being asked today, not only by soil scientists and conservationists (Larson et al., 1981) but also by the public. There are no simple answers.

Programs to reduce erosion and sustain productivity can be effective only if they are based on accurate appraisals of our soil resources. The major soil resource appraisals in the USA are the National Cooperative Soil Survey (NCSS) and the National Resources Inventory (NRI). The NCSS is a soil survey and interpretation program carried out by the Soil Conservation Service (SCS), other agencies of USDA, state agricultural experiment stations, and other state and local agencies. The SCS is the designated lead agency.

The NRI was conducted by SCS in 1977 (USDA, 1982b) and again in 1982. It provides information on the status, condition, and trends of soil erosion, land use, and many related natural resources issues. Both these appraisals are used for planning and carrying out action programs for conserving resources.

The principal soil and water conservation programs of USDA were established in the 1930s, amid the stark atmosphere of the Great Depression and the Dust Bowl. As people became aware of the droughts and dust storms of the period, they gave broad support for new programs of soil and water conservation. Because of the extent of erosion damage, there was little question of the need for Federal action. There was also little emphasis on monitoring the progress of the new programs.

Published in R. F. Follett and B. A. Stewart, ed. 1985. *Soil Erosion and Crop Productivity.*
© ASA-CSSA-SSSA, 677 South Segoe Road, Madison, WI 53711, USA.

How effective were those programs? Many studies have been conducted, including audits by the General Accounting Office (GAO) and evaluations by various USDA agencies and organizations and individuals outside the government (USDA, 1982a; GAO, 1977).

In February 1977, GAO issued a report on three major USDA conservation programs: Conservation Technical Assistance, Agricultural Conservation Program, and Great Plains Conservation Program. The GAO concluded that:

1. Each program fell short of achieving its erosion-control objective.
2. Program personnel and funds were directed toward broad objectives, such as water management and increased production, rather than specific goals for erosion control.
3. There was no indication that agency assistance was being targeted to farmers and ranchers with the most serious erosion problems.
4. None of the three programs was concentrating its resources on the most effective erosion control measures.

Both SCS and the USDA Agricultural Stabilization and Conservation Service made program changes in response to the GAO study. The changes included more emphasis on enduring conservation practices, revision of procedures for conservation planning, and redirection and targeting of funds and technical assistance. The results of the changes have yet to be evaluated, but a thorough evaluation of all USDA conservation programs was required by the Soil and Water Resources Conservation Act (RCA) of 1977 (Public Law 95-192). In response, SCS is now extensively evaluating its Conservation Technical Assistance Program. The RCA also required the Secretary of Agriculture to appraise the nation's nonfederal soil and water resources and to develop, in consultation with the public, a national program that would ensure the sustained productivity of these resources.

Much of the recent discussion of soil erosion and crop productivity has occurred in the context of the RCA program appraisal development. No longer are these topics of interest mainly to conservationists, scientists, and concerned groups. Instead, they form the basis for lively debate about government conservation policies and the proper role of universities and the private sector. "Everybody knows" that erosion reduces soil productivity. But the details—how severe is the effect, how lasting is it, where is it most serious, and many others—have important implications, not only for those of us who work in the agricultural sector but also for all who are sustained by it.

4-1 NATIONAL COOPERATIVE SOIL SURVEY

The National Cooperative Soil Survey has been describing and mapping soils in the USA since about the turn of the century. Early surveys were limited to important crop and range areas, and the scale of published maps was generally 1:63 360 (approximately 1.6 cm = 1 km; 1 in. = 1 mi). In the 1930s, soil survey methods were improved. Map scales increased to 1:15 840 (approximately 6.3 cm = 1 km; 4 in. = 1 mi), and soil information was more detailed.

Soil surveys delineate landscape areas that have similar soils, slope, erosion, and expected responses to management. In identifying and classifying different kinds of soils, the basis for consistency is a set of soil properties thought to be related to soil development. The soil classification system used by NCSS allows the placement of soil series into broader groups for progressively more general interpretations: soil families, subgroups, great groups, suborders, and orders (Soil Survey Staff, 1975). These broader groups are also used as mapping units of soil surveys in areas of low-intensity actual or expected land use.

Generalized small-scale soil maps have been made for the USA (Fig. 4-1, Table 4-1) and for each state. The general soil map of the USA identifies 27 suborders of soils that have been delineated in 61 areas.

Detailed modern soil surveys have been undertaken since the 1950's. Most are published on aerial base maps to help users locate soil boundaries. These surveys have been completed for approximately 65% (608 million ha; 1.5 billion acres) of the total U.S. land area and 90% of U.S. cropland.

The "once-over" detailed soil mapping of the country is expected to be completed in about 20 years, but some remapping will be required. Digitization of soil map unit information for computer storage is accelerating. It permits ready retrieval and manipulation of map data for planning and for producing custom maps.

About 13 000 kinds of soils (soil series) have been recognized in the USA. (More than twice as many kinds of map unit delineations exist when slope, erosion, rocky, and stony phases are considered.) Much information about many of these soils is stored in several separate computer data bases, but we need more information on the effects of soil properties on productivity and other aspects of soil quality. A coordinated computerized data system is needed to facilitate ready access to the soil information generated by the NCSS (Flach and Johannsen, 1981).

Interpretations of soils are based on experience and knowledge of the behavior of soils and the interaction of their properties when used for specific purposes. Many types of interpretations are possible, including suitability and limitations of use and the expected production of crops and forage.

Several methods are used to group the soils according to taxonomic properties or soil characteristics important for specific land uses. Two of the most important technical groupings and derived geographic classifications are major land resource areas (MLRA) (USDA, 1981a) and land capability classification (USDA, 1961).

4-1.1 Major Land Resource Areas

An MLRA is a geographically associated group of smaller areas called resource units. Each land resource unit is characterized by a particular

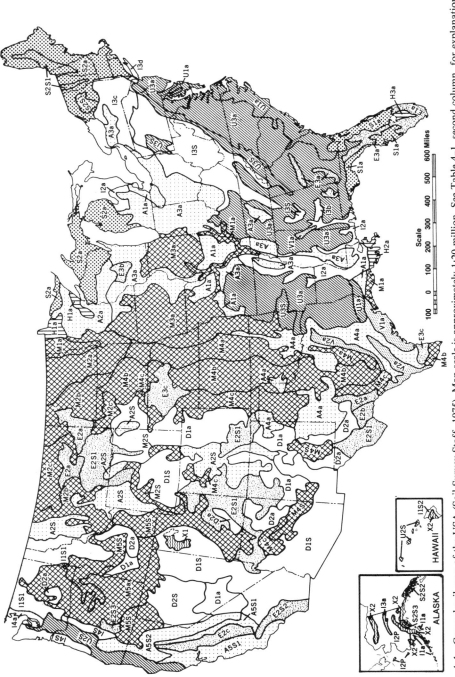

Fig. 4–1. General soil map of the USA (Soil Survey Staff, 1975). Map scale is approximately 1:20 million. See Table 4–1, second column, for explanation of the map unit symbols.

Table 4-1. Soil orders and suborders shown on the general soil map of the USA (Fig. 4-1).†

Order and suborder	Map symbol	Land area‡	Major land uses	Fertility§
		%		
Alfisols		13.4		
Aqualfs	A1	1.0	cropland	high
Boralfs	A2	3.0	forest land	high
Udalfs	A3	5.9	cropland	high
Ustalfs	A4	2.6	cropland	high
Xeralfs	A5	0.9	rangeland	high
Aridisols		11.5		
Argids	D1	8.6	rangeland	low
Orthids	D2	2.9	rangeland	low
Entisols		7.9		
Aquents	E1	0.2	wetland	moderate
Orthents	E2	5.2	range/forest	moderate
Psamments	E2	2.2	cropland	low
Histosols	H	0.5	wetland	moderate
Inceptisols		18.2		
Andepts	I1	1.9	forest/range	moderate
Aquepts	I2	11.4	cropland	moderate
Ochrepts	I3	4.3	crop/forest	moderate/low
Umbrepts	I4	0.7	forest land	low
Mollisols		24.6		
Aquolls	M1	1.3	cropland	high
Borolls	M2	4.9	cropland	high
Udolls	M3	4.7	cropland	high
Ustolls	M4	8.8	crop/range	high
Xerolls	M5	4.8	crop/range	high
Spodosols		5.1		
Aquods	S1	0.7	forest	low
Orthods	S2	4.4	forest	low
Ultisols		12.9		
Aquults	U1	1.1	forest	low/moderate
Humults	U2	0.8	forest	low
Udults	U3	10.0	forest/crop	low
Vertisols		1.0		
Uderts	V1	0.4	cropland	high
Usterts	V2	0.6	cropland	high
Areas with little soil	X	4.5	barren land	--

† Adapted from USDA (1981b), Table 20, p. 78–80.

‡ Percentages do not total because of rounding.

§ Terms for fertility are comparative and have no quantitative values. On all soils, fertilizer must be used for continued high production of cultivated crops.

pattern of soils, climate, water resources, and land use. Agriculture Handbook 296 (USDA, 1981a) describes major characteristics of each of the 204 MLRAs in the USA. The information includes extent and kinds of land use, relative extent of Federal or privately owned land, range of elevation, topography, annual precipitation and its seasonal distribution, average annual temperature, freeze-free period, water resources, major soils, and major plant species.

Major land resource areas are identified by descriptive geographic names. The smallest MLRA is the Subhumid Intermediate Mountain Slopes of Hawaii (285 km^2; 110 mi^2); the largest is the Northern Rocky Mountains (282 620 km^2; 109 130 mi^2). For most broad land-use planning, an MLRA can be treated as an agro-ecological zone with relatively homogeneous conditions.

The 204 MLRAs are geographically grouped into 24 land resource regions. The name of a region indicates its location and dominant land uses; for example, Central Great Plains Winter Wheat and Range Region. The major characteristics of each region are its physiographic setting, climatic range, dominant soils, major land uses, and important crops. A few MLRAs occur in more than one region. Many important statistical data are collected by political or administrative areas. To facilitate comparison and compilation of such data, a version of the map with boundaries adjusted to county lines is available. Data and maps for MLRAs can be associated with other information for planning and monitoring land use and conservation programs.

4–1.2 Land Capability Classification

The land capability classification system rates the degree of restrictions in the use of soils for most cultivated crops and pasture plants (USDA, 1961). The system assumes that modern farming methods are used. There are eight capability classes and four capability subclasses.

In general, soils in capability classes I to IV are considered arable, with limitations that range from slight to severe. Soils and land areas in classes V to VIII are generally unsuitable for cropland because they are steep, shallow, wet, stony, droughty, or have other severe limitations. Figure 4–2 shows the percentages of land in capability classes I to VII, by crop production region.

Capability subclasses are groups of soils that have the same dominant limitation for cropland. These limitations are erosion (subclass e), wetness (w), soil conditions in the root zone (s), or climate (c). (Class I has no subclasses.)

Of the nation's rural nonfederal land, about 58% (324 million ha; 802 million acres) is in capability classes I to IV. About one-fourth of this land is Class IV soils with marginal value as cropland (Fig. 4–3). The rest—soils in classes I to III—generally are suited for frequent cultivation and a wide range of other uses.

Land in classes V to VII generally is not suitable for cultivated crops but can be used for pastureland, rangeland, wildlife habitat, or recreation. Land in these classes makes up 42% of nonfederal rural land (USDA, 1981b).

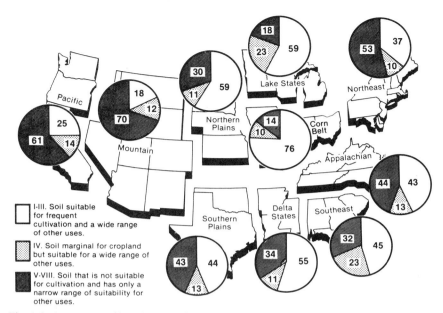

Fig. 4–2. Percentage of land in capability classes I to VIII, by crop production region (USDA, 1981b).

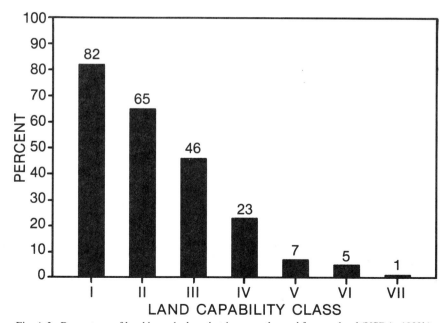

Fig. 4–3. Percentage of land in each class that is currently used for cropland (USDA, 1982b).

4-2 NATIONAL RESOURCES INVENTORY

The National Resources Inventory of 1977 conducted by the Soil Conservation Service has provided the most comprehensive data base on the status of nonfederal soil resources and their condition and trend (USDA, 1982b). It was not designed with support of the RCA appraisal in mind, but it has served that purpose well. The field work for the National Resources Inventory of 1982 has been completed. The data have been summarized and preliminary summaries were released in July, 1984. Availability of the 1982 NRI data will allow more accurate quantitative analyses of the trends and conditions of soil, water resources, and land use. (Detailed statistical tables are scheduled to be released by SCS in late 1985.)

For the 1977 NRI, SCS field personnel visited 70 000 primary sample units (PSUs). Most PSUs had an area of 65 ha (160 acres), and three points were examined per PSU of this size. A Graeco-Latin square design was used in the randomization procedure to spread the three points throughout the sample plot (Goebel and Schmude, 1980). Iowa State University Statistical Laboratory prepared the statistical design and expanded the point data for 11 natural resource elements.

The expanded 1982 NRI included 350 000 PSUs and about 1 million sampling points. Data were collected on 22 natural resource elements. Some of the 1982 data were collected to enable estimates of changes from 1977 to 1982. Other kinds of data were added at the request of potential users. The additions included data on crtically eroding areas, windbreaks, riparian vegetation, and cover conditions. In addition, more data were collected on soils, irrigation, wetlands, and water bodies.

The universe for the 1977 and 1982 NRI's consisted of the nonfederal land in the contiguous 48 states, Hawaii, Puerto Rico, and the Virgin Islands. (Alaska was excluded.) Each of the 3135 counties (or similar units) in these jurisdictions constituted a separate unit for sampling purposes.

Sampling procedures and the sampling rates varied somewhat from county to county, but all the county samples were stratified, two-stage, area samples. The first-stage sampling unit, a PSU, was an area of land. In most areas of the country, the PSU was a 65-ha (160-acre) square. Within each sample PSU, one or more sample points were selected for observation (Goebel and Baker, 1982).

4-3 THE RCA APPRAISAL: FOCUS ON EROSION AND PRODUCTIVITY

During the RCA appraisal, a large amount of data collected for the NRI and NCSS programs was analyzed to obtain a comprehensive view of resource conditions and land use on the nation's nonfederal land (USDA, 1981b). The discussion here focuses on those issues relating to soil erosion on cropland, crop productivity, and the supply of soil resources.

4–3.1   Soil Erosion on Cropland

Most of the soil materials lost by erosion come from cropland. One-half of all sheet and rill erosion and six-tenths of all wind erosion occur on cropland. Three-tenths of all sheet and rill erosion and four-tenths of all wind erosion occur on rangeland (USDA, 1981b). Sheet and rill erosion on all rural land accounts for four-fifths of all erosion by water.

Erosion is the dominant conservation problem on more than 84 million ha (208 million acres) of cropland—one-half of our total cropland. On this land, the average annual rate of wind and water erosion is 15.2 t/ha (6.8 tons/acre). On the other half of the cropland, the combined annual erosion rate is only 5.8 t/ha (2.6 tons/acre) (USDA, 1981b). Other forms of soil degradation include compaction, salinization, and subsidence, but these cause less total damage and affect a smaller land area. Cropland soils on which sheet and rill erosion is the dominant conservation problem are placed in *e* subclasses of the land capability classification system. Sheet and rill erosion on these soils is shown in Table 4–2.

A similar pattern appears in a comparison of wind erosion and cropland soils in *e* subclasses in the Great Plains states (Fig. 4–4). Figure 4–5 shows the rates of wind and water erosion on all cropland in each of these states.

Figure 4–6 provides a perspective on cropland sheet and rill erosion by crop production region. The volume of this erosion in the Corn Belt is about 625 million metric tons (690 million tons), twice the volume in any other region. The annual rate per hectare in the Corn Belt—15.7 t/ha (7 tons/acre)—is also higher than that in any other region, except the Appalachian states (about 20 t/ha; 9 tons/acre) and the Caribbean Area (63 t/ha; 28 tons/acre) (USDA, 1981b).

Only 10% of the nation's cropland accounts for 54% of all sheet and rill erosion. The same 10% constitutes 89% of all cropland that is eroding in excess of 22.4 t/ha (10 tons/acre) per year (USDA, 1982a).

Table 4–2. Area and annual rate of sheet and rill erosion on cropland soils in selected *e* subclasses.[†]

Capability class	Area of soils in *e* subclasses[‡]	Average annual soil loss[§]	
		e subclasses	Other subclasses
	million ha	t/ha	
II	36	11.0	5.8
III	32	15.5	6.3
IV	12	19.7	4.5
VI	4	33.8	5.2

[†] Data from USDA (1982b).
[‡] To convert figures to million acres, multiply by 2.47.
[§] To convert figures to tons per acre, multiply by 0.446.

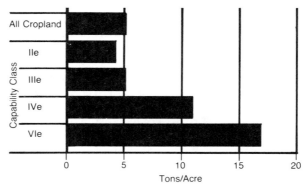

Fig. 4-4. Average annual wind erosion on erosion-prone soils in the Great Plains states (USDA, 1981b). To convert tons per acre to tons per hectare, multiply by 2.24.

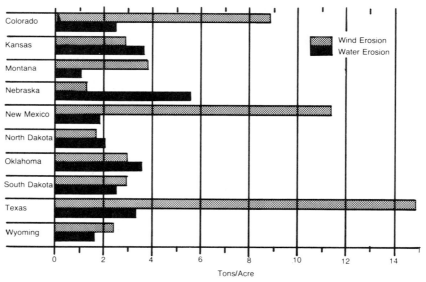

Fig. 4-5. Average annual rates of wind erosion and sheet and rill erosion on cropland in the Great Plains States (USDA, 1981b). To convert tons per acre to tons per hectare, multiply by 2.24.

4-3.1.1   Soil Loss Tolerance

Combined water and wind erosion exceeds tolerable levels on nearly 120 million ha (300 million acres) of nonfederal land in the USA. "Tolerable" refers to soil loss tolerance (T), which is defined as "the maximum rate of annual soil erosion that will permit a high level of crop productivity to be obtained economically and indefinitely" (McCormack et al., 1982). Deep soils with favorable subsoils have a T value of 5; that is, soil loss tolerance is 11.2 t/ha (5 tons/acre). Soils with thinner, less favorable subsoils have lower T values.

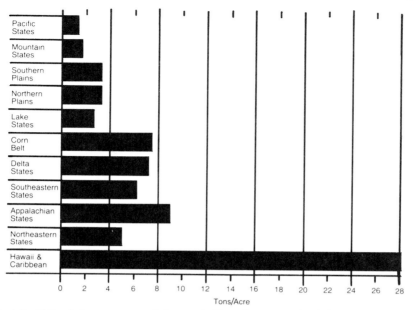

Fig. 4-6. Estimated average annual sheet and rill erosion on cropland, by crop production region (USDA, 1981b). To convert tons per acre to tons per hectare, multiply by 2.24.

However, T values need to be reevaluated. They imply that we know fairly precisely the degree to which erosion at a given rate affects the productivity of soils with assigned T values, but we do not know this "degree." We do know that a given erosion rate is not equally serious on all soils. On shallow soils with a T value of 2, erosion at a rate of 11.2 t/ha (5 tons/acre) could lead to relatively rapid productivity loss. In contrast, on some deep soils with a T value of 5, erosion at the same rate would not be expected to reduce soil productivity to the same extent.

Despite the imprecision of T values, they are useful in helping to gauge the relative sensitivity of a soil to continued erosion in excess of T. On the nation's cropland, the lower the T value, the more likely it is that erosion exceeds T (Table 4-3). On 30% of the cropland soils in e subclasses that have a T value of 5, the actual rate of sheet and rill erosion exceeds T. Of cropland with a T of 3, however, 54% is eroding in excess of T. The annual average rate of erosion on cropland with T of 5 is 13.2 t/ha (5.9 tons/acre), whereas the rate on cropland with T of 3 is 20.6 t/ha (9.2 tons/acre).

The T value and K, the soil erodibility factor of the universal soil loss equation (Wischmeier and Smith, 1978), can be used to identify soils that are potentially the most readily damaged by water erosion if used for cropland. The K value is closely related to soil erodibility, and the T value is related to topsoil thickness, the capability of the soil to regenerate itself, and characteristics of the root zone. Therefore, a soil with T of 2 and a high K factor is the most readily damaged. Soils with T of 4 or less comprise one-

Table 4-3. Average annual sheet and rill erosion on cropland soils in *e* subclasses, by soil loss tolerance value.†

T value‡	Extent	Average annual erosion	Extent of area with erosion greater than T value	
	million ha	t/ha	million ha	%
5	50.9	13.2	15.1	30
4	14.2	17.7	6.9	48
3	15.1	20.6	8.1	54
2	4.2	13.5	2.0	48

† Data from 1977 NRI. To convert hectares to acres, multiply by 2.47; to convert tons per hectare to tons per acre, multiply by 0.446.

‡ The T value is the soil loss tolerance in tons per acre per year. In metric units, erosion at the T level would be 11.2 t/ha for T = 5, 9.0 t/ha for T = 4, 6.7 t/ha for T = 3, and 4.5 t/ha for T = 2.

Table 4-4. Extent of cropland soils in various K-factor ranges with T value of 4 or less.†

T value	Total	K factor range		
		0.3	0.3-0.4	0.4
		million ha‡		
4	23.4	8.8	11.8	2.8
3	25.3	7.2	10.5	7.6
2	7.2	3.0	2.9	1.3
Total	55.9	19.0	25.2	11.7

† Data from 1977 NRI.

‡ For approximate equivalents in million acres, multiply figures by 2.47.

third of our nation's cropland (55.9 million ha; 138 million acres). Table 4-4 shows the extent of these soils in three K-factor ranges. The soils most susceptible to damage (T = 2, K ≥ 0.4) comprise 1.3 million ha (3.2 million acres).

4-3.1.2 Erosion and Productivity

In a recent symposium on soil loss tolerance, the consensus was that the present system has major weaknesses based on lack of sufficient data to predict the effects of soil erosion on productivity (Schmidt et al., 1982). In a review of these effects, Langdale and Shrader (1982) cite an urgent need for additional research to measure crop yield losses associated with erosion. They summarize previous work and describe the state of the art in modeling soil erosion and crop productivity.

In an analysis of this kind of modeling, Young (1984) cites a need not only for collecting more data but also for sorting out the yield effects of technology from those of erosion on various soils. For food and fiber production, knowing the rate of productivity loss—whatever it may be—would be more significant than knowing the actual volume of soil loss.

For conservation planning on erodible soils, more data collection is also needed to determine when the threshold to irreparable productivity loss may be crossed. Larson and colleagues (Larson et al., 1983, 1984; Pierce et al., 1984) have developed modeling procedures for estimating the decline in productivity. The procedures use erosion rates and selected soil properties and have been tested mainly in the Midwest.

4–3.2   Soil Resources and Productivity

Two-thirds of the USA is nonfederal land. This land consists of about 610 million ha (1.5 billion acres) of cropland, rangeland, pastureland, forest land, and other land. In 1977, three-fourths of the nonfederal land was about evenly divided among cropland, rangeland, and forest land (Fig. 4–7). Nine percent was pastureland, and twelve percent was urban and built-up areas and other nonagricultural land (USDA, 1981b).

Land use is not static, however. Cropland and forest land are decreasing in extent; pastureland and rangeland are increasing. Land in urban, transportation, and other nonagricultural uses is also increasing (USDA, 1981b). Along with soil degradation from erosion and other causes, the irreversible conversion of agricultural land to nonagricultural use is a major source of concern for the nation's long-term agricultural productivity.

4–3.2.1   Potential Cropland

Cropland extent in 1977 was about 167 million ha (413 million acres) (USDA, 1981b). The RCA analysis of NCSS and NRI data suggests that the

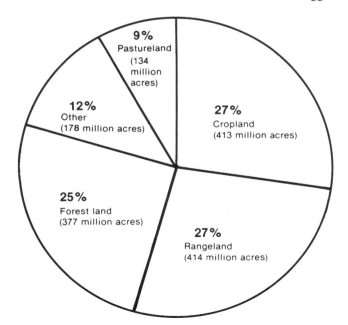

Fig. 4–7. Use of nonfederal land in the USA and the Caribbean Area, 1977 (USDA, 1981b). To convert acres to hectares, multiply by 0.4.

supply and productivity of U.S. cropland soils will be adequate through the next 50 years, provided that:

1. The rate of crop-yield increases due to technological advance equals or exceeds the present annual rate of 1.6%.
2. Improved conservation techniques, especially conservation tillage, are more widely applied.
3. Erosion-control efforts are targeted in areas where the soils are most vulnerable to erosion damage (USDA, 1981c).

World food needs are increasing. If more cropland is needed, where will it be located? Nearly all the best soils are already used for cropland. Many cropped soils appear to have reached the limit of their productive capacity at present levels of technology.

The National Resources Inventory of 1977 identified 51 million ha (127 million acres) of land with high or medium potential for cropland (USDA, 1982b). At present, 33 million ha (83 million acres) of this land are used for pastureland and rangeland, and about 10 million ha (25 million acres) are forest land.

The grazing lands that have high or medium potential for cropland account for 15% of all nonfederal pastureland and grazing land. Loss of meat production from grazing land could require the USA to import more meat, reduce meat consumption, or substitute grain feeding for grazing. Part of any loss of grazing land could be offset, however. At present, 7 million ha (18 million acres) of cropland are in classes V to VII (USDA, 1982b). This land could be converted to grazing land or hay production. Converting forest land to cropland would require similar tradeoffs. At present, more timber is being grown annually than is being cut. But future needs for lumber, fuel, and other forest products, and benefits such as wildlife habitat, must be kept in mind.

4-3.2.2 Prime Farmland

Prime farmland consistently produces high yields at relatively low economic cost and with the least damage to the environment. It has the best combination of physical and chemical characteristics for agriculture production. The USA has about 140 million ha (345 million acres) of prime farmland, about one-fourth of which is in the Corn Belt (Table 4-5). From 1967 to 1975, about 3.2 million ha (8 million acres) of prime farmland were lost to other uses (Schmude, 1977). Of this loss, 2.6 million ha (6.5 million acres) were converted to urban and built-up areas and 600 000 ha (1.5 million acres) to water areas.

Two-thirds of our prime farmland is used for cropland (93 million ha; 230 million acres), and most of the nation's crops are produced on prime farmland (USDA, 1981b). Export commodities produced on this land help to restrain our foreign trade deficit. The loss of prime farmland to nonagricultural uses puts more pressure on the nonprime lands, which generally are more erodible, droughty, difficult to cultivate, and generally less productive.

Table 4-5. Prime farmland, by crop production region.†

Region	Extent of prime farmland		
	million ha	million acres	%
Corn Belt	32.8	81.0	23.5
Northern Plains	24.5	60.5	17.6
Southern Plains	21.5	53.1	15.4
Lakes States	15.4	38.2	11.1
Delta States	13.3	32.8	9.5
Appalachian States	9.3	22.9	6.7
Southeast	8.3	20.5	6.0
Northeast	5.5	13.6	4.0
Pacific	4.9	12.2	3.5
Mountain	3.8	9.4	2.7
Total‡	139.3	344.2	100.0

† From USDA (1981b).
‡ Total does not include Hawaii, 92 000 ha (227 000 acres), and the Caribbean Area, 118 000 ha (291 000 acres). No data available for Alaska.

4-4 A GLOBAL VIEW OF SOIL RESOURCES

The area of the world's soil now under cultivation is estimated at 1.5 billion ha (3.8 billion acres) (Buringh, 1982). Estimates of the total area of cultivable soils range from about 3.3 billion ha (8.3 billion acres) (Buringh, 1982) to 9 billion ha (23 billion acres) (Clark, 1980). The latter figure appears excessively optimistic; about two-thirds of the world's reserve of potentially arable soils is in semiarid or arid climates. In addition, there are large areas of highly weathered and leached Oxisols and Ultisols, soils which require substantial inputs and high skills for maximum productivity.

The worldwide soil resource inventory is only about one-fourth completed at a level of detail adequate for analysis of productivity potential. Therefore, we must be cautious about predictions of sufficient availability of soils to support a growing world population (Van Wambeke, 1979). Buringh (1982) called for completion of soil surveys of potential agricultural land as soon as possible. The 1975 World Food and Nutrition Study (National Academy of Sciences, 1975) called for accelerated work on soil classification and land use of representative soils, with more readily available information systems. Much more site-specific information is needed to predict soil response to present and new technology (Van Wambeke, 1979).

The quality and degradation of soils are more serious problems in the developing countries than in the USA, and a number of these countries are already using most of their potentially arable soils (Dudal, 1982). Buringh (1982) estimates that 4 million ha (10 million acres) of the world's productive land are lost each year through soil degradation and that worldwide annual conversion of farmland to nonagricultural uses is 8 million hectares (20 million acres). The Global 2000 Report (Barney, 1979) estimates that urbanization is increasing twice as fast as population. A survey by the United Nations (1977) estimates that approximately one-fifth of the world's presently cultivated land is being degraded and that productivity has been reduced about 25% as a result. Lester Brown (1981) estimated that about one-third of the world's land is "being degraded." More precise estimates of land conversion are needed by decision-makers at national and international levels.

4-5 CONCLUSIONS

In late 1982, the USDA completed the development of a new national conservation program, as required by the Soil and Water Resources Conservation Act of 1977 (USDA, 1982a). The RCA studies rely heavily on data collected by the NRI and NCSS. This information indicates the relative severity of erosion on different soils, the relationship of erosion to land use and management, the relative hazards of continued erosion, and the availability of agricultural land. An analysis of this information and the RCA appraisal suggests several actions that would help maintain and increase the productivity of the nation's soils:

1. Increase the application of yield-raising technology. This will require more investment in research, extension, and education.
2. Obtain more specific information about the response of different soils to various types of technology.
3. Place a high priority on determining the separate effects of technology, weather, and soil erosion on crop yields.
4. Accelerate the application of proven and promising erosion control practices, especially conservation tillage, and extend the range of soils on which they can be applied. Additional research and development, and demonstration of conservation tillage systems are needed, especially on Udolls, Udalfs, and associated soils of the northern Corn Belt.
5. Monitor changes in land use to evaluate the need for policy, legislative, and zoning measures at state and local levels.
6. Give land users more assistance in soil conservation in areas where the soils are highly vulnerable to erosion damage.
7. Improve the quality and increase the use of NCSS and NRI data to identify and monitor resource status, conditions, and trends.

All these actions will require collecting more information on soil resources and making it more accessible to planners, land users, and others.

REFERENCES

Barney, G. O., (ed.). 1979. The global 2000 report to the President. U.S. Government Printing Office, Washington, DC.

Brown, L. R. 1981. World population growth, soil erosion and food security. Science 214: 995–1002.

Buringh, P. 1982. Potentials of world soils for agricultural production. p. 33–42. *In* Managing soil resources. Plenary Papers of 12th Int. Cong. of Soil Sci., New Delhi, India. Int. Soc. Soil Sci., Wageningen, the Netherlands.

Clark, C. 1980. Population and growth change in agriculture. Duckworth, London.

Dudal, R. 1982. Land degradation in a world perspective. J. Soil Water Conserv. 37:345–249.

Flach, K. W., and C. Johannsen. 1981. Land: Resource base and inventory. p. 11–13. *In* W. E. Larson et al. (ed.) Soil and water resources: Research priorities for the nation. Exec. Summ. Soil Sci. Soc. Amer., Madison, WI.

General Accounting Office. 1977. To protect tomorrow's food supply, soil conservation needs *priority* attention. Ed-22-30. U.S. Government Printing Office, Washington, DC.

Goebel, J. J., and H. D. Baker. 1982. The 1982 National Resources Inventory sample design and estimation procedure. Statistical Lab., Iowa State University, Ames.

––––, and K. O. Schmude. 1980. Planning the SCS National Resources Inventory. p. 148–153. *In* Arid land resources inventories workshop. USDA. For. Serv. Gen. Tech. Rep. WO-28. Washington, DC.

Heady, E. O. 1984. The setting for agricultural production and resource use in the future. p. 8–30. *In* Proceedings of RCA symposium on future agricultural technology and resource conservation (Washington, DC, December 1982). Center for Agric. and Rural Dev., Iowa State University, Ames.

Langdale, G. W., and W. D. Shrader. 1982. Soil erosion effects on soil productivity of cultivated cropland. p. 41–51. *In* B. L. Schmidt et al. (ed.) Determinants of soil loss tolerance. Spec. Pub. 45. American Society of Agronomy and Soil Science Society of America, Madison, WI.

Larson, W. E., F. J. Pierce, and R. H. Dowdy. 1983. The threat of soil erosion to long-term crop production. Science. 219:458–465.

––––, ––––, and ––––. 1984. Conservation of the nation's agricultural resources: Soil inventory and resource management. p. 40–59. *In* Proceedings of RCA symposium on future agricultural technology and resource conservation (Washington, DC, December 1982). Center for Agric. and Rural Dev., Iowa State University, Ames.

––––, L. M. Walsh, B. A. Stewart, and D. H. Boelter, (ed.). 1981. p. 21–40. *In* Soil and water resources: Research priorities for the nation. Soil Science Society of America, Madison, WI.

McCormack, D. E., K. K. Young, and L. W. Kimberlin. 1982. Current criteria for determining soil loss tolerance. p. 95–111. *In* B. L. Schmidt et al. (ed.) Determinants of soil loss tolerance. Spec. Pub. 45. American Society of Agronomy and Soil Science Society of America, Madison, WI.

National Academy of Sciences, 1975. World food and nutrition study. Washington, DC.

Pierce, F. J., W. E. Larson, R. H. Dowdy, and W. Graham. 1984. Productivity of soils— Assessing long-term changes due to erosion. J. Soil Water Conserv. 39:136–138.

Sampson, R. N. 1981. Farmland or wasteland—A time to choose. Rodale Press. Emmaus, PA.

Schmidt, B. L., R. R. Allmaras, J. V. Mannering, and R. I. Papendick (ed.). 1982. Determinants of soil loss tolerance. Spec. Pub. 45. American Society of Agronomy and Soil Science Society of America, Madison, WI.

Schmude, K. O. 1977. A perspective on prime farmland. J. Soil Water Conserv. 32:240.

Soil Survey Staff. 1975. Soil taxonomy: A basic system of soil classification for making and interpreting soil surveys. USDA-SCS Agric. Handb. 436. U.S. Government Printing Office, Washington, DC.

United Nations. 1977. Conference on desertification. Nairobi, Kenya. U.N. Food and Agricultural Organization, Rome, Italy.

USDA. 1961. Land capability classification. USDA-SCS Agric. Handb. 210. Washington, DC.

––––. 1981a. Land resource regions and major land resource areas of the United States. USDA-SCS Agric. Handb. 296. U.S. Government Printing Office, Washington, DC.

––––. 1981b. Soil, water, and related resources in the United States: Status, condition and trends. 1980 RCA Appraisal, Part I. USDA. Washington, DC.

––––. 1981c. Soil, water and related resources in the United States: Analysis of resource trends. 1980 RCA Appraisal, Part II. USDA, Washington, DC.

––––. 1982a. A national program for soil and water conservation: 1982 final program report and environmental impact statement. USDA, Washington, DC.

––––. 1982b. Basic statistics: 1977 National Resources Inventory. USDA Stat. Bull. 686. Washington, DC.

Van Wambeke, A. R. 1979. Land resources and world food issues. *In* World food issues. Center for Anal. of World Food Issues, Program in Int. Agric., N.Y. State College of Agriculture and Life Sciences, Cornell University, Ithaca, NY.

Wischmeier, W. H., and D. D. Smith. 1978. Predicting rainfall erosion losses—a guide to conservation planning. Agric. Handb. 537. USDA, Washington, DC.

Young, D. L. 1984. Modeling agricultural productivity impacts of soil erosion and future technology. p. 60–85. *In* Proceedings of RCA symposium on future agricultural technology and resource conservation (Washington, DC, December 1982). Center for Agric. and Rural Dev., Iowa State University, Ames.

5 Assessments of Soil Erosion and Crop Productivity with Process Models (EPIC)

J. R. Williams
Agricultural Research Service
U.S. Department of Agriculture
Temple, Texas

K. G. Renard
Agricultural Research Service
U.S. Department of Agriculture
Tucson, Arizona

Accurate estimates of future soil productivity are essential in agricultural decision making and planning from the field scale to the national level. Soil erosion depletes soil productivity, but the relationship between erosion and productivity is not well defined. Until the relationship is adequately developed, selecting management strategies that maximize long-term crop production will be impossible.

The Soil and Water Resources Conservation Act of 1977 (RCA) requires a report by 1985 that establishes the current status of soil and water resources in the USA. One important aspect of these resources is the effect of erosion on long-term soil productivity. The National Soil Erosion—Soil Productivity Research Planning Committee documented what is known about the problem, identified what additional knowledge is needed, and outlined a research approach for solving the problem (Williams et al., 1981). One of the most urgent and important needs outlined in the research approach was the development of a mathematical model for simulating erosion, crop production, and related processes. This model will be used to determine the relationship between erosion and productivity for the USA. Thus, a national ARS erosion-productivity modeling team was organized and began developing the model during 1981. This team consisted of J. R.

Published in R. F. Follett and B. A. Stewart, ed. 1985. *Soil Erosion and Crop Productivity.*
© ASA-CSSA-SSSA, 677 South Segoe Road, Madison, WI 53711, USA.

Williams, M. J. Shaffer, K. G. Renard, G. R. Foster, J. M. Laflen, L. Lyles, C. A. Onstad, A. N. Sharpley, A. D. Nicks, C. A. Jones, C. W. Richardson, P. T. Dyke, K. R. Cooley, and S. J. Smith. The model, called Erosion-Productivity Impact Calculator (EPIC), is composed of physically based components for simulating erosion, plant growth, and related processes and economic components for assessing the cost of erosion, determining optimal management strategies, etc.

Simultaneously and realistically, EPIC simulates the physical processes involved, using readily available inputs. Commonly used EPIC input data (weather, crop, tillage, and soil parameters) are available from a computer filing system assembled especially for applying EPIC throughout the USA. The model requires detailed soils information obtained from the SCS pedon descriptions of chemical and physical properties. The SCS Soils-5 database can be used by applying techniques for estimating missing data, although simulation accuracy is reduced.

Since erosion can be relatively slow, EPIC can simulate hundreds of years if necessary. EPIC is generally applicable, computationally efficient (operates on a daily time step), and capable of computing the effects of management changes on outputs. The model must be comprehensive to define adequately the erosion-productivity relationship.

Outputs from EPIC (crop inputs, costs, erosion rates, etc.) will be entered in the Center for Agricultural and Rural Development (CARD) linear programming model (Meister and Nicol, 1975) to accomplish the national RCA analysis.

The components of EPIC can be placed into eight major divisions for discussion—hydrology, weather, erosion, nutrients, plant growth, soil temperature, tillage, and economics. A detailed description of the EPIC model was given by Williams et al. (1982). An abbreviated description of the components of the EPIC model and results of model testing are presented here.

5-1 MODEL DESCRIPTION

Although EPIC is a fairly comprehensive model, it was developed specifically for application to the erosion-productivity problem. Thus, user convenience was an important consideration in designing the model. The computer program contains 53 subroutines but only 2700 FORTRAN statements. Since EPIC operates on a daily time step, computer cost for overnight turnaround is only about $0.15 per year of simulation on an AMDAHL 470 computer. The model can be run on a variety of computers, since storage requirements are only 210 kilobytes.

The drainage area considered by EPIC is generally small (ca. 1 ha) because soils and management are assumed to be spatially homogeneous. In the vertical direction, however, the model can work with any variation in soil properties. The soil profile is divided into a maximum of 10 layers (the top layer thickness is set at 10 mm and all other layers may have variable thickness). When erosion occurs, the second-layer thickness is reduced by the amount of the eroded thickness, and the top-layer properties are adjusted by interpolation (according to how far the top layer moves into the second layer). When the second-layer thickness becomes zero, the top layer starts moving into the third layer, etc.

Following are descriptions of EPIC's components and the mathematic relationships used to simulate the processes involved.

5-1.1 Hydrology

5-1.1.1 Surface Runoff

5-1.1.1.1. Surface runoff is predicted for daily rainfall using the SCS curve number equation (USDA-SCS, 1972):

$$Q = (R - 0.2s)^2/R + 0.8s \qquad [1]$$

where Q is the daily runoff, R is the daily rainfall, and s is a retention parameter. The retention parameter s is related to soil water content with the equation

$$s = s_{mx} (UL - SW)/UL \qquad [2]$$

where SW is the soil water content in the root zone minus the 15-bar water content, UL is the upper limit of soil water storage in the root zone (porosity minus 15-bar water content), and s_{mx} is the maximum value of s. Thus, the retention parameter s ranges from s_{mx}, the I (dry) moisture condition curve number, to zero. A depth weighting technique is used to express the effect of the soil water distribution on the retention parameter.

The EPIC model for simulating surface runoff is similar to option one of the CREAMS runoff model (Knisel, 1980; Williams and Nicks, 1982). The only difference is that EPIC accommodates variable soil layers. In addition to the CREAMS surface runoff component, EPIC includes a provision for estimating runoff from frozen soil. If the temperature in the second soil layer is less than 0°C, the curve number is assigned a value of 98 regardless of the soil water content.

5-1.1.1.2 Peak Runoff Rate. Peak runoff rate predictions are based on a modification of the rational formula:

$$q_p = (\alpha) (Q) (A)/360 (t_c) \qquad [3]$$

where q_p is the peak runoff rate in m^3/s, α is a dimensionless parameter that expresses the proportion of total rainfall that occurs during the watershed time of concentration (t_c in h), and A is the watershed area in ha.

The time of concentration can be estimated by adding the surface and channel flow times:

$$t_c = t_{cc} + t_{cs} \qquad [4]$$

where t_{cc} is the time of concentration for channel flow and t_{cs} is the time of concentration for surface flow. The t_{cc} can be computed using the equation

$$t_{cc} = \frac{1.1 \, (L) \, (n)^{0.75}}{(A)^{0.125} \, (\sigma)^{0.375}} \qquad [5]$$

where L is the channel length from the most distant point to the watershed outlet in km, n is Manning's roughness factor, and σ is the average channel slope. The equation for computing t_{cs} is

$$t_{cs} = \frac{(\lambda \cdot n)^{0.6}}{18 \, (S)^{0.3}} \qquad [6]$$

where λ is the surface slope length in m and S is the land surface slope in m/m.

Since the water erosion component of EPIC estimates the maximum 0.5-h amount of each daily rainfall, these estimates are used in calculating α. Besides the convenience of avoiding double calculation, it is important to assure that $\alpha 0._5$ and α are closely related for each storm. The relationship between $\alpha_{0.5}$ and α can be obtained from Hershfield (1961) by fitting a log function to the 10-year frequency rainfall distribution:

$$R_t = R_6 \, (t/6)^b \qquad [7]$$

where b is a parameter used to fit the TP-40 relationship at any location, R_t is the rainfall amount for any time t, and R_6 is the 6-h rainfall amount. The value of α is computed with the equation

$$\alpha = \alpha_{0.5} \, (R_{t_c}/R_{0.5}) \qquad [8]$$

Details of the procedure for estimating $\alpha_{0.5}$ are presented in section 5-1.3.1.

5-1.1.2 Percolation

The percolation component of EPIC uses a storage routing technique combined with a crack-flow model to predict flow through each soil layer in the root zone. Once water percolates below the root zone, it is lost from the watershed (becomes groundwater or appears as return flow in downstream basins). Percolation is computed using the equation

$$0 = SW_0 \, [1 - \exp(-\Delta t/TT)] \qquad [9]$$

where 0 is the percolation rate in mm/d, SW_0 is the soil water content at the beginning of the day in mm, Δt is the travel interval (24 h), and TT is the travel time through a soil layer in h.

The travel time, TT, is computed for each soil layer with the linear storage equation

$$TT_i = (SW_i - FC_i)/H_i \qquad [10]$$

where H_i is the hydraulic conductivity in mm/h and FC is the field-capacity water content in mm. The hydraulic conductivity varies from the saturated conductivity value at saturation to near zero at field capacity:

$$H_i = SC_i (SW_i/UL_i)^{\beta_i} \qquad [11]$$

where SC_i is the saturated conductivity for layer i in mm/h and β is a parameter that causes H_i to approach zero as SW_i approaches FC_i.

The saturated conductivity is estimated for each soil layer using the equation

$$SC_i = \frac{12.7\,(100 - CLA)}{100 - CLA + \exp[11.45 - 0.097(100 - CLA)]} + 0.25 \qquad [12]$$

where CLA is the percentage of clay in the soil layer.

If the layer immediately below the layer being considered is saturated, no flow can occur regardless of the results from Eq. [9]. The effect of lower-layer water content is expressed in the equation

$$0_{ci} = 0_i \sqrt{1 - SW_{i+1}/UL_{i+1}} \qquad [13]$$

where 0_{ci} is the percolation rate for layer i in mm/d corrected for the water content in layer i + 1 and o_i is the percolation computed with Eq. [9].

The crack-flow model allows percolation of infiltrated rainfall even though the water content of the soil is less than field capacity. When the soil is dry and cracked, infiltrated rainfall can flow through the cracks of a layer without becoming part of the layer's soil water. However, the portion that does become part of a layer's stored water cannot percolate until the water content exceeds field capacity.

Crack flow is simulated with relationships similar to those used to estimate percolation above field capacity. The amount of percolate caused by crack flow is estimated with the equation

$$O_i = O_{i-1}\,[1 - \exp(-\Delta t/TT_{ci})] \qquad [14]$$

where O_{i-1} is the flow from the layer above in mm/d (R − Q for the top layer) and TT_{ci} is the crack-flow travel time in h. Crack-flow travel time is estimated with the equation

$$TT_{ci} = o_{i-1}/(\varsigma_i)\,(SC_i) \qquad [15]$$

where ς_i is a dimensionless soil parameter with values from 0 to 1 that expresses degree of cracking.

Percolation is also affected by soil temperature. If the temperature in a particular layer is 0°C or below, no percolation is allowed from that layer. Water can, however, percolate into the layer as dictated by Eqs. [9] and [14].

Since the one-day time interval is relatively long for routing flow through soils, it is desirable to divide the water into several parts for routing. This is necessary because the flow rates are dependent on the soil water content which is continuously changing. For example, if the soil was extremely wet, Eqs. [9], [10], and [11] might greatly overestimate percolation if only one routing was performed using the entire amount $(SW_i - FC_i)$. To overcome this problem, EPIC divides each layer's inflow into 4-mm slugs for routing. Also, by dividing the inflow into 4-mm slugs and routing each slug individually through all layers, the lower-layer water-content relationship (Eq. [13]) is allowed to function.

5-1.1.3 Lateral Subsurface Flow

Lateral subsurface flow is calculated simultaneously with percolation. Each 4-mm slug is first given the opportunity to percolate and then the remainder is subjected to the lateral-flow function. Thus, lateral flow can occur when the soil water in any layer exceeds field capacity after percolation. The lateral-flow function (similar to Eq. [9]) is expressed in the equation

$$QR_i = (SW_i - FC_i)\,[1 - \exp(-\Delta t/TT_{Ri})] \qquad [16]$$

where QR_i is the lateral flow rate for soil layer i in mm/d and TT_{Ri} is the lateral-flow travel time in d (time required for subsurface flow to travel a distance equal to the land surface slope length λ).

The lateral-flow travel time is estimated for each soil layer using the equation

$$TT_{Ri} = \frac{1000\,CLA}{CLA + \exp(10.047 - 0.148\,CLA)} \qquad [17]$$

5-1.1.4 Drainage

Underground drainage systems are treated as a modification of the natural, lateral subsurface flow of the area. A drainage system is simulated by simply indicating which soil layer contains the drainage system. EPIC assigns a short travel time of one day to that layer. Since travel time depends on the soil properties and the drain spacing, the drainage travel time may require adjustment for certain applications.

5-1.1.5 Evapotranspiration

The evapotranspiration component of EPIC is Ritchie's ET model (Ritchie, 1972). To compute potential evaporation, the model uses the equation

$$E_0 = 1.28 (h_0)(\gamma) \qquad [18]$$

where E_0 is the potential evaporation rate in mm/d, h_0 is the net solar radiation, and γ is a psychometric constant. The value of h_0 is calculated with the equation

$$h_0 = (RA) (1 - AB)/2.44 \times 10^6 \qquad [19]$$

where RA is the daily solar radiation in J/m^2 and AB is the albedo. The albedo is evaluated by considering the soil, crop, and snow cover. If a snow cover exists with 5 mm or greater water content, the value of albedo is set to 0.8. If the snow cover is less than 5 mm and no crop is growing, the soil albedo is the appropriate value. When crops are growing, albedo is determined using the equation

$$AB = 0.23 (1.0 - EA) + (AB_s) (EA) \qquad [20]$$

where AB_s is the soil albedo, 0.23 is the albedo for plants, and EA is a soil-cover index. The value of EA ranges from 0 to 1.0 according to the equation

$$EA = \exp[-2.9 \times 10^{-5} (CV)] \qquad [21]$$

where CV is the sum of above-ground biomass and crop residue in kg/ha. The value of γ in Eq. [18] can be obtained from the equation

$$\gamma = \delta/\delta + 0.68 \qquad [22]$$

where δ is the slope of the saturation vapor-pressure curve at the mean air temperature. The expression for evaluating δ is

$$\delta = 5304/T_K^2 \exp(21.255 - 5304/T_K) \qquad [23]$$

where T_K is the daily average air temperature in K.
The model computes soil and plant evaporation separately. Potential soil evaporation is predicted with the equation

$$E_{s0} = E_0 \exp(-0.4 LAI) \qquad [24]$$

where E_{so} is the potential evaporation rate at the soil surface in mm/d and LAI is the leaf-area index defined as the area of plant leaves relative to the soil surface area. Actual soil evaporation is computed in two stages. In the first stage, soil evaporation is limited only by the energy available at the surface and, thus, is equal to the potential soil evaporation. When the accumulated soil evaporation exceeds the stage-one upper limit (6 mm), the stage-two evaporative process begins. Stage-two soil evaporation is predicted with the equation

$$E_s = 3.5 \, [t^{1/2} - (t - 1)^{1/2}] \qquad [25]$$

where E_s is the soil evaporation rate for day t in mm/d and t is the number of days since stage-two evaporation began.

Plant evaporation is computed with the equations

$$E_p = (E_0)(LAI)/3, \qquad 0 \le LAI \le 3 \qquad [26]$$

and

$$E_p = E_0 - E_s, \qquad LAI > 3 \qquad [27]$$

where E_p is the predicted plant evaporation rate in mm/d. If soil water is limited, plant evaporation will be reduced as described in section 5-1.6.

5-1.1.6 Irrigation

The EPIC user can simulate either dryland or irrigated agricultural areas. If irrigation is indicated, he must also specify the irrigation efficiency, a plant water-stress level to start irrigation, and whether water is applied by sprinkler or down the furrows. The plant water-stress factor ranges from 0 to 1.0 (1 means no stress and 0 means no growth). These stress factors are described in section 5-1.6. When the user-specified stress level is reached, enough water is applied to bring the root zone up to field capacity, plus enough to satisfy the amount lost if the application efficiency is less than one. The excess water applied to satisfy the specified efficiency becomes runoff and provides energy for erosion.

5-1.1.7 Snow Melt

The EPIC snow-melt component is similar to that of the CREAMS model (Knisel, 1980). If snow is present, it is melted on days when the maximum temperature exceeds 0°C using the equations

$$SML = 4.57 \, T_{mx}, \qquad SML < SNO \qquad [28]$$

and

$$SML = SNO, \qquad SML \ge SNO \qquad [29]$$

where SML is the snow melt rate in mm/d, T_{mx} is the daily maximum temperature in °C, and SNO is the water content of snow in mm before melt occurs. Melted snow is treated the same as rainfall for estimating runoff, percolation, etc.

5-1.2 Weather

The weather variables necessary for driving the EPIC model are precipitation, air temperature, solar radiation, and wind. If daily precipitation, air temperature, and solar radiation data are available, they can be input directly to EPIC. Rainfall and temperature data are available for many areas of the USA, but solar radiation and wind data are scarce. Even rainfall and temperature data are generally not adequate for the long-term EPIC simulations of 50 years or more. Thus, EPIC provides options for simulating temperature and radiation, given daily rainfall, or for simulating rainfall as well as temperature and radiation. If wind erosion is to be estimated, daily wind velocity and direction are simulated. (There is no option for inputing wind velocity and direction.) Following are descriptions of the models used for simulating precipitation, temperature, radiation, and wind.

5-1.2.1 Precipitation

The EPIC precipitation model developed by Nicks (1974) is a first-order, Markov chain model. Thus, the model must be provided with monthly probabilities of receiving precipitation if the previous day was dry and monthly probabilities of receiving precipitation if the previous day was wet. Given the initial wet-dry state, the model determines stochastically if precipitation occurs or not.

When precipitation occurs, the amount is determined by generating from a skewed, normal, daily-precipitation distribution. Inputs necessary to describe the skewed normal distribution for each month are the mean, standard deviation, and skew coefficient for daily precipitation. The amount of daily precipitation is partitioned between rainfall and snowfall using average daily air temperature. If the average daily air temperature is 0°C or below, the precipitation is snowfall; otherwise, it is rainfall.

5-1.2.2 Air Temperature and Solar Radiation

The model developed by Richardson (1981) was selected for use in EPIC because it simulates temperature and radiation that correlate properly with one another and with rainfall. The residuals of daily maximum and minimum temperature and solar radiation are generated from a multivariate normal distribution. The means and coefficients of variation for each variable must be input. Since rainfall affects most of the variables, the means and coefficients of variation must be input separately for wet and dry days. Fortunately, a simple cosine function with two parameters adequately fits both the means and coefficients of variation.

The multivariate generation model used implies that the residuals of maximum temperature, minimum temperature, and solar radiation are normally distributed and that the serial correlation of each variable may be described by a first-order linear autoregressive model. Details of the multivariate generation model were described by Richardson (1981). The dependence structure of daily maximum and minimum temperatures and solar radiation was described by Richardson (1982a).

5-1.2.3   Wind

Richardson (1982b) developed the wind simulation model for use with EPIC. The two wind variables considered are average daily velocity and daily direction. Average daily wind velocity is generated from a two-parameter gamma distribution of the dimensionless form

$$U = (V/V_p)^{\eta - 1} \exp[(\eta - 1)(1 - V/V_p)] \qquad [30]$$

where U is a dimensionless variable (0 to 1) expressing frequency of occurrence of wind velocity V in m/s, V_p is the wind velocity at the peak frequency, and η is the gamma-distribution shape parameter. The shape parameter is calculated with the equation

$$\eta = \overline{V^2}/SDV^2 \qquad [31]$$

where $\overline{V}$ is the annual average wind velocity in m/s and SDV is the standard deviation of daily wind velocity in m/s.

Values for the average annual wind velocity and the standard deviation of hourly wind are provided by the Climatic Atlas of the United States (U.S. Department of Commerce, 1968). By experimenting with standard deviations of hourly and daily wind, a correction factor of 0.7 was found to be appropriate for converting hourly to daily standard deviations. The base of the dimensionless gamma distribution (maximum V/V_p) can be determined using Newton's classical method of solving nonlinear equations. The objective function is to select the base to minimize the sum of ln(U) and 11.5. The value of V_p can be determined by differentiating the gamma function expressed in terms of V and setting the result equal to zero:

$$V_p = V_i (\eta - 1)/\eta \qquad [32]$$

where V_i is the mean daily wind velocity for month i. The rejection technique is used to generate a daily value of V/V_p. The daily wind velocity is then computed using the equation

$$V_j = (V_{pi})(V/V_p) \qquad [33]$$

where V_j is the generated velocity for day j, V_{pi} is the peak velocity for month i, and V/V_p is the generated value using the rejection technique.

Wind direction, expressed as radians from north in a clockwise direction, is generated from an empirical distribution specific for each location. The empirical distribution is simply the cumulative probability distribution of wind direction. The Climatic Atlas of the United States (U.S. Department of Commerce, 1968) gives monthly percentages of wind from each of 16 directions. Thus, to estimate wind direction for any day, the model draws a uniformly distributed random number and locates its position on the appropriate monthly cumulative probability distribution.

5-1.3 Erosion

5-1.3.1 Water Erosion

5-1.3.1.1 Rainfall. The water erosion component of EPIC uses the universal soil loss equation (USLE) (Wischmeier and Smith, 1978) as modified by Onstad and Foster (1975). The energy factor in the Onstad-Foster equation is composed of both rainfall and runoff variables. In contrast, the USLE energy factor contains only rainfall variables. The Onstad-Foster equation is

$$Y = [0.646\,EI + 0.45(Q)(q_p)^{0.333}](K)(CE)(PE)(LS), \qquad Q > 0 \quad [34]$$

$$Y = 0 \qquad Q \leq 0$$

where
  Y is the sediment yield in t/ha,
  EI is the rainfall energy factor in metric units,
  Q is the runoff volume in mm,
  q_p is the peak runoff rate in mm/h,
  K is the soil erodibility factor,
  CE is the crop management factor,
  PE is the erosion control practice factor, and
  LS is the slope length and steepness factor.
The value of K depends on the soil type and is assigned before the simulation begins. Similarly, the PE value is determined initially by considering the conservation practices to be applied. The value of LS is calculated with the equation (Wischmeier and Smith, 1978):

$$LS = (\lambda/22.1)^{\xi}\,(65.41\,S^2 + 4.56\,S + 0.065) \qquad [35]$$

where S is the land surface slope in m/m, λ is the slope length in m, and ξ is a parameter dependent on slope. To evaluate ξ, EPIC uses the equation

$$\xi = 0.6\,[1 - \exp(-35.835\,S)] \qquad [36]$$

The crop-management factor is evaluated for all days when runoff occurs, using the equation

$$CE = (0.8 - CE_{mn,j})\,\exp(-0.00115\,CV) + CE_{mn,j} \qquad [37]$$

where $CE_{mn,j}$ is the minimum value of the crop management factor for crop j and CV is the sum of above-ground biomass and surface residue.

The hydrology model supplies estimates of Q and q_p. To estimate the daily rainfall energy in the absence of time-distributed rainfall, one assumes that the rainfall rate is exponentially distributed. The exponential distribution allows easy substitution into and integration of the USLE energy equation for computing daily rainfall energy:

$$EI = \frac{R\,[12.1 + 8.9\,(\log r_p - 0.434)]\,(r_{0.5})}{1000} \qquad [38]$$

where R is the daily rainfall amount in mm, r_p is the peak rainfall rate in mm/h, and $r_{0.5}$ is the maximum 0.5-h rainfall intensity. The value of r_p can be obtained by integrating the exponential rainfall distribution. Since rainfall rates vary seasonally, $\alpha_{0.5}$ (the ratio of the maximum 0.5-h rainfall to the total storm rainfall) is evaluated for each month using U.S. Weather Service information (U.S. Department of Commerce, 1968 and 1979). To estimate the mean value of $\alpha_{0.5}$, the mean value of $R_{0.5}$ must be estimated. The value of $R_{0.5}$ can be computed easily if the maximum 0.5-h rainfall amounts are assumed to be exponentially distributed. From the exponential distribution, the expression for the mean maximum 0.5-h rainfall amount is

$$\overline{R}_{0.5,j} = R_{0.5R,j}/ - \ln F_j \qquad [39]$$

where $\overline{R}_{0.5,j}$ is the mean maximum 0.5-h rainfall amount, $R_{0.5F,j}$ is the maximum 0.5-h rainfall amount for frequency F, and subscript j refers to the month. The mean $\alpha_{0.5}$ is computed with the equation

$$\alpha_{0.5j} = \overline{R}_{0.5,j}/\overline{R}_j \qquad [40]$$

where $\overline{R}$ is the mean amount of rainfall for each event (average monthly rainfall/average number of days of rainfall) and subscript j refers to the month. Daily values of $\alpha_{0.5}$ are generated from a two-parameter gamma distribution in a similar manner to that described in generating wind velocity (section 5–1.2.3).

5–1.3.1.2   Irrigation. Erosion caused by applying irrigation water in furrows is estimated with the modified universal soil loss equation (MUSLE) (Williams, 1975):

$$Y = 11.8\,(Q \cdot q_p)^{0.56}\,(K)(CE)(PE)(LS) \qquad [41]$$

where CE, the crop management factor, has a constant value of 0.5 The volume of runoff is estimated by considering the irrigation efficiency:

$$Q = AIR\,(1.0 - EIR) \qquad [42]$$

where AIR is the volume of irrigation water applied in mm and EIR is the irrigation efficiency. The peak runoff rate is estimated for each furrow using Manning's equation and assuming that the flow depth is 0.75 of the row height and that the furrow is triangular. If irrigation water is applied to land without furrows, the peak runoff rate is assumed to be 0.00189 m³/s per m of field width. Erosion caused by sprinkle irrigation is not considered directly. However, erosion is usually greater than from nonirrigated areas because higher soil water content increased runoff.

5-1.3.2 Wind Erosion

The Manhattan, KS, wind-erosion equation (Woodruff and Siddoway, 1965) was modified by Cole et al. (1982) for use in the EPIC model. The original wind erosion equation is

$$WE = f(I, WC, WK, WL, VE)$$ [43]

where

WE is the wind erosion in t/ha,
I is the soil erodibility index in t/ha,
WC is the climatic factor,
WK is the soil-ridge roughness factor,
WL is the field length along the prevailing wind direction in m, and
VE is the quantity of vegetative cover expressed as small grain equivalent in kg/ha.

Equation [43] was developed for predicting average annual wind erosion. The main modification to the model was a conversion from annual to daily predictions to interface with EPIC.

Although two of the variables, I and WC, remain constant for each day of a year, the other variables are subject to change from day to day. The ridge roughness is a function of row height and row interval, according to

$$KR = 4\,HR^2/IR$$ [44]

where KR is the ridge roughness in mm, HR is the row height in mm, and IR is the row interval in mm. The ridge-roughness factor is a function of ridge roughness as expressed by the equations

$$WK = 1, \qquad KR < 2.27$$ [45]

$$WK = 1.125 - 0.153 \ln(KR), \qquad 2.27 \leq KR < 89$$ [46]

and

$$WK = 0.336 \exp(0.00324\,KR), \qquad KR \geq 89$$ [47]

Field length along the prevailing wind direction is calculated by considering field dimensions, field orientation, and wind direction:

$$WL = \frac{(FL)(FW)}{FL\,|\cos(\pi/2 + \theta - \phi)| + FW\,|\sin(\pi/2 + \theta - \phi)|}$$ [48]

where FL is the field length in m, FW is the field width in m, θ is the wind direction clockwise from north in radians, and ϕ is the clockwise angle between field length and north in radians.

The vegetative-cover equivalent factor is simulated daily as a function of standing live biomass, standing dead residue, and flat crop residue.

$$VE = 0.2533 \, (g_1 B_{AG} + g_2 SR + g_3 FR)^{1.363} \qquad [49]$$

where g_1, g_2, and g_3 are crop-specific coefficients, B_{AG} is the aboveground biomass of a growing crop in kg/ha, SR is the standing residue from the previous crop in kg/ha, and FR is the flat residue in kg/ha. Thus, all variables in Eq. [43] can be evaluated. To determine the wind-erosion estimate, however, requires a special combination of the factors as follows:

$$E2 = (WK) \, (I) \qquad [50]$$

$$E3 = (WK) \, (I) \, (WC) \qquad [51]$$

$$WL_0 = 1.56 \times 10^6 \, (E2)^{-1.26} \exp(-0.00156 \, E2) \qquad [52]$$

$$WF = E2 \, [1 - 0.1218 \, (WL/WL_0)^{-0.3829} \exp(-3.33 \, WL/WL_0)] \qquad [53]$$

$$E4 = (WF^{0.3484} + E3^{0.3484} - E2^{0.3484})^{2.87} \qquad [54]$$

$$E5 = \psi_1 \, E4^{\psi_2} \qquad [55]$$

$$WE = (E5) \, (DE)/(AE) \qquad [56]$$

where field lengths greater than WL_0 in m give no reduction in the erosion estimate,
WF is the field length factor,
ψ_1 and ψ_2 are parameters,
DE is the daily wind energy in KWH/m^2, and
AE is the average annual wind energy in KWH/m^2.
The parameters ψ_1 and ψ_2 are functions of the vegetative-cover factor described by the equations

$$\psi_2 = 1 + (8.93 \times 10^{-5} \, VE) + (8.51 \times 10^{-9} \, VE^2) - (1.59 \times 10^{-13} \, VE^3) \qquad [57]$$

and

$$\psi_1 = \exp[(-7.59 \times 10^{-4} \, VE)$$
$$- (4.74 \times 10^{-8} \, VE^2) + (2.95 \times 10^{-13} \, VE^3)] \qquad [58]$$

Daily wind energy is estimated with the equation

$$DE = 0.00617 \, (V)^{4.35} \exp(-0.0620 \, V) \qquad [59]$$

where V is the daily average wind velocity in m/s. Average annual wind energy is estimated by integrating the monthly gamma distributions of wind velocity.

$$AE = 30.4 \sum_{j=1}^{12} \frac{\int_{V_L}^{V_u} (DE)_j \, (U)_j \, dV}{\int_{V_L}^{V_u} U_j \, dV} \qquad [60]$$

where U is the frequency of occurrence of wind velocity V, V_u is the upper limit of wind velocity, and V_L is the lower limit of erosive wind velocity.

5–1.4 Nutrients

5–1.4.1 Nitrogen

5–1.4.1.1 Nitrate Loss in Surface Runoff. The amount of nitrate (NO_3–N) in runoff is estimated by considering the top soil layer (10 mm) only. The decrease in NO_3–N concentration caused by water flowing through a soil layer can be simulated satisfactorily using an exponential function. The average concentration for a day can be obtained by integrating the exponential function to give the NO_3–N yield and dividing by the volume of water leaving the layer:

$$VNO_3 = WNO_3 \, [1 - \exp(-QT/POR)] \qquad [61]$$

$$c_{NO_3} = VNO_3/QT \qquad [62]$$

where
WNO_3 is the weight of NO_3–N contained in the soil layer at the start of a day,
QT is the total water lost from the first layer (Q + O + QR),
POR is the porosity of the layer,
VNO_3 is the amount of NO_3–N lost from the first layer, and
c_{NO_3} is the concentration of NO_3–N in the first layer.
Amounts of NO_3–N contained in runoff, lateral flow, and percolation are estimated as the products of the volume of water and the concentration from Eq. [62].

5–1.4.1.2 Nitrate Leaching. Leaching and lateral subsurface flow in lower layers are treated with the same approach used in the upper layer, except surface runoff is not considered.

5–1.4.1.2 Nitrate Transport by Soil Evaporation. When water is evaporated from the soil, NO_3–N is moved upward into the top soil layer by mass flow. The equation for estimating this NO_3–N transport is

$$ENO_3 = \sum_{j=2}^{M} (E_S)_j \, (c_{NO_3})_j \qquad [63]$$

where ENO_3 is the amount of NO_3–N moved from lower layers to the top layer by soil evaporation E_S, subscript j refers to soil layers, and M is the number of layers contributing to soil evaporation (maximum depth is 300 mm).

5-1.4.1.4 Organic Nitrogen Transport by Sediment. A loading function developed by McElroy et al. (1976) and modified by Williams and Hann (1978) for application to individual runoff events is used to estimate organic nitrogen loss. The loading function is

$$\text{YON} = 0.001 \, (\text{Y}) \, (c_{\text{ON}}) \, (\text{ER}) \qquad [64]$$

where YON is the runoff loss of organic nitrogen (N) in kg/hg, c_{ON} is the concentration of organic N in the top soil layer in g/t, Y is the sediment yield in t/ha, and ER is the enrichment ratio. The enrichment ratio is the concentration of organic N in the sediment divided by that of the soil. Enrichment ratios are logarithmically related to sediment concentration as described by Menzel (Knisel, 1980). A relationship between enrichment ratio and sediment concentration for individual events was developed for EPIC considering upper and lower bounds. The upper bound of the enrichment ratio is the inverse of the sediment-delivery ratio. The lower limit of the enrichment ratio is 1.0—sediment particle size distribution is the same as that of the soil. The logarithmic equation for estimating the enrichment ratio is

$$\text{ER} = x_1 \, c_s^{x_2} \qquad [65]$$

where c_s is the sediment concentration g/m^3 and x_1 and x_2 are parameters set by the upper and lower limits.

5-1.4.1.5 Denitrification. As one of the microbial processes, denitrification is a function of temperature and water content. The equation used to estimate the denitrification rate is

$$\text{DN} = \text{WNO}_3 \, \{1 - \exp[(\text{CDN}) \, (\text{TF}_\text{N}) \, (\text{C})]\}, \qquad \text{SWF} \geq 0.9 \quad [66]$$

$$\text{DN} = 0.0, \qquad \text{SWF} < 0.9$$

where DN is the denitrification rate in kg/(ha·d). CDN is the denitrification constant, TF_N is the nutrient temperature factor, SWF is the soil water factor, and C is the percentage of organic carbon content. The temperature factor is expressed by the equation

$$\text{TF}_{\text{Nj}} = 0.1 + \frac{0.9 \, \text{T}_j}{\text{T}_j + \exp(9.93 - 0.312 \, \text{T}_j)}, \qquad \text{T}_j > 0 \quad [67]$$

$$\text{TF}_{\text{Nj}} = 0.0, \qquad \text{T}_j \leq 0$$

where T is soil temperature in °C and subscript j refers to the soil layers. The soil-water factor considers total soil water in the equation

$$\text{SWF} = (\text{SW} + \text{SW}_{15})/\text{POR} \qquad [68]$$

where SW_{15} is the 15-bar soil-water content in mm.

5-1.4.1.6 Mineralization. The nitrogen mineralization model is a modification of the PAPRAN (Production of annual pastures limited by rainfall and nitrogen) mineralization model (Seligman and van Keulen, 1981). The model considers two sources of mineralization: fresh organic N associated with crop residue and microbial biomass, and the stable organic N associated with the soil humus pool. Mineralization from the pool of fresh organic N is estimated with the equation

$$RMN = (DCR)(FON) \tag{69}$$

where RMN is the N mineralization rate for fresh organic N in kg/(ha•d), DCR is the decay-rate constant for the fresh organic N, and FON is the amount of fresh organic N present in kg/ha. The decay-rate constant is a function of the C:N ratio, the C:P ratio, the composition of crop residue, temperature, and soil water:

$$DCR = (CNP)(RC)(SWF)(TF_N) \tag{70}$$

where CNP is a C:N or C:P ratio factor and RC is a residue composition factor. The value of CNP is calculated with the equation

$$CNP = \min \begin{cases} \exp[-0.693\,(CNR - 25)]/25 \\ \exp[-0.693\,(CPR - 200)]/200 \end{cases} \tag{71}$$

where CNR is the C:N ratio and CPR is the C:P ratio. The value of RC is determined by the stage of residue decomposition.

Mineralization from the pool of stable organic N is estimated for each soil layer with the equation

$$HMN = (CMN)(SWF)(TF_N)(ON)(BD)^2/(BDP)^2 \tag{72}$$

where HMN is the mineralization rate for the humus pool in kg/(ha•d), CMN is the stable mineralization-rate constant (0.0001), ON is the amount of organic N in the soil layer in kg/ha, BD is the settled bulk density of the soil, and BDP is the current bulk density as affected by tillage.

5-1.4.1.7 Immobilization. Like mineralization, the immobilization model is a modification of the PAPRAN model. Immobilization is a very important process in EPIC because it determines the residue decomposition rate, which has an important effect on erosion. The daily amount of immobilization is computed by subtracting the amount of N contained in the crop residue from the amount assimilated by the microorganisms:

$$WIM = (DCR)(FR)(0.016 - c_{NFR}) \tag{73}$$

where WIM is the N immobilization rate in kg/(ha•d), c_{NFR} is the N concentration in the crop residue in g/g, and 0.016 is the result of assuming that C = 0.4 FR, the C:N ratio of the microbial biomass and their labile products

= 10, and 0.4 of C in the residue is assimilated. Immobilization may be limited by the availability of N or phosphorus (P). If the amount of N available is less than the immobilization predicted by Eq. [73], the decay rate constant is adjusted with the relationship

$$DCR' = 0.95 \ WNO_3/FR \ (0.016 - c_{NFR})$$ [74]

where DCR' allows 95% use of the available NO_3-N in a soil layer. A similar adjustment is made if P is limiting. The crop residue is reduced using the equation

$$FR = FR_o - (DCR') \ (FR_o)$$ [75]

where FR_o and FR are the amounts of residue in a soil layer at the start and end of a day in kg/ha.

5-1.4.1.8 Crop Uptake. Crop use of N is estimated using a supply and demand approach. The daily crop demand for N can be computed using the equation

$$UND_{IDA} = (c_{NB})_{IDA} \ (B)_{IDA} - (c_{NB})_{IDA-1} \ (B)_{IDA-1}$$ [76]

where UND_{IDA} is the N demand of the crop in kg/ha, c_{NB} is the optimal N concentration of the crop, and B is the accumulated biomass in kg/ha. The optimal crop concentration of N is computed as a function of growth stage using the equation

$$c_{NB} = bn_1 + bn_2 \exp(-bn_3 \ B_1)$$ [77]

where bn_1, bn_2, and bn_3 are crop parameters expressing N concentration and B_1 is a dimensionless (0 to 1) expression of accumulated thermal time.

Soil supply of N is assumed to be limited to mass flow of NO_3-N to the roots:

$$UNS_{IDA} = \sum_{i=1}^{M} (u_i)_{IDA} \left(\frac{WNO_{3i}}{SW_i} \right)_{IDA}$$ [78]

where UNS is the amount of N supplied by the soil in kg/ha, u_i is the water use in mm, and subscript i refers to the soil layers. Actual N uptake on IDA is the minimum of UNS and UND.

5-1.4.1.9 Fixation. Fixation of N is important for legumes. The EPIC model estimates fixation by adding atmospheric N to prevent N stress that constrains plant growth. Section 5-1.6 describes the determination of plant stress factors for N, P, water, and temperature. Plant growth is limited by the minimum for the four factors each day. If N is the active constraint, enough atmospheric N is added to the plant to make the N-stress factor equal the next most constraining factor. The amount of N needed is attributed to fixation.

5-1.4.1.10   Rainfall. To estimate the N contribution from rainfall, EPIC uses an average rainfall concentration of N for a location for all storms. The amount of N in rainfall is estimated as the product of rainfall amount and N concentration.

5-1.4.1.11   Fertilizer. The model provides two options for applying fertilizer. With the first option, the user specifies dates, rates, and depths of application of N and P. The second option is more automated—the only input required is a plant-stress parameter. At planting time, the model takes a soil sample and applies up to 15 kg/ha of N fertilizer if needed. The model also applies enough P to bring the concentration of labile P in the top two layers up to the concentration level at the start of the simulation. Additional N fertilizer may be applied during the growing season (at 25% and 50% of maturity). The amount of N applied with each of these two top dressings is determined by predicting the final crop biomass using a relationship derived from Eqs. [109] and [110] (the crop biomass-energy equations):

$$BF = B_j + (BE)(PRA - \sum_{k=1}^{j} RA_K)(0.03)(BFT) \qquad [79]$$

and

$$FN = (c_{NB})(BF) - UN_j - \sum_{i=1}^{M} WNO_{3_i} \qquad [80]$$

where BF is the crop biomass predicted for the end of the growing season in kg/ha, B_j is the accumulated biomass on day j, BE is the crop parameter for converting energy to biomass in kg/ha. PRA is the potential solar radiation for the growing season, BFT is the user-supplied plant stress parameter, FN is the amount of N fertilizer applied in kg/ha, c_{NB} is the plant concentration of N at the end of the growing season, and UN_j is the amount of plant N on day j. The value of BFT ranges from 0.0 to 1.0. Thus, the user can adjust the N fertilizer rate by assuming various stress levels (BFT) in predicting the final biomass. Obviously, before each fertilizer application, the model sums the NO_3–N content of each soil layer using Eq. [80].

5-1.4.2   Phosphorus

5-1.4.2.1   Soluble Phosphorus Loss in Surface Runoff. The EPIC approach is based on the concept of partitioning pesticides into the solution and sediment phases as described by Leonard and Wauchope (Knisel, 1980). Because P is mostly associated with the sediment phase, the equation for soluble P runoff can be expressed in the simple form:

$$YSP = 0.01 (c_{LP1})(Q)/k_d \qquad [81]$$

where YSP is the soluble P in kg/ha lost in runoff volume Q in mm, c_{LP1} is the concentration of labile P in soil layer 1 in g/t, and k_d is the P concentration in the sediment divided by that of the water in m^3/t. The value of k_d used in EPIC is 175.

5-1.4.2.2   Phosphorus Transport by Sediment. Sediment transport of P is simulated with a loading function as described in organic N transport. The P loading function is

$$YP = 0.001 \, (Y) \, (c_p) \, (ER) \qquad [82]$$

where YP is the sediment phase of the P loss in runoff in kg/ha and c_p is the concentration of P in the top soil layer in g/t.

5-1.4.2.3   Mineralization. The P mineralization model developed by Jones et al. (1982) is similar in structure to the N mineralization model. Mineralization from the pool of fresh organic P is estimated for each soil layer with the equation

$$RMP = (DCR) \, (FOP) \qquad [83]$$

where RMP is the mineralization rate of fresh organic P in kg/(ha·d). Mineralization from the pool of stable organic P associated with humus is estimated for each soil layer using the equation

$$HMP = (LF_M) \, (CMN) \, (TF_N) \, (SWF) \, (OP) \qquad [84]$$

where HMP is the mineralization rate of humus P in kg/(ha·d), LF_M is the mineralization factor for labile P, and OP is the organic P content of the soil layer in kg/ha. The mineralization factor for labile P is computed with the equations

$$LF_M = 5 - 0.16 \, c_{LP}, \qquad c_{LP} \le 25 \qquad [85]$$

$$LF_M = 1, \qquad c_{LP} > 25 \qquad [86]$$

Thus, when $c_{LP} > 25$, the mineralization of humus P is directly proportional to the mineralization of humus N. However, because of increasing phosphatase activity of soil microbes, the ratio of P to N mineralization increases when labile P is inadequate.

5-1.4.2.4   Immobilization. The P immobilization model also developed by Jones et al. (1982) is similar in structure to the N immobilization model. The daily amount of immobilization is computed by subtracting the amount of P contained in the crop residue from the amount assimilated by the microorganisms:

$$WIP = (DCR') \, (FR) \, (0.16 \, LF_I - c_{PFR}) \qquad [87]$$

where WIP is the P immobilization rate in kg/(ha·d), c_{PFR} is the P concentration in the crop residue, 0.16 is the result of assuming that carbon = 0.4 of fresh crop residue and 0.4 of the carbon in the residue is assimilated by soil microorganisms, LF_I is the immobilization factor by labile P allowing the P:C ratio of soil microorganisms to range from 0.01 to 0.02 as a function of labile P concentration. The immobilization factor for labile P is computed with the equations

$$LF_I = 0.01 + 0.0004\, c_{LP}, \qquad c_{LP} \le 25 \qquad [88]$$

$$LF_I = 0.02, \qquad c_{LP} > 25 \qquad [89]$$

5-1.4.2.5 Cycling of Mineral Phosphorus. The mineral P model was developed by Jones et al. (1982). Mineral P is transferred among three pools: labile, active mineral, and stable mineral. When P fertilizer is applied, it is labile (available for plant use). However, a fraction is quickly transferred to the active mineral pool according to the equation

$$MPR = 0.1\,(AP)\,(SWF)\,\exp(0.115\,T - 2.88) \qquad [90]$$

where MPR is the rate of flow from the labile pool (AP) to the active-mineral P pool in kg/(ha·d) for each soil layer. Simultaneously, P flows from the active mineral pool back to the labile pool (usually at a much slower rate) according to the equation

$$LPR = 0.1\,(MP_A)\,(SWF)\,\exp(0.115\,T - 2.88)\,(PSP/1\text{-}PSP) \qquad [91]$$

where LPR is the flow rate from the active mineral pool to the labile pool in kg/(ha·d), MP_A is the amount in the active-mineral P pool in kg/ha, and PSP is the P-sorption coefficient defined as the fraction of fertilizer P remaining in the labile pool after an initial rapid phase of P sorption is complete. The P-sorption coefficient is a function of chemical and physical soil properties as described by Eq. [92] for calcareous soils and Eq. [93] for non-calcareous soils.

$$PSP = -0.0577 - 0.00380\,CLA + 0.0682\,pH$$
$$- 0.00198\,CEC - 0.00624\,CAC \qquad [92]$$

$$PSP = -0.0682 - 0.00303\,AP - 0.00482\,CLA$$
$$+ 0.122\,pH - 0.00164\,CLA/C \qquad [93]$$

where PSP is the P-sorption coefficient for each soil layer, pH is the soil pH, CEC is the cation exchange capacity, C is the percentage of organic carbon, and CLA is the percentage of clay content of the layer (Jones et al., 1982). In either case, PSP is constrained within the limits $0.15 \le PSP \le$

0.75. Flow between the pools of active and stable mineral P is governed by the equation

$$ASPR = (\omega)(c_{MPA} - 5) \qquad [94]$$

where ASPR is the flow rate between the pools of active and stable mineral P in kg/(ha·d) for a particular soil layer, ω is the flow coefficient, and c_{MPA} is the concentration of active mineral P in the soil layer in g/t. The flow coefficient, ω, is a function of PSP as expressed (Jones et al., 1982) by Eq. [95] for noncalcareous soils and Eq. [96] for calcareous soils:

$$\omega = \exp(-1.77\,PSP - 7.05) \qquad [95]$$

$$\omega = 0.00076 \qquad [96]$$

5-1.4.2.6 Crop Uptake. Crop use of P is estimated with the supply and demand approach described in the N model. The daily plant demand is computed with Eq. [76] written in the form

$$UPD_{IDA} = (c_{PB})_{IDA}(B)_{IDA} - (c_{PB})_{IDA-1}(B)_{IDA-1} \qquad [97]$$

where UPD_{IDA} is the P demand for the plant in kg/ha and c_{PB} is the optimal P concentration for the plant. The optimal plant concentration of P is computed with Eq. [77] written in the form

$$c_{PB} = bp_1 + bp_2 \exp(-bp_3\,B_1) \qquad [98]$$

where bp_1, bp_2, and bp_3 are crop parameters expressing P concentration. Plant supply of P is estimated using the equation

$$UPS_{IDA} = 1.5\,UPD_{IDA} \sum_{i=1}^{M} (SWF)_i\,(LF_u)_i\,(RW_i/RWT_{IDA}) \qquad [99]$$

where UPS is the amount of P supplied by the soil in kg/ha, LF_u is the labile P factor for uptake, RW is the root weight in layer i, and RWT is the total root weight on day IDA in kg/ha. The labile P factor for uptake ranges from 0 to 1 according to the equations

$$LF_u = (c_{LP} - 0.5)/24.5, \qquad 0.5 \le c_{LP} \le 25 \qquad [100]$$

$$LF_u = 0.0, \qquad c_{LP} < 0.5 \qquad [101]$$

$$LF_u = 1.0, \qquad c_{LP} > 25 \qquad [102]$$

As with N, the actual P uptake is the minimum of UPD and UPS.

5-1.5 Soil Temperature

Daily average soil temperature is simulated at the center of each soil layer for use in nutrient cycling and hydrology. The basic soil-temperature equation is

$$T(Z,t) = \overline{T} + \frac{AM}{2} \exp(-Z/DD) \cos\left[\frac{2\pi}{365}(t - 200) - Z/DD\right] \qquad [103]$$

where
Z is depth from the soil surface in mm,
t is time in d,
$\overline{T}$ is the average annual air temperature in °C,
AM is the annual amplitude in daily average temperature in °C, and
DD is the damping depth for the soil in mm.

Equation [103] provides estimates of air temperature (Z = 0) as well as soil temperature. Since air temperature is provided by the weather component of EPIC, the soil-temperature model should be capable of using these air temperatures as drivers. Otherwise, Eq. [103] would predict the same temperatures for a given day each year. To allow simulated air temperature to be used as the soil-temperature driver, an equation was developed to estimate soil-surface temperature:

$$TG_{IDA} = (1.0 - AB)\left[\left(\frac{TMX + TMN}{2.0}\right)(1.0 - RA/3.35 \times 10^7)\right.$$
$$\left. + TMX\, RA/800\right] + (AB)(TG_{IDA-1}) \qquad [104]$$

where
TG is the soil surface temperature in °C,
AB is the surface albedo,
TMX is the maximum daily air temperature in °C,
TMN is the minimum daily air temperature in °C, and
RA is the daily solar radiation in J/m².

Besides providing a mechanism for using daily simulated air temperature, Eq. [104] also expresses the effect of solar radiation and cover (a function of AB) on soil temperature. The values of TG on the day of interest and the four days immediately preceding are averaged to adjust Eq. [103]. The adjustment is made by replacing T(0,t) with TG, which is a better estimate of the surface temperature than T(0,t) because current weather conditions are considered. Soil temperature at any depth is also corrected by damping the difference between TG and T(0,t) and adding it to the estimate from Eq. [103]. Thus, the final equation for estimating soil temperature at any depth is

$$T(Z,t)$$
$$= \overline{T} + \left\{\frac{AM}{2}\cos\left[\frac{2\pi}{365}(t - 200)\right] + TG - T(0,t)\right\}e^{-Z/DD} \qquad [105]$$

The damping depth is a function of soil-bulk density and water content as expressed in the equations

$$DP = 1000 + \frac{2500\ BD}{BD + 686\ \exp(-5.63\ BD)} \qquad [106]$$

$$\S = \frac{SW}{(0.356 - 0.144\ BD)\ Z_M} \qquad [107]$$

and

$$DD = DP \exp\left[\ln\left(\frac{500}{DP}\right)\left(\frac{1 - \S}{1 + \S}\right)^2\right] \qquad [108]$$

where DP is the maximum damping depth for the soil layer in mm, BD is the soil bulk density in t/m^3, Z_M is the depth of the lowest soil layer from the surface, and $\S$ is a scaling parameter.

5–1.6 Crop Growth Model

5–1.6.1 Potential Growth

A single model is used in EPIC for simulating all the crops considered (corn, grain sorghum, wheat, barley, oats, sunflowers, soybeans, alfalfa, cotton, peanuts, and grasses). Of course, each crop has unique values for the model parameters. Crop growth for both annual and perennial plants can be simulated. Annual crops grow from planting date to harvest date or until the accumulated heat units equal the potential heat units for the crop. Perennial crops (alfalfa and grasses) maintain their root systems throughout the year, although the plant becomes dormant after frost. They start growing when the average daily air temperature exceeds the base temperature of the plant.

Energy interception is estimated with the equation

$$PAR = 0.02092(RA) \left(\frac{HRLT}{12}\right)^3\left(1.0 + \frac{\Delta HRLT}{\Delta t}\right)^2$$
$$\{1.0 - \exp[-0.65(LAI + 0.05)]\} \qquad [109]$$

where PAR is the photosynthetic active radiation in MJ/m^2, HRLT is the daylight time during a 24-h period in h, and $\Delta HRLT/\Delta t$ is the change in daylight time during a 24-h period.

The potential increase in biomass for a day can be estimated with the equation

$$\Delta B_p = (BE)\ (PAR) \qquad [110]$$

where ΔB_p is the daily potential increase in biomass in kg/ha and BE is the crop parameter for converting energy to biomass in kg/MJ. The leaf area index (LAI), a function of biomass, is estimated with the relationships

$$LAI = \frac{(LAI_{mx})\,(WLV)}{WLV + 5512\,\exp(-0.000608\,WLV)}, \qquad B_i \le DLAI \qquad [111]$$

$$LAI = LAI_o \left(\frac{1 - B_i}{1 - DLAI}\right)^2, \qquad B_i > DLAI \qquad [112]$$

where LAI_{mx} is the maximum LAI potential for the crop, WLV is the above-ground biomass minus yield in kg/ha, DLAI is the fraction of the growing season when LAI starts declining, and LAI_o is the LAI value from Eq. [111] when $B_i = DLAI$.

Accumulated biomass after the yield-initiation stage of crop growth is designated as crop yield, using the equations

$$\Delta YLD = 0.0, \qquad \frac{\Sigma RA}{PRA} \le 0.8\,(1.0 - 1.0/GK) \qquad [113]$$

$$\Delta YLD = \Delta B_p, \qquad \frac{\Sigma RA}{PRA} > 0.8\,(1.0 - 1.0/GK) \qquad [114]$$

$$YLD \le B_p/GK \qquad [115]$$

where
  GK is the ratio of total biomass to crop yield under favorable growing conditions,
  ΔYLD is the amount of increase in yield during one day in kg/ha,
  YLD is the accumulated yield in kg/ha, and
  PRA is the potential solar radiation for the growing season in J/m^2.
Since yield is not allowed to occur until the later part of the growing season, late season stresses may reduce yield more than early season stresses. The amount of root growth for a day is estimated with the equation

$$\Delta RWT = \Delta B_P\,(0.4 - 0.2\,B_i) \qquad [116]$$

where ΔRWT is the daily change in root weight in kg/ha. Daily root sloughing is estimated with the equation

$$\Delta RWS = 0.1\,(\Delta B_P)\,(B_i) \qquad [117]$$

where ΔRWS is the amount of root sloughing during a day in kg/ha. The change in root weight through the root zone is simulated as a function of plant water use and accumulated root weight in each soil layer using the equation

$$RW_j = RW_{oj} + (\Delta RWL)\,(RWF)_j \qquad [118]$$

where RW_o and RW are the root weights in soil layer j at the start and end of a day in kg/ha, ΔRWL is the change in live root weight during a day in kg/ha, and RWF is a root-weight distribution factor. The daily change in live roots is computed by subtracting the sloughed roots from the total root growth:

$$\Delta RWL = \Delta RWT - \Delta RWS \qquad [119]$$

The root-weight distribution factor is determined by considering the sign of ΔRWL:

$$RWF = \frac{u_i}{\sum\limits_{i=1}^{M} u_i}, \qquad \Delta RWL > 0 \qquad [120]$$

$$RWF = \frac{RW_i}{\sum\limits_{i=1}^{M} RW_i}, \qquad \Delta RWL < 0 \qquad [121]$$

Equations [120] and [121] simply distribute root growth as a function of water use if the root weight is increasing and as a function of existing root weight if the root weight is decreasing. Rooting depth is simulated as a function of heat units and potential root-zone depth:

$$\Delta RD = 2(RZ)\,(HU)/PHU, \qquad RD \le RZ \qquad [122]$$

where ΔRD is the daily change in crop root depth in mm.

5-1.6.2   Growth Constraints

If one of the plant stress factors is less than 1.0, the potential biomass predicted with Eq. [110] is adjusted daily, using the equation

$$\Delta B = (\Delta B_P)\,(REG) \qquad [123]$$

where REG is the crop-growth regulating factor (the minimum stress factor). The water stress factor is computed by considering supply and demand in the equation

$$WS = \frac{E_P}{\sum\limits_{i=1}^{M} u_i} \qquad [124]$$

where WS is the water stress factor with values from 0 to 1. The value of E_P is predicted in the evapotranspiration model, and u_i is a function of depth and soil-water content:

$$u_{pi} = \left(\frac{E_P}{1 - \exp(-\Lambda)}\left(1 - \exp(-\Lambda)\,\frac{RD_i}{RZ}\right) - \sum\limits_{j=1}^{i-1} u_j \right)(RGF) \qquad [125]$$

where u_{pi} is the potential water use for layer i in mm/d, RGF is the root-growth stress factor with values from 0 to 1, and Λ is a parameter describing water-use rate as a function of root depth. The details of evaluating Λ are given by Williams and Hann (1978). The value of Λ used in EPIC (3.065) assumes that about 30% of the total water used comes from the top 10% of the root zone. Equation [125] allows roots to compensate for water deficits in certain layers by using more water in layers with adequate supplies. The potential water use must be adjusted for water deficits to obtain the actual use for each layer:

$$u_i = u_{pi}, \qquad SW > 0.25\,UL \qquad [126]$$

$$u_i = u_{pi}\,(SW/0.25\,UL), \qquad SW \le 0.25\,UL \qquad [127]$$

The temperature stress factor is computed with the equation

$$TS = \exp[\Omega(T_o - T/T)^2] \qquad [128]$$

where TS is the temperature stress factor with values from 0 to 1, Ω is the temperature stress parameter for the crop, T_o is the optimal temperature for the crop in °C, and T is the daily average air temperature in °C. The stress parameter is evaluated by appropriate substituting and rearranging of Eq. [128]:

$$\Omega = \frac{\ln(0.9)}{\left(\dfrac{T_o - [(T_o + T_b)/2]}{(T_o + T_b)/2}\right)^2} \qquad [129]$$

where T_b is the base temperature for the crop in °C. Equation [129] sets TS = 0.9 when the air temperature is halfway between T_b and T_o.

The nitrogen and phosphorus stress factors are based on the ratio of accumulated plant N and P to the optimal values. The stress factors vary nonlinearly from 1.0 at optimal levels of N and P to 0.0 when N or P is half the optimal level. The N stress factor is computed with the equation

$$SN_{IDA} = 1 - \frac{SN_{S,IDA}}{SN_{S,IDA} + 29.534\exp(-10.93\,SN_{S,IDA})} \qquad [130]$$

where SN_{IDA} is the N stress factor for day IDA and $SN_{S,IDA}$ is a scaling factor that allows SN to range from 0.0, when the ratio $UN/c_{NB}\cdot B$ is equal to 0.5, to 1.0 when the ratio is 1.0. The P stress factor is computed with Eq. [130] written in P terms. Finally, the value of REG is determined as the minimum of WS, TS, NS, and PS. REG is used to adjust YLD, RWT, and RWL with equations similar to Eq. [123].

The root-growth stress factor is the minimum of stresses caused by soil strength, temperature, and aeration. Temperature stress for each soil layer is computed using Eq. [128]. Stress caused by poor aeration is estimated with the equations

$$AS = \exp[23\,(0.85 - SWF)], \qquad SWF > 0.85 \qquad [131]$$

$$AS = 1.0, \qquad SWF \leq 0.85 \qquad [132]$$

where AS is the aeration stress factor and SWF is the soil-water factor computed from Eq. [68]. Stress from soil strength is estimated as a function of soil texture and bulk density using the equation

$$SS = 0.1 + \frac{0.9\,BDP}{BDP + \exp[bt_1 + (bt_2)\,(BDP)]} \qquad [133]$$

where SS is the soil strength factor, BDP is the soil bulk density, and bt_1 and bt_2 are parameters dependent on soil texture. The values of bt_1 and bt_2 are obtained by considering boundary conditions for stress. The lower boundary where stress is essentially nil is given by the equation (Jones, 1983):

$$BD1 = 1.15 + 0.00445\,SAN \qquad [134]$$

where BD1 is the bulk density that gives no stress (SS = 1.0) for a particular percentage of sand, SAN. The upper boundary is given by the equation (Jones, 1983):

$$BD2 = 1.5 + 0.005\,SAN \qquad [135]$$

where BD2 is the bulk density that gives SS ~ 0.2 for a particular percentage of sand, SAN. The equations for estimating bt_1 and bt_2 are

$$bt_2 = \frac{\ln\,(0.0112\,BD1) - \ln\,(8\,BD2)}{BD1 - BD2} \qquad [136]$$

$$bt_1 = \ln\,(0.0112\,BD1) - (bt_2)\,(BD1) \qquad [137]$$

Equations [136] and [137] assure that Eq. [133] gives SS values of 1.0 and 0.2 for BDP equal BD1 and BD2. Finally, the root-growth stress factor, RGF, is the minimum of AS, SS, and TS. Besides constraining water use as defined in Eq. [125], RGF also constrains rooting depth. Combining RGF with Eq. [122] gives

$$\Delta RD = 2\,(RZ)\,(HU)\,(RGF)/PHU, \qquad RD \leq RZ \qquad [138]$$

5-1.7 Tillage

The EPIC tillage component was designed to mix nutrients and crop residue within the plow depth, simulate the change in bulk density, and convert standing residue to flat residue. Each tillage operation is assigned a mixing efficiency with values from 0 to 1. The tillage mixing equation is

$$X_i = (1 - EF) X_{oi} + \left(\frac{Z_i - Z_{i-1}}{PD}\right) EF \sum_{j=1}^{M} X_{oj} \qquad [139]$$

where X_i is the amount of the material in layer i after mixing in kg/ha, EF is the mixing efficiency of the tillage operation (0-1), X_{oj} is the amount of the material in layer j before mixing in kg/ha, and M is the number of soil layers in the plow depth (PD) in mm.

The change in bulk density in the plow layer is simulated for each tillage operation using the equation

$$BDP_i = BDP_{oi} - (BDP_{oi} - 2/3\ BD_{oi})\ (EF) \qquad [140]$$

where BDP_o is the bulk density in soil layer i before tillage in t/m^3, BD_o is the bulk density of the soil when it has completely settled after tillage, and BDP is the bulk density after tillage. Between tillage operations, the soil settles with each rainfall according to the equations

$$SZ_i = \frac{O_{i-1}}{Z_i^{0.6}} \left[1.0 + \frac{2.0\ SAN_i}{SAN_i + \exp(8.597 - 0.075\ SAN_i)}\right] \qquad [141]$$

and

$$BDP_i = BDP_i + (BD_i - BDP_i) \left[\frac{SZ_i}{SZ_i + \exp(3.735 - 0.008835\ Z_i)}\right] \qquad [142]$$

where SZ_i is a scaling factor for soil layer i, O_{i-1} is the amount of water that percolates into the layer in mm (R-Q for the top layer), and SAN is the percentage of sand in the layer.

Another important function of the tillage model, converting standing residue to flat residue, is done with the equation

$$SR = (SR_o) \exp[-0.0569\ (PD)\ (EF)^2] \qquad [143]$$

where SR_o and SR are the standing residue weights before and after tillage in kg/ha and PD is the plow depth in mm.

Other functions of the tillage component include simulating row height and surface roughness. Both these variables are specified for each tillage implement. The user also specifies the date and depth for each tillage operation. The tillage operation is carried out on the specified date if the soil is dry enough; if not, on the next suitable day.

The EPIC model can simulate three kinds of harvest: (i) traditional harvest that removes seed, fiber, etc. (multiple harvests are allowed for crops like cotton); (ii) hay harvest (multiple harvests are allowed); and (iii) no harvest (green manure crops, etc.). The traditional harvest partitions the crop stover into 10% residue on the top soil layer and 90% standing residue. A shredder is often applied after harvest, which further partitions the residue according to the height before and after cutting. Yield from a hay harvest is also estimated for height before and after cutting.

5-1.8 Economics

The crop budgets are calculated with components from the crop budget generator developed at Oklahoma State University (Kletke, 1979). Budgets can be calculated for each year in the EPIC simulation or for the average yields and resource requirements for the period of simulation.

Inputs are divided into two categories: fixed and variable. Fixed inputs include depreciation; interest or return on investment; insurance; and taxes on equipment, land, and capital improvements (terraces, drainage, irrigation systems, etc.). Variable inputs are defined as machinery repairs, fuel and other energy, machine lubricants, seed, fertilizer, pesticides, labor, and irrigation water.

Total variable cost is expressed with the equation

$$TVC = \sum_{i=1}^{M} (PR_i)(QA_i) \qquad [144]$$

where TVC is the total variable cost in dollars, PR is the price of the variable input (i) in dollars, and QA is the quantity of the input used per ha.

The machinery complement file is a list of 100 pieces of equipment for use in simulating user-specified tillage operations. This file contains equipment information like purchase price, size, expected life, and repair cost. With this information, fixed costs for the machinery are allocated to each crop on an hours-of-use basis. Equations [145] through [149] are used for the fixed costs of machinery.

$$DEPC = (PP - SV)/(HUA)(YRO) \qquad [145]$$

where DEPC is the depreciation cost in dollars/hr, PP is the purchase price in dollars, SV is the salvage value in dollars, HUA is the annual use in h, and YRO is the time the equipment is owned in yr.

$$AI = (PP + SV)/2 \, HUA \qquad [146]$$

where AI is the average investment in dollars/h.

$$IC = (AI) \, (IT) \qquad [147]$$

where IC is the interest cost in dollars/h and IT is the interest rate.

$$INSC = (AI) \, (INSR) \qquad [148]$$

where INSC is the insurance cost in dollars/h and INSR is the insurance rate.

$$MTAX = (PP)(TR)/HUA \qquad [149]$$

where MTAX is the machinery tax in dollars and TR is the tax rate.
   Machine variable cost is also calculated on a per-hour basis.

$$PL = (HUA)(YRO)/HOL \qquad [150]$$

where PL is the percentage of machinery life used in a given year and HOL is the total machine life in h.
   Total accumulated repair cost is calculated with the equation

$$TAR = (rc_1) \, (PP) \, (rc_2) \, (PL) \, (rc_3) \qquad [151]$$

where TAR is the total accumulated cost of repairs in dollars and rc_1, rc_2, and rc_3 are repair cost coefficients (American Society of Agricultural Engineers, 1971). Repair cost is placed on an hourly basis using the equation

$$rc = TAR/(HUA)(YRO) \qquad [152]$$

Fuel-consumption cost is given by the equation

$$CF = (DH) \, (CFM) \, (PF) \qquad [153]$$

where CF is the fuel-consumption cost in dollars/h, DH is the drawbar horsepower, CFM is the fuel-consumption coefficient, and PF is the price of the fuel in dollars/liter. Lubrication cost is estimated as 15% of the fuel cost.

Actual annual cost of operating machinery can be estimated by converting from cost/h to cost/ha using the equation

$$HPA = 10/(VT)(WD)(FE) \tag{154}$$

where HPA is the time required to till 1 ha in h, VT is the tractor velocity in km/h, WD is the implement width in m, and FE is the field efficiency of the equipment. Finally, total annual machinery cost is estimated by summing the costs of the individual operations.

Fixed costs like rent, land taxes, and management are also charged on a per-ha basis. Total cost is the sum of the variable and fixed costs.

Gross income from the crop is simply the market price of the crop times the yield minus any marketing or harvest cost not accounted for in the machinery cost. Net profit, of course, is the difference between gross income and total cost. Simulated annual net profits are useful in illustrating the effects of erosion.

5-2 MODEL TESTS

Simulations of EPIC have been performed on 150 test sites in the continental USA and 13 sites in Hawaii. Table 5-1 shows runoff and sediment yield results from three small watersheds in Falls County, TX. More tests are planned for these components of the model, although they have been tested extensively (Knisel, 1980; Williams, 1982). Crop yield results are shown in Tables 5-2 to 5-4. Table 5-2 shows comparisons of simulated and recently measured yields for 12 research plots. Older, long-term average yields from 8 research plots are compared with simulated yields in Table 5-3. The estimated yields in Table 5-4 (county averages, local experts' estimates, etc.) are compared with simulated yields. Table 5-5 shows results of simulated wind erosion. Although there are no measurements of wind

Table 5-1. Comparisons of simulated and measured runoff and sediment yield in Falls County, TX.

Water-shed	Yr	Annual runoff				Annual sediment yield			
		Mean		SD		Mean		SD	
		Meas-ured	Simu-lated	Meas-ured	Simu-lated	Meas-ured	Simu-lated	Meas-ured	Simu-lated
		mm				t/ha			
W-10	5	246	264	126	131	0.082	0.082	0.098	0.044
SW-11	4	150	140	139	125	1.33	1.11	0.93	0.82
Y-14	4	204	245	138	164	0.82	1.04	1.11	1.81

erosion with which to compare the simulations, the results are similar to estimates obtained using the annual wind-erosion equation.

To determine the model's sensitivity to erosion, crop yields were related to accumulated erosion for the 50-yr simulation periods. The resulting linear regression equations were used to compare expected yields at the start and end of the 50-yr period. Generally, the regression analysis indicated a reduction in crop yield depending on soil and climatic characteristics and fertilization rate. In some areas with high erosion rates and unfavorable subsoil, crop yield was reduced as much as 40%.

Table 5-2. Comparisons of simulated and recently measured crop yields.

State	County	Yr	Crop	Yield		Standard deviation	
				Measured	Simulated	Measured	Simulated
				——— kg/ha ———			
IA	Monona	5	Corn	6996	7653	1110	1035
IA	Monona	5	Oats	1755	2225	774	1000
IA	Monona	10	Corn	6162	7325	1908	1895
IA	Ringold	7	Corn	7270	7235	1702	798
IA	Ringold	7	Soybean	1910	2065	284	531
IA	Ringold	10	Corn	6593	7095	1296	1075
IA	Story	5	Corn	6664	7580	815	790
IA	Story	5	Corn	6575	7265	922	1215
IA	Story	5	Corn	6077	7250	1279	1210
IA	Story	4	Corn	7033	7205	1010	1175
MO	Boone	10	Corn	7833	7632	2077	1635
OH	Coshocton	3	Corn	8399	7460	2665	2020

Table 5-3. Comparisons of simulated and long-term average measured crop yields.

State	County	Yr	Crop	Yield	
				Measured	Simulated
				——— kg/ha ———	
AL	Escambia		Soybeans	1893	1911
			Corn	5290	5350
			Wheat	1231	1584
AL	Escambia		Corn	5278	5325
			Cotton	2470	1415
ND	Morton	29	Corn	1625	2060
			Wheat	1022	835
ND	Stark	40	Corn	879	1015
			Wheat	908	625
MT	Hill	31	Corn	659	1040
			Wheat	800	565
WY	Sheridan	30	Corn	910	1185
			Wheat	1204	835
WY	Laramie	32	Corn	816	1165
			Wheat	693	825
KS	Ellis	31	Grain sorghum	1280	3450
			Wheat	1480	1550

Table 5-4. Comparisons of simulated and estimated crop yields.

State	County	Crop	Yield Estimated	Yield Simulated
			———— kg/ha ————	
AL	Conecuh	Peanut	3 500	2 885
AZ	Cochise	Pasture	1 000	1 290
AZ	Maricopa	Cotton	2 500	2 505
AZ	Yuma	Cotton	3 760	3 065
		Alfalfa	15 700	21 685
CO	Washington	Corn	2 511	3 135
		Wheat	982	1 570
GA	Emanual	Corn	6 275	6 155
		Wheat	3 700	1 665
		Cotton	2 520	2 410
GA	Oconee	Soybean	2 020	2 395
		Grain sorghum	4 710	5 605
IL	Marion	Corn	4 269	6 875
		Soybean	1 567	2 410
KS	Ellis	Wheat	1 930	2 250
KS	Finney	Corn	8 440	8 280
KS	Greeley	Corn	9 500	9 785
KS	Sherridan	Corn	10 044	11 265
KS	Thomas	Corn	2 084	2 240
		Wheat	1 765	1 066
KY	Caldwell	Soybean	1 850	2 390
		Corn	4 960	6 455
		Wheat	2 333	2 270
KY	Fayette	Corn	5 660	
MN	Polk	Wheat	2 450	3 000
		Alfalfa	5 160	5 420
MS	Hinds	Soybean	2 200	2 305
MS	Oktibbeha	Corn	2 373	4 225
		Soybean	1 338	1 420
MS	Sunflower	Cotton	1 929	2 085
MT	Gallatin	Wheat	2 470	2 465
MT	Judith Basin	Wheat	1 470	1 110
		Corn	2 260	2 350
MT	Richland	Wheat	1 790	2 295
NC	Craven	Corn	4 457	6 280
		Peanut	2 330	2 290
ND	Morton	Wheat	1 880	2 315
NE	Cheyenne	Wheat	2 340	2 065
NE	Red Willow	Wheat	2 640	2 500
NM	Curry	Wheat	1 000	1 390
		Alfalfa	10 760	7 525
NM	Eddy	Corn	6 025	9 125
		Wheat	2 970	3 410
		Cotton	2 240	2 775
OH	Coshocton	Corn	5 900	6 540
		Wheat	2 170	3 245
OK	Canadian	Wheat	1 855	2 070
OK	Comanche	Grain sorghum	1 505	1 855
		Wheat	1 010	1 215
SD	Bennett	Wheat	2 020	2 440
SD	Lyman	Wheat	1 960	2 350
TN	Marshall	Soybean	1 540	1 245
		Corn	3 365	4 415

(continued on next page)

Table 5-4. Continued.

State	County	Crop	Yield Estimated	Yield Simulated
			kg/ha	
TX	Bell	Grain sorghum	4 620	4 215
		Cotton	1 655	1 960
		Wheat	2 220	2 485
TX	Hartley	Corn	940	1 151
TX	Howard	Cotton	1 315	1 095
TX	Potter	Grain sorghum	6 100	6 350
		Wheat	2 800	3 025
WA	Whitman	Wheat	4 500	3 930

Table 5-5. Simulated wind erosion.

State	County	Crop	Soil erodibility factor	Climate factor	Soil loss
					t/ha
AZ	Yuma	Corn	300	495	46
		Alfalfa			
CO	Prowers	Wheat	150	75	0
CO	Washington	Corn	193	75	31
		Wheat			
IA	Harrison	Corn	193	16	3
IA	Monona	Corn	193	18	1
		Soybean			
KS	Finney	Corn	193	100	47
KS	Greeley	Corn	193	97	43
KS	Sherman	Corn	108	94	6
		(irrigated)			
ND	McLean	Wheat	108	58	0
ND	Morton	Corn	193	37	26
		Wheat			
NE	Cheyenne	Wheat	125	54	1
NE	Red Willow	Wheat	108	32	0
		Grain sorghum			
NM	Curry	Wheat	300	93	34
		Grain sorghum			
NM	Quay	Cotton	125	116	73
		Grain sorghum			
NV	Churchill	Oats	300	36	0
		Alfalfa			
OH	Auglaize	Corn	108	4	0
		Soybean			
OK	Comanche	Grain sorghum	193	20	16
SD	Bennett	Wheat	108	48	0
SD	Lyman	WHeat	193	50	1
TX	Bailey	Cotton	300	105	125
TX	Carson	Wheat	193	105	1
TX	Deaf Smith	Wheat	193	105	14
		Grain sorghum			
TX	Gaines	Cotton	695	202	717
TX	Howard	Cotton	193	120	65

5–3 CONCLUSIONS

The EPIC model is operational and has produced reasonable results under a variety of climatic conditions, soil characteristics, and management practices. It has also demonstrated sensitivity to erosion in terms of reduced crop production.

More extensive testing is planned for EPIC. Although some components of the model such as hydrology and erosion are based on accepted technology, other components require rigorous testing for validation. The two components that most need testing are crop growth and nutrients, because they are newly developed and are extremely important to the success of the EPIC model.

The model has many potential uses beyond the RCA analysis, including: (i) national conservation policy studies, (ii) national program planning and evaluation, (iii) project-level planning and design, and (iv) as a research tool.

REFERENCES

American Society of Agricultural Engineers. 1971. Agricultural machinery management data. p. 287–294. *In* Agricultural Engineers Yearbook. ASAE, St. Joseph, MI.

Cole, G. W., L. Lyles, and L. J. Hagen. 1982. A simulation model of daily wind erosion soil loss. Paper 82-2575. Winter Meeting 1982. American Society of Agricultural Engineers, Chicago, IL.

Hershfield, D. M. 1961. Rainfall frequency atlas of the United States for durations from 30 minutes to 24 hours and return periods from 1 to 100 years. Technical Paper 40. U.S. Department of Commerce, Washington, DC.

Jones, C. A. 1983. Effect of soil texture on critical bulk densities for root growth. Soil Sci. Soc. Am. J. 47(6):1208–1211.

----, C. V. Cole, and A. N. Sharpley. 1982. A simplified soil phosphorus model, I. Documentation. (Submitted to Soil Sci. Soc. of Am. J.)

Kletke, D. D. 1979. Operation of the enterprise budget generator. Agric. Exp. Stn. Res. Rep. P-790. Oklahoma State University, Stillwater.

Knisel, W. G. 1980. CREAMS, A field scale model for chemicals, runoff, and erosion from agricultural management systems. USDA Conservation Res. Rep. 26. Tucson, AZ.

McElroy, A. D., S. Y. Chiu, J. W. Nebgen, A. Aleti, and F. W. Bennett. 1976. Loading functions for assessment of water pollution from nonpoint sources. EPA-600/2-76-151, Environmental Protection Technology Series, Washington, DC.

Meister, A. D., and K. J. Nicol. 1975. A documentation of the national water assessment model of regional agricultural production, land and water use, and environmental interaction. Misc. Report. Center for Agric. and Rural Dev., Iowa State University, Ames.

Nicks, A. D. 1974. Stochastic generation of the occurrence, pattern, and location of maximum amount of daily rainfall. p. 154–171. *In* USDA-ARS Misc. Publ. 1275. Proc. Symp. on Statistical Hydrology, Tucson, AZ. August–September 1971. USDA-ARS.

Onstad, C. A., and G. R. Foster. 1975. Erosion modeling on a watershed. Am. Soc. Agric. Eng. Trans. 18(2):288–292.

Richardson, C. W. 1981. Stochastic simulation of daily precipitation, temperature, and solar radiation. Water Resour. Res. 17(1):182–190.

----. 1982a. Dependence structure of daily temperature and solar radiation. Am. Soc. Agric. Engr. Trans. In press.

----. 1982b. A wind simulation model for wind erosion estimation. Paper 82-2576. Winter Meeting 1982. American Society of Agricultural Engineers, Chicago, IL.

Ritchie, J. T. 1972. A model for predicting evaporation from a row crop with incomplete cover. Water Resour. Res. 8(5):1204–1213.

Seligman, N. G., and H. van Keulen. 1981. PAPRAN: A simulation model of annual pasture production limited by rainfall and nitrogen. p. 192–221. *In* M. J. Frissel and J. A. van Veen (ed.) Simulation of nitrogen behaviour of soil-plant systems. Proceedings of a Workshop on Models for the Behavior of Nitrogen in Soil and Uptake by Plant: Comparison between Different Approaches, Wageningen, The Netherlands. 28 Jan.–1 Feb. 1980. CABO.

USDA–SCS. 1972. National engineering handbook, Hydrology Section 4, Ch. 4–10. USDA, Washington, DC.

U.S. Department of Commerce. 1968. Climatic atlas of the United States. Enivronmental Science Services Administration, Environmental Data Service, Washington, DC.

----. 1979. Maximum short duration precipitation. *In* Climatological Data National Summary, U.S. Department of Commerce, Asheville, NC.

Williams, J. R. 1975. Sediment yield prediction with universal equation using runoff energy factor. p. 244–252. *In* USDA, ARS-S-40.

----, R. R. Allmaras, K. G. Renard, Leon Lyles, W. C. Woldenhauer, G. W. Langdale, L. D. Myer, and W. J. Rawls. 1981. Soil erosion effects on soil productivity: A research perspective. J. Soil Water Conserv. 36(2):82–90.

----. 1982. Testing the modified universal soil loss equation. p. 157–165. *In* Proc. Workshop on Estimating Erosion and Sediment Yield on Rangelands, Tucson, AZ, 7–9 March 1981. USDA-ARS, Agric. Reviews and Manuals, ARM-W-26/June 1982, Tucson, AZ.

----, P. T. Dyke, and C. A. Jones. 1982. EPIC—A model for assessing the effects of erosion on soil productivity. Proc. Third Int. Conf. on State-of-the-Art in Ecological Modelling, Colorado State University, Fort Collins, CO, 24–28 May 1982. Int. Soc. for Ecological Modeling.

----, and R. W. Hann. 1978. Optimal operation of large agricultural watersheds with water quality constraints. Tech. Rep. No. 96. Texas Water Resources Inst., Texas A&M University, College Station.

----, and A. D. Nicks. 1982. CREAMS hydrology model—option one. p. 69–86. *In* V. P. Singh (ed.) Applied modeling catchment hydrology. Proc. Int. Symp. on Rainfall-Runoff Modeling, Mississippi State University, Mississippi State, 18–21 May 1981.

Wischmeier, W. H., and D. D. Smith. 1978. Predicting rainfall erosion losses, a guide to conservation planning. USDA Agricultural Handbook 537. USDA, Washington, DC.

Woodruff, N. P., and F. H. Siddoway. 1965. A wind erosion equation. Soil Sci. Soc. Am. Proc. 29(5):602–608.

6 Assessment of Soil Erosion and Crop Productivity with Economic Models

Paul T. Dyke
Economic Research Service
U.S. Department of Agriculture
Temple, Texas

Earl O. Heady
Iowa State University
Ames, Iowa

"Erosion is destroying our lands." "We must protect our soil for the next generation." "We are losing our productive capacity down the Mississippi River." "We are exporting our good farm land by our fence-to-fence farming and cheap grain export policies." "If we don't protect our soil and water resources today, our grandchildren won't eat tomorrow." "I don't understand why we pay farmers not to grow grain and then subsidize irrigation projects."

These are the kinds of statements that have been coming from U.S. citizens for decades. Who is right? Who is wrong? What difference does it really make? Public Law 95-192, passed by Congress on 18 November 1977, directed the Secretary of Agriculture to provide quantitative answers to similar questions. The Soil and Water Resources Conservation Act of 1977 (known as RCA) requires that a report be submitted to Congress every five years containing the following: (i) a physical assessment of the nation's soil- and water-related resources, (ii) an economic assessment of the benefits and costs of alternative soil and water conservation practices, and (iii) a recommended program to be administered by the Secretary of Agriculture which "shall be responsive to the long-term needs of the Nation, as determined under the provisions of this Act" (95th Congress, 1977).

Another section of the Act instructs: "The Secretary shall establish an integrated system capable of using combinations of resource data to de-

Published in R. F. Follett and B. A. Stewart, ed. 1985. *Soil Erosion and Crop Productivity.*
© ASA-CSSA-SSSA, 677 South Segoe Road, Madison, WI 53711, USA.

termine the quality and capabilities for alternative uses of the resource base. . . ." The Erosion-Productivity Impact Calculator (EPIC) described in the previous paper and the National Linear Programming model (NLP) described in this paper combine to provide a substantial part of that "integrated system capability" to be used to produce the 1985 RCA report and to design long-term national resource policies. Although not totally correct, it may be useful to think of the EPIC program and its data bases as the physical integrator for questions related to soil erosion and crop productivity, and the linear programming model and its data bases as the economic integrator.

The linear program not only keeps track of the quantity and value of the principal physical resources used in crop production but also provides a mechanism by which to evaluate the physical and economic effects on the nation of alternative long-range conservation policies that Congress might adopt.

The discussion to follow will be presented in four parts: (i) a general description of the economics of soil erosion, (ii) the methodology of linear programming when used as economic integrators, (iii) a study of the specific characteristics of the NLP model plus the way it interfaces with the EPIC model, and (iv) a compilation of the potential uses of the NLP model in the 1985 RCA and its ability to address questions about soil erosion and crop productivity.

6-1 ECONOMICS OF SOIL EROSION

The economics of soil erosion are influenced by two major forces: the demand for food and fiber, and the supply of food and fiber. The interaction of these separate but related pressures will ultimately determine how important or unimportant erosion is when compared to the other economic concerns of a society.

6-1.1 Demand for Food and Fiber

To address properly the economies of soil erosion, we must concern ourselves with food and fiber demands in at least two dimensions—across space and across time. Let us first look at the space dimension.

Because political institutions control space allocations and ownership of our world's natural resources, and since we are addressing U.S. policies regarding management of soil and water resources under U.S. jurisdiction, we will divide demand once again into domestic demand and export or foreign demand. Domestic demand for food and fiber at any given time is determined by summing all the needs and desires of the national population for each of many pricing policies of the various mixes of food and fiber. Without digressing into a review of basic supply and demand theory, let it suffice to point out that many kinds of domestic influences will alter the mix and the absolute amounts of food and fiber demanded by the domestic population: the population, the amounts and distribution of purchasing power, the ages of the population, their food preferences (meat vs. plant), and numerous other factors. Although we are looking at the space dimension and therefore at only one point in time, we must remember that a different mix is required for each given time. Any analysis to address resource policy must take this factor into consideration. We will see that linear programming has this capability to a limited degree.

Export demands are those demands from outside the national boundaries for food and fiber produced with the resources owned and controlled within the national boundaries. They are important demands because they are more easily controlled by policy decisions than are domestic demands. Although influenced by pressures similar to those of domestic demands, export demands behave somewhat differently because of the availability of alternative sources of food and fiber from other suppliers (countries). Internal national policy can influence but seldom control these changing world market conditions; therefore, assumptions dealing with export demand must often be taken as "givens" for policy purposes. For these and other reasons, it is useful to keep domestic and export demand separate.

The RCA did not ignore the other dimension of demand—time. In fact, this legislation and the concern with erosion and land protection have no meaning without the time consideration. Sufficient information is available to conclude that the present land base can support the demands of the present population. The question Congress asked is, "How are the present policies and practices affecting the capacity of our soil and water resources to meet the food and fiber needs of future generations?" The next question is, "Which future generations?" The addition of this time dimension immensely increases the complexity of the economics. Since there are no right answers to this question, a somewhat arbitrary decision was made to observe the demands in three years—1990, 2000, and 2030.

6-1.2 Supply of Food and Fiber

The other side of the economic market for food and fiber is the supply. Traditionally, supply is subdivided into land, labor, and capital. To a limited degree, any one of the three components can substitute for the other two. However, this substitution becomes increasingly more difficult as increased amounts of one are substituted for another. Within the economics of soil erosion, land supply is the most critical. Labor and capital supply, especially in the USA, are much more flexible since only a small fraction of available labor and capital is used in the production of food and fiber. Therefore, when pressures get strong enough, the necessary quantities of labor and capital can be pulled away from other industries for use in food production. The same is not true for land. When timber production is included with food and fiber, most of the land area is used for these three purposes. Land dedicated to other purposes such as urban, transportation, water reservoirs, etc., is of such specialized and highly valued uses that the food and fiber industry is not an effective competitor. As a result, the supply of land becomes quite rigid.

This being the case, the next two economic questions become critical: What influences the quality of the land for food and fiber production? and How can labor and capital be substituted for land? This substitution can be accomplished in many ways. Given a fixed supply of food and fiber, the various mixes of food stuffs, livestock, fiber production, and timber push and shove among themselves in competition for the nutrients, light, space, and moisture available upon a given volume of soil and water resources. Labor and capital can be used to drain, clear, irrigate, fertilize, till, control weeds, reduce disease and insects, provide plant breeding, educate managers, improve harvesting, minimize storage loss, and improve marketing, as well as numerous other nonland supply activities, all of which will increase the amounts and/or decrease the prices of foods and fibers available to the consumer.

When these labor and capital resources are applied and on which lands brings us to the time and space dimensions for supply. If we are to understand how soil erosion fits into this maze of interactions, systematic procedures must be devised to track, monitor, account, and report these interacting influences in a form that can be understood and translated into economic impacts, policy analyses, and alternative programs. This need led to the call for the "integrated system" directive as specified in the RCA.

The land, or soil, supplies many resources used in food and fiber production. For this discussion, we will restrict our comments to those soil-related resources that are altered by soil erosion. To what extent can capital and labor substitute for these resources? What is the additional cost incurred when these substitutions are made? When do these substitutions become physically impossible or uneconomical? The EPIC model has been designed to help track and answer these questions.

Soil erosion influences crop production and, therefore, the supply of food and fiber in several ways. When soil is carried away by wind or water erosion, valuable nutrients go with the particles. The economic loss of nutrients is compounded by the presence of more plant-available nutrients near the soil surface than deeper in the profile. Therefore, under most conditions the near-surface layers are the most valuable with respect to soil nutrients. In many soils the surface layers are also more hospitable to root growth than are the deeper layers. For example, many soils have aluminum levels that are toxic to plant roots. These toxic concentrations most frequently occur in the deeper profile of the soil, since natural processes and past inputs, such as liming, have corrected the surface layers for crop production. As erosion peels away the upper layers, the roots are increasingly exposed to the toxic area. Most of the organic material containing the plant-available nitrogen is near the surface, and most of the plant-available phosphorus is contained in the top 150 mm (6 in.). These and other nutrients combine to make the surface layer far more economically important than it first appears.

A second major resource provided by the soil and used in plant growth is the water-holding capacity of the soil. Since most soil's properties vary with depth, the water-holding capacity varies with soil layers. Removal of surface soil by erosion has a wide variation on the water-holding capacity of the soil. Only water that can be reached and extracted by the plant can be called a water resource for plant growth. This portion is often called plant-available water. Three dominant forces control the soil-water supply to the plant: (i) the volume of water that can be stored in each volume of soil, often referred to as the drained upper limit; (ii) the volume of soil that can be reached or influenced by the roots, also known as the rooting zone; and (iii) the fraction of the water that cannot be pulled away from the soil particles by the plant, identified as the wilting point of the plant. Soil texture and organic matter seem to influence this point. When unrestricted by soil and soil environment, the rooting zone is controlled by the genetic characteristics of the plant. However, this genetic potential is often not reached because of characteristics of the soil environment. Thus, rooting can be restricted by high bulk density, low aeration, cold temperatures, toxic elements, or lack of plant-available water. All these limiting factors are altered by soil erosion and, therefore, are economically important to a study of erosion.

We have discussed the ability of soils to provide water to the plant once the water is in the soil profile. Soil erosion may also alter the properties of the surface soil and, therefore, the ability of the soil to take in the water that falls or is applied to the soil surface. This infiltration rate is controlled by the soil texture and other chemical and physical properties of the soil (for example, the clay mineralogy and soil aggregation properties). For a majority of soils, the small soil particles (clays) are more numerous in the subsoils than in the surface soil. The tighter the soil, the slower the infiltration rate. As soil erosion exposes the tighter soils, the amount of water entering the soil decreases and the amount of runoff increases. This phenomenon has a compounding effect. Less water is stored in the root zone for plant growth and more runoff energy is created on the surface, thereby increasing the erosion resulting from a given amount of rainfall. Although other primary and secondary effects of erosion influence the economics of crop production, those discussed above are the dominant ones.

6-2 MODELS THAT FUNCTION AS ECONOMIC INTEGRATORS

Numerous models can serve as economic integrators for addressing changing supply and demand conditions within a region or country. The most widely used of these national models are the macro-economic models. These are time-series (econometric) models which express supply and demand conditions in a system of simultaneous and/or recursive, statistically derived equations. However, this type of model has at least three major limitations which, without augmentation, make it ineffective and inappropriate for studying erosion economics.

First, time-series models operate on "trend" equations. One of the underlying assumptions of trend models is that relationships creating the trend do not change appreciably during the time of projection. This is a reasonable assumption for time periods of 1–5 yr; however, it is a very weak assumption when long-term projections of 25–50 yr are needed. The objective here is to establish long-term resource policies that will affect future generations.

Second, econometric models use aggregated relationships. They are not conducive to monitoring the level of detail about specific resource use needed to trace properly the impacts of erosion on the supply functions. As discussed earlier, erosion is soil and tillage dependent. The better econometric models may have regional supply equations, but they do not have land-quality accounting.

A third reason why econometric models have limited use in evaluating erosion economics is that many of the phenomena to be considered when designing long-range resource policy have never been observed. Therefore, there are no samples from which to derive statistical estimate of the coefficients to use in the models. For example, the combination of future domestic population and export demand gives a future demand level never before experienced. What marginal resources will be drawn upon to meet the demand?

Types of economic models much more suited for studying erosion economics are the linear programming (LP) models plus related models that have an associated LP subsystem. As we will see shortly, conventional LP models can handle the aforementioned difficulties, but they are limited by a fixed demand level (which includes export demand) for each final commodity in each market region as defined by a geographic area of production.

Generally, we believe that this specification is adequate for analyzing soil loss and conservation possibilities. Of course, in certain cases, the tightening up of the system would cause restrained production and increased prices. Linear programming models can estimate changes in commodity "shadow" or supply prices. However, with fixed demands specified for each market region, they do not allow the quantity consumed to respond to commodity price levels. This response can be attained, however, in several types of demand-responsive variations. The linear programming model makes a snapshot in time, such as for the year 2000, 2030, etc. Another snapshot-in-time model is a quadratic programming model which does incorporate demand functions and permits the quantities consumed to be responsive to prices. A third alternative for a single year's results in the tatonnement model, which also incorporates demand functions allowing quantities consumed to respond to commodity prices. Finally, the linked hybrid model combines the market characteristics of an econometric model with the characteristics of resource use and supply of a linear programming model. It provides a recursive analysis with a path through time of market response and input allocations with specified levels of total resource use.

Current plans call for using the conventional linear programming model with fixed demands and the linked hybrid model as the two primary integrating economic models for the 1985 RCA analysis. Of these two, the conventional LP model provides the greatest detail for tracing economic impacts of erosion. For this reason, much of the remainder of this paper will be dedicated to describing how a conventional LP model is set up to evaluate erosion economics for national policy design.

Linear programming models, when designed for national policy analysis, become powerful tools for tracing economic impacts of changing economic conditions. These LP models include regional and interregional characteristics so that impacts can be analyzed at substate, state, regional, and national levels. The models allow a detailed analysis of the use of land, water, and other resources. They allow a normative evaluation of future potentials in resource use for agriculture and their impacts on food supplies, commodity prices, farm income, and resource income. Most importantly, they trace the interdependence among regions and among land classes within them throughout the nation.

For example, the LP models will show the change in cropping and livestock systems and the change in land and water use necessary on level lands of western Kansas and Nebraska if soil erosion is reduced to "t" levels in areas several thousand kilometers away in Mississippi, western Tennessee, or the Palouse area of Washington. They can trace the most economic combination of cropping systems, conservation practices, and tillage methods on thousands of land classes in hundreds of producing regions over the nation if specified goals in soil-loss abatement, environmental improvement, or rates of water use are to be attained. In connection with land and water use, cropping systems, conservation programs, technologies, and other alternatives, the models can indicate where major types of livestock can be produced most economically and the rations they should use to meet with particular configurations of national agriculture. When LPs have transportation submodels, they report the route that commodities should take for export when moving from producing regions to consuming or market regions. The models can assess different levels of exports, different future technologies, alcohol production for fuel use or other scenarios on soil loss, land and water use, commodity supply prices, net farm income, or resource returns. Linear programming models designed in this way have a rich history of providing analysis for national policy questions.

The Center for Agriculture and Rural Development (CARD) at Iowa State University, in cooperation with USDA and the National Science Foundation, has over a period of many years developed the most widely used of these linear programming models. One configuration of this model was used to provide solutions for the 1980 RCA evaluation by SCS (USDA, RCA Appraisal, 1981). Earlier versions of such models have been used to evaluate erosion and environmental impact potentials, conservation alternatives, and land- and water-use possibilities (Boggess and Heady, 1980; Campbell and Heady, 1979; English and Heady, 1980; Heady et al., 1972). They have also been used for numerous other policy applications and by the Food and Fiber Commission, the President's National Water Commission, the Water Resource Council's 1975 National Water Assessment, the Midwest Governors' Conference Land Use Studies, and other entities.

The components of the LP model are tailored for each specific application, but the structure is basic. To understand this basic structure, we need only to look at the matrix of a very simple LP (Table 6–1).

A linear program describes a system of simultaneous equations which is overidentified (that is, the system has a large number of solutions). On the left-hand side (LHS) of the equations are recorded the amounts of resources used, the amounts of products resulting for each combination of resources, and the cost of producing the products. Each of the values on the right-hand side (RHS) shows the total amounts of a resource available for use, the total amount of a resource used, the total amount of a crop produced, or the total cost of producing all products.

Each column in the production activity section of the matrix represents an activity. All resources and products of interest associated with that activity are recorded in that column. For example, activity A1 in Table 6–1 is an irrigated corn-wheat rotation which would use the following resources: 1 ha of soil of quality SOIL1, w1 cm of irrigation water, h1 hours of labor, f1 kg of fertilizer, j1 MJ of energy, plus unrecorded quantities of other re-

Table 6–1. Example of linear programming matrix designed for erosion policy analysis.

	Production activities†							Right hand side
	A1	A2	A3	A4	A5	A6	A7	
Objective Function								
Total Cost	$A1	$A2	$A3	$A4	$A5	$A6	$A7	= $
Constraints								
Land								
SOIL1	1	1	1					≤ × ha irrigated
SOIL2				1	1	1		≤ y ha
SOIL3							1	≤ z ha
Irrigation Water	w1	w2						≤ d m³
Commodity Demands								
Corn	y1c‡			y4c				≥ a kg
Wheat	y1w	y2w		y4w	y5w			≥ b kg
Hay		y2h	y3h		y5h	y6h	y7h	≥ c kg
Accounting Rows								
Erosion	e1	e2	e3	e4	e5	e6	e7	> 0 t
Labor	h1	h2	h3	h4	h5	h6	h7	> 0 hr
Fertilizer	f1	f2	f3	f4	f5			> 0 kg
Energy	j1	j2	j3	j4	j5	j6	j7	> 0 MJ
Water Runoff	r1	r2	r3	r4	r5	r6	r7	> 0 m³
Chemical Runoff	c1	c2		c4	c5			> 0 kg

† This table shows production activities for one producing area, one conservation practice, and one tillage level. The complete matrix is created from multiples of this sub-matrix.
‡ y1c = corn yield per hectare of activity 1.

sources not of immediate interest. This specific combination of resources would produce y1c kg of corn, y1w kg of wheat, e1 tons of soil erosion, r1 cm of runoff, c1 g of chemical runoff, plus unrecorded quantities of other products and by-products not of immediate interest. If one has the proper data, any resource or product associated with activity A1 can be recorded and monitored in this manner. This example is set up to manage three soil qualities and three crops (corn, wheat, and hay). Activities A1 and A4 are corn-wheat rotations, with A1 irrigated on SOIL1 and A4 a dryland rotation on SOIL 2. SOIL1 and SOIL2 can grow any one of three rotations (corn-wheat, wheat-hay, or continuous hay), and SOIL3 is permitted to grow only continuous hay.

Even though SOIL1 is irrigable, the continuous hay rotation (A3) is dry farmed as shown by the blank in the irrigation row. Chemical inputs are not used on the hay activities. Each of the seven activities produces a different level of soil erosion (e1 to e7) and requires different levels of labor, fertilizer, and energy. Activities can be stratified into various levels of detail. Since erosion is affected by location, tillage, and conservation practices, stratification by these categories is important. Therefore, one would repeat the relevant activities for each geographic producing area (PA), tillage practice (T) (for example, fall plow, spring plow, conservation tillage, or zero tillage), and conservation practice (for example, straight row, contouring, strip cropping, or terracing). Each geographic producing area will, of course, have different soils and climates and, therefore, different yields and byproducts. Of necessity, new land rows must be added to the matrix to accommodate each geographic area.

In the land section, each row represents one type of land, with the total hectares of land in the producing area recorded on the right-hand side (RHS). In the final solution of the linear program, the sum of all hectares used by all activities for that quality of land cannot exceed the total hectares of that land. This is called a constraint row and is illustrated by the less than or equal to signs. The LHS of the commodity section records the yields as described above. On the RHS the total demand for each commodity is recorded as the sum of the domestic and export demand for a given point in time (present or future). For example, when the corn yield for each activity is multiplied by the number of hectares assigned to that activity and all these quantities of corn are summed for all activities growing corn in the rotation, the total must, at minimum, equal the total corn demand in the nation. Where irrigation water is limited, the amount of water used by all irrigation activities cannot exceed the total supply of water available for irrigation. Rows with a greater than zero for the RHS are called accounting rows. These rows, unlike the constraint rows, will not restrict activity resource assignments. They are used to record the total quantities used for resources of interest.

The top row is the objective row. The LHS records the dollar cost of all resources used in an activity. As was stated previously, a linear program is a system of simultaneous equations with a large number of solutions. One solution is achieved when limited resources are assigned so that none of the constraints is violated. Therefore, a restricted quantity of land will be assigned to specified (not necessarily all) activities. When hectares are multiplied by their activity price and these quotients are summed, the total cost of producing all the commodities demanded in the model is determined. The linear programming algorithm uses this objective function to choose one solution from the multiple selections.

The preferred solution is the one that minimized the total cost of producing and transporting all agricultural commodities considered by the model. In effect, this minimum-cost solution gives a long-run market equilibrium where all resources receive their market rate of return. However, since returns to land, water, and other limiting resources are determined endogenously (internal to the model), their return will depend on the supply and demand constraints placed on the model. In summary, the LP model can assign resources in an optimum manner, simulate market equilibrium, record each limiting resource's market rate of return, and provide accounting for all other resources and byproducts of interest. This tool is exceedingly effective for integrating the economic relationships associated with soil erosion.

6-3 SPECIFICS OF THE NATIONAL LINEAR PROGRAMMING MODEL

Let us now proceed to some of the specific characteristics of the national policy model maintained at CARD, its relationship to EPIC, and the configuration as it will likely be used for the erosion policy assessments of the 1985 RCA.

The data needed to run an LP model must be assembled from many sources. Since the output is a minimum-cost solution, the resource data must have consistent quality across all geographic regions, and the collection process must be subject to the same set of rules and definitions. The soil-inventory data to be used in the RHS of the 1985 model will be compiled from the SCS 1982 National Resource Inventory (NRI) to be released in the fall of 1983. This inventory will report the hectares (acres) of soil in each land quality group by irrigation (or dryland) and land use for each major land resource area (MLRA) in the USA. The national LP model will have conversion activities (not shown in Table 6-1), which will allow the model to change land use—for example, from rangeland to cropland, or from dryland to irrigated land. This change in use sustains some capital cost of conversion, which is charged to the cost of producing the crop. This type of land conversion will occur when the total demand for commodities is exceedingly high and will likely move more fragile soil into cultivation. These soils often will have higher erosion rates and require better management if they are to remain productive as cultivated lands.

Water inventories are used to determine the total amount of ground and surface water available for irrigation in each geographic region. These regions in the national LP model were, in fact, chosen to represent major drainage basins in the USA. This water supply information was collected for and is available from the 1975 National Water Assessment (U.S. Water Resources Council, 1978). The land, water, and other resource inventories must be adjusted to the expected conditions for the time under consideration. For example, if the LP run is to represent the year 2030, the supply of land available for food and fiber production must be adjusted from the 1982 level to the land supply in 2030. The adjustment should represent the resources leaving agriculture as lands are converted to nonagricultural uses, but the conversion from range to cropland is provided for the model.

The projected regional and national demands for each commodity in future time periods are provided by national econometric models such as the hybrid model mentioned above. Various levels of sophistication are used to determine these demands. One method is to predict per capita consumption as a function of income, taste, geographic distribution of the population, age distribution, etc. Future population projections are provided by the Bureau of Census. The per capita consumption is then multiplied by the projected population for each future time period to obtain the domestic demand. Export demand is added to the domestic demand to give the total commodity demand. This total is then used as a constraint in the RHS of the LP model.

Right-hand side constraints are often provided, or existing values are adjusted, by policy analysts who desire to consider possible policy alternatives. Examples are a constraint on the total amounts of sediment concentration permitted in the streams of a region, or the total tons of eroded materials leaving the land. Model adjustments for these types of policies will be discussed in the last section.

A large portion of the LHS information is provided by the EPIC model. For any given activity (rotation and conservation-tillage practice), one run of the EPIC model will provide values for one LHS column. Those values will include yield estimates for each crop grown in the rotation, annual tons of soil eroded by wind and water, the amount of irrigation water required if irrigated, the amount of fertilizer applied for the rotation, the amount of lime required to maintain the soil pH as erosion occurs, the quantity of nitrogen fixed by legumes in the rotation, the estimates of nutrients leaving the land with the sediments, the quantity of runoff from the hectare, and numerous other bits of input and output information. Although EPIC provides this information for each year of a simulation run, only the expected values for each point in time are passed to the LP model. Because of specific management practices and changing soil properties due to erosion, these expected values will change through time (that is, they are not the average values over the 50 years of simulation). Therefore, for the LP run representing year 2030, the yields, fertilizer, etc. inputs into the model will be different from those values used for a 1990 LP run.

Other LHS values, such as requirements for energy, labor, pesticides, equipment, management, etc., are obtained from input and output information of the Firm Enterprise Data System (FEDS) of the Economic Research Service. These budgets are compiled by USDA from farm surveys conducted nationwide to determine the farm management practices and their associated costs. Budgets reporting cost and returns by crop and sub-state regions are published annually. These same budgets are used to provide the tillage information required as input to the EPIC program. Costs for the individual crops are combined by rotation and tillage practices, adjusted for fertilizer, irrigation, and lime requirements from the EPIC output, and summed to provide the activity cost ($A1) for the objective row of the LP output. Other LHS information, such as the cost of converting land from range to cultivation, dryland to irrigated, or wetland to drained land, is obtained from other federal agencies, such as the Agricultural Stabilization and Conservation Service (ASCS) and SCS.

The expected configuration of the national model for the 1985 RCA Assessment will consider internal to the model 105 producing areas, each with 8 soil groups, 4 tillage practices, 4 conservation practices, dryland or irrigated, and 10 to 25 crop rotations selected from combinations of 10 crops. The crops internal to the model will include corn, grain sorghum, barley, oats, wheat, soybeans, cotton, peanuts, legume hay, and nonlegume hay. The model will have 28 market regions and 54 water-supply regions. When all the relevant combinations are formulated into activities, the LP will have 40 000 to 60 000 LHS production activity columns. In addition to these production columns transfer columns will account for land conversion, transportation between market regions, transport of water between water-supply regions, and other specialized columns designed for policy analyses and resource accounting purposes. The model will likely have 3000 to 5000 constraint and accounting rows in the matrix.

6-4   USES OF THE NATIONAL LINEAR PROGRAMMING MODEL FOR EROSION POLICY ANALYSES

After the matrix has been formulated to include all inputs and outputs of interest, it is solved for the preferred resource allocations. This is referred to as a baseline solution. One base line is obtained for each time of interest. At this point the model is ready to be used to address various policies. A few examples will be given to illustrate how the matrix can be adjusted to address policy issues.

Policy Question: What would be the price of food and fiber and the regional distribution of crops in the year 2000 if a national program were established to encourage farmers not to use any rotation or practices which exceed three times the soil tolerance level (t)? As a result, how many acres would be shifted out of cultivation? How many acres of additional irrigation would be required? How much extra energy would be consumed each year? To address these questions, the LHS of the erosion row would be searched for all activities with erosion levels exceeding 3t. These activity columns would be removed from the matrix and a new preferred solution would be obtained. Results from this new solution would be compared to the baseline solution.

Policy Question: What would be the effect on U.S. agriculture in the year 2030 if the USA were to adopt an export policy in 1985 which would increase exports of wheat and corn by 30% of the 1985 export level? The RHS of the national demand rows for the wheat and corn would be adjusted by calculating 30% of the 1985 export demand and adding this to the demand levels shown in the baseline solution for 2030.

Policy Question: What would be the benefit to society in the year 2030 if we were to reduce the total erosion in the nation to 70% of the 1990 baseline level? What lands and regions would it affect? This analysis would be achieved by recording the RHS solution from the accounting row for total national erosion shown in the 1990 baseline solution. The value would be reduced 30%, and the resulting number would be entered as a constraint in the 2030 matrix row which reports the national erosion total. The constraint would be achieved by changing the "greater than zero" RHS to a "less than or equal to" equation. Many more examples could be described, but we feel these should suffice to illustrate how the LP model can be used for policy analysis.

6-5 CONCLUSION

Linear programming has been and will continue to be used to address many types of resource policy analyses. It is now possible to provide the LP data from physical simulation models such as EPIC. These data can greatly enhance the information extracted from linear programming for policy analysis. Their marriage gives us powerful analytical capabilities not previously available to policy decision makers.

REFERENCES

Boggess, W. G., and E. O. Heady. 1980. A separable programming analysis of U.S. agriculture export, price and income, and soil conservation policies in 1985. CARD Report 89. Center for Agric. and Rural Dev., Iowa State University, Ames.

Campbell, J. C., and E. O. Heady. 1979. Potential economic and environmental impacts of alternative sediment control policies. CARD Report 87. Center for Agric. and Rural Dev., Iowa State University, Ames.

English, B. C., and E. O. Heady. 1980. Short- and long-term anlaysis of the impacts of several soil loss control measures on agriculture. CARD Report 93. Center for Agric. and Rural Dev., Iowa State University, Ames.

Heady, E. O., H. C. Madsen, K. J. Nicol, and S. H. Hargrove. 1972. Agricultural and water policies and the environment: An analysis of national alternatives in natural resource use, food supply capacity and environmental quality. CARD Report 40T. Center for Agric. and Rural Dev., Iowa State University, Ames.

USDA. 1981. Soil, water, and related resources in the United States: Analysis of resource trends. 1980 RCA Appraisal, part II. USDA, Washington, DC.

U.S. Water Resources Council. 1978. The nation's water resources, 1975-2000. Second national water assessment.

7 Economic and Social Perspectives on T Values Relative to Soil Erosion and Crop Productivity

Peter J. Nowak and John Timmons
Iowa State University
Ames, Iowa

John Carlson
University of Idaho
Moscow, Idaho

Randy Miles
University of Missouri
Columbia, Missouri

Controlling excessive soil erosion is a challenging but not insurmountable task. Recent USDA actions indicate that we are committed to understanding how, where, and the extent to which soils erode. Soil scientists are beginning to quantify the complex relationship between erosion and soil productivity. Agronomists and agricultural engineers are introducing new production systems designed with conservation as an explicit criterion. Further, social scientists are beginning to move beyond simplistic or unrealistic explanations of the behavior that induces excessive erosion. A concentrated, interdisciplinary effort toward synthesizing our understanding of the relationship between soil erosion and crop productivity has been lacking.

This paper examines the following social and economic perspectives on the relation of soil loss tolerance (T values) to soil erosion and crop productivity in the USA: (i) why soil erosion is of concern; (ii) why soil erosion will likely continue to be a problem; (iii) some of the limitations of using T

Published in R. F. Follett and B. A. Stewart, ed. 1985. *Soil Erosion and Crop Productivity.*
© ASA-CSSA-SSSA, 677 South Segoe Road, Madison, WI 53711, USA.

values in the formulation of conservation policy; and (iv) how some of these limitations can be resolved through development of multiple soil-loss tolerance values.

7-1 WHY SOIL EROSION IS OF CONCERN

Soil is an essential natural resource for growing plants for food and fiber. The suitability of a soil for crop production is based on the quality of the soil's physical, chemical, and biological properties. One naturally occurring process that may detrimentally affect these soil properties and subsequent crop production is soil erosion.

Although erosion is a natural process, man's activities have greatly increased the imbalance between soil loss and soil formation. Destruction or removal of vegetation or natural cover resulting from tillage, overgrazing, or clear-cutting exposes the soil to a greater possibility of detachment. Soil losses resulting from sheet, rill, and wind erosion from cropland, pastureland and forest land averaged 15.3, 5.9, and 2.7 Mg/ha respectively (6.8, 2.6, and 1.2 T/acre) in 1977 (USDA, 1981).

Water and wind erosion account for an average of over 5.8 billion Mg (6.4 billion tons) of soil loss annually (USDA, 1981). Although wind erosion accounts for 1.4 billion Mg (1.5 billion tons), discussion will focus on water erosion.

Water can erode soil material through sheet, rill, gully, and streambank erosion (Troeh et al., 1980). Although gullies represent a large soil loss from specific areas of the landscape, sheet and rill erosion are the major types of water erosion on cropland. Sheet and rill erosion are estimated to account for 78% of erosion by water on all nonfederal land (USDA, 1981), thus having a profound effect on not only cropland, but also on pasture, range, and forest land.

As shown in Table 7-1, land capability classes that are either marginal or not suitable for cropland (classes IV, VI, and VII) have relatively high rates of erosion, but relatively small areas are affected. Erosion rates are relatively moderate on Class II and Class III lands (8.1 and 11.4 Mg/ha or 3.6 and 5.1 T/acre, respectively), but approach or exceed the generally accepted maximum tolerance values of 9 to 11.2 Mg/ha (4 to 5 T/acre) for

Table 7-1. Average erosion rates by land capability class (USDA, 1977).

Capability class	Average annual erosion rate	
	Mg ha^{-1}	1000 ha
I	6.3	12 770
II	8.1	76 049
III	11.4	53 336
IV	14.8	17 757
V	4.0	936
VI	25.1	5 240
VII	31.8	1 232
VIII	0.0	57
Average	10.5	167 377

moderately deep to deep soils. The alarming fact about these Class II and Class III areas is their suitability and widespread use (129.4 million ha or 319.5 million acres) for cultivated crops, and thus their high potential for a widespread effect of soil erosion on productivity.

7-1.1   Soil Erosion and Soil Productivity

Soil productivity is the capacity of a soil, in its normal environment, to produce a particular plant or sequence of plants under a defined set of management practices (Soil Science Society of America, 1975). Historically, decreases in soil productivity have been attributed to soil erosion (Smith, 1941; Lowdermilk, 1948; Dregne, 1982; Langdale and Scrader, 1982). This relationship is difficult to study, however, because yield decreases resulting from erosion are poorly defined, cumulate in a nonlinear fashion, and may be recognized only after much of the land has been rendered economically unfit for crop production (USDA, 1981). Moreover, technological inputs, such as fertilization or irrigation, may mask the productivity losses due to erosion. The nonlinearity of the erosion also adds to this masking effect (USDA, SEA–AR, 1981).

7-1.2   Research

Many research studies have been conducted to assess the influence of soil erosion on productivity (USDA, SEA–AR, 1981; Langdale and Scrader, 1982; Pierce et al., 1983; Schertz, 1983). Reviews of these studies (Council for Agricultural Science and Technology, 1982; USDA, SEA–AR, 1981) have consistently concluded that evidence of productivity losses due to erosion exists throughout the USA.

Many factors contribute to the decrease in soil productivity from soil erosion. The most frequently mentioned factor is the reduction in the soil's plant-available, water-holding capacity. The crop is subject to greater moisture stress through the interaction of decreased soil depth (rooting volume) and the increase in the percentage of the crop root zone in the subsoil. The subsoil usually contains less desirable physical properties because of more clay and poorer structure.

Degradation of surface structure is a second factor induced by erosion. This less desirable structure creates greater bulk density which restricts seedling emergence and root penetration.

A third factor is the loss of plant nutrients. Nutrients such as nitrogen, phosphorus, and potassium can be solubilized in surface runoff or attached to soil particles that are moved during erosion.

Fourth, losses in organic matter are also prevalent with erosion, thus further contributing to the loss of nutrients, the decrease of structural stability, and a decline in the effectiveness of herbicides in certain crop production systems. In addition, tillage and seedbed preparation are usually made more difficult by erosion. More energy is needed and soil-seed contact is less complete when less topsoil and greater amounts of denser, more clayey subsoil are mixed in the plow layer.

Finally, confounding the influence of soil erosion on productivity by all the previously mentioned factors is the selective nature of soil erosion. Soil erosion is varaible for particular soil components and within a particular field or landscape. This variability is usually a result of topographic and soil nonuniformity. One area of a field may experience great losses of runoff, soil particles, and nutrients. Other segments of the same field may experience smaller losses or may even have net additions of water, soil particles, and nutrients as a result of deposition from other areas. The result is a relatively large loss of productivity in some areas of the field and a relatively small loss or even a gain of productivity in other segments of the field. The variability of erosion on productivity within a field not only impedes scientific investigation, but also makes difficult the selection of the best suited fertilizers, herbicides, or cropping systems.

7-2 WHY SOIL EROSION WILL CONTINUE
TO BE OF CONCERN

As research results accumulate on the relationship between soil erosion and crop productivity, the need for actions to reduce soil-erosion losses appears imperative. However, the precise actions to be taken remain less clear and are debatable for two reasons. First, because of the tremendous variability of the impact of erosion on different soils, no one agronomic or engineering action is completely appropriate for a particular field, let alone for an entire farm, watershed, or county.

A second reason for uncertainty on appropriate corrective actions is the diversity of factors affecting soil erosion. Any future policy decision must necessarily be directed toward those contributing factors responsible for soil-erosion losses. Yet social, economic, technological, and political causes of soil erosion are frequently impervious to policy decisions. Both these reasons imply that soil erosion will continue to be problematic in the immediate future. We will briefly examine four related reasons for this prediction.

7-2.1 Demand and Supply Process

Beginning in the 1930s, the nation responded to the threat of erosion to agricultural productivity, and considerable progress was achieved from the 1940s through the 1960s (Cory and Timmons, 1978). But serious, intransigent obstacles remained. After about 5 decades of agricultural surpluses and low farm prices, interrupted briefly by World War II, abrupt and important changes arose in the early 1970s. These changes, whose causes run deep into the past, appear likely to persist into the future (Frey, 1952; Held and Timmons, 1958; Blase and Timmons, 1961; Hauser, 1976).

In the early and middle 1970s, domestic and international developments substantially increased the demand for agricultural products and at the same time taxed agriculture's capacity to respond to their demands. This increased demand in conjunction with a reduced capacity to respond aggravated the degradation of agricultural resources.

The factors associated with increased demand for agricultural products included increased per capita income in developed and developing countries; population growth; and the continuing need for foreign exchange by the USA to purchase petroleum, vehicles, electronics, textiles, and other products that until recently had been produced domestically.

Another concomitant set of natural and man-made factors adversely affected the ability of U.S. agriculture to respond to the increased demand for its products. One of these factors, weather conditions (in particular, inadequate moisture during the middle 1970s) interfered with crop production in large sections of the nation and the world. Second, petroleum-based inputs, particularly fertilizers, pesticides, and energy, which were responsible for much of the increase in crop yields in recent decades, became increasingly expensive. Third, environmental quality constraints, especially nonpoint water pollution including suspended silt and chemical residuals, discouraged continuation of some agricultural practices. Fourth, the expansion of urban and industrial development into productive farming areas frequently conflicted with and aggravated crop-production practices. Fifth, per acre productivity increases appeared to level off, due to the above factors and possibly to inadequate basic and applied production research. Sixth, the resulting reduction in yield increases per acre, increased crop prices, and increased production costs necessitated bringing additional acres into cultivation. These additional acres usually were less productive than current cropland, so per acre output tended to be less, particularly in terms of net value productivity. Also, the additional acres converted to cropland frequently were more fragile and more susceptible to water and wind erosion, with consequent adverse effects on environmental quality and continuing productivity of soil.

These macro demand and supply factors will probably continue to accelerate erosion processes in U.S. agriculture in the immediate future.

7-2.2 Economic Incentives

Soil erosion is also likely to continue being a problem because of the difficulty in designing a conservation policy based on economic incentives.

From studies of obstacles to soil-erosion control, economists conclude that people are most interested in increasing their net income over their respective planning horizons. This process involves minimizing costs and maximizing revenues in using labor, capital, and land. Compounding this motivation to maximize net income are institutions that also affect people's behavior in regard to accelerated soil erosion. Such institutions include land tenure (including tenancy), credit, tax, and market systems. Thus, from an economic perspective, maintaining soil productivity will involve manipulating these market and institutional processes so that erosion control becomes profitable to the land user.

Unfortunately, a number of other social and economic conditions interfere with this rational allocation of benefits. We are operating in an imperfect or incomplete market in that social benefits and costs do not match private benefits and costs. This imperfection is reflected in the lack of consistent criteria around which scarce economic resources are distributed. Currently, cost-sharing funds are distributed according to various rationales. A partial list includes:

1. Distribution to the states based on the political consideration that all states, regardless of need, should receive some of the federal funds.

2. Distribution within a district based on the logic of "first come—first served." Although some argue that this method is fair, in reality the distribution of funds is highly correlated to one's position in the social community and weakly correlated to the need for erosion control.

3. Distribution to a state or district based on gross erosion rates. Current targeting efforts are being reoriented around the fact that only 3.5% of the nation's cultivated land accounts for 58% of the nation's gross soil loss, by weight (Ogg et al., 1982). Yet, no direct, linear relationship exists between rates of soil erosion and losses in soil productivity. Targeting based on soil loss, by weight, also fails to account for the variability in off-site damages.

These popular distribution schemes illustrate the power of economic motivations—we know *how* to motivate land users. Nevertheless, we still need to determine *who* needs to be motivated while also assessing the political feasibility (i.e., who pays) of the specific motivational strategy. Ambiguity and controversy regarding these decisions prevent a quick, economic solution to our erosion problems. This failure is largely due to the political inertia surrounding current distribution schemes and to the anticipation that the amount of real dollars allocated to conservation efforts will remain constant at best.

7–2.3 Technical Assistance Programs

Soil conservationists, soil technicians, and others who actively promote conservation on a day-to-day basis frequently become frustrated and cynical because of their inability to get more practices and structures on the land. Although they begin as highly motivated individuals well-trained in soils, engineering, and agronomic principles, they quickly learn they must practice these skills in a world ruled also by social, economic, and political realities. Technically accurate recommendations must be modified in accord with social, economic, and political constraints. Assistance is often requested by those with the least need, while those with serious erosion problems fail to see a need for assistance or are unaware of how to obtain help.

Technical assistance is but one in a triad of local conservation efforts. Economic assistance and informational or educational programs must also be considered. Educational efforts can create a demand for technical assistance which, in turn, is modified by the availability of cost-sharing funds. Continued educational efforts will then be needed to ensure that the technical recommendations are carried out in an agronomically sound manner. This sequence implies the need for coordination of technical, economic, and educational efforts at the local level (Nowak, 1982). Although a formal, local coordinating body was rejected in the RCA process (USDA, 1981), we should not ignore coordination efforts between existing organizations.

Regardless of the policy or program designed to control excessive erosion, it must be implemented at the local level. How effectively it is carried out will depend on the training and leadership qualities of the district, state, and regional personnel. More effort needs to be directed, not so much at *what* we are doing, but *how* we are carrying out local conservation efforts. Until we begin to respond to this situation, we can expect our soil-erosion problems to continue.

7–2.4 Technological Innovations

Most experts agree that another reason erosion will continue to be a significant problem is that the "technological fix" is not forthcoming. We cannot anticipate any major breakthroughs in conservation technologies (Office of Technology Assessment, 1982).

Some might dispute this claim based on the evolution in U.S. agriculture relative to reduced tillage systems. However, these claims ignore the difference between adoption and adaptation. Any technology has both an equipment and a managerial component. The equipment refers to the actual tool or implement, whereas management refers to the knowledge and skills necessary to use the equipment. For example, some reduced tillage systems do not require a significant change in equipment, for example, shifting from a moldboard plow to a chisel or disk as the primary tillage tool. However, the managerial skills required for a reduced tillage system are much more complex than for conventional tillage. Thus, a number of operators who have adopted the equipment of a reduced tillage system have yet to fully adapt their management skills to realize teh system's full economic and ecological potential. More importantly, some will never adapt, resulting in little difference between it and conventional tillage systems.

The critical difference between adoption and adaptation can be seen in Table 7–2. Where actual crop residue was calculated for those operators claiming to have adopted a reduced tillage system. The data were collected during 1980 and 1981 in three watersheds in the Iowa–Cedar River Basin of Iowa (Nowak and Korsching, 1983). The population (N = 193) consisted of all farmers who operated land within the three watersheds. The universal soil loss equation (USLE) was calculated on a field-by-field basis. The residue figures used in Table 7–2 were partially derived from this information, as well as crop yields and machinery information.

If we use one suggested definition of conservation tillage—2242 kg/ha of corn residue or 1121 kg/ha of soybean residue—we have clearly overesti-

Table 7–2. Reported versus actual use of conservation tillage. †

Claimed use of conservation tillage	
Using Conservation Tillage	78.2% (151)
Not Using Conservation Tillage	21.8% (42)

Actual residue levels for those claiming use of conservation tillage

Corn residue
73.3% (99) maintaining less than 1680 kg/ha (1499 lb/acre)
19.3% (26) maintaining at least 1682 kg/ha (1500 lb/acre)
7.4% (10) maintaining at least 2242 kg/ha (2000 lb/acre)

Soybean residue
26.4% (29 maintaining less than 840 kg/ha (749 lb/acre)
47.3% (52) maintaining at least 841 kg/ha (740 lb/acre)
26.4% (29) maintaining at least 1121 kg/ha (1000 lb/acre)

† Average for all fields with same rotation. Estimated residue levels based on calculations prior to spring planting.

mated the amount of reduced tillage on the land. Instead of a projected 25 to 35% adoption rate for conservation tillage in 1980 (Crossen, 1981; USDA, 1975), based on these data a more realistic assessment would be 10 to 15%. Also, recent surveys of reduced-tillage technologies have probably overestimated its use because the figures were based on farmers' perceptions rather than actual residue levels (e.g., Pioneer Hi-Bred International, Inc., 1982). Overall, this realization implies that we can expect only a slow accumulation of the benefits of reduced tillage or any other conservation-related technological innovations.

7–3 POLICY LIMITATIONS OF THE T VALUE

7–3.1 Agronomic Limitations

One of the ultimate purposes in predicting soil loss and assessing its effect on soil productivity under a specific management system is to correlate this loss to a permissable agronomic loss for each soil. The arbitrarily set criteria for a permissable loss of soil have been the center of considerable discussion (Schertz, 1983; Van Doren and Bartelli, 1956).

The T value is based primarily on topsoil thickness, physical properties of the soil, gully prevention, organic-matter reduction, and plant nutrient losses (Wischmeier and Smith, 1978). Most T values range from 2.2 to 11.2 Mg/ha (1 to 5 T/acre). The smaller values are correlated with shallow soils and soils with restrictive layers within particular depths. The larger T values are assigned to deeper soils and soils with restrictive layers at greater depths.

Correlation of T values to the combination of soil, climatic, and management factors that influence soil erosion in a specific cropping system is accomplished with the USLE (Wischmeier and Smith, 1978). Problems may occur, however, when comparing soil-loss values obtained from the USLE to T values for a specific soil. Wischmeier (1976) stated that "the equation is used for purposes that it was not designed for, simply because it seems to meet the need better than any other tool available" (p. 5).

Inherently, the USLE does not distinguish between soil loss and sediment yield (Wischmeier, 1976). The USLE estimates the soil moved from a specific landscape component but does not account for the deposition of soil material on the toe- and foot-slope positions. Other variables that can influence the accuracy of the USLE are variability of soil properties for a particular soil, a lack of detailed soil surveys, variability of rainfall, and lack of uniformity, in some cases, of plant and residue cover.

Some soil scientists believe that T values are consistent with soil-development rates. Soil-formation rates commonly found in the research literature and popular press are 30 to 100 years to form a 25-mm (1-in.) thick layer of organic enriched topsoil. If an erosion rate of 11.2 Mg/ha (5 T/acre) were assumed, this 25-mm (1-in.) layer of topsoil would be eroded in approximately 30 years. In summary, at the maximum T value for this soil, erosion *may* be in equilibrium with soil development. Although many soil scientists find this soil renewal quite large, this value has been validated in specific cases (Hallberg et al., 1978).

However, total soil renewal does not arise only from the development of an A horizon from the subsoil or, in some cases, an underlying, unconsolidated, parent material. Generation of unconsolidated soil material from a consolidated parent material may take place 1 to 2 m or greater below the soil surface, thereby tempering the major impact of the active soil-forming factors of climate and biotic activity. Compared to the soil-renewal rates from consolidated bedrock suggested by Smith and Stamey (1965), the tolerance levels of 2.2 to 11.2 Mg/ha (1 to 5 T/acre) exceed development of soil material from consolidated parent materials. Soil-renewal values are probably best approximated on weighted values of topsoil formation from subsoil balanced by subsoil formation from an underlying parent material. Undoubtedly, it seems that the rate of soil formation at the interface of subsoil and parent material is the controlling factor in long-term soil renewal and could be better utilized in assessing the long-term influence of erosion on soil productivity.

The assignment of a T value for a particular soil series leads to a wide variation of soil properties that correspond to a tolerable erosion rate. Much of this variation is manifested by the soil-series concept. An example would be a soil series that has a solum thickness range of 50 to 100 cm (20 to 40 in.) and an assigned T value of 9 Mg/ha (4 T/acre). The productivity of a shallow solum member of this series range (e.g., 54 cm or 24 in.) would seem to be impaired more by the tolerable range than would a deeper-end member (e.g., 91 cm or 36 in.) of the same series. The same analogy could be used for the surface-soil thickness of the series, though the subsoil's physical and chemical properties would have an input in this situation.

Another important factor currently having little input into the assignment of the T value is slope gradient. Many series have a wide range of slope-gradient classes, thus producing different degrees of surface runoff and erosion susceptibility, which in turn could affect long-term soil productivity. The assignment of either multiple T values or one T value with a standard deviation to cover particular segments of a soil-series range would be appropriate for more accurate assessment of soil erosion/productivity influences.

7–3.2 Economic Limitations

Loss limits for soil erosion to maintain productivity and reduce non-point pollution of water are becoming established throughout the nation and are assuming the character of compliance tests for land use. Yet these current values for erosion tolerances and water quality standards need to be treated as provisional and imperfect.

T values are based on estimated erosion losses calculated from the USLE. Yet the T value is essentially a physical on-site concept devoid of interpretations and does not include off-site effects of erosion, or what economists term externalities. Thus, T values require validation in terms of off-site effects as well as productivity, output, and cost effects. Results of such validation would probably mean a wider range in T values to reflect off-site effects as well as on-site soil productivity.

Nonpoint pollution control standards are more complex than soil-erosion T values for productivity because pollution control standards must embrace the off-site consequences, including amounts of suspended sediment and the residuals for which sediment serves as a transport agent in water and possibly air. Therefore, despite the fact that more is known about setting soil-erosion limits for productivity than for nonpoint pollution effects, overall policies and programs should be considered provisional.

If we accept the current formulation of T values, then we can gauge probable costs of using this factor directly in the policy arena. Wade and Heady (1977) estimated that meeting the tolerance limits would cost approximately $931 million per year nationwide. This estimate suggests that the aggregate effect of soil-erosion control programs on farmers' incomes would be less significant than the effect of weather patterns, pest problems, nitrogen prices, government policies, or export markets—none of which is under direct control of the farmer. However, soil-erosion control programs could significantly affect those farmers with large acreages of erosive land.

Others have investigated the income consequences of using T values for policy parameters on both a regional and individual farm basis (Taylor and Frohberg, 1977; Seitz et al., 1978). The effects of T values on net farm income were estimated by extending and adjusting a programming model currently used in estimating conversion of noncropland to cropland in Iowa (Amos and Timmons, 1983). Income and erosion effects were estimated from 14 various T values under several combinations of crop prices, factor costs, cropping practices and discount rates (Timmons and Armor, 1982). Effects of various T values on present value of net income ranged from reductions of $178.90 to $1705.77 per ha ($72.40 to $690.39 per acre) under one set of variables and reductions of $2075.69 to $6752.16 per ha ($840.00 to $2732.50 per acre) under another set of variables. Results suggest that, in general, conservation programs at the farm level will have a negative effect on farm income (Kasal, 1976; Boggess et al., 1980).

Establishment and enforcement of T values will affect farm income, U.S. capacity to export crops, availability and prices of food for domestic and foreign consumers, future soil productivity, and water and air quality. Therefore, all these factors and effects must be considered when examining the economic consequences of establishing particular T values.

7–3.3 Social Limitations

Although most research has focused on agronomic or economic limitations, there are also social constraints in using T values as a policy instrument.

The definition of what constitutes *excessive* soil loss remains a social interpretation of the interaction between the natural processes of soil formation and the outcome of agricultural production practices. Actual erosion and soil formation rates are important, but what ultimately constitutes excessive erosion is determined by the socioeconomic and political consequences of that erosion. Erosion considered excessive in one context may not be considered so in another. For example, assume that the erosion rates in two similar regions of cropland are exceeding the T values. However, in the first region this land is an insignificant proportion of the nation's overall agricultural productivity, whereas in the second case it represents the major share. It is unlikely that both would be considered instances of excessive erosion. Or, as another example, the label of "excessive erosion" is often based not on the erosion rate, but on the off-site damages of that erosion. This tells us that even though the T value has been based on the principles of soil science, that value is interpreted in a social, economic, and political setting.

Cook (1982) also reminds us that changing T values to reflect soil properties more accurately is a political process. This is especially true when raising the T values. On the one hand, soil scientists have revised their estimates of how long a soil will maintain its productivity. On the other hand, the practitioners and proponents of conservation are now being told that they have overinvested in erosion control.

Caught in the middle of this dialogue are two additional groups. First, the administrators of natural resource agencies find themselves having to choose between the scientific results and political support at the local level. Needless to say, science does not always win. Second are the land users for whom the policy-based T values have no meaning. These individuals are being told to change their agricultural practices on the basis of an abstract concept, the T value.

An additional social consideration in using T values as a policy instrument arises when we consider whose land is being eroded. Poor lands with high susceptibility to erosion (with a low T value) are more likely to be farmed by operators with few of the necessary management skills and economic resources. Defining excessive erosion solely on the basis of T values would bear relatively heavily upon the operators of these poor lands, not only because of the magnitude of the erosion problems they face, but also because some will lack the capital and the knowledge to apply appropriate conservation technologies. Further confounding the situation is the widespread belief that these operators receive less technical and financial assistance from the responsible government agencies.

7-4 EXPANDING THE SOIL LOSS TOLERANCE VALUE

Clearly, we need to match the complexity of our conservation policy to the complexity of the processes affecting soil-loss tolerances. The T values, as currently formulated, are incapable of this task. Therefore, we suggest that the formulation of soil-loss tolerance values be modified to minimize the noted agronomic, soil science, social, and economic limitations. One way of capturing this complexity is to create multiple T values (Larson, 1981). Initial consideration will be given to a T_1 and a T_2 value.

A T_1 value would be similar to the current T value. This value (T_1) would be the maximum rate of annual soil erosion allowable while economically maintaining current levels of crop productivity within a specified planning period.

T_1 differs from the current T value in that it is dynamic. As the planning period changes, say between 5 years, 50 years, and an infinite time line, then, except for a few unique soils, the T_1 value would also change. In most cases the T_1 value would decrease as the planning period increases.

The T_1 value would also change as our technological ability to economically maintain crop productivity changes. If advances in technology allow us profitably to produce more from less, then the T_1 values would be increased.

One also assumes that work would continue on current modeling efforts (e.g., the Erosion-Productivity Impact Calculator) to specify the correlation between soil loss, soil generation, and crop productivity. The accuracy of T_1 would also be enhanced by establishing measures of standard deviation to capture variation in solum thickness and slope gradient.

Once the T_1 value is established, then it becomes possible to formulate T_2. It is the resulting rate of annual soil erosion after accounting for the social, economic, and political costs of changing current erosion rates to the recommended T_1 level. This soil-loss value explicitly acknowledges that what may be appropriate from a soil-science perspective is not possible because of social, economic, or political constraints. In essence, T_2 is the resulting tolerable soil losses after these trade-offs have been made.

The T_2 value can be either higher or lower than the T_1 value. It would be higher where costs (economic, social, or political) are excessive to reduce current erosion rates to the T_1 value. For example, the cost may be too high if the only way T_1 can be achieved is through buying cropping rights, requiring drastic shifts in the enterprise mix, or extensively using expensive conservation structures. In such instances, the soil-loss tolerance values used in formulating local conservation programs would be higher than those based on soil-science principles. Although this would be a temporary situation—tied to the length of the planning period used in formulating

T_1—it explicitly recognizes that excessive erosion would be tolerated if social, political, or economic constraints required it.

Of course, T_2 may be lower than T_1. Concern over off-site damages (e.g., water, pollution, sedimentation, etc.) could mandate an erosion rate lower than the T_1 value. In other instances social or ethical demands would mandate certain behaviors that would also put the T_2 value below that of T_1. For example, a landlord could require certain conservation treatments as part of the lease. Another example would be family or community norms that strongly encourage the maintenance of conservation traditions.

Both T_1 and T_2 must be dynamic referents because the factors influencing crop productivity are dynamic. Although excessive soil erosion is likely to continue to reduce crop productivity in the future, we will be in a better position to control these damages for two reasons. First, we will be better able to specify how and where soil erosion is damaging crop productivity. This will allow us to target accurately whatever policy options are available. Second, both the design of policies and their implementation are beginning to acknowledge the need for integrated and interdisciplinary efforts. Both these reasons are incorporated in multiple values for soil-loss tolerance. What remains to be accomplished is further interdisciplinary work on the design and implementation of such values.

REFERENCES

Amos, C. M., and J. F. Timmons. 1983. Iowa crop production and soil erosion with cropland expansion. Am. J. Agric. Econ. 65(2):486–492.

Blase, M. G., and J. F. Timmons. 1961. Soil erosion control: Problems and progress. J. Soil Water Conserv. 16(4).

Boggess, W., J. Miranowski, K. Alt, and E. Heady. 1980. Sediment damage and farm production costs: A multiple objective analysis. North Cent. J. of Agric. Econ. 2(2):107–112.

Cook, K. 1982. Soil loss: a question of values. J. Soil Water Conserv. 37(2):89–92.

Cory, D. C., and J. F. Timmons. 1978. Responsiveness of soil erosion losses in the Corn Belt to increased demands for agricultural products. J. Soil Water Conserv. 33:221–226.

Council for Agricultural Science and Technology. 1982. Soil erosion: Its agricultural, environmental and socioeconomic implications. Report 92. CAST, Ames, IA.

Crosson, Pierre. 1981. Conservation tillage and conventional tillage: Comparative assessment. Soil Conserv. Soc. of Am., Ankeny, IA.

Dregne, H. E. 1982. Historical perspective of accelerated erosion and effect on world civilization. p. 1–14. In Determinants of soil loss tolerance. Spec. Pub. 45, American Society of Agronomy and Soil Science Society of America, Madison, WI.

Frey, J. C. 1952. Some obstacles to soil erosion control in western Iowa. Iowa Agric. Home Econ. Exp. Station Bull. 391. Ames.

Hallberg, G. R., N. C. Wollenhaupt, and G. A. Miller. 1978. A century of soil development in soil derived from loess in Iowa. Soil Sci. Soc. of Am. J. 42:339–343.

Hauser, W. R. 1976. Soil erosion control in western Iowa. M.S. thesis. Iowa State Univ., Ames.

Held, R. B., and J. F. Timmons. 1958. Soil erosion control in process in western Iowa. Iowa Agric. and Home Econ. Exp. Station Res. Bull. 460. Ames.

Kasal, J. 1976. Trade-offs between farm income and selected environmental indicators: A case study of soil loss, fertilizer and land use constraints. USDA Tech. Bull. 1550. Washington, DC.

Langdale, G. W., and W. D. Scrader. 1982. Soil erosion effects on soil productivity of cultivated cropland. p. 41-52. *In* Determinants of soil loss tolerance. American Society of Agronomy and Soil Science Society of America, Madison, WI.

Larson, W. E. 1981. Protecting the soil resource base. J. Soil Water Conserv. 36(1):13-16.

Lowdermilk, W. C. 1948. Conquest of land through seven thousand years. USDA Misc. Pub. 32. Washington, DC.

Nowak, P. J. 1982. Phase one final report: The selling of soil conservation. Dep. of Sociology and Anthropology. Iowa State University, Ames.

----, and P. F. Korsching. 1983. Sociological factors in the adoption of best management factors: A final report to the Environmental Protection Agency. Dep. of Sociology and Anthropology, Iowa State University, Ames.

Office of Technology Assessment, U.S. Congress. 1982. Impacts of technology on U.S. cropland and rangeland productivity. U.S. Govt. Printing Office, Washington, DC.

Ogg, C. W., J. D. Johnson, and K. C. Clayton. 1982. A policy option for targeting soil conservation expenditures. J. Soil Water Conserv. 37(2):68-72.

Pierce, F. T., W. E. Larson, R. H. Dowdy, and W. A. P. Graham. 1983. Productivity of soils: Assessing long-term changes due to erosion. J. Soil Water Conserv. 38(1):39-44.

Pioneer Hi-Bred International, Inc. 1982. Soil conservation attitudes and practices: The present and the future. Jefferson Davis Associates, Inc., Cedar Rapids, IA.

Schertz, D. L. 1983. The basis for soil loss tolerance. J. Soil Water Conserv. 38(1):10-14.

Seitz, W., D. Gardner, S. Gove, K. Guntermann, J. Karr, R. Spitze, E. Swanson, C. Taylor, D. Uchtmann, and J. van Es. Alternative policies for controlling nonpoint agricultural sources of water pollution. U.S. Environ. Protection Agency, Athens, GA. EPA-600/5-78-005.

Smith, D. D. 1941. Interpretation of soil conservation data for field use. Agric. Eng. 22:173-175.

Smith, R. M., and W. L. Stamey. 1965. Determining the range of tolerable erosion. Soil Sci. 100:414-424.

Soil Science Society of America. 1975. Glossary of soil science terms. Madison, WI.

Taylor, C. R., and K. K. Frohberg. 1977. The welfare effects of erosion controls, banning pesticides, and limiting fertilizer application in the corn belt. Am. J. Agric. Econ. 59:13-24.

Timmons, J. F., and A. M. Armor, Jr. 1982. Economics of soil erosion control with application to T values. p. 139-153. *In* B. L. Schmidt et al. (ed.) Determinants of soil loss tolerance. American Society of Agronomy and Soil Science Society of America, Madison, WI.

Troeh, F. R., J. A. Hobbs, and R. L. Donahue. 1980. Soil and water conservation for productivity and environmental protection. Prentice-Hall, Englewood Cliffs, NJ.

USDA. 1975a. Minimum tillage: a preliminary technology assessment. USDA, Office of Planning and Evaluation, Washington, DC.

----. 1977. National Resources Inventory. U.S. Government Printing Office, Washington, DC.

----. 1981. Soil, water, and related resources in the United States: Status, condition, and trends. 1980 RCA Appraisal, part I. USDA. Washington, DC.

USDA—Science and Educational Administration, Agricultural Research. 1981. Soil erosion effects on soil productivity: A research perspective. J. Soil Water Conserv. 36(2):82-90.

Van Doren, C. A., and L. T. Bartelli. 1956. A method for forecasting soil loss. Agric. Eng. 37:355-361.

Wade, J., and E. Heady. 1977. Controlling nonpoint sediment sources with cropland management: A national economic assessment. Am. J. of Econ. 59(1):13-24.

Wischmeier, W. H. 1976. Use and misuse of the universal soil loss equation. J. Soil Water Conserv. 31:5-9.

----, and D. D. Smith. 1978. Predicting rainfall erosion losses—a guide to conservation planning. Agric. Handb. 537. USDA, Washington, DC.

8 Assessment: A Farmer's Perspective

Charles McLaughlin
Britt, Iowa

If soil erosion is controlled to assure sustenance for posterity, it will be because those who own and operate the land are motivated to conserve. This symposium brings together an awesome compilation of the latest data, technology, and expertise. We have the technology, I am confident, to control erosion at a tolerable level for every acre of soil this nation needs for productive purposes. What we must now explore is the form of motivation. Initially, we must contend with historic disincentives to conservation. A significant body of our tax laws, our public policies and related economic incentives, are those of a young nation promoting exploitation rather than the rationale of a mature nation promoting civilization.

If our goal is a sustainable society, academic discussion of the sciences of soil and water management is talking about the cart while ignoring the horse. While this symposium comprises learned Societies, I am convinced that the impact of these choices upon the broad interests of society is of paramount importance to each of us as a citizen. Any nation which expends such a disproportion of its time, money, and natural resources on systems to destroy the planet, has not opted for a sustainable society. Our soil may last as long as humans do, without our intervention, unless society changes direction and allocates meaningful amounts of thought, money, and resources to peaceful resolution of the stresses arising from irresponsibly increasing the numbers of the human species using this finite planet.

If we are committed to protecting a heritage for our posterity, and assuming that we are sufficiently "civilized" to project one, we must develop an ethic based on acknowledgment of the source of our wealth. We must understand that it does not "trickle down" in spite of the bungling of some bureaucratic entity, but "percolates up" from the earth, by our labor. Products of the soil, the waters, and the mines are the only source of real wealth.

Published in R. F. Follett and B. A. Stewart, ed. 1985. *Soil Erosion and Crop Productivity.*
© ASA-CSSA-SSSA, 677 South Segoe Road, Madison, WI 53711, USA.

Mitigation of soil erosion is not the primary, nor even a principal, thrust of our current laws and government policies. The Office of Management and Budget (OMB) has for several years attempted to delete all assistance for soil conservation from the Federal budget. Administrative procedures circumvent law in ominous ways. Perhaps the OMB is of the same persuasion as those persons who feel we don't need farmers as long as we have supermarkets. Price support programs have historically rewarded the farmer exploiting the land, and penalized the diversified farmer who had a good conservation program in place. The forthcoming PIK program may prove to be an exception.

Tax laws are fueling conversion of good range land to poor farm land. Those of us who remember the "Dirty Thirties" may have yet to see the *great* Dust Bowl. Tax incentives have effectively removed cattle feeding from midwestern farm feedlots to industrial, concentrated feedlots where profit may depend upon tax breaks rather than upon feed efficiency or prudent buying and selling. The absence of cattle on farms indicates more row-crop tillage and less rotation, with conserving ground cover crops.

Investment credit rewards the purchase of bigger equipment which disturbs more ground faster and fits less conveniently into terraces and contours.

The traditional tax deduction for dependent children lingers from a time when infant mortality was high and a frontier was yet to be settled. Implicitly, it favors increased pressure on the land.

In the 1970s, when the bins were emptied by a stroke of luck, farmers were admonished to plow and plant to feed a hungry world, only to find their credibility as suppliers scuttled by embargo and talk of embargo from officials incapable of handling a political crisis in a constructive, statesman-like manner.

In the cost-price squeeze thus created, the only recourse for the farmer was to become more efficient, until he is now floundering in a sea of grain produced at a morally indefensible cost to the soil and water resources inventory.

Soil erosion has been permitted, sometimes inadvertently, by earlier generations. The benefits of conservation will accrue to unborn generations who will not vote in 1984. It is, therefore, indisputable logic that the public should assume a share of the investment in the soil resource base. It is entirely appropriate that Federal funds be used to pay, in part, for protection of the resource base which provides food for the public and a significant portion of the exchange medium for international trade. This investment is a modest insurance premium for the goose that lays the golden egg.

In feeding the nation, employing a significant proportion of the labor and industrial force in the supply, processing, and distribution of food, the agricultural sector provides the tax base to support the several levels of government. Our perception of priorities has not yet recognized society's obligation to protect the soil base with a just share of the taxes thus generated.

It is sometimes easier to analyze objectively a conservation dilemma which is miles from our door. To cite the rape of the Ogallala aquifer as a crime, which cries out for intervention by society through government, may only stir emotions. To describe an unprecedented government initiative, to save the tigers of India, may inform us without arousing controversy. Volume 11, No. 6 of *Ambio*, a journal of the Royal Swedish Academy of Sciences, relates in an article by H. S. Panwar how India secured a 62% increase in population of endangered tigers, through a 10-year program.

During the colonial era, tigers were hunted for sport by the affluent, but for balance, large forest tracts were left as wilderness. In this natural habitat the tiger was an essential factor within the ecosystem. With Indian nationhood, "development" projects to reclaim forestland, irrigate marginal lands, and both cultivate and graze those lands, exposed tigers to ominously rapid depopulation. By 1972, this all-too-familiar disregard for balance among species had caused the new country's tiger census to shrink from 40 000 to only 1827. These few specimens were endangered by poachers who slaughtered them for the valuable, faddish skin, and by farmers whose livestock they took as substitute for the deer and other natural prey vanishing with them from the ecosystem. Control of hunting, and restoration of habitat, was urgent if the elegant animals were to be saved as a part of the cultural heritage. To control hunting, the Indian Board for Wildlife secured a ban on tiger hunting, and the poachers themselves became the hunted. The milestone Wildlife Protection Act was passed in 1972.

In advocating this long-range conservation program. Prime Minister Gandhi said, "It (the tiger) is at the apex of a large and complex biotope." To provide habitat, human competitors for space had to be resettled. To create "core" areas of forest reserve, 27 entire agricultural villages were removed to better land, supplied with fertilizer and improved seed, and offered low-cost financing. So greatly improved was life for the first group of villagers to be resettled, that by their own request 17 additional villages were relocated. Buffer areas, in which agriculture was restricted and closely regulated, surrounded the "core" wilderness reserves from which all human activity had been removed. The parallel here is that there are circumstances under which society, through government, must say to occupants of a mis-used area:" You are in the wrong place, doing the wrong thing. Financial incentives will be given, but desist you must." There are no tigers in the Great Plains, but there is encroachment upon the water reserves of a sub-continent.

It is said that mankind is the only species to live upon the face of the earth which can change and modify that face at will. I believe he can improve his habitat if he will use with wisdom and discretion the accumulated knowledge at his disposal.

We must promulgate and examplify an ethic which demands the conservation-with-change required for a sustainable society. We have made progress in recent years, but we are easily distracted from long-range goals.

Our situation requires mature judgment in establishing priorities, and rigorous self-discipline in adhering to them. Knee-jerk, adolescent over-reaction to bogey-man emergencies has often vitiated sound policy.

Obsolete laws and policies need fundamental re-thinking and adjustment. Informed voters must elect public officials who have intestinal fortitude, and must vigilantly monitor their performance. Only public support rooted and grounded in a conservation ethic can enforce legislative and executive perseverance.

Production must be controlled. Excess food, feed and fiber is waste. Along with the more obvious import, this is an aspect of the warning of Chief Seattle, who with intuitive wisdom told his audience of 125 years ago, "You (white people) will someday suffocate in your own waste." To afford the long-range investment required for conservation, we must have a profitable agriculture. A wasteful agriculture cannot be profitable. We farmers did not make enough money in 1982 to service our debt of over $215 billion. During 1982, farmers' equity in their holdings fell by $35 billion, the steepest drop ever recorded by USDA. We cannot provide all the food for a hungry world, nor do I believe that we should ravage our non-renewable resources in trying to. Rather, we should offer our technology, our expertise, and, in some circumstances, financial surety to help developing countries grow food, feed and fiber suited to their specific needs. Local production is constructive in morale and self-respect, saves transportation cost, and is a permanent infusion of skill and initiative.

Writing in the November, 1982, issue of *Natural History* magazine, Peter D. Moore, a senior lecturer in the Department of Plant Sciences at King's College in London, described his long-range research into the origin of the large areas of non-productive wetlands in Wales, England, and Scotland. He has bored numerous bogs and employed pollen analysis, thus revealing how the activities of prehistoric peoples caused and spread formation of bogs and mires. His research dates the origin of the bogs as early as 2800 B.C., and even 5000 B.C. in the Pennine Range of northern England. Neolithic peoples, although relatively few in number, by burning forests and by shallow "plowing" with sticks, set in motion erosive and decay processes which formed the bogs. I have walked on those springy, sphagnum, bog surfaces, and I vouch that even a sheep finds it poor habitat. Moore concluded: "The lesson of the blanket bogs demonstrates that from the earliest agricultural and even preagricultural days, people have rarely planned beyond immediate and obvious economic gain. As far as the spread of the bogs is concerned, Stone Age man had some excuse, for they could not read the fossil record. We have none, for we can."

If the human species, by self-discipline and a little luck, survives heedless waste of natural resources and the portent of nuclear devastation, it will find its nurture in the soil. I am a tiller of the soil. If posterity finds barren infertile rock where I have been the steward, I have no excuse.

Unless the expertise available in this symposium is made manifest in action, neither do you.

9 Processes of Soil Erosion by Water

G. R. Foster
Agricultural Research Service
U.S. Department of Agriculture
West Lafayette, Indiana

R. A. Young
Agricultural Research Service
U.S. Department of Agriculture
Morris, Minnesota

M. J. M. Römkens
Agricultural Research Service
U.S. Department of Agriculture
Oxford, Mississippi

C. A. Onstad
Agriculture Research Service
U.S. Department of Agriculture
Morris, Minnesota

9–1 FUNDAMENTAL EROSION PROCESSES

Fundamental erosion processes include detachment, transport, and deposition of soil particles by raindrops and surface flow. Detachment is the dislodging of soil particles from the soil mass. Many soils are cohesive in that primary soil particles are bound together by physical and chemical bonding forces. For example, soil having some clay is usually cohesive because clay is a bonding agent. In contrast, pure sand is noncohesive, loose, and already detached. No forces other than gravity and friction hold these particles in place. The force required to dislodge particles from a cohesive soil is greater than the force required to move the same particles lying loose on the soil surface. By definition the latter particles are detached.

Transport is the movement of soil particles over the soil surface. Raindrops striking bare soil detach and splash soil particles in all directions, with some particles landing on nearby vegetation. Some of these remain there indefinitely; others are washed from the vegetation back to the soil surface to be splashed again. Some particles are splashed directly into surface runoff

Published in R. F. Follett and B. A. Stewart, ed. 1985. *Soil Erosion and Crop Productivity.*
© ASA-CSSA-SSSA, 677 South Segoe Road, Madison, WI 53711, USA.

that can transport them long distances before they are deposited. After deposition, these particles may remain in place until runoff increases later in that storm or until a subsequent storm, perhaps years later, moves them further downslope.

Transport capacity is the capacity of the erosive agents to transport sediment. Sediment load is the actual transport rate and may be either greater or less than the transport capacity. If sediment load exceeds transport capacity, deposition occurs at a rate proportional to the difference between transport capacity and sediment load. Deposition reduces the sediment load, and sediment accumulates on the soil surface. Conversely, if sediment load is less than transport capacity, detachment by flow may occur.

This concept for sediment transport by surface runoff is expressed by the equations (Foster and Meyer, 1975; Foster et al., 1981a):

$$D_d = \alpha(T_c - g) \qquad [1]$$

where D_d is the deposition rate (mass/area•time), α is a parameter describing the likelihood of a given sediment class being deposited (length^{-1}), T_c is the transport capacity (mass/width•time), and g is the sediment load (mass/width•time); and

$$D_e = (D_c/T_c)(T_c - g) \qquad [2]$$

where D_e is the detachment rate (mass/area•time), and D_c is the detachment capacity (mass/area•time). Equations [1] and [2] are identical, except that the proportionality coefficient α is defined differently for deposition and detachment. Mathematically, deposition and detachment are the same except for sign—negative for deposition, positive for detachment. In both cases, deposition and detachment decrease as sediment load approaches transport capacity.

Some sediment may travel only a few millimeters after detachment before being deposited. One example is the detachment of sediment from protruding soil clods after primary tillage and its deposition in an adjacent depression. Another example is detachment from a row-ridge sideslope and immediate deposition in low-gradient furrows between the ridges. Sediment may also travel several meters down steep surfaces having no depressional storage before being deposited on a significantly flatter slope. Sediment may not be deposited anywhere on some convex land profiles during some storms, but may be deposited on the upper end of these profiles during other storms.

Sediment movement downslope is described by the continuity equation for mass transport. During a rainstorm, erosion is unsteady during the initial accumulation of surface storage of water and during other times as rainfall, infiltration, and runoff rates change. The continuity equation for steady conditions is (Foster, 1982a):

$$dg/dx = D_f + D_i \qquad [3]$$

where x is the distance along a slope, D_f is the deposition or detachment rate by flow, and D_i are the lateral additions of sediment, either from detachment by raindrop impact or from laterally entering flow. According to Eq. [3], detachment adds sediment to the flow and deposition takes it away in a downslope direction. The unsteady continuity equation is similar to Eq. [3] but includes additional terms for temporary storage of sediment in the flow (Foster, 1982a).

Sometimes Eq. [1] and [2], expressions for an interaction between deposition or detachment and sediment load, are ignored (Meyer and Wischmeier, 1969). In that case, sediment load is limited either by the amount of sediment produced by detachment or by transport capacity. When detachment produces more sediment than the flow can transport, the sediment load is said to be limited by transport capacity. At other times, it is said to be limited by detachment.

9–2 SEDIMENT SOURCES

Several sediment sources and sinks occur within a field (Foster and Meyer, 1977). A sediment source loses sediment while a sink gains sediment. On a microscale, a source area may be no larger than a protruding soil clod on a rough surface after primary tillage and a sink no larger than an adjacent depression. On a macroscale, an entire field may be a source area, or deposition may be concentrated in a few relatively large areas like the toe of concave slopes. Other macroscale sediment sinks are grass buffer strips, terrace channels, grassed waterways, backwater areas near field outlets, and natural and constructed ponds.

Sediment source areas are defined according to flow patterns. Overland flow, though often considered to be of uniform depth across the slope, tends to concentrate in numerous small channels. These channels, resulting from either erosion or tillage, are the main pathway of downslope runoff, and they are called rills regardless of whether erosion occurs in them. Any erosion that occurs in them is defined as rill erosion. Erosional rills are numerous small channels, usually about 100 to 300 mm wide by 50 to 150 mm deep uniformly distributed across the slope. Removal of a single rill has little effect on the hydrologic and erosional responses of a hillslope. Furthermore, rills are traditionally defined as small channels that can be obliterated by tillage. Flow on areas between the rills tends to be lateral toward the rills and is much shallower than flow in the rills. These sediment sources are called interrill areas, and erosion that occurs on them is called interrill erosion.

Overland flow on many landscapes converges in a few major channels before leaving field-sized areas (Foster, 1982c). This flow, called concentrated flow, will often erode channels a few meters wide but still so shallow that they can be tilled across. Most runoff leaves fields through this type of channel, and the removal of a single concentrated flow would have a noticeable effect on the hydrologic and erosional responses of an area.

When concentrated flow erodes channels deeper than can be crossed with farm equipment, the eroded channels are called gullies. Another sediment source is erosion by stream channels. In this paper, however, we do not discuss gully or stream-channel erosion.

9-3 INTERRILL EROSION

Impacting raindrops are the most important erosive agent on interrill areas. Raindrops range in size from about 0.2 to 6.0 mm, and impact at about 9 m/s (Bubenzer, 1979). Raindrop impact on bare soil creates intense local shear and pressure forces (Al-Durrah and Bradford, 1982), which can detach large quantities of sediment (Ellison, 1944). Some flow in the drop impact is a high-velocity radial jet along the soil surface, which detaches and carries sediment. A second flow component is the film of water that subsequently rises vertically and breaks into splash droplets. These droplets may carry sediment as high as 1.0 m and as far as 0.8 m in a single splash (Ellison, 1944; Martinez, 1979). Sediment may be splashed directly to flow in rills and be immediately transported downslope, or it may be transported laterally to rills by interrill surface flow (Meyer et al., 1975b; Young and Wiersma, 1973).

Interrill erosion can be described by the function (Foster, 1982a):

$$D_i = r_i K_i [2.96(\sin \theta)^{0.79} + 0.56] C_i \qquad [4]$$

where D_i is the interrill erosion rate (mass/area•time), r_i is an erosivity factor for interrill erosion, K_i is a soil erodibility factor for interrill erosion, θ is the angle of the interrill surface with the horizontal, and C_i is a factor for the effect of cover and management on interrill erosion.

9-3.1 Interrill Erosivity

Fluid forces on a bare soil surface from the impact of a single drop increase asymptotically toward a maximum as drop diameter increases. The presence of a surface layer of water affects the magnitudes of these forces. As water depth increases from zero to about 0.3 drop diameters, the forces increase (Mutchler and Hansen, 1970), but at greater depths the forces decrease. When water depth exceeds approximately 3 drop diameters, drop impact forces on the soil surface and subsequent soil detachment are negligible. Raindrops of all sizes occur in most storms, but median drop diameter varies with the 0.2 power of intensity (Laws and Parsons, 1943). However, data from natural rainfall are highly scattered for the relation of erosion to rainfall characteristics such as intensity.

Kinetic energy of raindrops, the product of drop mass and the square of impact velocity, is frequently used as an indicator of rainfall erosivity. Energy per unit depth of rainfall varies approximately with the 0.14 power of rainfall intensity (Wischmeier and Smith, 1978; Foster and Meyer, 1975). Total energy for a storm is the summation:

$$E = \sum_{m=1}^{n} e_m \Delta V_m \qquad [5]$$

where E is the total energy for the storm; e_m is the energy per unit depth of rainfall for the intensity of each depth increment of rainfall, ΔV_m; and n is the number of increments used to describe the rainstorm. Another frequently used indicator of rainfall erosivity is maximum 30-min intensity. It is multiplied by E to obtain storm EI, the erosivity index in the universal soil loss equation (USLE) (Wischmeier and Smith, 1978).

Rainfall simulator studies show that interrill erosion varies with the square of rainfall intensity (Meyer, 1981), but drop size is seldom varied with intensity in these studies. Therefore, the erosivity factor r_i, for a storm of constant intensity is

$$r_i = i^2 T \qquad [6]$$

where i is the rainfall intensity and T is the storm duration. Since storm amount, V, is i•t, Eq. [6] becomes

$$r_i = Vi \qquad [7]$$

Storm energy, E, is almost directly proportional to storm amount, V, because unit energy, e, varies only slightly with intensity, i. Therefore, if the maximum 30-min intensity is assumed to be a characteristic intensity for a storm, the EI erosivity factor is almost directly proportional to the V•i erosivity factor of Eq. [7]. Thus, volume of rainfall and maximum 30-min intensity are the two most important general measures of rainfall erosivity. These two variables were as good as EI for describing erosion from fallow plots (Foster et al., 1982c).

Wind affects raindrop diameter, impact velocity, and impact angle. Although Lyles (1977) showed that soil detachment by raindrops can be 73% greater with a 13 m/s wind than with no wind, this effect is often ignored because of the unpredictability of wind.

9–3.2 Interrill Soil Erodibility

Soils, even when managed alike, vary in their susceptibility to interrill erosion. Soil characteristics that significantly affect detachment by raindrop impact include degree of aggregation of the in situ soil, stability of soil aggregates, and resistance of soil aggregates to breakdown from fluid shear and pressure forces (Al-Durrah and Bradford, 1982). In turn, these characteristics are affected by aggregate size and density, amount and type of clay, organic carbon, and such inorganic soil constituents as iron, sodium, calcium, and magnesium. Stability of aggregates usually increases with organic carbon content up to about 7% (Luk, 1979), but it may decrease with greater carbon content because of acquired hydrophobic properties. Large aggregates are less stable than small ones because of increased organic carbon content in the smaller aggregates (Edwards and Bremner, 1967; Garey, 1954). Microbial activity, because of its relation to soil organic-matter content, affects soil erodibility. Increased erosion following oilseed crops (Laflen and Moldenhauer, 1979) may be related to microbial activity.

Clay in combination with organic matter increases aggregate stability. Its effects are greatest at about 6% organic carbon content. High aggregate densities generally indicate high clay content and increased aggregate strength. Soil chemistry affects soil erodibility because high sodium content increases soil erodibility and high iron content reduces soil erodibility by their effects on aggregate stability (Römkens et al., 1977).

Soil texture is often used as a general indicator of soil erodibility. Generally, silt soils are considered the most erodible, and clay and sand soils the least erodible. Clay soils are less erodible because of the binding characteristics of clay and its associated organic and inorganic constituents. The classification of sand soils as not erodible may be valid only within the context of an erosion equation like the USLE. The USLE soil-erodibility factor represents the net effect of both detachment and transport capacity in controlling erosion. Particles from sandy soils are easily detached; therefore, these soils should be classed as highly erodible when detachment alone is considered. However, infiltration also tends to be rapid on sandy soils, leaving little runoff to transport sediment from the interrill area. Therefore, reduced transport capacity of runoff on sandy soils may limit erosion.

9-3.3 Interrill Length and Steepness

Since all detachment on interrill areas is assumed to be by raindrop impact, detachment over an interrill area should be uniform and unaffected by slope length of an interrill area, except as length affects flow depth. This effect is not great according to the relation of flow depth to discharge rate in flow equations (Chow, 1959). However, discharge rate, and thus length, affects transport capacity, which could affect delivery of sediment from interrill areas to the rills. This effect is discussed later in the section on sediment transport (section 9-7).

Interrill erosion increases with slope steepness of the interrill area as indicated by the $[2.96 (\sin \theta)^{0.79} + 0.56]$ term in Eq. [4] (Foster, 1982a). According to this term, erosion from a 20% interrill slope is about twice that from a 1% interrill slope, whereas this ratio is about 30 according to the USLE. This difference reflects the USLE slope factor lumping the effects of interrill detachment and transport and rill detachment and transport with the effect of slope on runoff amount. Even in the slope-related term of Eq.

[4], transport capacity may reduce interrill erosion for slopes less than about 6% (Foster, 1982a; Fertig et al., 1982). On steep slopes, where transport capacity is great, interrill slope may hardly affect detachment by raindrops.

9–3.4 Interrill-Cover Management

Complete cover at the soil surface fully protects the soil from raindrop impact, and management that changes aggregate stability affects interrill erosion. These types of effects are described by factor C_i in Eq. [4] and can be divided into three classes: (i) canopy effects, (ii) ground-cover effects, and (iii) within-soil effects (Wischmeier, 1975).

9–3.4.1 Canopy Effects

Canopy intercepts raindrops before they reach the ground. Some of the intercepted water evaporates before it reaches the ground, some flows down the plant stems and adds to runoff but causes no detachment by drop impact, and some re-forms drops and falls to either lower plant surfaces or to the ground. Drops falling from plant surfaces are usually larger than natural raindrops. Therefore, when they fall from a tall canopy like trees which has its lowest leaves high above the ground, their erosivity is greater than that of raindrops (Chapman, 1948). However, the canopy of most agricultural crops is so close to the ground that the overall effect is to reduce the erosivity of rainfall. Even though interception increases drop diameter, fall height is too short for drops to reach a velocity to produce energy greater than that of natural rainfall. Obviously, a dense, short canopy leads to the greatest reduction in erosion.

9–3.4.2 Ground-Cover Effects

Cover in direct contact with the ground protects more against interrill detachment than does canopy, since drops intercepted by ground cover have no fall height to regain velocity. Interrill detachment is at least reduced by the percentage of ground cover, which includes crop residue, stones, and live vegetation. However, ground cover also increases hydraulic roughness, causing reduced flow velocity and increased flow depth to protect the soil from impacting raindrops. Increased flow depth on 0.7-m long interrill plots covered with straw mulch decreased interrill erosion by 20% for a 25% ground cover, by 44% for 61% cover, and by 77% for a 90% cover over and above the reduction from interception by cover alone (Foster, 1982a).

The fraction of a soil surface covered by water deeper than 3 raindrop diameters ranges from very little for a densely rilled slope having smooth interrill areas to 100% where ground cover is dense or land is nearly flat. Water depth on interrill areas also depends on discharge rate and slope of the interrill areas. The effect of water depth is minimal on 100-mm long,

50% steep interrill slopes, but probably is significant on 10-m long, 0.5% steep interrill slopes. Murphree and Mutchler (1981) reduced USLE estimates by up to 58% on very flat slopes in the Mississippi Delta to account for the increased water depth reducing the erosivity of raindrops.

Water ponded in the depressions of rough surfaces like those left by primary tillage eliminates interrill detachment in these areas. However, soil protrusions or clods above the waterline are still exposed to interrill detachment. Interrill sediment is deposited in adjacent depressions, and eventually the depressions become filled with sediment. Erosion significantly increases because deposition is eliminated (Römkens and Wang, 1985).

Initial storage volume in tillage depressions depends on initial soil bulk density, soil texture, moisture content at the time of tillage, type of tillage implement, and speed at which the implement is pulled (Voorhees et al., 1981). A rough surface following moldboard plowing can store up to 10 mm of water, but wetting by rainfall, detachment, deposition, and soil settlement reduces this roughness.

9–3.4.3   Within-Soil Effects

Factors other than canopy and ground cover must be considered to estimate monthly, seasonal, and annual variations in interrill erosion. The susceptibility of a soil to interrill detachment varies greatly depending on the soil's condition and recent history. These effects are actually soil-erodibility effects. In equations like the USLE, however, soil erodibility is defined for convenience as erosion at some baseline condition. The within-soil effect describes erosion as actual soil conditions depart from the baseline.

For example, cropped land is more susceptible to detachment by raindrop impact in the spring than in late summer (Meyer, 1985). Freshly plowed meadowland has good soil aggregation, making the soil less erodible than continuously row-cropped soils. Furthermore, soils where crop residue, manure, and other organic matter are regularly incorporated are less erodible than soils without such incorporations. Organisms, including earthworms, can improve soil aggregation and reduce erosion. Certainly, most or all these effects could be incorporated in a basic soil-erodibility relationship, but ignoring them may lead to large overestimates of erosion.

9–4   RILL EROSION

Rill erosion is a hydraulic process where shear of the flow at the soil-water interface detaches and transports soil particles (Foster, 1982 and 1982c). Detachment is nonuniform along a rill. Some rills are formed by headcuts moving upslope, with much localized erosion at the headcuts. The likelihood of headcuts depends on rill grade, soil texture, soil condition, and discharge rate. Headcuts seem much more likely on freshly tilled soil than on the same soil untilled and on a granular soil than on a highly cohesive soil. Miniature grade controls within the soil, like plant roots and undecomposed crop residue, can greatly reduce formation of headcuts and rill erosion. For example, 1.0 kg/m² of incorporated straw reduced rill erosion by 80% on a tilled soil compared to the same soil with no incorporated straw (van Liew and Saxton, 1983). Also, plant roots may physically bind and hold soil in place.

Erosion can also occur all along a rill from shear stress exerted on the channel boundary by the flow in the rills. This erosion can undercut the rill sidewalls, and, when the weight of the overhanding soil exceeds soil strength, it sloughs into the rill to be carried away by flow in the rill.

A relationship typically used to describe detachment capacity in a single rill is (Foster, 1982a)

$$D_c = K_{rs} (\tau - \tau_c) \qquad [8]$$

where D_c is the detachment capacity (mass/unit area of wetted perimeter time), K_{rs} is the soil erodibility for rill erosion, τ is shear stress, and τ_c is critical shear stress. Except in well-defined, uniform, ridge-furrow surface configurations, discharge rates can vary greatly across slope from rill to rill. Therefore, rather than analyze each rill as a separate channel, uniform flow across slope is assumed, and the following equation is used to estimate rill-detachment capacity (Foster, 1982a):

$$D_c = K_{ru} q s_r C_r \qquad [9]$$

where D_c is detachment capacity per unit land area, K_{ru} is soil erodibility for rill erosion, q is the discharge rate per unit width, s_r is the sine of the grade angle along the rill, and C_r is a cover-management factor for rill erosion. Equation [9] distorts the soil erodibility factor K_{ru}, because flow is actually concentrated in rills rather than uniformly distributed across the slope (Foster et al., 1982b). Flow velocities in the rills are much greater than those given by the assumption that flow is uniform across the slope.

9-4.1 Flow Erosivity

Shear stress, τ, the hydraulic variable frequently used for flow erosivity, is given by

$$\tau = \gamma R s_r \qquad [10]$$

where γ is the weight density of the flow and R is the hydraulic radius (Chow, 1959). Shear stress increases as both discharge rate in the rill and grade along the rill increase. Shear stress is also nonuniformly distributed around the wetted perimeter, increasing from zero at the water surface to a maximum at the bottom of the rill. Local shear stress at this point is about 1.3 times the average for the entire cross section of the rill (Graf, 1971). If local shear stress exceeds the soil's critical shear stress, soil is detached. If a constant discharge is maintained and any nonerodible soil layer is deep, a narrow incised channel will erode to an equilibrium cross section (Foster and Lane, 1983). Soil-detachment rates will vary during the initial incisement of the rill, but they become steady as the channel's shape approaches an equilibrium.

If a resistant layer develops in the bottom of the rill or the rill reaches nonerodible soil or buried grade controls like roots or residue, downward erosion of the rill stops and the rill begins to widen. As a rill widens under steady discharge, the erosion rate decreases (Foster and Lane, 1983). Eventually, rill width approaches a final width characteristic of the given discharge rate, rill grade, critical shear stress, and hydraulic roughness of the soil (Foster and Lane, 1983). If a subsequent discharge rate in the rill is less than the discharge rate that initially formed the rill, erosion will be slight. Therefore, rill erosion for a given discharge rate depends on the extent of erosion from previous storms.

The product qs_r in Eq. [9] is the flow erosivity term for the assumption of uniform flow across slope. It was also derived from a shear-stress concept. Localized erosion at headcuts, usually not considered in erosion equations, is generally a function of these same flow-erosivity variables.

Total shear stress of the flow is distributed between that acting on the cover and that acting on the soil. Ground cover, by increasing hydraulic roughness, increases total shear stress of the flow, but shear stress on the soil, the part responsible for detachment and sediment transport, is reduced (Foster et al., 1982b). If the shear stress acting on the cover exceeds a critical value, it can move unanchored surface material like wheat straw and corn stalks, leaving the soil bare and susceptible to rill erosion (Foster et al., 1982b). The ratio of shear stress acting on soil with cover to that without cover is approximately equal to the ratio of values for Manning's hydraulic-roughness factor for the bare and covered soil conditions, respectively. The shear stress acting on the soil having a 1.0-kg/m² mulch of corn stalks is about 0.15 of that with no cover (Foster et al., 1982b).

The effect of ground cover in Eq. [9] is described by the C_r factor. It is proportional to the cube of the ratio of flow velocity with cover to that without cover. For 1.0-kg/m² corn stalk mulch, the velocity ratio is 0.3, which gives a C_r value of 0.03 (Neibling and Foster, 1977). Therefore, surface cover greatly reduces rill erosion, provided that rill erosion does not occur beneath the cover or that the cover is not washed away. Surface ground cover reduces rill erosion much more than interrill erosion for a given percentage of cover.

Raindrop impact should have minimal effect on detachment by flow in rills because this flow is usually much deeper than 3 raindrop diameters. However, when flow in rills was protected from direct raindrop impact of a 64-mm/h rain, rill erosion of a freshly tilled soil was about one half of that when the flow was unprotected (Meyer et al., 1975a).

9–4.2 Rill Erodibility

Susceptibility of a soil to rill erosion is indicated by the soil erodibility factor K_{rs} and the critical shear stress τ_c in Eq. [8] or by the soil erodibility factor K_{ru} in Eq. [9]. Many of the same soil and management factors affecting detachment by raindrop impact also affect detachment by surface flow, but in a different way. Two soils equally susceptible to interrill erosion differed greatly in susceptibility to rill erosion (Meyer et al., 1975b).

Tillage greatly increases the susceptibility of some soils to rill erosion. Rill erosion from a range of discharge rates was 3 to 15 times more on a freshly tilled soil than on the same soil not tilled for a year (Foster et al., 1982a). Similarly, the critical shear stress of a typical soil that has not been tilled for some time is about 10 to 15 times that of the same soil freshly tilled (Foster et al., 1980). Tillage and soil consolidation may have an even greater effect on rill erosion than does soil texture. Since untilled soil has a large critical shear stress relative to the tilled zone, soil just below the tilled zone can act as a nonerodible layer restricting eroded depth of rills.

9-5 EROSION BY CONCENTRATED FLOW

Detachment by concentrated flow is similar to detachment in a single rill except that untilled soil acting as a nonerodible layer below the tilled zone has a greater influence. In rill erosion, a typical ratio of eroded channel width to depth to the nonerodible layer is about 1, whereas this ratio may be 30 for concentrated flow (Foster, 1982c). Therefore, relatively little erosion occurs by concentrated flow before the influence of the nonerodible layer is noticed. The sidewalls of concentrated flow channels are usually low relative to the width of the channel and may hardly exist as the channel bottom blends into them. Headcuts in concentrated flow areas generally are not important in the erosion process.

9-6 GULLY EROSION

High sidewalls and headcuts are characteristic of many gullies, the strength of soil in the head and sidewalls being important in maintaining gully stability. Strength of gully walls depends greatly on the water content of the soil. A saturated soil has reduced soil strength, and thus may be susceptible to sloughing. If the discharge rate in a gully is too low to remove sloughed material, then the soil at the toe of the walls will tend to stabilize the gully (Piest et al., 1975). Many gullies elongate at their upper ends as a headcut moves upstream, much like rills.

9-7 SEDIMENT TRANSPORT CAPACITY

If detachment produces a sediment load greater than the runoff can transport, deposition occurs. The sediment load leaving a depositional area is closely related to the transport capacity of the runoff. Even when transport capacity is greater than sediment load, transport capacity regulates detachment by flow according to Eq. [2].

Transport on interrill areas is by thin-film flow toward rills and direct splash of sediment to rill flow. Raindrop impact greatly enhances sediment transport by thin flow. Without raindrop impact, this flow has little transport capacity and moves little sediment. Raindrop impact is of lesser importance for transport by flow in rills and larger channels.

Most downslope movement of sediment is by surface runoff. Transport capacity of flow depends on hydraulic variables (primarily discharge rate, velocity, and flow depth) and sediment characteristics (primarily particle diameter and density). Many sediment transport equations exist for stream channels, but no particular equation is universally accepted. Usually, one of these equations is used to estimate sediment transport capacity on field-sized areas (Alonso et al., 1981). However, the surface flow and sediment characteristics for erosion on most field-sized areas differ so much from channel conditions that few if any of the channel equations apply. Furthermore, hydraulics of overland flow are often described with broad sheet-flow assumptions, without considering the nonuniformity in flow depth and velocity across a slope (Foster and Meyer, 1972). This variation distorts parameter values in the equations. A sediment-transport equation is needed specifically for shallow flow and sediment characteristic of field-sized areas. The equation should consider mixed, nonuniform sediment ranging from clay to small sand-sized particles and specific gravities ranging from about 1.5 to 2.7 (Foster, 1982a). It should also consider the effect of turbulence from raindrop impact on transport capacity of shallow flow. Surface-tension effects may also be important in sediment transport by shallow flow (Young and Mutchler, 1969).

9-8 DEPOSITION

Deposition often occurs where transport capacity is reduced—on a flattening grade at the toe of a concave slope, in a grass strip, in a low-gradient furrow or terrace channel, or in backwater near a field outlet where a ridge or vegetation around a field has slowed runoff. Where sediment load exceeds transport capacity, sediment is deposited at a rate proportional to the difference between transport capacity and sediment load according to Eq. [1].

Deposition is a selective process. Large and dense particles are readily deposited, whereas fine and lightweight particles travel much further downstream. As a result, the gradation of particles in a depositional area, for example on a concave slope, is from coarse to fine in a downstream direction.

Equation [1] is written for individual particle classes in the sediment because of the great nonuniformity and nonlinearity in transport and depositional characteristics over a range of particle classes. The transport-capacity equation must also account for this nonuniformity and must shift transport capacity among particle classes as deposition removes some classes more rapidly than others. Therefore, Eq. [1] is usually written as (Foster et al., 1981a)

$$D_k = \alpha_k(T_{ck} - g_k) \qquad [11]$$

where the subscript k designates the particle class. The proportionality constant α is defined as

$$\alpha_k = \beta \, V_{fk}/v \, y \qquad [12]$$

where β is a factor to account for degree of flow turbulence, V_f is the particle-fall velocity, v is the flow velocity of the runoff, and y is the flow depth. Particles are deposited slowly by highly turbulent flow, especially flow on interrill areas highly disturbed by raindrop impact. Therefore, β is small for this type of flow and equals 1.0 for deep flow where raindrop disturbance is minimal, as in concentrated flow areas (Einstein, 1968). The difference in deposition rate among particles is described by particle-fall velocity, V_f. Fall velocity is about $2 \cdot 10^{-2}$ m/s for medium-sized sand versus about $2 \cdot 10^{-6}$ m/s for clay. When all other factors are equal, particles are deposited more rapidly in shallow flow because of the short distance they must settle to reach the channel bottom. Conversely, particles are less likely to be deposited in deep, high-velocity flow.

The significance of Eq. [11] can be illustrated with the example of deposition along a uniform, low-gradient furrow that has much interrill erosion on adjacent ridge sideslopes. Deposition occurs along the entire furrow length except near the outlet, where the grade may steepen near the furrow discharge into a tail ditch. The coefficient, α, is so large for the sand-sized particles that they are deposited immediately upon entering flow in the furrow. A characteristic of Eq. [11] when α is large is that sediment load at the outlet of the furrow nearly equals transport capacity for these particles.

Sediment load of clay-sized particles at the outlet of the furrow is only slightly less than the amount added to the flow by interrill erosion on the ridge sideslopes. This result is from very low values for α, which calculate a near-zero deposition rate by Eq. [11]. Depositional characteristics of other particle classes are intermediate between the extremes for clay- and sand-sized classes. The detachment-transport-deposition processes on very rough, primary-tilled surfaces with deposition in nearby depressions are also similar to those of the ridge-furrow configuration.

9–9 SEDIMENT CHARACTERISTICS

Sediment detached from most soils is a nonuniform mixture of primary particles (clay, silt, and sand) and aggregates (conglomerates of primary particles) (Foster, 1982a). The transportability of sediment is primarily related to the diameter and density of the particles, with density being more important than size for particles larger than 0.1 mm (Davis et al., 1983). Specific gravity of aggregates ranges from about 1.5 to 2.7, depending on their composition, size, and degree of saturation of aggregate pore space (Young, 1980; Allmaras, 1967; Yoder, 1936; Long, 1964). Aggregate density decreases as the silt content of the soil or the degree of saturation of the aggregates increases (Young, 1980; Rhoton et al., 1983). Density also decreases as size of the aggregates increases (Voorhees et al., 1966). The size distribution of sediment at its point of detachment is related to soil texture

(Foster et al., 1982e). Other factors, such as clay mineralogy; the amount, type, and availability of organic matter; chemical composition of the soil; soil structure; and previous land use also affect sediment size, but their effects have not been quantified.

The minimum and specific gravity (s.g.) classes needed to route sediment by particle classes in erosion models are (i) primary clay (0.002 mm, 2.65 s.g.), (ii) primary silt (0.01 mm, 2.65 s.g.), (iii) small aggregate (approx. 0.03 mm, 1.8 s.g.), (iv) large aggregate, (approx. 0.2 mm, 1.6 s.g.), and (v) primary sand (0.2 mm, 2.65 s.g.) (Foster et al., 1982e). At most, these classes represent broad, general classes of particles. Because clay is an important bonding agent for soil aggregation, in situ clay is a key variable in the equations that estimate the distribution of these classes and the diameters of the aggregates at the point where sediment is detached. The fraction of the sediment composed of primary clay is about 25% of the soil-clay fraction. The fractions of the primary-silt and small-aggregate classes combined about equal the silt fraction of the in situ soil. The small-aggregate class is composed of clay and silt, and the large-aggregate class is composed of clay, silt, and sand.

9–10 CHANGES IN SOIL PROPERTIES BY EROSION

Erosion changes soil properties in several ways. Primarily, erosion reduces thickness of the root zone and changes the composition of the soil by selectively removing fine soil particles, thereby affecting nutrient content, and soil water-holding characteristics.

9–10.1 Particle Selectivity

Sediment from large erosion plots and small watersheds is frequently enriched in fines, that is, the fraction of clay and silt particles is greater in the sediment than in the in situ soil. Therefore, erosion increases the fraction of coarse primary particles in the in situ soil while selectively removing fine primary particles, organic matter, nutrients, and agricultural chemicals associated with fine particles. In terms of basic erosion processes, sediment can be enriched during either detachment or deposition. The textural composition of sediment changes little during transport unless deposition occurs.

Selectivity during interrill detachment is minimal as shown by analysis of interrill erosion data from 19 soils (Foster et al., 1982e). Sediment from interrill plots showed enrichment of fines at low slopes, but such enrichment was completely eliminated as slope steepness was increased. This suggests that transport and deposition were the enriching processes (Fertig et al., 1982). Therefore, factors that reduce transport capacity and cause deposition on interrill areas, like low runoff rates and increased ground cover, enrich the sediment in fines. Also, soil detachment in rills removes soil in bulk, with little selectivity of particles (Foster et al., 1982d). Therefore, detachment produces a distribution of primary particles in sediment that is more or less the same as that in the surface soil subject to erosion.

Deposition, on the other hand, is a highly selective process. It is this process, rather than detachment, that seems primarily responsible for changes in the textural and chemical composition of the sediment. Raindrops may dislodge larger particles than interrill flow or splash can transport to rills. Even if the particles do enter rills, flow in the rills may be unable to transport them. Large or dense particles will be deposited, causing the sediment leaving the field to be enriched in fines.

Obviously, soil may contain gravel particles so large that neither raindrops nor flow can move them even though these particles may be "loose." Detachment can remove finer particles, leaving these coarse particles to armor the soil surface and the bottom of rills. The smallest primary particle that would not be moved during detachment is likely the largest primary particle in the aggregates. According to this concept, all larger primary particles are noncohesive and already detached.

The extent to which erosion selectively removes fines depends on the degree of aggregation of the sediment. Most of the clay and silt in sediment detached from sandy soils is in fine particles not easily deposited. Therefore, any sediment deposited will be mainly sand particles and will leave the sediment load greatly enriched in fines. Clay and silt are much more evenly distributed across particle classes in highly aggregated sediment eroded from heavy-textured soils. When this sediment is deposited, aggregates are deposited containing significant amounts of clay and silt. Therefore, little selectivity will be observed for this condition. An analysis with CREAMS (a field-scale model for chemicals, runoff, and erosion from agricultural management systems) (Knisel and Foster, 1981) showed that if 90% of the sediment is deposited before leaving a field, the enrichment ratio for the sediment load is 5.5 for sediment detached from a sandy soil but only 1.6 for sediment detached from a clay soil.

Selectivity is minimized by reducing deposition. Therefore, selectivity is less on smooth than on rough surfaces. For example, for a given soil loss, selectivity will be greater on a rough-plowed surface than on a smooth surface covered by a heavy mulch. A heavy cover of mulch apparently reduces detachment more than transport capacity so that detachment controls soil loss. However, a light mulch cover can enrich sediment in fines because of deposition behind mulch pieces (Brenneman and Laflen, 1982).

9–10.2 Surface Sealing

Raindrops striking a soil surface rupture soil aggregates. Intense fluid pressures during drop impact can wash fine particles into the surface pores, forming a seal up to 10 mm thick. This seal greatly restricts infiltration, thereby substantially increasing runoff and decreasing soil moisture. As a seal dries, it forms a crust, a dense surface layer of soil particles that can

impede seedling emergence and reduce plant population. During a rainstorm, the seal is somewhat fluid, which allows detachment of soil particles
from beneath it (Gabriels and Moldenhauer, 1978). Surface sealing can be
reduced by using mulch or dense, low canopies to protect the soil from raindrop impact. Soil can also be managed to improve aggregate stability, thereby reducing susceptibility of the soil to aggregate rupture. Since fewer fines
are produced, sealing is reduced. Soils also vary in their tendency to seal.

9–10.3 Nutrient Loss

Nutrients attached to sediment are lost during erosion in proportion to
their concentration on the sediment at the point of detachment. Most of
these nutrient losses are associated with the removal of fine, inorganic and
organic, colloidal material where the nutrients are adsorbed. With a
reduction in soil colloidal content over time, the productive capacity of the
soil is reduced. Dissolved nutrients are also lost in the runoff.

9–10.4 Loss of Soil Depth

When erosion removes soil more rapidly than new soil is being formed,
soil depth is reduced. If a soil profile is shallow, erosion reduces rooting
depth and soil moisture storage. Both losses potentially reduce crop yield.
Erosion must occur for several years on deep soils before rooting depth is
affected. When the erosion rate exceeds the rate of topsoil formation,
tillage eventually will mix subsoil with topsoil in the surface-tilled zone.
Since properties of subsoil are usually less desirable for crop growth than
those of topsoil, this mixing often degrades soil for crop production. These
changes, combined with selective removal of fine soil particles and
deterioration of soil structure, can result in significant yield reductions.

9–10.5 Secondary Changes

Some of the secondary effects of erosion on soil properties and profile
characteristics include reduced water and nutrient availability, deteriorated
soil tilth, increased susceptibility to compaction, and increased draft of
farm implements. The degree of these effects usually depends on past erosion and on the properties of the original soil.

9–10.5.1 *Water and Nutrient Availability*

Topsoil mixed with subsoil, or the subsoil itself after the topsoil is
eroded, may have poor water-storage capacity and transmissibility from
reduced pore space. Thus, less water can be stored for later plant use. Likewise, nutrient availability may be reduced from the mixing of topsoil and
subsoil. Also, nutrient availability is usually less in subsoil, which may be

exposed by erosion, than in topsoil. However, most nutrients can be re-placed by fertilization, although this increases crop-production costs. Ferti-lizer is more effective on soils having adequate organic-matter content soils than on soils with depleted organic matter (Barrows and Kilmer, 1963). Like plant nutrients, organic matter is lost with sediment and is usually less in the residual subsoil than in the original topsoil.

9–10.5.2   Soil Temperature

Because soil temperature is a function of soil properties as well as climate, it can be changed when those soil properties are altered, directly or indirectly, by erosion. However, modification of the soil-temperature regime may be much more significant to crop growth in some regions than in others. In the temperate (Allmaras et al., 1964) and tropic (Lal, 1976) regions, change in soil temperature is especially important.

9–10.5.3   Soil Tilth

Soil tilth, a qualitative measure of a soil's physical condition for tillage and crop growth, is related to soil structure and aggregate stability. Selec-tivity of the erosion process and mixing of topsoil with subsoil often result in deteriorated soil tilth because of the generally poorer tilth of subsoil.

9–10.5.4   Soil Compaction and Implement Draft

Erosion does not directly compact soil or increase implement draft, but, since erosion forces tillage into subsoil, the mixed topsoil-subsoil or subsoil may be susceptible to compaction and increased implement draft.

9–10.6   Role of Management

Degradation of some soil properties may be more directly related to poor soil management than to erosion. For example, increased crusting, though caused by breakdown of aggregates, is more a result of poor soil management than of erosion. Crusting may be related to a failure to main-tain adequate levels of organic matter by incorporation of crop residue or by continuous row cropping. Even if erosion were significant, physical changes in the soil would probably be minimal if topsoil were maintained, unless erosion had significantly reduced total soil depth.

9–11   HYDROLOGY

The hydrologic processes of rainfall and runoff drive erosion pro-cesses. Rainfall amount and the maximum 30-min. intensity are two im-portant rainstorm characteristics affecting erosion. Rainfall erosivity varies widely across the country, being 10 to 20 times greater in the Southeast than in many parts of the West (Wischmeier and Smith, 1978). Also, rainfall erosivity is not uniformly distributed throughout the year in many areas of the USA. The erosion hazard is greater if peak rainfall erosivity occurs during periods when land is bare than if it occurs during full crop canopy.

Surface runoff is directly related to rainfall, infiltration, surface storage, and plant interception (Brakensiek and Rawls, 1982). Infiltration depends on soil texture, soil-surface condition, soil porosity, and antecedent soil moisture (Rawls et al., 1982). Infiltration tends to be greater in sandy soils than in clay soils. Surface sealing greatly reduces infiltration and increases runoff. With rough surface conditions or dense vegetation, all rain can be stored from small storms, resulting in no runoff or erosion.

Infiltration is greater for initially dry soil than for wet soil. Soil moisture at any particular time is a function of several simultaneously occurring processes, including infiltration, percolation to deep soil layers, and water extraction by soil evaporation and plant transpiration. A dense, deep-rooted crop can remove large quantities of water from the soil profile between storms, thereby substantially increasing infiltration from subsequent rainfall because of low antecedent soil moisture. Nonliving ground cover also can increase infiltration by reducing surface sealing. It also reduces soil evaporation, thereby increasing soil moisture and decreasing infiltration because of high antecedent soil moisture. The net effect, however, is generally increased infiltration and reduced runoff. Therefore, not only do growing ground cover and crop residue reduce the erosive forces of rainfall and runoff, but they also reduce runoff itself, thereby providing a double effect.

9–12 PRINCIPLES OF EROSION CONTROL

Erosion is controlled by reducing the erosivity of the erosive agents, by reducing the susceptibility of soil to erosion, or both. Erosivity can be reduced by protecting the soil surface with canopy or ground cover, reducing runoff amount and rate, reducing slope length and steepness, and increasing surface roughness. Of these, protecting the soil surface with increased ground cover, for example with conservation tillage, is often the most effective. Measures to reduce erosivity can usually be accomplished quickly.

The susceptibility of soil to erosion can be reduced over time by improved soil management. These practices include incorporation of crop residue and manure to build up the soil's organic matter and rotation of row crops with sod crops.

Contouring, terracing, strip-cropping, and grassed waterways are structural conservation practices that support cultural practices such as conservation tillage. The effectiveness of these support practices results primarily from control of runoff. For example, contouring causes runoff to flow along a much reduced grade than when it flows directly downhill. Terraces shorten slope length, which reduces runoff rate. Terraces and grassed waterways both dispose of runoff from fields at nonerosive velocities, thus preventing concentrated flow erosion. Densely vegetated strips in strip-cropping reduce the transport capacity of runoff and cause deposition. These dense strips also protect their respective areas from erosion.

Not all measures control erosion equally across a range of conditions. Reducing runoff amount and rate is more effective on soils highly suscepti-ble to rill erosion than on soils resistant to rill erosion. Also, since moderate mulch rates have a greater effect on rill erosion than on interrill erosion, the percentage reduction in erosion from moderate mulch rates will be greater on those slopes where the potential for rill erosion is great, such as soils susceptible to rill erosion or soils on steep slopes.

The effectiveness of surface cover on rill and interrill erosion combined is (Laflen et al., 1980)

$$F = \exp(-\zeta M) \tag{13}$$

where F is the ratio of erosion with mulch to erosion without mulch and M is the fraction of the surface covered. Experimentally, the empirical coef-ficient ζ has varied from less than 2 to greater than 7, with most values near 3.5 (Laflen et al., 1984). When M equals 0.5 and ζ equals 2, F equals 0.37. Thus, a 50% cover reduces erosion by 63%. If ζ equals 7, the reduction is 97%. This range in effectiveness of cover is partly related to the sus-ceptibility of a soil to rill erosion relative to its susceptibility to interrill erosion.

Management practices control the amount of sediment leaving a field by controlling detachment or by inducing deposition. Roughness from moldboard plowing can reduce erosion by 50% by inducing deposition in depressional areas (Wischmeier and Smith, 1978). Over time, however, the depressions will fill and lose their effectiveness. Roughness is less effective when the percentage of cover is high because the effect of cover dominates (Laflen et al., 1984).

Another example is impoundment terraces, which can reduce the amount of sediment leaving an area by 95% by depositing sediment in a pond (Wischmeier and Smith, 1978). However, deposition can significantly enrich the remaining suspended sediment in fines, so that its nonpoint source pollution may be as great as that from another practice that controls sediment yield by limiting detachment. Sediment yield from the latter practice may be greater than from the impoundment, but its sediment may be hardly enriched in fines (Knisel and Foster, 1981).

A problem with impoundments is that they provide little or no protec-tion of the slope against rill and interrill erosion. Consequently, a system of practices is usually best. For example, interrill and rill erosion may be con-trolled with conservation tillage, concentrated flow erosion may be con-trolled with terraces and grassed waterways, and sediment yield may be con-trolled with impoundments. The extent to which any practice or combina-tion of practices is used depends on the nature and severity of the particular situation.

9–13 EROSION PREDICTION

Methods for estimating erosion are important tools in soil conservation. Since erosion cannot be measured easily, estimation methods are used
to assess the magnitude of erosion, to identify areas of excessive erosion,
and to project long-term changes in crop production from soil erosion. Procedures for estimating erosion are also useful in selecting practices to control site-specific erosion and nonpoint source pollution. The universal soil
loss equation (USLE), by far the most widely used erosion equation, is
(Wischmeier and Smith, 1978)

$$A = R \, K \, L \, S \, C \, P \qquad\qquad [14]$$

The factor R represents effects of climatic erosivity; K, soil erodibility; LS,
slope length and steepness; and CP, land use including cover, managem-
ment, and supporting conservation practices. Soil loss A is the long-term
average annual value, the one most often used in assessing long-term effects
of erosion on productivity. This value is compared to a soil-loss tolerance
value previously assigned to each soil (Wischmeier and Smith, 1978). As
long as erosion is less than this value, erosion is not considered excessive.

The USLE was derived from statistical analysis of 10 000 plot-years of
data from natural runoff plots, plus the equivalent of 1000 to 2000 plot-
years of data from rainfall simulators. Most of the data were from east of
the Rocky Mountains, but the data uniformly covered that area (Meyer,
1982; Foster, 1982b). Consequently, although the developers of the USLE
were mainly associated with the Midwest, the equation is not a midwestern-
based one. Plot lengths producing the USLE data set ranged from 11 to 180
m, not just the 22.1-m length of the plot definition for a USLE unit (Foster,
1982b). Slope steepness ranged from 0.5 to 25%, not just the 9% slope of
the unit plot (Foster, 1982b; Murphree and Mutchler, 1981). Even though
the USLE soil-erodibility factor K is defined for a unit plot on a tilled, continuous fallow soil on 22.1-m-long, 9% slope, the USLE is by no means
limited to that condition. The L, S, C, and P factors account for differences
between the given situation and the unit plot.

Although most of the data for the USLE came from agricultural conditions, limited data from construction sites, surface mines, pasture, rangeland, and forestland have been used to adapt the USLE to nonagricultural
situations (Foster et al., 1981b). Although the USLE is used in the West, infrequent storms causing great erosion, spatial variability of rainfall, and
soil and cover differences from data used to derive the USLE cause special
problems in such use. Furthermore, research is just beginning to develop a
full set of parameter values for range and forest land.

The effect of frozen soil on erosion in the Northwest is not well described by the USLE. Since the USLE predicts interrill-rill erosion occurring
during rainfall and the runoff generated by rainfall, the USLE does not
estimate erosion from snowmelt or surface irrigation. However, measured

erosion from snowmelt was only 7% of annual soil loss at Morris, MN (Meyer, 1980).

The USLE includes rill erosion but not erosion from concentrated flow. When do rills approach concentrated flow areas and become so large that the USLE no longer applies? This issue troubles many users of the USLE, and it requires considerable judgment during application of the USLE. Since the USLE is data based, it does not apply when conditions depart too far from those of the plots generating the data. Furthermore, since rill erosion occurred on these plots, the USLE includes rill erosion in its estimate. However, being a lumped equation, the USLE cannot separately estimate interrill and rill erosion.

Questions continually arise concerning the accuracy of the USLE. Occasionally, studies conducted at a single site or only a few locations show the USLE to predict either quite well or quite poorly. The USLE is meant only to describe the major trends of a complete, widely ranging data set. As such, it does well at some sites and not so well at others. Wischmeier and Smith (1978) describe the accuracy of the USLE. The adequacy of a model or equation must be judged on the basis of how well it meets its intended purpose. Indeed, more than degree of accuracy needs to be considered in selecting an equation. Furthermore, if two models consistently result in the same conclusion, then they do not differ in performance. Given the alternatives, which are practically none, the USLE remains a powerful tool for rather easily estimating average annual erosion (Foster et al., 1981b).

While the USLE is not an accurate estimator of erosion from single storms, it can be modified to improve its use for this purpose. Two things are usually done (Laflen et al., 1984; Foster et al., 1983). First, a soil-loss ratio appropriate for the crop and soil conditions on the day of the storm is used rather than a general value for a crop stage that covers several weeks. The cover-management factor C, used for average, annual soil-loss estimates with the USLE, is the average soil-loss ratio weighted according to the annual distribution of erosivity. Second, a runoff term is added to the USLE rainfall-erosivity factor. The best known modification is Williams' (1975) use of the product of volume of runoff and peak runoff rate raised to a power. The equation resulting from this modification, known as the modified universal soil loss equation (MUSLE), is frequently used to estimate sediment yield. Use of MUSLE eliminates the need for sediment-delivery ratios, which are required to estimate sediment yield with the USLE. This modification reflects the high correlation of erosion and sediment yield with runoff.

Another modification to the erosivity factor of the USLE is the Onstad-Foster (1975) modification, where the modified erosivity factor includes the standard USLE individual storm erosivity factor, storm energy times maximum 30-min. intensity, plus a runoff erosivity term based on volume of runoff and peak runoff rate. This modification estimates erosion on slopes where the USLE applies and should be used in conjunction with a sediment-transport-capacity equation to prevent computation of erosion for little or no runoff. This modification is an attempt to include both runoff and rainfall erosivity effects.

A third modification to the erosivity factor of the USLE (Foster et al. 1977 attempts to further separate interrill erosion from rill erosion. This modification uses the USLE single-storm erosivity factor with terms from the USLE thought to represent interrill erosion, plus the runoff erosivity factor of Onstad and Foster (1975) with USLE slope length and steepness terms thought to represent rill erosion. This modification, used in CREAMS (Knisel and Foster, 1981), also requires a sediment transport equation to prevent overestimation of erosion for low runoff.

A fourth modification to the USLE erosivity factor (Foster et al., 1982b; Lombardi, 1979) that worked well for natural erosion plot data uses as an erosivity factor the product of the volume of runoff and the maximum 30-min intensity for the rainstorm, raised to a power. This factor overcomes the overprediction by the Onstad-Foster erosivity factor when little or no runoff occurs by combining a key rainfall factor, intensity, with a key runoff factor, runoff amount. All these modified erosivity factors require a hydrology model to estimate volume of runoff, and some require an estimate of maximum 30-min rain intensity and/or peak runoff rate.

Hydrologic modeling permits the calculation of runoff hydraulics at many points in time and space over an area during a storm. This capability permits the use of fundamentally based erosion equations to describe detachment, transport, and deposition processes. Models based on at least some of these fundamentals include CREAMS, KINEROS, SEDLAB, ANSWERS, CSU, and ARM, each having unique features that make it strong in specific applications (Foster, 1982a). Unfortunately, data are quite limited for defining coefficients, exponents, and parameters in the governing relationships for several of these models. Until more data become available, values are assumed from limited data, from the USLE, or from calibrating the models using observed data.

Each of these models requires a moderate-sized computer, substantial input data, and significant computer time compared to the USLE. Each can be much more powerful than the USLE, however, in special cases such as estimating erosion and sediment yield from given storms on specific complex areas.

The basic governing equations for fundamental erosion processes can be analytically solved for special cases. Foster and Meyer (1975) gave steady-state solutions for a range of slope shapes. The unsteady continuity equation and other governing equations have been analytically solved for steady rainfall and infiltration for broad sheet flow (Singh, 1983; Lane and Shirley, 1978), and for a ridge-furrow surface configuration (Croley, 1982). The general case of varying rainfall intensity, however, requires numerical solution on a computer. Even the empirical USLE can be manipulated with the steady-state continuity equation, Eq. [3], to make it applicable to irregular slopes, except for those portions where deposition occurs (Foster and Wischmeier, 1974).

9-14   FUTURE DIRECTIONS IN EROSION PREDICTION

One direction sure to be followed is continued development of hydrologically based erosion models that use both rainfall and runoff variables. Also, simple crop-growth models will generate canopy, ground cover, roughness, and residue decomposition values needed to compute erosion on a storm-by-storm basis. Increased computer power in calculators and personal computers and easier access to central computers will make these models increasingly practical for field conservationists. In spite of its limitations, the USLE will continue to be used because it is simple and gives satisfactory estimates for many applications. Refinement and development of the USLE will continue, but new developments will be increasingly based on fundamental erosion mechanics rather than entirely on data.

Current erosion assessments are now based on a relatively few selected points. Since erosion is highly variable, more sample points are desirable. Remote sensing and large data bases managed by computers will be increasingly important in increasing the density of erosion assessment.

Erosion, including deposition, varies nonlinearly even over a uniform slope profile or small watershed. If soil productivity is nonlinearly related to erosion, an average erosion rate for a slope or watershed may give a poor estimate of average loss of productivity. Relationships for landscape form over a watershed will be developed, as will companion erosion-deposition models that will allow computation of the effect of erosion and deposition on productivity at all points over a watershed. Thus, not only will the average productivity loss over the watershed be estimated accurately, but also the distribution of productivity loss over the watershed will be determined.

REFERENCES

Allmaras, R. R. 1967. Soil water storage as affected by infiltration and evaporation in relation to tillage induced soil structure. *In* Tillage for greater crop production. American Society of Agricultural Engineers, St. Joseph, MI.

----, W. C. Burrows, and W. E. Larson. 1964. Early growth of corn as affected by temperature. Soil Sci. Am. Proc. 28(2):271-275.

Alonso, C. V., W. H. Neibling, and G. R. Foster. 1981. Estimating sediment transport in watershed modeling. Am. Soc. Agric. Eng. Trans. 24(5):1211-1200, 1226.

Al-Durrah, M. M., and J. M. Bradford. 1982. The mechanism of raindrop splash on soil surfaces. Soil Sci. Soc. Am. J. 46(5):1086-1090.

Barrows, H. L., and V. J. Kilmer. 1963. Plant nutrients losses from soils by water erosion. Adv. Agron. 16:303-316.

Brakensiek, D. L., and W. J. Rawls. 1982. An infiltration based rainfall runoff model for SCS Type 2 distribution. Am. Soc. Agric. Eng. Trans. 25(9):1607-1611.

Brenneman, L. G., and J. M. Laflen. 1982. Modeling sediment deposition behind corn residue. Am. Soc. Agric. Eng. Trans. 25(5):1245-1250.

Bubenzer, G. D. 1979. Rainfall characteristics important for simulation. p. 22–34. *In* Proc. of the Rainfall Simulator Workshop. USDA–SEA. ARM-W-10.

Chapman, G. 1948. Size of raindrops and their striking force at the soil surface in a red pine plantation. Am. Geophys. Union Trans. 29(5):664–670.

Chow, V. T. 1959. Open channel hydraulics. McGraw-Hill Book Co., New York.

Croley, T. E. 1982. Unsteady overland sedimentation. J. Hydrol. 56(3/4):325–346.

Davis, S. S., G. R. Foster, and L. F. Huggins. 1983. Deposition of nonuniform sediment on concave slopes. Am. Soc. Agric. Eng. Trans. 26(4):1057–1063.

Edwards, A. P., and J. M. Bremner. 1967. Microaggregates in soils. Soil Sci. 18(1):64–73.

Ellison, W. D. 1944. Studies of raindrop erosion. Agric. Eng. 25(4):131–136.

Einstein, H. A. 1968. Deposition of suspended particles in a gravel bed. Am. Soc. Civil Eng. Proc. J. Hydr. Div. 94(HY5):1195–1205.

Featig, L. H., E. J. Manke, and G. R. Foster. 1982. Characterization of eroded soil particles from interrill areas. Paper 82-2038. American Society of Agricultural Engineers, St. Joseph, MI.

Foster, G. R. 1982a. Modeling the erosion process. p. 297–380. *In* Hydrologic modeling of small watersheds. American Society of Agricultural Engineers, St. Joseph, MI.

————. 1982b. Relation of USLE factors to erosion on rangeland. p. 17–35. *In* Proc. Workshop on Estimating Erosion and Sediment Yield on Rangelands, Tucson, AZ. 7–9 March 1981. USDA–ARS. ARM-W-26.

————. 1982c. Channel erosion within farm fields. Preprint 82-007. Am. Soc. Civil Eng., New York.

————, C. B. Johnson, and W. C. Moldenhauer. 1982a. Critical slope lengths for unanchored cornstalk and wheat straw residue. Am. Soc. Agric. Eng. Trans. 25(4):935–939.

————, ————, and ————. 1982b. Hydraulics of failure of unanchored cornstalk and wheat straw mulches for erosion control. Am. Soc. Agric. Eng. Trans. 25(4):940–947.

————, and L. J. Lane. 1983. Erosion by concentrated flow in farm fields. p. 9.65–9.82. *In* Proc. of Symp. to Honor D. B. Simons. Colorado State University, Ft. Collins. 28–30 June 1983.

————, ————, and J. D. Nowlin. 1980. A model to estimate sediment yield from field sized areas: Application to planning and management for control of nonpoint source pollution. p. 193–281. *In* CREAMS: A field scale model for chemicals, runoff, and erosion from agricultural management systems. USDA-Conserv. Res. Rep. 26.

————, ————, ————, J. M. Laflen, and R. A. Young. 1981a. Estimating erosion and sediment yield on field sized areas. Am. Soc. Agric. Eng. Trans. 24(5):1253–1262.

————, F. Lombardi, and W. C. Moldenhauer. 1982c. Evaluation of rainfall-runoff erosivity factors for individual storms. Am. Soc. Agric. Eng. Trans. 25(1):124–129.

————, and L. D. Meyer. 1972. Transport of soil particles by shallow flow. Am. Soc. Agric. Eng. Trans. 15(1):99–102.

————, and ————. 1975. Mathematical simulation of upland erosion by fundamental erosion mechanics. p. 190–207. *In* Present and prospective technology for predicting sediment yields and sources. USDA–ARS. ARS-S-40. USDA.

————, and ————. 1977. Soil erosion and sedimentation by water—An overview. p. 1–13. *In* Proc. Natl. Symp. on Erosion and Sedimentation by Water, Chicago, IL. ASAE Publ. 4-77. American Society of Agricultural Engineers, St. Joseph, MI.

————, ————, and C. A. Onstad. 1977. A runoff erosivity factor and variable slope length exponents for soil loss estimates. Am. Soc. Agric. Eng. Trans. 20(4):683–687.

————, W. R. Osterkamp, L. J. Lane, and D. W. Hunt. 1982d. Effect of discharge rate on rill erosion. Paper 82-2572. American Society of Agricultural Engineers, St. Joseph, MI.

————, J. R. Simanton, K. G. Renard, L. J. Lane, and H. B. Osborn. 1981b. Discussion of "Application of the Universal Soil Loss Equation to Rangelands on a Per-Storm Basis" by Trieste and Gifford in *Journal of Range Management* 33:66–70, 1980. J. Range Manage. 34(2):161–165.

————, R. E. Smith, W. G. Knisel, and T. E. Hakonson. 1983. Modeling the effectiveness of on-site sediment controls. Paper 83-2092. American Society of Agricultural Engineers, St. Joseph, MI.

----, and W. H. Wischmeier. 1974. Evaluating irregular slopes for soil loss prediction. Am. Soc. Agric. Eng. Trans. 17(2):305-309.

----, R. A. Young, and W. H. Neibling. 1982e. Comosition of sediment for nonpoint source pollution analysis. Paper 82-2580. American Society of Agricultural Engineers, St. Joseph, MI.

Gabriels, D., and W. C. Moldenhauer. 1978. Size distribution of eroded material from simulated rainfall: Effect over a range of textures. Soil Sci. Soc. Am. J. 42(6):954-958.

Garey, C. L. 1954. Properties of soil aggregates. I: Relation to size, water stability, and mechanical composition. Soil Sci. Am. Proc. 18(1):16-18.

Graf, W. H. 1971. Hydraulics of sediment transport. McGraw-Hill Book Co., New York.

Knisel, W. G., and G. R. Foster. 1981. CREAMS: A system for evaluating best management practices. p. 177-194. In Economics, ethics, ecology: Roots of productive conservation. Soil Conserv. Soc. Am., Ankeny, IA.

Laflen, J. M., G. R. Foster, and C. A. Onstad. 1985. Simulation of individual-storm soil loss for modeling impact of soil erosion on crop productivity. In Proc. of Preserve the Land, 1983 Int. Conf. on Soil Conserv., Honolulu, HI. Jan. 1983. Soil Conserv. Soc. Am., Ankeny, IA.

----, and W. C. Moldenhauer. 1979. Soil and water losses from corn-soybean rotations. Soil Sci. Soc. Am. J. 43(6):1213-1215.

----, ----, and T. S. Colvin. 1980. Conservation tillage and soil erosion on continuously row-cropped land. p. 121-133. In Crop production with conservation in the 80's. ASAE Publ. 7-81. American Society of Agricultural Engineers, St. Joseph, MI.

Lal, R. 1976. No-tillage effects on soil properties under different crops in Western Nigeria. Soil Sci. Soc. Am. J. 40(5):762-768.

Lane, L. J., and E. D. Shirley. 1978. Mathematical simulation of erosion on upland areas. USDA-Agr. Res. Service, Tucson, AZ.

Laws, J. O., and D. A. Parsons. 1943. The relation of raindrop size to intensity. Am. Geophys. Union. Trans. 24:452-460.

Lombardi, F. 1979. Universal soil loss equation (USLE), runoff erosivity factor, slope length exponent, and slope steepness exponent for individual storms. Ph.D. diss. Purdue Univ., West Lafayette, IN.

Long, D. C. 1964. The size and density of aggregates in eroded soil material. M.S. thesis. Iowa State University, Ames.

Luk, S. H. 1979. Effect of soil properties on erosion by wash and splash. Earth Surf. Process. 4(3):241-255.

Lyles, L. 1977. Soil detachment and aggregate disintegration by wind-driven rain. p. 152-159. In Soil erosion: Prediction and control. Spec. Publ. 21. Soil Conservation Society of America, Ankeny, IA.

Martinez, M. 1979. Erosion modeling for upland areas. Ph.D. diss. University of Arizona, Tucson.

Meyer, L. D. 1980. Adding erosion from snowmelt to an erosion prediction equation. p. 444-445. In CREAMS: a field scale model for chemicals, runoff, and erosion from agricultural management systems. USDA-Conserv. Res. Rep. 26.

----. 1981. How intensity affects interrill erosion. Am. Soc. Agric. Eng. Trans. 25(6):1472-1475.

----. 1982. Soil erosion research leading to development of the Universal Soil Loss Equation. p. 1-16. In Proc. Workshop on Estimating Erosion and Sediment Yield on Rangelands, Tucson, AZ. 7-9 March 1981. USDA-ARS. ARM-W-26. USDA.

----. 1984. Sediment eroded from interrill sources. In Proc. of Preserve the Land, 1983 Int. Conf. on Soil Conserv., Honolulu, HI. Jan. 1983. Soil Conservation Society of America, Ankeny, IA.

----, G. R. Foster, and S. Nikolov. 1975a. Effect of flow rate and canopy on rill erosion. Am. Soc. Agric. Eng. Trans. 18(5):905-911.

----, ----, and M. J. M. Romkens. 1975b. Source of soil eroded by water from upland slopes. p. 177-189. In Present and prospective technology for predicting sediment yields and sources. USDA-ARS. ARS-S-40.

----, and W. H. Wischmeier. 1969. Mathematical simulation of the process of soil erosion by water. Am. Soc. Agric. Eng. Trans. 12(6):754–758, 762.

Monke, E. J., and G. R. Foster. 1982. Characterization of eroded soil particles from interrill areas. Paper 82-2038. American Society of Agricultural Engineers, St. Joseph, MI.

Murphree, C. E., and C. K. Mutchler. 1981. Verification of the slope factor in the universal soil loss equation for low slopes. J. Soil Water Conserv. 36(5):300–302.

Mutchler, C. K., and L. M. Hansen. 1970. Splash of a waterdrop at terminal velocity. Science 169:1311–1312.

Neibling, W. H., and G. R. Foster. 1977. Estimating deposition and sediment yield from overland flow processes. p. 75–86. In Proc. of the Int. Symp. on Urban Hydrology, Hydraulic, and Sediment Control, Lexington, KY. July 1977. University of Kentucky, Lexington.

Onstad, C. A., and G. R. Foster. 1975. Erosion modeling on a watershed. Am. Soc. Agric. Eng. Trans. 18(2):288–292.

Piest, R. F., J. M. Bradford, and R. G. Spomer. 1975. Mechanisms of erosion and sediment movement from gullies. p. 162–176. In Present and prospective technology for predicting sediment yields and sources. USDA-ARS. ARS-S-40. USDA.

Rawls, W. J., D. L. Brakensiek, and K. E. Saxton. 1982. Estimation of soil water properties. Am. Soc. Agric. Eng. Trans. 25(5):1316–1320.

Rhoton, F. E., L. D. Meyer, and F. D. Whisler. 1983. Densities of wet aggregated sediment from different textured soils. Soil Sci. Soc. Am. J. 47(3):576–578.

Römkens, M. J. M., C. B. Roth, and D. W. Nelson. 1977. Erodibility of selected clay subsoils in relation to physical and chemical properties. Soil Sci. Soc. Am. J. 41(5):954–960.

----, and J. Y. Wang. 1985. Soil translocations by rainfall. In Proc. of Preserve the Land, 1983 Int. Conf. on Soil Conserv., Honolulu, HI. Jan. 1983. Mississippi Academy of Sciences, Jackson.

Singh, V. J. 1983. Analytical solutions of kinematic equations for erosion on a plane. Adv. Water Resour. 6(June):88–95.

Van Liew, M. W., and K. E. Saxton. 1983. Slope steepness and incorporated residue effects on rill erosion. Trans. Am. Soc. Agric. Eng. 26:1738–1743.

Voorhees, W. B., R. R. Allmaras, and C. E. Johnson. 1981. Alleviating temperature stress. p. 217–266. In Modifying the root environment to reduce crop stress. American Society of Agricultural Engineers, St. Joseph, MI.

----. ----, and W. E. Larson. 1966. Porosity of surface aggregates at various moisture contents. Soil Sci. Am. Proc. 30(2):163–167.

Williams, J. R. 1975. Sediment-yield prediction with Universal Equation using runoff energy factor. p. 244–252. In Present and prospective technology for predicting sediment yields and sources. USDA-ARS. ARS-S-40.

Wischmeier, W. H. 1975. Estimating the soil loss equation's cover and management factor for undisturbed land. p. 118–125. In Present and prospective technology for predicting sediment yields and sources. USDA-ARS. ARS-S-40.

----, and D. D. Smith. 1978. Predicting rainfall erosion—a guide to conservation planning. USDA Agriculture Handbook 537.

Yoder, R. E. 1936. A direct method of aggregate analysis of soils and a study of the physical nature of erosion losses. Agron. J. 8(5):377–351.

Young, R. A. 1980. Characteristics of eroded sediments. Am. Soc. Agric. Eng. Trans. 23(5):1139–1142, 1146.

----, and C. K. Mutchler. 1969. Soil and water movement in small tillage channels. Am. Soc. Agric. Eng. Trans. 12(2):543–545.

----, and J. L. Wiersma. 1973. The role of raindrop impact in soil detachment and transport. Water Resour. Res. 9(6):1629–1639.

10 Wind Erosion: Processes and Prediction

Leon Lyles, G. W. Cole, and L. J. Hagen
Agricultural Research Service
U.S. Department of Agriculture
Manhattan, Kansas

Conditions conducive to wind erosion exist when the soil is loose, dry, and finely granulated; the soil surface is smooth and vegetative cover is sparse or absent; the susceptible area is sufficiently large; and the wind is strong and turbulent enough to move soil. Those conditions often prevail in semiarid and arid climates, for example west of the 99th meridian in the USA.

The many problems researchers face in determining effects of erosion on crop productivity are represented by Fig. 10-1. A host of interactions determines crop production or yield. One way to isolate the effects of erosion in this complex crop-production system is to translate the problem from practical terms to basic concepts at the process level, where research is performed.

Even though wind erosion damages soil, crops, and the environment, we will omit any discussion of environmental damages, other than to note that future experimental research and modeling should include such impacts on crop yields. In this paper, wind erosion will be linked to productivity through its alteration of surface-soil properties and/or soil depth. This alteration must then be linked to other factors that influence the growth and development of crops.

10-1 WIND EROSION AND CROP PRODUCTION

Wind erosion is a set of processes that contribute to the motion of soil from its initiation until final deposition. Neglecting the abrasive effect of aggregate impact on crops, the interface between wind erosion and the crop-

Published in R. F. Follett and B. A. Stewart, ed. 1985. *Soil Erosion and Crop Productivity.*
© ASA-CSSA-SSSA, 677 South Segoe Road, Madison, WI 53711, USA.

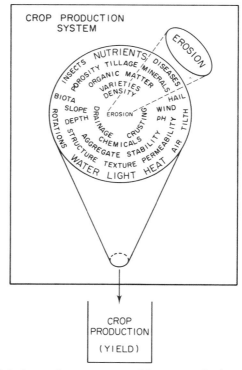

Fig. 10–1. Interacting components of the crop-production system.

production system (CPS) must be with the soil component of the CPS (Fig. 10–2). Furthermore, the soil system can only be affected by a change in soil depth, which is a measure of the size of the system, and/or a change in soil properties.

Soil properties refer to those characteristics that depend on the soil but not to any great extent on the size of a soil sample, such as, hydraulic conductivity as opposed to hydraulic conductance. This description, therefore, excludes the total soil mass of the CPS, which is a state of the system similar to the aggregate-size distribution. If the rate of soil loss exceeds the rate of soil generation, eventually the CPS will terminate and the yield will decline to zero. (The total loss is indicated by process A in Fig. 10–2). The effect of selective loss (process B in Fig. 10–2) on the CPS is not quite as obvious. To understand it, we must consider both wind erosion and the soil system.

The literature indicates that some soil properties can be correlated directly to the primary particle-size distribution (PSD) (Gupta and Larson, 1979; Arya and Paris, 1981). Because wind erosion is generally a selective soil-loss process, which moves aggregates of various size fractions at different mass-flow rates (Chepil, 1951), one also needs to understand how the aggregate-size distribution (ASD) is related to soil properties. Selective soil loss appears to be linked to soil properties primarily through the ASD

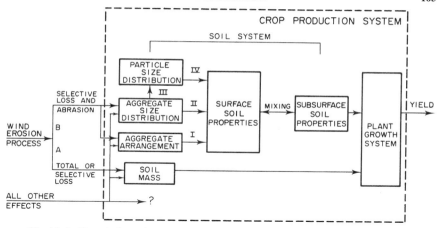

Fig. 10–2. The wind-erosion process and its interface with the crop-production system.

and the PSD. This linkage is illustrated in Fig. 10–2 as lines II, III, and IV, which represent functional relationships. Line III suggests that a change in ASD due to abrasion and sorting may affect the PSD as well. In general, when the aggregates are not homogeneous with respect to their PSD's, this path will exist. Line IV, between PSD and surface-soil properties, is that cited previously. Line II, between ASD and surface-soil properties, indicates the effect on properties associated with fluid and energy transport through the surface, such as gas and liquid permeabilities. Except for fluid and energy transport properties, all these effects occur at or near the soil surface. Unless these effects are extended to subsurface layers, the plant-growth system will be relatively unaffected. Unfortunately, soil mixing accomplishes this extension.

The spatial arrangement of surface aggregates, of which nothing quantitative is known at present, also affects surface-soil properties (line I, Fig. 10–2). Aggregate arrangement can also affect soil loss by sheltering the smaller aggregates. Other factors, such as water erosion, tillage, and weather, also affect the same variables that are affected by wind erosion and are included in Fig. 10–2 as "all other effects."

10–2 WIND EROSION PROCESSES

The most comprehensive summaries on the movement of surface material by wind action have been prepared by Bagnold (1941) for desert sands and by Chepil and Woodruff (1963) for agricultural lands. Wind erosion consists of initiation, transport (suspension, saltation, surface creep), abrasion, sorting, avalanching, and finally deposition of soil aggregates/particles (A/P). We will limit our viewpoint here to periods when erosion is actually occurring. Soil transport by wind is commonly described in three distinct modes: suspension, saltation, and surface creep.

10-2.1 Suspension

Suspension refers to the vertical and (eventual) horizontal transport of A/P that are generally removed from the local source area. Chepil (1945) reported that 3 to 38% of total transport could be carried in suspension, depending on soil texture. Generally, the vertical transport is less than 10% of the horizontal (Gillette, 1977 and 1978). Suspendible A/P's range in size from 2 to 100 μm in diameter, with a mass median diameter of about 50 μm in an actively eroding field (Chepil, 1957a; Gillette and Walker, 1977). This size range excludes the fine, medium, coarse, and very coarse sand particles and aggregates of corresponding size, which remain in the local area. Because organic matter and some plant nutrients are usually associated with the finer soil fractions, suspension samples are enriched in such constituents compared with the bulk soil source. Furthermore, the enrichment ratio increases as the amount of sand particles too large to be suspended by common winds increases in the bulk soil. Consequently, suspension indirectly impacts productivity through removal of organic matter and plant nutrients or, conversely, by leaving behind the less-fertile soil constituents. It directly affects the surface-layer ASD during erosion, but, as previously noted, ASD in the new surface layer, depending on PSD, may be unchanged when erosion ceases.

Because of our present definition of the wind-erosion process, we exclude subsequent deposition of suspension-sized A/P's. On an expanded treatment of wind-erosion processes, deposition contributes to soil renewal (Smith et al., 1970) and might even need to be included in determining the effects of erosion on crop productivity.

10-2.2 Saltation

The characteristics of saltation (jumping) A/P's in wind have been described (Bagnold, 1941; Chepil, 1945; Free, 1911; White and Schultz, 1977). Roughly 50 to 80% of total wind transport is by saltation. During saltation, individual A/P's lift off the surface (eject) at 50 to 90° angles, rotate at 115 to 1000 r/s, and follow distinctive trajectories under the influence of air resistance and gravity (Chepil, 1945; White and Schulz, 1977). Those A/P's 100 to 500 μm in diameter (too large to be suspended by the flow) return to the surface at impact angles of 6 to 14° from the horizontal, either to rebound or to embed themselves, thus influencing the breakdown and movement of other A/P's. The size range for saltation excludes the coarse and very coarse sand particles, which remain in the local area. During erosion, saltating aggregates may shift to the suspension mode because of abrasion and may cause other aggregates at the surface to shift modes. Saltation is the major cause of aggregate breakdown during erosion. Its role is to initiate and sustain suspension, drive the creep transport, and influence ASD of the soil surface. Therefore, linkage through those factors must be established to determine the impact of saltation on crop productivity. As with suspension, ASD in the new surface layer may be unchanged when erosion ceases.

10–2.3 Surface Creep

Coarse, sand-sized, mineral-soil A/P's 500 to 1000 μm in diameter, too large to leave the surface in ordinary erosive winds, can be set in motion by the impact of saltating A/P's. Reportedly, surface creep constitutes 7 to 25% of total transport (Bagnold, 1941; Chepil, 1945; Horikowa and Shen, 1960). In high winds, the entire surface appears to be creeping slowly forward at speeds much less than 2.5 cm/s pushed and rolled (driven) by the saltation flow. Surface creep normally excludes very coarse sand particles and gravels greater than 2000 μm in diameter which, if contained in the bulk soil, must remain near their current location during wind erosion. Creep appears nearly passive in the erosion process, but creep-sized aggregates may abrade into the size ranges of saltation and suspension and thus shift modes of transport. The impact of surface creep on productivity appears to be linked primarily to ASD effects.

10–2.4 Abrasion

Many aggregation and deaggregation processes affect the soil-surface layer between erosion events. These processes generally produce log-normal, surface soil ASD's (Gardner, 1956). The log-normal distributions often approach limits at the extremes, however, because the maximum size of the aggregates may be controlled by processes such as tillage, and the minimum aggregate size may be controlled by the size of the primary particles themselves, which usually do not have log-normal distributions. Suspension of the particles less than 100 μm in diameter during wind erosion may also change the lower limit of the surface ASD.

The percentage of erodible soil (i.e., less than 1000 μm in diameter) in the surface layer is highly correlated with the mass of soil one can remove from that surface in wind-tunnel tests (Chepil, 1958). On long fields, the amount of soil that passes from a control volume on the soil surface because of saltation and creep increases nearly linearly with field length (Chepil, 1957b). Such a result implies abrasive breakdown of both erodible and nonerodible aggregates. Indeed, on long, erosion-susceptible fields, the total amount of soil that can be lost is usually several times the amount of erodible material initially present at the surface. Thus, both initial ASD and resistance to abrasive breakdown of surface aggregates are important in wind erosion.

An abrasion susceptibility term (w) can be defined as the mass of material abraded from target aggregates per unit mass of impacting abrader. To determine how various factors affect w, large soil aggregates (50–100 mm in diameter) have been abraded with sand particles and soil aggregates using a calibrated nozzle (Hagen, 1984). The results show that

$$w = f(V_a, \alpha, d_a, S_t, S_a, \varrho_a) \tag{1}$$

where V_a is the average velocity of the impacting A/P's; α is the A/P impact angle with the surface plane; d_a is the average diameter of abrading A/P's; S_t and S_a are dry mechanical stabilities of the target aggregates and abrading aggregates, respectively; and ϱ_a is the A/P density of the abrader. Aggregate abrasion affects the soil system through ASD and aggregate arrangement (Fig. 10–2).

10–2.5 Sorting

Unless surface-layer A/P's are homogeneous in physical properties (size, shape, density), which is highly unlikely in agricultural soils, sorting will occur during erosion. Sorting here refers to the selective removal during erosion of A/P's, because various sizes move at different mass-flow rates. The impact of sorting on crop productivity would ultimately be expressed through changes in ASD associated with discrete erosion events. Changes in ASD are contingent on initial PSD and ASD, homogeneity with depth, aggregate stability, erosive wind duration, presence or absence of erosion-resistant "layers", and arrangement of nonerodible aggregates with depth. In most cases envisioned, ASD would change during discrete erosion events. The most common case where ASD would remain the same would involve A/P homogenity with depth, no particles greater than 1000 μm in diameter, an erosion-resistant layer below the soil surface caused by binding agent(s) or water, and all the A/P's above the resistant layer being removed by the erosion event.

10–2.6 Process Alteration

In general, wind erosion can be decreased only by reducing wind forces on erodible A/P's or by creating aggregates or surfaces more resistant to wind forces. Nonerodible elements reduce wind-drag forces on erodible A/P's (Fig. 10–3). Various components of the erosion process might be altered by reducing field length, increasing dry-aggregate stability, changing ASD, altering the path of the wind, providing trapping surfaces, or reducing wind forces.

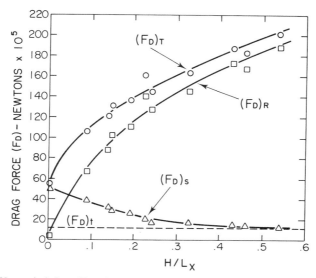

Fig. 10-3. How wind drag (F_D) changes as an eroding surface stabilizes by exposing nonerodible elements. Subscripts T, R, S, and t refer to total drag forces, drag forces on the nonerodible roughness elements, drag forces on the intervening surface, and threshold drag of erodible aggregates/particles, respectively. H is height and L_x is downwind distance between nonerodible elements (adapted from Lyles et al., 1974).

10-3   PREDICTION OF WIND EROSION

10-3.1   Present Methods

Currently, prediction of wind erosion is largely associated with the wind-erosion equation (WEE) originally reported by Woodruff and Siddoway (1965):

$$E = f(I, K, C, L, V) \qquad [2]$$

where E is the potential annual soil-loss flux, I is the soil erodibility, K is the soil-ridge-roughness factor, C is the climatic factor, L is the unsheltered "weighted" distance that wind travels across a field, and V is the equivalent vegetative cover. Cole et al. (1982) discussed two "weighting" methods that have been used to determine L, and Skidmore (unpublished data) has proposed another.

Procedures have been developed for applying Eq. [2] to periods shorter than 1 year, which involves partitioning erosion amounts over time with erosive wind-energy distribution as the criterion (Bondy et al., 1980). Recently, a similar approach has been used in Erosion-Productivity Impact Calculator (EPIC) (Williams et al., 1984), which operates on a daily time step. Details of the daily, wind-erosion, soil-loss model have been reported by Cole et al. (1982).

Regardless of the modifications, the E term in WEE predicts only total soil removal. Hence, it can be viewed only in terms of accumulated loss of topsoil depth when applied to the problem of determining erosion effects on crop productivity.

10–3.2 Future Methods

Classically, prediction of wind erosion has been linked to the idea that any soil loss is bad and that the appropriate measure of "badness" is the potential average, annual, soil loss (E). Because of the soil erosion/crop productivity problem, however, the measure must now be expanded to include wind-erosion effects on surface-soil properties. This is represented in Fig. 10-2 as the wind-erosion processes of sorting and abrasion, which affect ASD. A relationship between E and ASD is also implied in Fig. 10-4, where a mass balance model is depicted for n soil-size classes. Also shown are two ASD's, representing possible initial and final states of the surface-soil mass.

The final ASD of Fig. 10-4 represents the solution generated by the model after running for a specified time, having started with the condition implied by the initial ASD. Therefore, given the initial state of the system and a description (i.e., equations) of the soil-loss ratios ($\dot{m}_i$) and abrasion rates ($\dot{a}_i$), the final ASD is predictable.

Because the model illustrated is for a field of size A, E could be computed by taking the average sum over time of all $\dot{m}_i$ and dividing by A. Hence, the information needed to generate our present soil-loss measure is

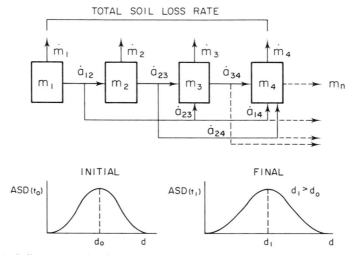

Fig. 10–4. Soil transport-abrasion model for an homogeneous field surface, in which $\dot{m}_i$ and $\dot{a}_{ij}$ are mass soil-loss rates and abrasion rates by aggregate/particle size class, respectively.

inherent in this new model. Another way to view the relationship between Fig. 10–4 and E is simply to lump all the $\dot{m}_i$ compartments into one. All the lines representing the abrasion process thus disappear, and the sum of $\dot{m}_i$ becomes $\dot{m}_{total}$, which relates to E as previously noted.

Probably of more significance in the new model than the inclusion of abrasion rates is that we are now interested in the resulting state of that portion of the system that is *not* lost. This contrasts with predicting E, which says nothing about what remains behind in a given size class. Although the abrasion rates were not previously required, they must be included here because we now need to predict the final ASD and because abrasion is the only other wind erosion process that affects the ASD.

Clearly, Fig. 10–4 represents an expanded model capable of predicting both ASD and E, although this new model requires equations for $\dot{m}_i$ and $\dot{a}_i$ which, unfortunately, are unknown at present. Obviously, for prediction of ASD, development of such equations is required.

10–4 FUTURE RESEARCH

As indicated previously, Fig. 10–2 portrays the wind-erosion processes and fundamental relationships involved in changing soil properties and size distributions. In particular, the crucial processes for describing wind-erosion effects on the CPS are labeled A and B, and the functional relationships, which are not unique to the wind-erosion process, are labeled I to IV. The distinction between processes and functions provides a clear image of the direction of future research. For example, part of process A represents the state-of-the-art of wind-erosion research: prediction of the total average soil-loss flux. However, the subdivision of the soil-loss flux by aggregate-size class has not yet been accomplished (except for rough generalizations among transport modes). That subdivision, in conjunction with abrasion rates by size class, will be required to describe process B. Both of the above involve field, wind-tunnel, and theoretical studies.

In contrast, functional relationships I and II do not involve the processes of wind erosion. They are in the domain of soil physics and are independent of the process that caused the ASD or aggregate arrangement to change. Relationship IV is being studied. Relationship III is proposed for future study because it is an important link in interactions between wind erosion and the crop-production system.

REFERENCES

Arya, L. M., and J. F. Paris. 1981. A physicoempirical model to predict the soil moisture characteristics from particle-size distribution and bulk density data. Soil Sci. Soc. Am. J. 45: 1023–1030.

Bagnold, R. A. 1941. The physics of blown sand and desert dunes. Methuen, London.

Bondy, E., L. Lyles, and W. A. Hayes. 1980. Computing soil erosion by periods using wind-energy distribution. J. Soil Water Conserv. 35(4):173–176.

Chepil, W. S. 1945. Dynamics of wind erosion: I. Nature of movement of soil by wind. Soil Sci. 60(4):305–320.

––––. 1951. Properties of soil which influence wind erosion: IV. State of dry aggregate structure. Soil Sci. 72:387–401.

––––. 1957a. Sedimentary characteristics of dust storms: III. Composition of suspended dust. Am. J. Sci. 255:206–213.

––––. 1957b. Width of field strips to control wind erosion. Kans. Agric. Exp. Stn. Tech. Bull. 92.

––––. 1958. Soil conditions that influence wind erosion. USDA Tech. Bull. 1185.

––––, and N. P. Woodruff. 1963. The physics of wind erosion and its control. Adv. Agron. 15: 211–302.

Cole, G. W., L. Lyles, and L. J. Hagen. 1983. A simulation model of daily wind erosion soil loss. Trans. Am. Soc. Agric. Eng. 26(6):1758–1765.

Free, E. E. 1911. The movement of soil material by wind. USDA Bureau of Soils Bull. 68.

Gardner, W. R. 1956. Representation of soil aggregate-size distribution by a logarithmic-normal distribution. Soil Sci. Soc. Am. Proc. 20:151–153.

Gillette, D. A. 1977. Fine-particle emissions due to wind erosion. Trans. Am. Soc. Agric. Eng. 20(5):890–897.

––––. 1978. A wind tunnel simulation of the erosion of soil: Effect of soil texture, sandblasting, windspeed, and soil consolidation on dust production. Atmos. Environ. 12:1735–1743.

––––, and T. R. Walker. 1977. Characteristics of airborne particles produced by wind erosion of sandy soil, high plains of West Texas. Soil Sci. 123:97–110.

Gupta, S. C., and W. E. Larson. 1979. A model for predicting packing density of soils using particle-size distribution. Soil Sci. Soc. Am. J. 43:758–764.

Hagen, L. J. 1984. Soil aggregate abrasion by impacting sand and soil particles. Trans. Am. Soc. Agric. Eng. 27(3):805–806, 816.

Horikowa, K., and H. W. Shen. 1960. Sand movement by wind action. Beach Erosion Bull., Corps Engin. Tech. Memo. No. 119. Washington, DC.

Lyles, L., R. S. Schrandt, and N. F. Schmeidler. 1974. How aerodynamic roughness elements control sand movement. Trans. Am. Soc. Agric. Eng. 17(1):134–139.

Smith, R. M., P. C. Twiss, R. K. Krauss, and M. J. Brown. 1970. Dust deposition in relation to site, season, and climatic variables. Soil Sci. Soc. Am. Proc. 34(1):112–117.

White, B. R., and J. C. Schulz. 1977. Magnus effect in saltation. J. Fluid Mech. 81(3):497–512.

Williams, J. R., P. T. Dyke, and C. A. Jones. 1982. EPIC—a model for assessing the effects of erosion on soil productivity. Presented at the Third Int. Conf. on the State-of-the-Art in Ecological Modeling, Colorado State University, Fort Collins. 24–28 May.

Woodruff, N. P., and F. H. Siddoway. 1965. A wind erosion equation. Soil Sci. Soc. Am. Proc. 29:502–608.

11 Criteria for Determining Tolerable Erosion Rates

G. F. Hall and T. J. Logan
The Ohio State University
Columbus, Ohio

K. K. Young
Soil Conservation Service
U.S. Department of Agriculture
Washington, DC

11-1 HISTORICAL PERSPECTIVE

In the mid-1930s, the total sediment carried to the sea from our land was estimated at about 3.3 billion t (3.6 billion tons) annually (Bennett, 1939). By the 1940s, it had become apparent that soil loss research was needed to differentiate geologic erosion, a phenomenon we must accept, from destructive man-made erosion, which we should not accept. However, specifying how much erosion can be tolerated is meaningful only if erosion rates can be estimated and the effects of erosion determined.

11-1.1 Early Estimates

Hays and Clark (1941) concluded that average annual erosion should be kept under 6.7 t/ha (3 tons/acre). They reasoned that at this rate it would take about 50 years to remove 2.5 cm (1 in.) of surface soil, so that the top 20 cm (8 in.) would be removed in 400 years. From a practical standpoint, they believed that farmers would regard such a loss as reasonably safe. Smith (1941) concluded that annual soil loss of 9 t/ha (4 tons/acre) might be excessive for maintaining fertility of Shelby soils if the surface and subsoil layers were mixed. He suggested, however, that the addition of organic matter might offset the loss of productivity due to erosion. He even speculated that the dilution of the surface soil with a small amount of subsoil could be beneficial because the subsoil contained larger amounts of certain nutrients than the surface layer.

Published in R. F. Follett and B. A. Stewart, ed. 1985. *Soil Erosion and Crop Productivity.*
© ASA-CSSA-SSSA, 677 South Segoe Road, Madison, WI 53711, USA.

Browning et al. (1947), in a thorough review of soil erosion research in the Midwest, identified for 12 soils the "maximum average annual permissible soil loss without decreasing productivity" (p. 66). They stated that tolerance levels for soil loss vary with soil type. Their suggested values ranged from a low of 4.5 t/ha (2 tons/acre) per year on soils with a restrictive layer such as a claypan to a high of 13.5 t/ha (6 tons/acre) per year on thick loess soils with a deep, favorable root zone.

Maintenance of soil fertility as a soil-conservation objective, or the concept that soil loss is excessive if fertility declines, was emphasized by Smith and Whitt (1948). Using organic matter as an indicator of fertility, they plotted the annual change in soil organic matter content against the annual soil loss for Shelby, Marshall, and Putnam soils to determine the upper limit of soil loss without a concurrent loss in soil fertility. They found that an average annual loss of 9 t/ha (4 tons/acre) could be tolerated on Marshall and Shelby soils, but only 6.7 t/ha (3 tons/acre) on Putnam soils. For other soils in Missouri they suggested soil loss tolerances ranging from a high of 9 t/ha (4 tons/acre) on the better soils to 4.5 t/ha (2 tons/acre) on claypan or shallow soils.

11-1.2   Identification of T Value

During the 1950s and early 1960s, regional workshops were held on developing a procedure to estimate allowable soil loss. The latest soil erosion research was used in these workshops to develop the factors in a predictive equation for soil loss. Each workgroup report stated that a soil loss tolerance of 11.2 t/ha (5 tons/acre) was considered a reasonable maximum. By 1962, SCS conservationists in states east of the Rocky Mountains were using the soil loss equation in conservation planning.

The definition of soil loss tolerance developed during that period remains essentially unchanged. Wischmeier and Smith (1978) defined the soil loss tolerance (T), or allowable soil loss, as denoting "the maximum level of soil erosion that will permit a high level of crop productivity to be maintained economically and indefinitely" (p. 2). The T value is operationally defined by the SCS in terms of long-term average annual soil losses as estimated by the universal soil loss equation (USLE). Guidelines for establishing tolerance values have been developed over the years. In 1973, SCS issued Advisory Notice Soils- (USDA-SCS, 1973), requesting each state to update soil-loss tolerances based on the guidelines in Table 11-1. T values of 2.2 to 11.2 t/ha (1 to 5 tons/acre) were used. The numbers represent the permissible average annual soil loss on agricultural soils. T values are not applicable to construction or other nonfarm land uses.

A maximum T value of 11 t/ha was chosen in the above case, with lower values based primarily on the depth of soil favorable for plant root growth. McCormack et al. (1982) indicated that one reason for selecting an upper value of 11 t/ha was the estimation, from sketchy data, that an A horizon can develop in permeable, medium-textured material in well-managed cropland at about this rate. Also, erosion above this rate was felt

Table 11-1. Guide for assigning soil loss tolerance values to soils having different rooting depths (USDA–SCS, 1973).

Rooting depth	Soil-loss tolerance values	
	Renewable soil†	Nonrenewable soil‡
cm	t/ha	
<25	2.2	2.2
25–51	4.5	2.2
51–102	6.7	4.5
102–152	9.0	6.7
>152	11.2	11.2

† Soils that have a favorable substratum and can be renewed by tillage, fertilizer, organic matter, and other management practices.
‡ Soils that have an unfavorable substratum, such as rock, and cannot be renewed economically.

to cause gullying (Klingebiel, 1961), but the SCS, in its 1977 Technical Advisory on T values, based its values only on the maintenance of soil productivity. Favorable crop rooting depth was the sole criterion used to set values in the range of 2 to 11 t/ha.

11-1.3 Soil Productivity and T Values

Since the middle to late 1970s, considerable attention has been given to soil erosion in the USA, its effects on soil productivity, and offsite damages due to sediments and the nutrients and other chemicals transported by sediment. Logan (1977 and 1982), McCormack et al. (1982), and others have discussed the need to separate soil loss tolerance values from soil loss limits. Soil-loss tolerances would be set for the sole purpose of maintaining crop productivity, whereas soil loss limits would have the objective of reducing sediment loads from a watershed. While T values can be assigned on the basis of a soil series or soil phase, soil limits should be assigned on an areally distributed basis for the control of sediment loads from critical areas. Such areas could either be those that have high erosion rates, or those that, because of their location with respect to the watershed drainage system, have high sediment-delivery ratios. It may be advantageous to keep these two soil loss objectives separate if different incentive programs (cost sharing, set-aside payments, soil-loss taxes, etc.) are to be used for their adoption.

To assign T values more accurately, soil productivity must be defined more specifically. Early research (Hays and Clark, 1941; Stallings, 1957; Uhland, 1949) focused on the loss of favorable rooting depth as well as loss of plant nutrients, but, since the widespread use of chemical fertilizers in the late 1940s and early 1950s, the emphasis has been placed entirely on maintenance of rooting depth with suitable physical properties (Larson et al., 1983; Pierce et al., 1983). Although the value of plant nutrients lost by erosion may exceed the profits of many grain farmers (Langdale et al., 1979), and although it has been estimated (Larson et al., 1983) that the national loss of nitrogen, phosphorus, and potassium is $1 billion/year, most farmers do not see nutrient loss as sufficient reason to adopt soil conservation measures. Costs of structural measures, for example, may exceed the value of lost nutrients. If we assume that all factors of crop production are maximized and that the only factor limiting long-term crop productivity is an adequate rooting depth (Pierce et al., 1983), then the term productive soil potential rather than productivity is probably more appropriate. Larson et al. (1984) recently used this term in discussing the effects of various physical properties of soil on soil productivity. However, as recently pointed out by Lal (1984), the perspective of Third World subsistence farming is quite different. In these areas, fertilizer is difficult to obtain, its price is extremely high, so most crops rarely receive fertilizer. The subsistence farmer must rely on native fertility or other sources of nutrients, such as manures or legumes, and for him loss of nutrients by erosion results in reduced crop yields. In this instance, the concern is the loss of actual productivity rather than potential productivity. As Lal (1984) has shown, the higher concentrations of organic matter and nutrients in selectively eroded, fine particle-size fractions result in a loss of nutrients per ton of eroded sediment that is greater than the nutrients in an equivalent mass of the original soil. If soil loss is 10 t/ha and the phosphorus enrichment ratio of the sediment is 3.0, then the phosphorus loss is equivalent to a loss of 30 t/ha of topsoil. Lal points out that this latter figure should be used in determining soil loss tolerance if maintenance of inherent soil fertility is the major factor determining soil productivity.

11-2 RATES OF SOIL FORMATION

One parameter that must be used in determining an acceptable rate of soil loss from erosion is the rate at which the soil will rejuvenate or form. If the erosion rate exceeds that rate of rejuvenation or formation, the subsurface horizons or subsoil layers must be evaluated to determine if they will provide a suitable medium for plant growth. Only then can tolerable erosion rates be determined, and after that cultural practices for agricultural production can be selected.

Soil formation is the result of a complex interaction of vegetation and parent material, as conditioned by climate, acting on topography over time. Because each of these parameters can be considered an independent variable, the rate and type of soil formation is infinitely variable. Even within a small area or region, the rates of soil formation are highly variable.

Soil should not be equated only with the A horizons, as was done formerly when the concept of soil and subsoil (beneath the soil) prevailed. Although the primary rooting zone of many agricultural row crops is in the A horizon or upper 10 to 20 cm, the underlying horizons are important because of their limitations to rooting and water movement. With soil erosion, these underlying horizons become more significant because they become the rooting zone.

When evaluating the soil, all soil horizons and their physical and chemical differences must be considered. The horizons underlying the A are important because they differ from the A with respect to content of organic matter, available K and P levels, available water-holding capacity, bulk density, presence of toxic material (particularly aluminum), and other characteristics.

11-2.1   Soil Formation

The literature states that a certain thickness of soil develops in a given period of time. This suggests that, somehow, a finite amount of material is added to the surface of the soil. Though this process does occur on certain portions of the landscape or as the result of certain geologic processes, it is usually the result of deposition resulting from erosion from another portion of the landscape. Occasionally, it is the result of slow aeolian deposition. Rock weathering does occur, but the rates are highly variable. Furthermore, such rates decrease as the overlying soil develops and limits the amount of water moving downward to the bedrock (Colman, 1981). In most cases, rock weathering adds little soil material to the rooting zone. Soil-forming processes well above the rock weathering zone often create a partial barrier to rooting.

Soils do not form by simply starting at the surface and progressing downward through the original geological material. Instead, soil formation consists of many facets, with gains, losses, transformations, and translocations of different components taking place at different rates in different horizons (Simonson, 1959). The first step in soil formation is the accumulation of organic matter at the surface (Fig. 11-1). Simultaneously with or shortly after the accumulation of organic matter, soluble salts may be leached starting at the surface and proceeding downward. The rate and depth of leaching is determined largely by the quantity and distribution of rainfall. Once leaching has taken place, other processes, such as clay migration, occur throughout a relatively uniform depth (Fig. 11-2) (Stobbe and

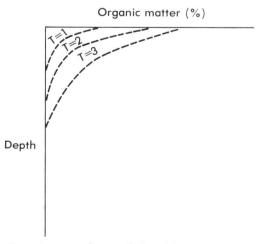

Fig. 11-1. General pattern of accumulation of the organic matter with time.

Clay content (%)

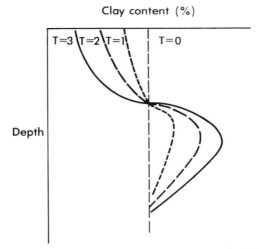

Fig. 11-2. General pattern of clay translocation and accumulation with time.

Wright, 1959). The degree of soil formation is the extent of departure in properties at various depths from the properties of the original material.

The early phases of soil formation, such as accumulation of organic matter and leaching of soluble salts, are considered desirable with respect to crop production. However, once the genetic processes advance beyond these initial stages, horizons less suitable for crop production may form. These less desirable horizons are important considerations in all crop production but are even more important when accelerated erosion brings them closer to the soil surface.

11-2.2   Diagnostic Horizons Important to Crop Production

Soil horizons categorized according to physical and chemical properties have been identified as diagnostic horizons (Soil Survey Staff, 1975). Other classification schemes use other terms for these surface and subsurface horizons, but there is nearly universal agreement that the horizons discussed below are significant in the differentiation of soils.

The kind of soil in a given location is a function of the five soil forming factors: parent material, climate, biota, relief, and time. The diagnostic horizons in a given soil are thus the result of soil processes dictated by these five factors.

Of the diagnostic horizons, some are considered to have a positive effect on crop production, many are thought to have a negative effect, and a few are thought to have no effect.

11-2.2.1   Desirable horizons

Table 11-2 lists those diagnostic horizons considered to enhance production of most crops under normal management. Historically, soils and

Table 11-2. Diagnostic horizons that enhance root growth.

Horizon	Location	Properties
Mollic	Surface	High organic matter content, base-rich, friable
Histic	Surface	Organic horizon, high water holding capacity, release of nutrients on mineralization
Cambic	Subsurface	Structural development, leaching of soluble salts

horizons rich in organic matter have been considered desirable for crop production. The mollic epipedon is a surface mineral horizon that is thick, high in organic matter, and relatively highly saturated with such bases as calcium, magnesium, and potassium. Approximately 25% of the soil area in the USA has a mollic epipedon as a surface horizon. Its inherent natural fertility resulted in the high production observed initially as U.S. croplands were brought into production.

The cambic horizon is a subsurface horizon in which leaching, structural development, and some mineral weathering has taken place. In situ materials prior to soil formation are often massive or laminated, which tends to restrict root growth. Formation of a cambic horizon is usually accompanied by formation of vertical and horizontal planes of weakness that are identified as soil structure. At least partial leaching of soluble salts is a prerequisite for a cambic horizon. The removal of salts of sodium and sometimes of calcium and magnesium results in a medium more suitable for root growth of most agricultural crops.

A histic epipedon is a surface organic horizon with properties dominated by organic matter, both decomposed and undecomposed. High water holding capacity, good tilth, and mineralization of the organic matter make this horizon a desirable rooting medium.

11-2.2.2 Limiting horizons

Table 11-3 lists the diagnostic horizons that may be considered limiting to root growth in many cases. The most common limiting horizon is the argillic, which is the result of accumulation of clay in the subsoil. This accumulation fills voids and pores and therefore provides a less permeable medium for root growth and water movement. The increased clay also increases shrink and swell characteristics. In sandy soils the argillic horizon may be considered desirable because it decreases the permeability and increases the plant-available water capacity. In some additional situations the argillic is considered an improved rooting medium compared with the original material.

A natric horizon has the same properties as the argillic, plus sodium on the exchange complex. Sodium tends to disperse the clay, which makes this horizon even less permeable than an argillic with the same amount of clay. The sodium also restricts root growth of many crops.

Spodic horizons are also horizons of accumulation that result in a decrease in porosity and permeability compared to the original material. Spodic horizons are often weakly to strongly cemented by amorphous organic matter and aluminum, with or without iron. In very sandy soils the

Table 11-3. Diagnostic horizons that can limit root growth.

Horizon	Location	Properties
Argillic	Subsurface	High clay content, low porosity and permeability
Natric	Subsurface	High clay content, low porosity and permeability, exchange sodium
Spodic	Subsurface	High in amorphous organic matter, aluminum and/or iron, low porosity and permeability
Oxic	Subsurface	Low CEC, absence of weatherable minerals, high iron oxide content
Duripan	Subsurface	Cemented by silica, nearly impermeable
Fragipan	Subsurface	Weak cementation, low porosity and permeability
Albic	Surface or subsurface	Low nutrient supply
Calcic	Subsurface	Secondary carbonate enrichment
Petrocalcic	Subsurface	Cemented by carbonates, nearly impermeable
Lithic contact†	Surface or subsurface	Bedrock

† An interface rather than a horizon.

spodic horizon may be desirable because of its higher exchange capabilities and its ability to retain moisture for crop growth.

In tropical soils the oxic horizon is associated with very old portions of the landscape and is identified by the absence of weatherable minerals and 2:1 clay minerals. As a result of intense weathering, the oxic horizon has a low exchange capacity and low inherent fertility. It is often high in iron which inhibits uptake of fertilizer supplied nutrients, particularly phosphorus. A desirable feature of the oxic is its usually well developed structure that enhances water movement and root growth.

Duripans and petrocalcic horizons are subsurface layers cemented by silica and carbonates, respectively. The cementation produces a zone almost impermeable to water and roots.

The fragipan is a weakly cemented subsurface horizon resulting from pedogenesis in humid regions. It is characterized by dense polygons more than 10 cm across with clay- and/or silt-rich vertical faces. Root growth and water movement take place in the space between the polygons.

The lithic contact, although not a diagnostic horizon, is the contact between the soil and underlying bedrock. The limits to water movement and root growth depend on the type and fracture pattern of the bedrock.

The way in which many earthen materials were deposited created undesirable zones for root growth. An example of one of these materials is glacial till. Many till deposits, emplaced directly by the ice, were compacted to some extent in the process. This compaction, combined with a wide range in size of particles, results in values for bulk density as high as 1.8 to 2.0 g/cc. This high density greatly restricts root growth and water movement. Alluviation is another example of a depositional process that sometimes leads to root-restricting horizons. As a result of variations in energy levels of the water, material deposited in floodplains often has strata of widely varying particle size, ranging from clay to gravel. Water movement and rooting habit tend to be restricted by this gross stratification. However, some of the best cropland in the world is on soils developed in alluvium.

11–2.3 Time for Horizon Formation

Diagnostic horizons require various times to form. The time required for a given horizon to form may also depend on the impact of the various soil-forming factors.

Using isotopic dating techniques, historic records, stratigraphic relationships, and other methods, researchers have determined typical amounts of time required for the formation of diagnostic horizons (Table 11–4). Most of these vlaues represent minimum times. Longer times may be required in other settings, depending on the impact of the various soil forming factors.

11–2.3.1 Mollic Epipedon

Incorporation of organic matter into the mineral surface is usually considered the first indication of soil formation. Thin A horizons enriched with organic matter have reportedly formed in as little as 30 to 50 years (Ugolini, 1968; Simonson, 1959). Ruhe et al. (1975) reported steady-state organic carbon levels at 30 years in Iowa. A study by Hallberg and associates (1978) showed that a mollic epipedon will form in loess under Iowa climatic conditions in about 100 years. The soil studied had 2.6% organic carbon in a 31-cm surface zone. In Oregon, Parsons et al. (1970) reported formation of a mollic epipedon in 120 years. These and other studies suggest that organic matter can accumulate very rapidly, particularly under grass vegetation. Horizons that can be considered A or A1 can form in tens of years, and under ideal conditions a mollic epipedon with a steady-state carbon level can be reached in a little more than 100 years.

Table 11–4. Typical rates of diagnostic horizon formation.

Horizon	Age	Location	Reference
	years		
A1	24–30	Alaska	Ugolini, 1968
OC-enriched†	30	Iowa	Ruhe et al., 1975
A	50	North Dakota	Simonson, 1959
Mollic	100	Iowa	Hallberg et al., 1978
Mollic	120	Oregon	Parsons et al., 1970
Cambic	200	Pennsylvania	Bilzi and Ciolkosz, 1977
Cambic	450	Pennsylvania	Cunningham et al., 1971
Cambic	550	Oregon	Balster and Parsons, 1968
Argillic	<2000	Iowa	Dietz and Ruhe, 1965
Argillic	2350	Oregon	Parsons and Herriman, 1976
Fragipn-like	2000	Pennsylvania	Bilzi and Ciolkosz, 1977
Bir	250	Alaska	Ugolini, 1968

† OC = Organic carbon.

11-2.3.2   Cambic Horizon

Identification of a cambic horizon is based on the leaching of soluble salts, mineral weathering, and the formation of structural units (peds) in a subsurface zone. As in other genetic horizons, the rate of formation depends on the soil forming factors. Ugolini (1968) identified a color B horizon that would probably qualify as cambic which had formed in only 55 years. Studies in Pennsylvania and Oregon (Bilzi and Ciolkosz, 1977; Cunningham et al., 1971; Balster and Parsons, 1968) showed that identifiable cambic horizons could develop in 200 to 500 years.

11-2.3.3   Argillic Horizon

The argillic horizon is one of the most important and extensive diagnostic subsurface horizons in the major crop production areas of the USA. It is estimated that about 40% of U.S. land area, or 3.7 million km^2, is underlain by argillic horizons.

Argillic horizons are generally believed to result from the translocation and accumulation of clay in subsurface horizons. Formation of a cambic horizon is assumed to be an intermediate step in the development of an argillic horizon.

Dietz and Ruhe (1965), although not designating it an argillic horizon, reported that a strong, textural B horizon was formed on a landscape no older than 2080 years. Work in Oregon by Parsons and Herriman (1976) and by Balster and Parsons (1968) showed that clay films and identifiable argillic horizons were formed in 2350 to 5250 years.

11-2.3.4   Fragipan

Dense, subsoil layers identified as fragipans are common in the eastern and southern sections of the USA. These pans, with surfaces 40 to 75 cm below the soil surface, are very restrictive to water movement and root penetration. Many of these pans are located in late Wisconsin-age material. Bilzi and Ciolkosz (1977) described a fragipan-like horizon in 2000-year-old alluvium in Pennsylvania. Thick, strongly expressed fragipan horizons occur in Mississippi soils that are between 5000 and 10 000 years old (Hall et al., 1982).

11-2.3.5   Other Diagnostic Horizons

Other diagnostic horizons have not been studied in great detail, so the time required for their formation is not well documented. Many spodic horizons in northeastern U.S. soils are formed in Late Wisconsin-age material and are thus assumed to have been formed in the past 10 000 to 15 000 years. Ugolini (1968) identified a Bir horizon in a soil that was approximately 250 years old. Although the Bir does not necessarily translate to a spodic horizon, it does have many similar properties.

11-2.4 Influence of Restrictive Horizons

In much of the crop producing area of the USA, the soil material originally deposited did not restrict root growth. Much of this same area now has horizons or zones in the subsoil that restrict roots and water. The amount of restriction depends on the type and degree of development of the restrictive layer. For example, the argillic horizon developed in the loessial materials of western Iowa is much less restrictive than the argillic horizon in loess of similar age overlying a paleosol in southern Ohio.

The tops of restrictive layers form tens of centimeters below the soil surface. As erosion or land shaping removes the soil surface, these restrictive layers are brought closer to the surface and, therefore, affect crop growth more. Problems associated with plant growth in these diagnostic horizons as they become incorporated in the plow zone are discussed by Larson et al. (1984).

11-3 PRESENT POLICY

In 1978, an SCS Soil Loss Tolerance Committee was established to produce a draft revision in the concept of T factors. The draft was reviewed by SCS state office personnel, who also solicited reviews by other knowledgeable people in their states. Two changes were proposed. The first proposed change was to remove sediment damage as a consideration in establishing T values, because an erosion rate derived from the USLE does not represent sediment yield, although the two are related. The amount of sediment that reaches a stream is more closely related to the proximity of the field to the stream, the size of sediment particles, sediment entrapped along the way, and the size and other characteristics of the watershed. Also, sediment damage is more related to water use and land use in the area receiving the sediment than to soil loss tolerance. For example, a bottomland farm may tolerate large amounts of sediment in the nongrowing season. In fact, the productivity of some soils is enhanced by annual sediment deposition. However, even a small amount of sediment in a trout stream can be harmful. This proposed change met with almost unanimous agreement, and it has been incorporated in the SCS National Soils Handbook.

The other proposed change was more controversial: to raise the soil loss tolerance for certain soils having a thick, favorable root zone. A maximum annual loss of 22.4 t/ha (10 tons/acre) was suggested for soil with 203 cm (80 in.) or more of favorable rooting depth. Progressively lower tolerances would be assigned as the favorable root zone becomes thinner (Fig. 11-3). For example, if no soil renewal is assumed, a soil with a favorable rooting depth of about 230 cm (90 in.) would be reduced to 100 cm (40 in.) in 1000 years and to 25 cm (10 in.) in 2500 years. Assuming an annual soil renewal rate of about 1 t/ha (0.5 ton/acre), however, the 230 cm (90 in.) of soil depth would decrease to 25 cm (10 in.) in approximately 3000 years. The soil loss tolerance at this point would be set at 2.2 t/ha (1 ton/acre).

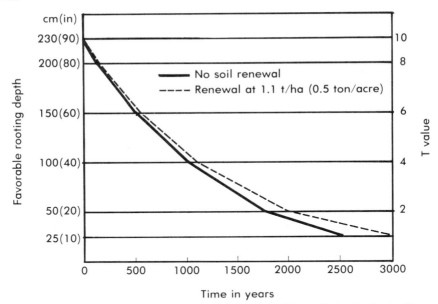

Fig. 11-3. Illustration of a proposal to vary the T value of soil according to the depth of favorable soil remaining (Young, 1980).

This proposed change received mixed comments. Most respondents accepted the idea of an increased T value. Of those that opposed it, however, most were strongly opposed. The SCS has concluded that the maximum soil loss tolerance should remain at 11.2 t/ha (5 tons/acre). Conclusive evidence has not been assembled to indicate that long-term productivity can be maintained at a higher rate of erosion. SCS personnel have been asked to encourage scientists at experiment stations to assemble current research results and initiate new research on this question. The SCS Soil Loss Tolerance Committee has also proposed several symposia on soil loss tolerance.

11-4 TESTING NEW IDEAS

Recent reviews of soil loss tolerance have recognized that the 2 to 11 t/ha annual soil loss range used in the USA does not maintain absolute solum thickness, since estimates of annual rock weathering are 1 t/ha or less. The allowable rates of erosion, however, assume that an A horizon can develop with good management at a much faster rate (perhaps as fast as 11 t/ha/yr), and that an A horizon depth will be maintained as long as subsoil characteristics are not unfavorable to A horizon development. The presence of unfavorable subsoil material at shallow depth has, therefore, been the main criterion for establishing T values less than the maximum of 11 t/ha. While this criterion has provided a reasonable goal for soil conservation planning, several authors (McCormack et al., 1982; Larson et al., 1983) have indicated the need to develop specific soil parameters that can quantify the reduction in soil productivity resulting from long-term erosion.

McCormack et al. (1982) identified soil rooting depth, topsoil thickness, available water holding capacity, plant nutrition storage, surface runoff, soil tilth, and soil organic matter content as soil and site characteristics important in determining potential soil productivity. Pierce et al. (1983) cite the work of Neill (1979), who considered available water holding capacity, bulk density, aeration, pH, and electrical conductivity as the parameters most important for root growth. Pierce et al. (1983), based on the approach of Neill (1979), used bulk density, available water capacity, and pH to develop a model for quantifying losses in soil productivity resulting from erosion. They used water permeability rates to adjust the bulk density factor to account for continuity of pores and resistance to root penetration.

Soil loss tolerance is now established at the soil series level on the basis of pedon characteristics. However, many of the parameters previously discussed are not readily available from profile descriptions or from routine characterization in a soil-survey laboratory. Grossman and Berdanier (1982) proposed a procedure for assigning T values on the basis of taxonomic features indicative of root growth limitation, including horizons with undesirable physical properties and horizons high in extractable aluminum and low in calcium, or both.

While T values are currently assigned to soil series or their eroded phases, soils are spatially distributed in the field by mapping unit, and agricultural land is often uniformly farmed on fields that may contain several mapping units. The soil distribution may represent toposequences where erosion is occurring on one or more units and the sediment is being deposited on other units. Langdale et al. (1979), measuring corn yields on eroded and depositional areas within a watershed in the Southern Piedmont, found that yields were higher in the depositional areas than on the eroded areas of the watershed, with the difference in yield being a function of depth to the argillic horizon. They also found no yield increase due to soil deposition that would compensate for the yield reduction due to erosion. This finding means that the overall productivity of the watershed should continue to decline with continued erosion. In such situations, a farm field must be managed to control erosion on the eroding area if crop production is to be maintained.

Langdale et al. (1979) also found that differences in corn yields between eroded and depositional areas of a watershed were affected by rainfall distribution. This finding means that soil loss tolerance values determined for soil series on the basis of available water holding capacity must be adjusted for regional differences in rainfall distribution and specific crop requirements. If eroded soils are assumed to have a greater susceptibility to drought stress than noneroded soils, local adjustments in the T value could be based on an analysis of drought frequency for the area. Such adjustments might be accomplished with the new Erosion-Productivity Impact Calculator (EPIC) (Williams, 1984).

The rate at which soil productivity declines with continued erosion varies considerably among soils. Soils that are already severely eroded, or that are shallow to bedrock or to some other undesirable substratum, decline in productivity more rapidly than soils that are thicker or whose substrata do not greatly restrict the development of A horizon characteristics under good management (McCormack et al., 1982; Larson et al., 1983). Soil productivity, rather than potential soil productivity, declines more rapidly with Third World farming systems, which cannot rely on continued use of fertilizers. Pierce et al. (1983) define the rate of soil productivity decline with continued erosion as soil *vulnerability*. Categorizing soils on the basis of their relative vulnerabilities provides a mechanism for short-term allocation of soil conservation funding. Soil vulnerabilities established for soil series should be adjusted for localized effects on soil productivity of climate, cropping patterns, and other factors.

11-5 FUTURE CRITERIA

Several authors (Logan, 1982; McCormack et al., 1982; Larson et al., 1983; Grossman and Berdanier, 1982) have recognized the need to establish an upper limit to allowable soil loss as a reflection of a national commitment to preserving our soil resources and maintaining long-term soil productivity. An upper limit of 11 t/ha is generally accepted, since it approximates the maximum rate of A horizon development under optimum conditions. Future research on the factors affecting soil productivity may allow some refinement of the rate of allowable soil loss. As parameters are identified that best describe the loss of soil productivity with erosion, they can be used to set T values for individual soil series. As Grossman and Berdanier (1982) have suggested, these decisions should be made by a national team of research scientists most familiar with the research and with the soil survey data base. A detailed set of standard instructions should then be provided to state and local SCS staff to adjust the factors for local conditions. These conditions should include frequency of drought stress and desirable soil characteristics for specific crops. Soil loss limits to meet objectives for watershed sediment loads should be independent of such T values and should be determined by a protocol different than that used to determine T values.

11-6 RESEARCH NEEDS

Further extensive field study of effects of erosion on crop yields is needed. McCormack and Moorman (1984) recently proposed that yield plots be established throughout the USA on eroded and noneroded areas of the same soil. Such pairings would eliminate many factors affecting crop yield, such as local climate, soil and crop management, insect and weed problems. The standardized experimental design would be replicated on major soils, in major regions, and with the important economic crops of the region. Various parameters would be measured and used in the calibration and testing of models such as the root-growth models of Larson et al. (1983) and Pierce et al. (1983), and the EPIC model of Williams (1984). Measured parameters would be correlated with taxonomic data to aid the use of soil survey information in setting soil loss tolerances. Field plot research should also investigate the ability of various soil management practices to improve the productivity of eroded soils. More reliable data are needed on the development of A horizon characteristics under a wide range of conditions and on the costs of soil renovation. A nationwide network of paired erosion plots should be immediately established, with funding adequate for their continued maintenance. Data obtained from these studies should be periodically published in report and computer formats.

REFERENCES

Balster, C. A., and R. B. Parsons. 1968. Geomorphology and soils, Willamette Valley, Oregon. Oregon State Agric. Exp. Stn. Spec. Rep. 265:8–22.

Bennett, H. H. 1939. Soil conservation. McGraw-Hill, Inc., New York.

Bilzi, A. F., and E. J. Ciolkosz. 1977. Time as a factor in the genesis of four soils developed in recent alluvium in Pennsylvania. Soil Sci. Soc. Am. J. 41:122–127.

Browning, G. M., G. L. Parish, and John Glass. 1947. A method for determining the use and limitation of rotation and conservation practices in the control of soil erosion in Iowa. J. Am. Soc. Agron. 39:65–73.

Colman, S. M. 1981. Rock-weathering rates as a function of time. Quat. Res. 15:250–264.

Cunningham, R. L., E. J. Ciolkosz, R. P. Matelski, G. W. Peterson, and R. W. Ranney. 1971. Characteristics, interpretations and uses of Pennsylvania soils: Armstrong county. Pennsylvania Agric. Exp. Stn. Prog. Rep. 316.

Dietz, W., and R. V. Ruhe. 1965. Soil development in less than 2,000 years in Iowa. Agron. Abstr. American Society of Agronomy, Madison, WI. p. 104.

Grossman, R. B., and C. R. Berdanier. 1982. Erosion tolerance for cropland: Application of the soil survey data base. p. 113–130. In B. L. Schmidt et al. (ed.) Determinants of soil loss tolerance. Spec. Pub. 45. American Society of Agronomy, Madison, WI.

Hall, G. F., R. B. Daniels, and J. E. Foss. 1982. Rates of soil formation and renewal in the U.S.A. p. 23–29. In B. L. Schmidt et al. (ed.) Determinants of soil loss tolerance. Spec. Pub. 45. American Society of Agronomy, Madison, WI.

Hays, D. E., and N. Clark. 1941. Cropping systems that help control erosion. Bull. 452, State Soil Conserv. Comm. in cooperation with USDA-SCS and the Wisconsin Agric. Exp. Stn., University of Wisconsin-Madison.

Hallberg, G. R., N. C. Wollenhaupt, and G. A. Miller. 1978. A century of soil development in spoil derived from loess in Iowa. Soil Sci. Soc. Am. J. 42:339–343.

Klingebiel, A. A. 1961. Soil factor and soil loss tolerance. In Soil loss prediction, North and South Dakota, Nebraska, and Kansas. USDA-SCS, Lincoln, NE.

Lal, R. 1984. Soil erosion and crop productivity relationship for a tropical soil. In Proc. Int. Conf. on Soil Erosion and Conserv., Honolulu, HI. Jan. 1983. Soil Conservation Society of America, Ankeny, IA.

Langdale, G. W., J. E. Box, Jr., R. A. Leonard, A. P. Barnett, and W. G. Fleming. 1979. Corn yield reduction on eroded Southern Piedmont soils. J. Soil Water Conserv. 34:226–228.

Larson, W. E., F. J. Pierce, and R. H. Dowdy. 1983. The threat of soil erosion to long-term crop production. Science 219:458–465.

––––, ––––, and ––––. 1984. Loss in soil productivity from erosion in the United States. In Proc. Int. Conf. on Soil Erosion and Conserv., Honolulu, HI. Jan. 1983. Soil Conservation Society of America, Ankeny, IA.

Logan, T. J. 1977. Establishing soil loss and sediment yield limits for agricultural land. p. 59–69. In Soil erosion and sedimentation. Proc. Natl. Symp. Soil Erosion and Sedimentation by Water. American Society of Agricultural Engineers, St. Joseph, MI.

––––. 1982. Improved criteria for developing soil loss tolerance levels for cropland. p. 131–139. In B. L. Schmidt et al. (ed.) Determinants of soil loss tolerance. Spec. Publ. 45. American Society of Agronomy, Madison, WI.

McCormack, D. E., K. K. Young, and L. W. Kimberlin. 1982. Current criteria for determining soil loss tolerance. p. 95–111. *In* B. L. Schmidt et al. (ed.) Determinants of soil loss tolerance. American Society of Agronomy, Madison, WI.

————, and F. R. Moorman. 1984. Sequential testing—within field data collection to determine erosion impacts. *In* Proc. Int. Conf. on Soil Erosion and Conserv., Honolulu, HI. Jan. 1983. Soil Conservation Society of America, Ankeny, IA.

Neill, L. L. 1979. An evaluation of soil productivity based on root growth and water depletion. M.S. thesis. University of Missouri, Columbia.

Parsons, R. B., C. A. Balster, and A. O. Neas. 1970. Soil development and geomorphic surfaces, Willamette Valley, Oregon. Soil Sci. Soc. Am. Proc. 34:485–491.

————, and R. C. Herriman. 1976. Geomorphic surfaces and soil development in the Upper Rogue River Valley, Oregon. Soil Sci. Soc. Am. J. 40:933–938.

Pierce, F. J., W. E. Larson, R. H. Dowdy, and W. A. P. Graham. 1983. Productivity of soils: Assessing long-term changes due to erosion. J. Soil Water Conserv. 38:39–44.

Ruhe, R. V., T. E. Fenton, and L. L. Ledesma. 1975. Missouri River history, floodplain construction, and soil formation in southwestern Iowa. Iowa Agric. Home Econ. Exp. Stn. Res. Bull. 580:738–791.

Simonson, Roy W. 1959. Outline of a generalized theory of soil genesis. Soil Sci. Soc. Am. Proc. 23:152–156.

Smith, D. D. 1941. Interpretation of soil conservation data for field use. Agric. Eng. 22:173–175.

————, and D. M. Whitt. 1948. Evaluating soil losses from field areas. Agric. Eng. 19:394–396, 398.

Soil Survey Staff. 1975. Soil taxonomy: a basic system of soil classification for making and interpreting soil surveys. USDA Agriculture Handb. 426. U.S. Government Printing Office, Washington, DC.

Stallings, J. H. 1957. Soil conservation. Prentice Hall, Englewood Cliffs, NJ.

Stobbe, P. C., and J. R. Wright. 1959. Modern concept of the genesis of podzols. Soil Sci. Soc. Am. Proc. 23:161–164.

Ugolini, F. C. 1968. Soil development and alder invasion in a recently deglaciated area of Glacier Bay, Alaska. *In* J. M. Trappe et al. (ed.) Biology of alder. USDA, Forest Service, Pacific Northwest Forest and Range Exp. Stn., Portland, OR.

Uhland, R. E. 1949. Crop yields lowered by erosion. USDA–SCS–TP–75. Washington, DC.

USDA–SCS. 1973. Advisory notice, Soils-6. Washington, DC.

Williams, J. R. 1983. The physical components of the EPIC model. *In* Proc. Int. Conf. on Soil Erosion and Conserv., Honolulu, HI. Jan. 1983. Soil Conservation Society of America, Ankeny, IA.

Wischmeier, W. H., and D. D. Smith. 1978. Predicting rainfall erosion losses—a guide to conservation planning. USDA Agriculture Handb. 537. Washington, DC.

Young, K. K. 1980. The impact of erosion on the productivity of soils in the United States. p. 295–303. *In* M. DeBoodt and D. Gabriels (ed.) Assessment of erosion. John Wiley and Sons, New York.

12 Effects of Soil Erosion on Soil Properties as Related to Crop Productivity and Classification

W. E. Larson
University of Minnesota
St. Paul, Minnesota

T. E. Fenton
Iowa State University
Ames, Iowa

E. L. Skidmore
Agricultural Research Service
U.S. Department of Agriculture
Manhattan, Kansas

C. M. Benbrook
National Academy of Sciences
Washington, DC

Soil erosion is the detachment and movement by wind or water of soil particles from their place of origin. Soil productivity is the capacity of a soil, in its normal environment, to produce a particular plant or sequence of plants under a specified management system. Productivity is generally expressed in terms of yields.

Two recent committee reports (The National Soil Erosion-Soil Productivity Research Planning Committee, 1981; Council for Agricultural Science and Technology, 1982) have discussed in detail the soil erosion and soil productivity problem. Their discussions will not be reported here.

Whether it is called the topsoil, surface soil, A horizon, or epipedon, the part of the soil affected first by erosion is the surface few centimeters, which is generally highest in organic matter and plant nutrients.

Published in R. F. Follett and B. A. Stewart, ed. 1985. *Soil Erosion and Crop Productivity.*
© ASA-CSSA-SSSA, 677 South Segoe Road, Madison, WI 53711, USA.

This article reviews the effects of erosion on soil properties, how these changes in properties can cause problems in classification and mapping, the effects of erosion on hydraulic properties, and the relation of soil properties to productive potential.

12-1 DESCRIPTION OF SOILS

Twelve soil profiles representing eight soil series were selected from the Soil Conservation Service's current listing of benchmark soils of the USA for discussion here. They were selected to provide a range of soils used for row-crop production and subject to water erosion. The soils and some selected information are listed in Table 12-1.

12-1.1 Mollisols

A property common to all Mollisols is the presence of a relatively thick, dark-colored, humus-rich horizon or horizons in which bivalent cations are dominant on the exchange complex. In soils with sola that are 76 cm or greater thick, the mollic epipedon must be 25 cm or more thick to be classified as a Mollisol. Many sloping Mollisols used for intensive row-crop production have dark-colored surface horizons which erosion has decreased in thickness. The Seymour and Marshall profiles discussed in this paper illustrate the problem associated with these soils (Table 12-2).

Table 12-1. Classification, erosion factors, and organic matter contnet of selected soil series.†

Soil series	Classification	Erosion factors		Organic matter
		K	T	
				%
Mollisols				
Seymour	Fine, montmorillonitic, mesic, Aquic Argiudoll (3 profiles)	0.37	3	2–4
Marshall	Fine-silty, mixed, mesic, Typic Hapludoll (3 profiles)	0.32–0.43	5	3–4
Tillman	Fine, mixed, thermic, Typic Paleustoll	0.32	5	1–3
Alfisols				
Fayette	Fine-silty, mixed mesic, Typic Hapludalf	0.37	4 or 5	1–2
Grenada	Fine-silty, mixed, thermic, Glossic Fragiudalf	0.37–0.43	3	0.5–2
Ultisols				
Cecil	Clayey, kaolinitic, thermic, Typic Hapludult	0.15–0.28	4	0.5–2
Tifton	Fine-loamy, siliceous, thermic Plenthic Paleudult	0.05–0.24	4	1–2
Fullerton	Clayey, kaolinitic, thermic Typic Paleudult	0.20–0.28	5	0.5–2

† Data from Soil Interpretation Record of named series, Soil Conservation Service.

Table 12–2. Selected properties of slightly (s.e.) and moderately eroded (m.e.) Seymour and Marshall soils.†

Soil	Color 10YR moist value/chroma	Color 10YR dry value/chroma	O.C.‡ %	Depth cm	Clay max. %	D.B.§ Mg/m³	Clay max. %	O.C. >0.58 cm	D.B. ≥1.5 cm	pH ≥6.0 cm	B/A/ clay ratio	% of max C %
Seymour												
(93-6) s.e.	3/1	5/1	2.06	28.0	22.5	1.33	65.0	71.0	114.0	91.0	2.27	51.0
(93-5) m.e.	3/1	5/1	1.68	15.0	33.0	1.37	29.0	36.0	81.0	64.0	1.46	48.1
(93-4) m.e.	3/1	5.4/1	1.77	15.0	27.6	1.36	39.0	25.0	48.0	81.0	1.63	45.1
Marshall												
(15-1) s.e.	2/1	4/1	2.46	58.0	28.6	1.34	65.0	91.0	180.0	71.0	1.19	34.0
(15-2) s.e.	2/1-2/2	5/2	2.14	46.0	30.9	1.42	39.0	66.0	180.0	33.0	1.07	33.2
(15-3) m.e	2/2-3/2	5/2	2.05	24.0	31.7	1.39	20.0	46.0	180.0	2.5	1.05	33.3

Column group headers: Surface horizon (Color 10YR moist/dry, O.C.‡, Depth, Clay max., D.B.§); Subsurface horizon — Depth to: (Clay max., O.C. >0.58, D.B. ≥1.5, pH ≥6.0), B/A/ clay ratio, % of max C.

† Data from Soil Survey Investigation Report 31. USDA–SCS.
‡ Organic carbon.
§ Bulk density.

12–1.1.1 Seymour Series

The Seymour soils were sampled in a transect within a total distance of 46 m on 2 to 5% slope gradients. Profile 93-6 was on an interfluve summit, 93-4 was on a sideslope, and 93-5 was on a less stable interfluve summit. Profile 93-6 classifies as an Aquic Argiudoll. The other profiles have a mollic epipedon that is 15 cm thick and qualify as Udollic Ochraqualfs rather than Mollisols (USDA, 1975). In soil-survey field operations, they are correlated as taxadjuncts of the Seymour series. Trends shown in Table 12–2 for the eroded Seymour profiles compared to the slightly eroded profile are decreasing thickness of mollic epipedon, decreasing organic carbon content in the surface horizon and profile, increasing clay content in the surface horizon, greater bulk density, shallower depth to the clay maximum, shallower depth to less than 0.58% organic carbon, and bulk density equal to or greater than 1.5 Mg/m³. The maximum clay percentage in the profile also decreases. However, the thickness of the zone of 40% or more clay is quite similar—43 cm for 93-6 and 48 cm for 93-4 and 93-5. The argillic horizons in these soils are considered limiting to plant root growth.

12–1.1.2 Marshall Series

Marshall soils were also sampled in a transect approximately 100 m long. The data in Table 12–2 for these soils have trends similar to those for Seymour soils. The three Marshall profiles qualify as Mollisols, but profile 15-3 has the minimum thickness of mollic epipedon required, 25 cm. All have a cambic horizon. Marshall soils have more organic carbon, lower bulk densities in the subsurface horizons, and better natural drainage than Seymour soils.

12-1.1.3   Tillman Series

This extensive series occurs in Texas and Oklahoma. These are deep, well-drained, slowly permeable soils formed in clayey alluvium derived from redbed clays and shales. Slopes range from 0 to 5%. The A horizon is silty clay loam or clay loam. The Bt horizon is clay or clay loam ranging from 35 to 55% clay. The subsoil extends to depths below 200 cm. Calcium carbonate in some form is usually present in the lower part of the B horizon.

12-1.2   Alfisols

12-1.2.1   Fayette Series

Fayette soils, formed under forest vegetation, have a thin (15 cm or less), dark-colored, surface horizon underlain by a lighter-colored E horizon. Before cultivation, a thin zone of partially decomposed organic material is commonly present on the surface. Cultivation mixes these materials with the surface and E horizon, resulting in an Ap horizon with composite properties. Table 12-3 documents some of these changes. Compared to the moderately eroded sites, the color of the surface horizons of the slightly eroded profiles is darker, an E horizon is present, depth to the argillic (Bt) horizon is greater, clay content in the upper 33 to 38 cm is less, and organic-carbon content is higher.

Table 12-3. Characteristics of eroded phases of Fayette soils, 5 to 9% slopes.

Location	Erosion	Horizon	Thick-ness	Color 10YR	Clay	Organic carbon
			cm	value/chroma	%	
Illinois†	slight	A1	15.2	3/2	14.9	1.62
		E1	7.6	4/2	17.1	0.68
		E2	10.2	4/3–5/3	22.7	0.44
		Bt1	7.6	4/3–5/3	27.3	0.34
		Bt2	15.2	4/4	31.7	0.27
Iowa‡	slight	A1	10.2	3/2	14.1	2.70
(13689–		E	15.2	4/2	15.9	0.53
13692)		Bt1	12.7	4/3	23.1	0.30
		Bt2	12.7	4/4	30.0	0.19
Iowa‡	moderate	Ap	17.8	3/3–4/3	24.5	0.88
(13693–		Bt1	10.2	4/4	31.4	0.31
13696)		Bt2	7.6	4/4	31.9	0.25
		Bt3	15.2	4/4	31.1	0.21
Iowa‡	moderate	Ap	17.8	3/1 & 4/2	24.2	0.84
(16828)						
Iowa‡	severe	Ap	15.2	4/3	29.9	0.62
(16827)						

† Data from Soil Survey Investigation Report 19. USDA–SCS.
‡ Data from Soil Survey Investigation Report 31. USDA–SCS.

12–1.2.2   Grenada Series

Grenada soils formed in loess and have a fragipan at a depth of about 60 cm. As indicated by the classification in Table 12-1, the texture of these soils is similar to Fayette soils, but these soils differ by having a fragipan and a thermic temperature regime. Similar trends to those listed for Fayette soils are expected as the degree of erosion increases. However, rooting volume becomes more limiting as erosion decreases the depth to the fragipan.

Accelerated erosion does not affect the taxonomic classification of Fayette or Grenada soils since the A and E horizons are not diagnostic horizons for Alfisols.

12–1.3   Ultisols

As a group these soils have a horizon or horizons with an accumulation of translocated silicate clays but few bases. The release of bases by weathering is usually equal to or less than the removal by leaching. Most of the bases are held in the vegetation and the upper 20 to 30 cm of the soil. Base saturation normally decreases with depth and is less than 35% (by sum of cations) at specified depths, depending on the sequence of horizons present in a profile. Also, in an uneroded state, most Ultisols have an E horizon.

12–1.3.1   Cecil Series

These soils are extensive in the southeastern USA. The major limiting factor for crop production is the low base status throughout the profile. The clayey Bt horizon is not considered a major limiting factor for root penetration and water movement. The A and E horizons typically are sandy loam, fine sandy loam, or loam or their gravelly counterparts. Eroded phases are sandy clay loam or clay loam. Except for the severely eroded units, these soils respond well to good management. Depth to bedrock is 180 cm or more.

12–1.3.2   Tifton Series

These extensive soils occur in Alabama, Florida, Georgia, and South Carolina. The siliceous mineralogy class indicates the presence of more than 90% by weight of silica minerals (quartz, chalcedony, or opal) and other extremely durable minerals resistant to weathering. Ironstone nodules are present in most horizons. Clay content of the upper part of the Bt horizon averaged 20 to 35%. Plinthite makes up 5 to 15% of the lower part of the subsoil. Low organic matter and low fertility limit crop production. Areas used for several years for row-crop production may develop pressure pans at the bottom of the plow layer.

12-1.3.3   Fullerton Series

This series occurs widely in Alabama, Georgia, and Tennessee. They are cherty soils formed in residuum weathered from dolomite. Chert content (by volume) of each horizon ranges from 15 to 35%. Slopes range from 2 to 45%. The horizons, except where limed, are strongly or very strongly acid. Clay content in the argillic (Bt) horizon ranges from 40 to 75% and base saturation is low, generally less than 20%. Texture of the surface horizon is cherty silt loam, cherty loam, or cherty fine sandy loam except for severely eroded phases which may be cherty silty clay loam or finer in texture.

12-2 EFFECT OF SOIL EROSION ON SOIL CLASSIFICATION AND MAPPING

In soil mapping, the effect of erosion on the epipedon is described in terms of erosion classes. In the Iowa Cooperative Soil Survey program, erosion classes are defined quantitatively:

Erosion Class 1—None or slight erosion.
   Little or no mixing of the subsoil with the plow layer. The plow layer consists mainly of the A horizon or A + E horizons. Dark-colored material is greater than 18 cm thick.
Erosion Class 2—Moderate erosion.
   Only 7 to 18 cm of A or (A + E) horizon remaining. Some of the B or AB are mixed with the plow layer.
Erosion Class 3—Severe erosion.
   Less than 7 cm of A or (A + E) horizon remaining. Most of the plow layer is B (or AB) horizon.

In this system, soil properties can be used to estimate the degree of accelerated erosion and the amount of A horizon that has been removed.

Erosion classes can be mapped accurately. Based on 240 randomly selected profiles of three soil series, Wilding et al. (1965) reported that the series was mapped accurately 42% of the time and erosion class 94% of the time. Dideriksen (1966) summarized data from the Conservation Needs Inventory Statistical Sample utilized in a corn yield study in Iowa. Study sites were drawn randomly from the 2% statistical sample used in the Conservation Needs Study. At each sample site, a detailed description of the soil was made, the delineation of the area by soil-mapping unit was recorded, and the slope was measured at the point. These data were then compared with data previously mapped for that point. In general, the mapping unit delineations were most accurate in expressing slope group, erosion class, and soil series, in that order (Table 12-4). In this study the soil series over all slope groups was mapped accurately from 63 to 83% of the time, slope from 83 to 100%, and erosion class from 79 to 100%.

Table 12-4. Accuracy of mapping soil series, soil slope, and soil erosion classes in Iowa (Dideriksen, 1966).†

Soil	Slope group	Average percent correct		
		Series	Slope	Erosion class
Ida	5–9%	91	100	91
(Typic	9–14%	80	70	60
Udorthent)	14–20%	71	100	86
	weighted mean	83	90	79
Monona	0–2%	66	100	100
(Typic	2–5%	60	100	90
Hapludoll)	5–9%	69	100	69
	9–14%	100	100	94
	14–20%	57	86	71
	weighted mean	76	98	84
Marshall	0–2%	60	80	100
(Typic	2–5%	83	66	83
Hapludoll)	5–9%	79	100	71
	9–14%	70	80	80
	weighted mean	75	83	81
Sharpsburg	0–2%	100	100	100
(Typic	2–5%	63	100	78
Argiudoll)	5–9%	55	90	73
	weighted mean	63	100	81
Tama	0–2%	100	100	100
(Typic	2–5%	100	100	100
Argiudoll)	5–9%	100	100	75
	weighted mean	100	100	92
Shelby	5–9%	100	100	100
(Typic	9–14%	77	89	100
Argiudoll)	14–18%	50	100	100
	weighted mean	75	92	100

† Based on 161 profile descriptions: Ida, 29; Monona, 49; Marshall, 41; Sharpsburg, 22; Tama, 8; and Shelby, 12.

The effect of accelerated erosion on Mollisols is a major problem in soil classification. The criteria for classificaiton at the highest category, the order level, is linked directly to surface-soil thickness (mollic epipedon). Smith (1978, p. 13) stated:

> In general, we tried throughout taxonomy to use the characteristics of the subsurface horizon rather than the surface horizon because we wanted to keep the eroded and uneroded soils in the same series, as has been our practice in mapping. The use of the mollic epipedon as a diagnostic horizon violated the general principles that we started with, but we could find no escape from it.

In soils with sola thicker than 75 cm, the minimum thickness of the mollic epipedon for the soil to be classified as a Mollisol is 25 cm. Failure to meet the thickness criterion for a mollic epipedon results in a classification of Mollic Hapludalf, if the soils are well drained and have an argillic horizon. Without an argillic horizon but with a cambic horizon, the soils would be classified as Inceptisols. Because of the emphasis given to the mollic epipedon, a slight change in thickness can result in a change at the highest level in *Soil Taxonomy* (USDA, 1975). The Mollisols and associated mollic subgroups are the only groups of soils where accelerated erosion results in such a dramatic change in classificaiton.

The only other order that depends on a diagnostic epipedon is the Histosols. Because of the unique characteristics of Histosols, accelerated erosion is generally not a problem. Other orders are defined in terms of diagnostic subsurface horizons which are not commonly removed completely by accelerated erosion. The major classification-related change resulting from accelerated erosion in these soils is at the phase level, where the textural class may change, for example, from silt loam on the uneroded phase to silty clay loam on the eroded phases.

While acclerated erosion may have different effects on the formal classification of different soils, some changes appear to be universal among eroded soils. Erosion usually lowers soil fertility, but the permanent loss in productivity following erosion usually results from poorer tilth with associated reduced infiltration rates, soil crusting, increased power requirements, poorer stands, and decreased water-holding capacities. In general, soils with less desirable subsoils (that is, with limiting rooting volumes) or less desirable substrata are more adversely affected by erosion than soils with optimal subsoil or substrata. The changes listed above are consistent with the changes in soil properties discussed for the Seymour, Marshall, and Fayette soils.

Langdale and Shrader (1982) discussed the problem of obtaining reliable crop-yield data on eroded soils compared to noneroded soils. Technology advances have masked soil-productivity decline due to erosion and have rendered of little value the extensive soil-erosion and soil-productivity research completed between 1935 and 1950. Technology has also affected the design of mapping units in soil surveys. Boundaries on soil maps are justified on the basis of differences in use, management, and behavior. Technology has reduced the significant statements that could be made to differentiate eroded and uneroded units. Initially, the problem was one of correlation, that is, eroded units shown on the field sheets were combined with uneroded units or vice versa and were not shown on the published soil maps. Mapping-unit descriptions were broadened to treat the unit combined as an inclusion. However, the eroded and uneroded units were identified on the field sheets. Soil scientists soon lose interest in making separations that are not published, so gradually, erosion differences in many cases were not identified on the field sheets. These trends are based on the authors' experience in the Midwest and in a report by Ditzler (1981, p. 15), a soil survey party leader from Tennessee, who stated ". . .the trend in many of the more recent surveys has been to map slope phases and not try to map out eroded phases."

This period of evolving technology also coincides with the evolution and application of the soil classification system, published as *Soil Taxono-*

my (USDA, 1975). This system stresses the classification of soils based on a set of observed properties. A letter of transmittal attached to USDA-SCS National Soils Handbook Notice 63 dated 11 September 1980 stated:

> Eroded phases are used only if accelerated erosion has not caused a change in the taxonomic class. If the taxonomic class has changed, the map unit name is based on the name of the new taxon and is not considered to be an eroded phase of that new taxon. Naming the new taxon as a taxadjunct of an eroded phase of the original soil is not appropriate because taxadjuncts imply similar use, management, and behavior.

Strict application of this procedure eliminates the identification of erosion classes for many soils, especially in the Mollisols order and Mollic great groups.

For example, Tama, Downs, and Fayette soils form a biosequence. Tama is classified as a Typic Argiudoll, Downs as a Mollic Hapludalf, and Fayette as a Typic Hapludalf. Classification criteria require that Tama soils have 25 cm or more of surface horizon that qualifies as a mollic epipedon. Downs soils have 18 to 25 cm of mollic material, and Fayette soils have 18 cm or less of mollic-like material at the surface. Strict application of this criteria results in moderately eroded Tama being classified as slightly eroded Downs, and severely eroded Tama as Fayette. Also accelerated erosion of Downs results in a classification as Fayette. In many survey areas, these rules have not been enforced, and erosion phases have been correlated as taxadjuncts of the series.

As pointed out by Ditzler (1981), the use of eroded phases in designing mapping units seems about to complete a full cycle. Because of the current interest in the National Resources Inventory and a renewed interest in the effects of soil erosion on productivity, there is a renewed effort to document the extent of soil erosion in published soil surveys.

The above discussion serves as a warning to those researchers who plan to conduct studies relating soil productivity to named soils and erosion phases. A qualified soil scientist should aid in an on-site evaluation of the research area. This evaluation will preclude complications arising from the soil classification and erosion problems discussed in this paper. In addition, a natural variability associated with soil map units may influence research results (Beckett and Webster, 1971).

12-3 EFFECTS OF SOIL EROSION ON HYDRAULIC PROPERTIES

Erosion often degrades soil hydrologic conditions and decreases plant-available water capacity. Runoff is increased and the amount of water available for plant growth is decreased, thus lowering crop productivity. The following discussion shows how crop production can be affected by erosion-induced change in soil hydraulic properties.

12-3.1 Hydrologic Condition and Runoff

Runoff from daily rainfall is predicted by the equation (USDA–SCS, 1972):

$$Q = (P - 0.2S)^2/(P + 0.8S) \quad [1]$$

where Q is daily runoff, P is rainfall, and S is a retention parameter. Williams et al. (1983) showed that the retention parameter is related to soil water content by

$$S = S_{mx}(UL - SM)/UL \quad [2]$$

where S_{mx} is the maximum value of S, UL is the upper limit of soil water storage in the root zone (porosity minus water content at 1.5 MPa), and SM is the soil water content in the root zone minus water content at 1.5 MPa. The influences of the maximum value of the retention parameter and the upper limit of soil water storage on the retention parameter for SM equal to 0.1 m^3/m^3 are shown in Fig. 12-1. The maximum value of the retention parameter for different management conditions is shown in Table 12-5. To calculate S_{mx}, we obtained runoff curve numbers for hydrologic soil-cover complexes from Table 9.1 in USDA–SCS (1972) for antecedent moisture

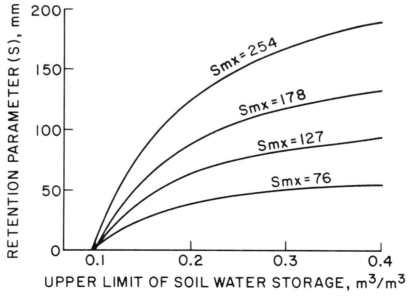

Fig. 12-1. Influence of upper limits of soil water storage on retention parameter at indicated levels of maximum.

Table 12-5. Maximum values for retention parameter (S) for specified conditions on row-cropped land.

Practice	Hydrologic condition	Hydrologic soil group			
		A	B	C	D
		mm			
Contoured	Poor	254	155	114	84
Contoured	Good	310	201	132	99
Contoured and terraced	Poor	297	208	150	132
Contoured and terraced	Good	351	244	170	142

condition II. We then used the relationship given by Smith and Williams (1980) to convert from curve numbers for antecedent moisture condition II to curve numbers for antecedent moisture condition I. Then

$$S_{mx} = 1000/CN_I - 10 \qquad [3]$$

where CN_I is the curve number for antecedent moisture condition I.

Erosion degrades the hydrologic conditions and thus lowers the maximum retention parameters. Loss in organic matter and resulting poor soil structure could easily degrade hydrologic condition from good to poor, or even degrade from a higher to a lower hydrologic soil group. The USDA–SCS (1972) has classified Cecil, Fullerton, Tifton, Fayette, and Marshall soils into hydrologic soil group B; and Grenada, Tillman, and Seymour soils into hydrologic group C. Hydrologic groups B and C are soils having moderate and slow rates of water transmission, respectively. If the soils in hydrologic group B with contoured row crops degraded from good to poor hydrologic condition (Table 12-5), the maximum value of the retention parameter would change from 20.1 to 15.5 cm, and, if they further degraded to hydrologic soil group C, the resulting retention parameter is 11.4 cm.

The lower retention parameter caused by degradation of hydrologic conditions increases runoff as shown in Eq. [1] and Eq. [2] and Fig. 12-2. This reduces the quantity of water available for crop production.

The influence of soil removal from the surface on the upper limit of soil water in the root zone (UL) and on available water capacity (AWC) was evaluated. The UL and AWC for each horizon of Cecil, Fayette, Marshall, and Seymour soils down to 100 cm was calculated, where UL is porisity minus water content at 1.5 MPa and AWC is water content at .033 MPa minus 1.5 MPa water content. The input data were taken from Soil Survey Laboratory work sheets for one pedon of each soil.

A depth-weighted factor (W_i) was computed for each depth increment using the Smith and Williams (1980) method:

$$W_i = 1.016\, e^{-4.16\,(D_{i-1}/RD)} - e^{-4.16\,(D_i/RD)} \qquad [4]$$

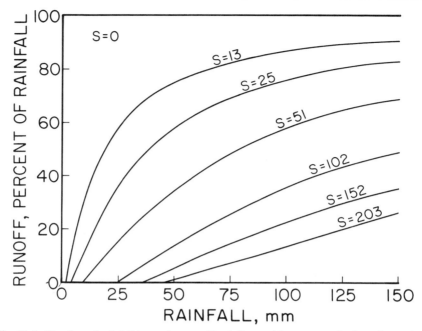

Fig. 12-2. Fraction of rainfall becoming runoff as influenced by amount of rain and retention parameter(s).

where D_i is the depth to the bottom of storage depth and RD is the total rooting depth. Using the weighting factor, the mean weighted UL and AWC were calculated from

$$UL = \sum_{i=1}^{n} W_i \cdot UL_i$$

and

$$AWC = \sum_{i=1}^{n} W_i \cdot AWC_i$$

where

$$\sum_{i=1}^{n} W_i = 1$$

Calculations were repeated by separately disregarding the upper 2.5, 5.1, 7.6, 12.7, 20.3, and 30.5 cm, and including a corresponding depth at the lower end of the soil depth considered. Results are shown in Table 12-6.

The $\overline{UL}$ for the Cecil, Fayette, and Seymour soils gradually decreases from the present condition as soil is removed, whereas for the Marshall soil, $\overline{UL}$ gradually increases with depth of soil removal. The $\overline{AWC}$ for Marshall is essentially constant with soil removal, whereas for Seymour and Fayette it decreases with soil removal. For Cecil, $\overline{AWC}$ changes as soil is removed from the surface. The $\overline{UL}$ and $\overline{AWC}$ would decrease drastically with surface soil removal if the lower soil depth were more restricting.

Table 12–6. The influence of soil removal on mean weighted upper limit water capacity ($\overline{UL}$) and mean weighted available water capacity ($\overline{AWC}$).

Soil	Soil property	Surface soil removed (cm)						
		0	2.5	5.1	7.6	12.7	20.3	30.5
		m^3/m^3						
Cecil	$\overline{UL}$	0.311	0.299	0.286	0.272	0.261	0.238	0.252
Cecil	$\overline{AWC}$	0.157	0.154	0.150	0.147	0.148	0.151	0.168
Fayette	$\overline{UL}$	0.364	0.359	0.355	0.349	0.337	0.313	0.282
Fayette	$\overline{AWC}$	0.233	0.229	0.226	0.222	0.212	0.200	0.190
Marshall	$\overline{UL}$	0.331	0.337	0.344	0.351	0.368	0.378	0.376
Marshall	$\overline{AWC}$	0.165	0.166	0.166	0.167	0.169	0.168	0.164
Seymour	$\overline{UL}$	0.285	0.282	0.279	0.276	0.268	0.257	0.240
Seymour	$\overline{AWC}$	0.162	0.160	0.158	0.155	0.149	0.142	0.140

12–3.2 Available Water Capacity and Relative Yield

A National Soil Erosion-Soil Productivity Research Planning Committee (1981) reported that erosion reduces productivity primarily through loss of plant-available soil water capacity.

We prepared some hypothetical situations to illustrate how changes in AWC can affect crop production and then to show how UL and AWC are affected in Cecil, Fayette, Seymour, and Marshall soils.

The connecting link between AWC and production is the manner in which AWC influences relative evapotranspiration. The generalized influence of water deficit on crop production is

$$1 - Y_a/Y_m = ky\,(1 - ET_a/ET_m) \qquad [5]$$

where Y_a and Y_m are actual and maximum yield, ET_a and ET_m are actual and maximum evapotranspiration, and ky is a yield-response factor (Doorenbos and Kassam, 1979). Maximum yield is the harvested yield of a high-producing variety well-adapted to the growing environment under conditions where water, nutrients, pests and diseases do not limit yield. The relationship of Eq. [5] is illustrated by Fig. 12–3, which shows four groups of crops according to their sensitivity to water stress. Crops with ky < 1 are less sensitive to water stress than those with ky > 1. The stage of plant growth also affects sensitivity to water stress and may be represented graphically as general crop sensitivity in Fig. 12.3. Generally, the order of sensitivity according to growth stage is flowering > yield formation > vegetation > ripening.

Actual evapotranspiration equals ET_m when available water (AW) is greater than (1 − p) AWC, but, when AW is less than (1 − p) AWC,

$$ET_a = ET_m\,[AW/(1 - p)\,AWC] \qquad [6]$$

where p is the soil-water-depletion fraction, or the fraction to which the available water can be depleted while maintaining ET_a equal to ET_m. The value of p varies with crop, stage of crop development, and ET_m.

We assumed three levels of AWC (6, 14, and 20 cm) which correspond to low, medium, and high AWC for a 100-cm root zone, and the other conditions given in Table 12-7. Then we calculated relative evapotranspiration and relative yield decrease.

The relative yields decreased 60, 20, and 6% when AWC was 6, 14, and 20 cm, respectively, and initial available water equaled available water

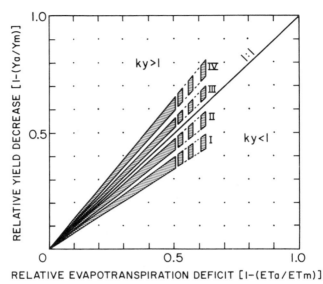

Fig. 12-3. Generalized relationship between relative yield decrease $(1 - Y_a/Y_m)$ and relative evapotranspiration $(1 - ET_a/ET_m)$ (redrawn from Doorenbos and Kassam, 1979).

Table 12-7. The influence of available water capacity (AWC) on relative yield decrease for specified conditions.

Condition†	Available water capacity	Initial available water	ET_a/ET_m‡	Relative yield decrease
	cm			
1	6	6	0.40	0.60
	14	14	0.80	0.20
	20	20	0.94	0.06
2	6	6	0.40	0.60
	14	6	0.36	0.64
	20	6	0.32	0.68
3	6	6	0.92	0.08
	14	14	1.00	0.0
	20	20	1.00	0.0
4	6	6	0.77	0.23
	14	6	0.63	0.37
	20	6	0.53	0.47

† Calculations are for a 21-day period and are based on a soil depletion fraction (p) of 0.5. Conditions 1 and 2 received no additional water, whereas 3 and 4 had 3.0 cm of water added to available water after every seventh day. ET_m was assumed to be 0.7 cm/day.
‡ ET_a is actual evapotranspiration; ET_m is maximum evapotranspiration.

capacity. Yields started declining 5, 11, and 15 days after full soil reservoir (Fig. 12–4). However, if each soil initially contained 6 cm of available water, a yield-decreasing situation already existed at the beginning of the period for the 14-cm and 20-cm AWC soils (Fig. 12–5). For this condition (condition 2) relative yield was reduced 60, 64, and 68% when AWC was 6, 14, and 20 cm, respectively.

The sufficiency of the AWC of a soil for maximum crop production can be interpreted best with respect to the adequacy of soil water to supply the needs of the crop. Rainfall distribution, potential evaporation, runoff, and plant water requirements must all be evaluated simultaneously. A model is needed that incorporates all these important variables for specified rainfall probability levels and length of accounting periods to explain more clearly the influence of erosion-induced change in AWC on crop production for different climatic regimes and initial soil properties.

12–4 RELATION OF SOIL PROPERTIES TO PRODUCTIVE POTENTIAL FOR CORN

Using an approach described by Pierce et al. (1983), we calculated the relative productive potential of selected soils. The method determines the relative productive potential of soil in terms of the environment the soil provides for root growth based on the soil's available water capacity (AWC),

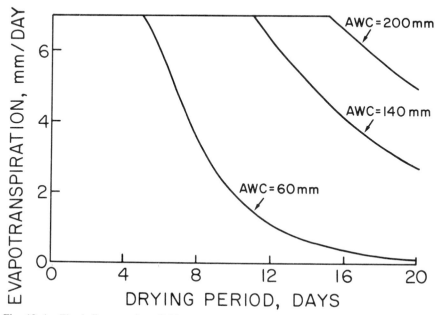

Fig. 12–4. The influence of available water capacity (AWC) on actual evapotranspiration during a drying cycle with potential evapotranspiration at 7 mm/day and initial full soil reservoir.

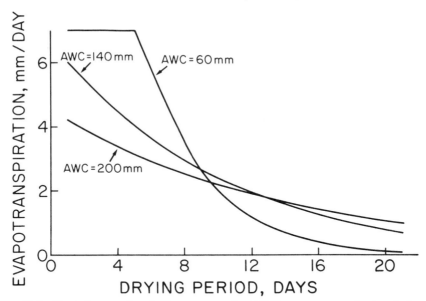

Fig. 12–5. The influence of available water capacity (AWC) on evapotranspiration during a drying cycle with potential evapotranspiration at 7 mm/day and initial available water content of 60 mm for each soil.

resistance to root growth and development, and adequacy of pH to a depth of 100 cm. The equation used by Pierce et al. (1983) is

$$PI = \sum_{i=1}^{r} (A_i \cdot B_i \cdot C_i \cdot WF_i)$$ [7]

where A is the sufficiency of AWC, B is the sufficiency of bulk density (DB), C is the sufficiency of pH, WF is a weighting factor, and r is the number of 10-cm increments in the rooting. The productivity index (PI) increases directly with productive potential from 0 to 1.0, as does each sufficiency factor. The weighting factor (WF) epresses an ideal corn root distribution to 100 cm depth. It is normalized so that the area under the WF-depth curve is equal to 1.0.

The AWC, DB, and pH data by horizons used in the PI calculations for the Seymour and Marshall pedons were taken from Soil Survey Staff (1978). Data representative of the central concept of the other soils were taken from SOILS-5 (The SOILS-5 data base contains soil descriptions, physical and chemical properties, crop yields, and capabilities and limitations for every soil series, and their variants established in the USA. USDA–SCS 1983, SOILS-5 data base. USDA–SCS, Ames, IA). The PI indices for these selected soils are shown in Table 12–8.

Table 12–8. Productivity Index (PI) for eight benchmark soils.

| Soil Series | Erosion | |
	Slight	Moderate
	PI	
Seymour (3 profiles)	0.79	0.64, 0.59
Marshall (3 profiles)	0.89, 0.73	0.82
Tillman	0.43	--
Fayette	0.86	--
Grenada	0.74	--
Cecil	0.51	--
Fullerton	0.41	--
Tifton	0.40	--

12–4.1 Mollisols

The characteristics of the three Seymour soils from which PI was cal-culated are given in Fig. 12–6. For all three pedons, bulk density was suf-ficient while pH and AWC were somewhat limiting to productive potential. Available water capacity was related to the depth of maximum clay content in the profile, and pH increased to a more neutral level deeper in the profile. For the Seymour soil, the slightly eroded site (93-6) had a PI of 0.79, where-as the moderately eroded sites (93-5 and 93-4) had PIs of 0.64 and 0.59. Productive potential in this soil declines as erosion moves the zone of high clay closer to the surface.

The characteristics of the three Marshall pedons are given in Fig. 12–7. The PI for the two slightly eroded sites of the Marshall soil (15-1 and 15-2) was 0.89 and 0.73, and, for the moderately eroded site (15-3), the PI was 0.82. The somewhat lower PI for site 15-2 (slightly eroded) reflects a lower AWC, although no reason is apparent for it being less than the other two pedons. However, AWC, DB, and pH all reflect generally good root-growth conditions to depths of more than 1 m.

To simulate the effects of erosion on productive potential, PI was cal-culated after successive removals of 5 cm of soil from the surface. As soil was removed, the rooting depth was maintained at 1 m by moving the root-ing function an equal distance down the soil profile.

The Tillman series are deep, well-drained, slowly permeable soils formed in clayey alluvium and have a computed PI of 0.43. While the soils have a relatively high AWC (approximately 0.16 cm/cm in the surface and decreasing to 0.14 cm/cm below 1.2 m), they have root-restricting bulk densities (1.55 Mg/m³) at depths below 15 cm. Soil pH below 15 cm ap-proaches 8.0. The PI of 0.43 reflects the relatively high DB in horizons be-low 15 cm. Erosion as simulated herein slightly reduces the PI of the Till-man soils as restricting DB in subsurface horizons moves closer to the sur-face (Fig. 12–8).

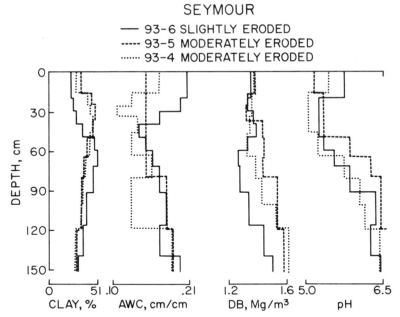

Fig. 12-6. Soil characteristics of three Seymour soil profiles as influenced by erosion. (AWC is available water capacity, and DB is bulk density).

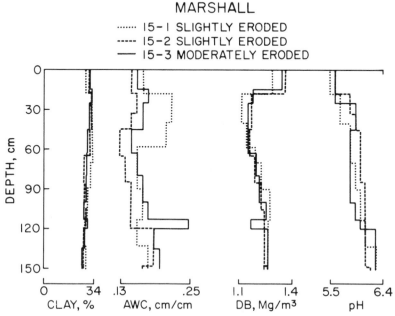

Fig. 12-7. Soil characteristics of three Marshall soil profiles as influenced by erosion. (AWC is available water capacity, and DB is bulk density).

12-4.2   Alfisols

The two Alfisols studied have PIs of 0.86 (Fayette) and 0.74 (Grenada). The data in Fig. 12-8 show that the PI for Fayette does not decrease appreciably as erosion occurs, whereas the PI for the Grenada series decreases appreciably. The Fayette soils are developed from deep loess and have high AWC and nonrestricting DB to 120 cm. The Grenada soils are also developed from deep loess but, unlike the Fayette, have root-restricting fragipans (DB from 1.45 to 1.60 Mg/m³) at about 60 cm. During simulated erosion (Fig. 12-8), the PI decreases as the fragipan comes closer to the soil surface.

12-4.3   Ultisols

The three Ultisols studied all have relatively low PIs ranging from 0.40 (Tifton) to 0.51 (Cecil). The Cecil soils have moderate AWC throughout the top 125 cm, moderate DBs (1.3 to 1.5 Mg/m³), and low pHs (4.5 to 6.0). The low pH of the subsoil is reflected in the PI of 0.51. With simulated erosion, PI decreases slightly (Fig. 12-8). Fullerton soils have low but nearly constant AWC (0.10 to 0.15 cm/cm) to a depth of 200 cm. Bulk densities are somewhat root restricting (1.45 to 1.65 Mg/m³), and subsoil pH values are severely restricting (4.5 to 5.5). These characteristics result in a PI of 0.41. The PI decreases with simulated erosion to 0.31 with 50 cm of the soil removed.

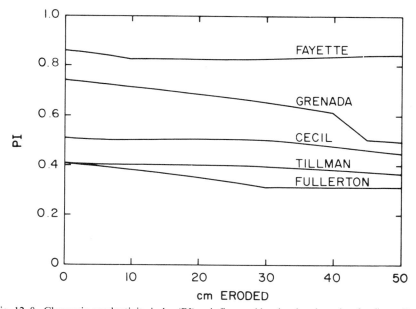

Fig. 12-8. Change in productivity index (PI) as influenced by simulated erosion for five soils.

12-5 RELATIONSHIP BETWEEN PI AND CORN YIELD

Englestad and Shrader (1961) compared the effects of surface soil thickness on corn yields for Marshall soils using artificially exposed subsoil and normal soil. In two years, yields on subsoil control plots (zero nitrogen) averaged 3.07 Mg/ha lower than those from corresponding normal soil plots. However, corn yields were equal when adequate nitrogen fertilizer was supplied.

Odell (1950) reported that each 2.5 cm of surface soil added 0.072 to 0.075 Mg/ha to yield on a soil with a favorable rooting zone in the subsoil and substratum. On a soil with an unfavorable rooting zone in the subsoil and substratum, each 2.5 cm of surface soil added 1.4 to 2.0 Mg/ha to yield of corn. Seymour soils are in this category.

Fenton (1980) used data reported by Swanson and MacCallum (1969) to calculate changes in corn yields related to surface soil thickness. Values calculated ranged from a change of 0.033 Mg/ha per 1 cm of surface soil for soils with favorable rooting zones to 0.080 Mg/ha per 1 cm of surface soil for soils with unfavorable rooting zones.

The estimated row-crop yields expected for various erosion phases of the Marshall and Seymour soils are given in Table 12-9. Even with high-level management, the effect of topsoil loss is greater on Seymour than on Marshall soils.

The relationship between corn yield and PI for the soils discussed here is given in Fig. 12-9. The symbols with a circle are the soils for which corn yields are given in SOILS-5. The soil data used as input in Eq. [7] were also taken from SOILS-5. Compare the point marked M ' (calculated from SOILS-5) in the upper right hand corner with the three squares. The points marked as squares are the three Marshall pedons discussed earlier. The data used to calculate the points marked as squares were taken from measured data in Soil Survey Staff (1978). Estimated corn yields were taken from Table 12-9. The squares near the line represent a slightly and moderately eroded Marshall pedon. The one removed from the line is a slightly eroded pedon. Likewise, compare the point marked S ' (Seymour) in the center of the diagram (calculated from SOILS-5) with the three triangles. The triangles represent PIs calculated from measured data reported by Soil Survey Staff (1978) and corn yields from Table 12-9. The triangle to the right

Table 12-9. Estimates of crop yields for selected phases of the Marshall and Seymour series.

Soil	Erosion phase	Corn	Soybeans
		——— Mg/ha ———	
Marshall	Slight	6.69	2.73
(silty clay loam, 2-5% slopes)	Moderate	6.50	2.67
	Severe	6.13	2.47
Seymour	Slight	5.50	2.20
(silt loam, 2-5% slope)	Moderate	5.19	2.13
	Severe	4.56	1.87

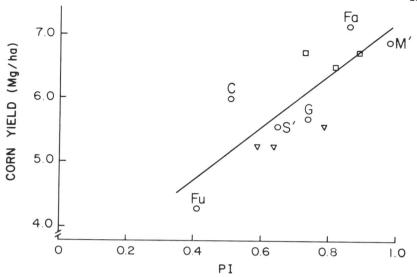

Fig. 12-9. Corn yield relative to the productivity index, PI, for various soils (C = Cecil, Fa = Fayette, G = Grenada, S = Seymour, M = Marshall, Fu = Fullerton).

represents the slightly eroded site, and the triangles to the left represent the moderately eroded sites.

From the data, we conclude that PI is generally related to corn yields. However, PI is sensitive to the actual measured characteristics of the pedon within a soil series. Since these characteristics vary within a mapped series, a general yield loss for that series is difficult to estimate. Last, the corn-yield values used in Fig. 12-9 are not precise estimates and, hence, probably contribute considerable error to the general relationship.

12-6   SUMMARY

We have pointed out some of the difficulty in classification of eroded soils, particularly Mollisols. Degree of erosion can be mapped with a high degree of accuracy if quantitative definitions are provided to the soil surveyor. However, eroded phases of a soil series are not always displayed on the published map, erosion phases often being combined into one mapping unit. A major problem in mapping Mollisols is that the eroded and uneroded phases may result in a different classification if *Soil Taxonomy* (USDA, 1975) is strictly adhered to. In practice, however, the eroded phases are often identified in the soil reports as a taxadjunct of the uneroded phases. Obviously, strict guidelines and adherence to the guidelines are needed for quantitative estimation of the characteristics of the soil and the inerpretations made therefrom.

Probably the most serious loss in long-term productivity from soil erosion is loss in plant-available soil water capacity. Using several simple models, we have shown that root-zone water-storage capacity can be changed materially by erosion. Losses in available water capacity are most serious on soils with unfavorable subsoils. On soils with deep favorable sola, losses in available water capacity may not be significant. Losses in available water-storage capacity can result in significant reductions in crop yields.

The relative productive potential of selected soils was estimated using a model that calculates the environment the soil provides for root growth and available water capacity. Using this model, relative reductions in crop yield are estimated using simulated erosion. Erosion is shown to reduce the productive potential on soils with unfavorable subsoils.

We conclude that losses in crop production potential are related to the characteristics of the soils, and that improved mapping procedures are needed to retain mapping units that reflect erosion's effect on sola characteristics to the user of soil-survey information.

REFERENCES

Beckett, P. H. T., and R. Webster. 1971. Soil variability: A review. Soils Fert. 34:1–15.

Council for Agricultural Science and Technology. 1982. Soil erosion: Its agricultural, environmental, and socio-economic implications. Report 92. Ames, IA.

Dideriksen, R. I. 1966. An evaluation of soil mapping units by erosion groups in Iowa. Mimeograph, Soil Conservation Service-Iowa Soil Survey. Ames, IA.

Ditzler, C. 1981. Mapping eroded phases of soil. Soil Surv. Horiz. 22(3):15–17.

Doorenbos, J., and A. H. Kassam. 1979. Yield response to water. FAO Irrigation and Drainage Paper 33. FAO, Rome.

Englestad, O. P., and W. D. Shrader. 1961. The effect of surface soil thickness on corn yields: II. As determined by an experiment using normal surface soil and artificially exposed subsoil. Soil Sci. Am. Proc. 25:497–499.

Fenton, T. E. 1980. Soil erosion in the corn belt—problems and solutions. Proc. of Quaternary Geology and Geomorphology Symp.: Soil Erosion. Geol. Soc. Am., Atlanta, GA. 17–20 Nov. 1980.

Langdale, G. W., and W. D. Shrader. 1982. Soil erosion effects on soil productivity of cultivated cropland. p. 41–51. In Determinants of soil loss tolerance. Spec. Pub. 45. American Society of Agronomy, Madison, WI.

National Soil Erosion-Soil Productivity Research Planning Committee. 1981. Soil erosion effects on soil productivity: A research perspective. J. Soil and Water Conserv. 36:82–90.

Odell, R. T. 1950. Measurement of the productivity of soils under various environmental conditions. Agron. J. 42:282–292.

Pierce, F. J., W. E. Larson, and R. H. Dowdy. 1983. Productivity of Soils: Assessing long-term changes due to erosion. J. Soil Water Conserv. 38:39–44.

Smith, G. D. 1978. Conversations in Taxonomy. Soil Surv. Horiz. 19(2):11–14.

Smith, R. E., and J. R. Williams. 1980. Simulation of the surface water hydrology. p. 13–35. In CREAMS, a field scale model for chemicals, runoff, and erosion from agricultural management systems. USDA, Conserv. Res. Rep. 26.

Soil Survey Staff. 1978. Soil survey laboratory data and descriptions for some soils of Iowa. soil Survey Investigation Report 31. USDA-SCS and Iowa Agriculture and Home Economics Experiment Station.

Swanson, E. R., and D. E. MacCallum. 1969. Income effects of rainfall on erosion control. J. Soil Water Conserv. 24:56–59.

USDA. 1975. Soil taxonomy—a basic system of soil classification for making and interpreting soil surveys. USDA Agric. Handbk. 436.

USDA–SCS. 1972. National Engineering Handbook, Hydrology Section 4. USDA–SCS.

Wilding, L. P., R. B. Jones, and G. M. Shafer. 1965. Variations of soil morphological properties within Miami, Celina, and Crosby mapping units in west-central Ohio. Soil Sci. Soc. Am. Proc. 29:711–717.

Williams, J. R., C. A. Jones, and P. T. Dyke. 1984. A modeling approach to determining the relationship between erosion and soil productivity. Trans ASAE 27:129–144.

13 Experimental Approaches for Quantifying the Effect of Soil Erosion on Productivity

L. D. Meyer
Agricultural Research Service
U.S. Department of Agriculture
Oxford, Mississippi

A. Bauer
Agricultural Research Service
U.S. Department of Agriculture
Mandan, North Dakota

R. D. Heil
Colorado State University
Fort Collins, Colorado

Experiments to determine the effects of soil erosion on crop yields were an important part of soil- and water-conservation research from the 1930s through the early 1950s. Some studies were rather elaborate and others were rather simple, but they confirmed the observations of farmers and soil conservationists that severe soil erosion generally decreased crop production.

Both observations and results of studies of the consequences of excessive erosion on crop production were so conclusive that most research was shifted to other aspects of the erosion problem, such as (i) quantifying the rates of soil erosion for different soils, topographic conditions, climates, and crops, and (ii) developing erosion-control practices to manage storm runoff or dissipate wind energy to prevent excessive soil erosion. These pursuits have dominated most soil-erosion research until recently.

During recent years, questions similar to those asked during the earlier years of erosion awareness have reemerged with special emphasis on the need for quantifying erosion-induced productivity losses, their economic

Published in R. F. Follett and B. A. Stewart, ed. 1985. *Soil Erosion and Crop Productivity.*
© ASA-CSSA-SSSA, 677 South Segoe Road, Madison, WI 53711, USA.

consequences for modern agriculture, and their short- and long-term impacts on our nation's ability to produce food, fiber, and forage. (We use "productivity" as a measure of crop-production potential attributable to inherent soil properties, and crop "production" as a measure of crop yield.) Considerable effort is being directed toward evaluating the effects of erosion on productivity, often at the expense of research on methods to control soil losses. The primary impetus for this redirection has been Public Law 95-192, the Soil and Water Resources Conservation Act of 1977 (RCA). To respond to research needed for RCA, the Agricultural Research Service initiated a coordination effort with the Soil Conservation Service, state agricultural experiment stations, and other interested agencies. This effort began with a planning workshop at Lafayette, IN, in September 1981. This paper summarizes experimental research discussions during and after the workshop, including early research in this field, alternative experimental approaches, current and planned research activities, and related issues.

13-1 COMPLEXITY OF THE PROBLEM

During the past half century, increased fertilization, improved cultivars, selective pesticides, and supplemental irrigation have increased production so much that any yield reductions due to erosion have been easy to overlook. During the same period, drastic changes in farming systems and cropping practices have greatly affected soil losses and necessitated major changes in erosion-control practices.

Probably the most perplexing aspects of designing experiments to quantify the effect of soil erosion on productivity are the complex interrelationships among the many factors involved and the variability of their characteristics with time and space. Soils subject to erosion differ greatly in texture, profile characteristics, position on the landscape, and chemical composition. Furthermore, the extent of erosion varies from negligible to so extensive that farm operations are almost impossible. Yields of some crops are severely affected by erosion-caused changes, such as reduced soil water and thinner root zones. Yet, other crops are less affected. Erosion also affects crop yields indirectly; for example, sealing of the soil surface may decrease infiltration and impede seedling emergence. Erosion events vary in number and magnitude from year to year and from season to season, making their average annual and accumulated impacts difficult to assess for short research periods. Finally, erosion may selectively remove certain portions of the soil material, and then may selectively deposit the eroded sediment at distant locations.

With these complexities in mind, important questions arise: Can we accurately assess the effects of soil erosion on productivity? What are the major questions to resolve? Which solutions are critical in ensuring that we adequately protect our soil resources from excessive exploitation and still maintain an adequate productive capacity?

Regardless of the priorities, experimental research will have a major role since qualitative knowledge is not adequate for RCA and other uses. Appropriate quantitative data are essential for improved assessment of the effects of erosion on productivity. Mathematical models must have quantitative values from pertinent experiments, and model results need validation by experiments. Major variables must be experimentally evaluated to determine their relative importance.

13-2 PAST EXPERIMENTAL RESEARCH TO QUANTIFY THE EFFECTS OF SOIL EROSION ON PRODUCTIVITY

Many early studies evaluating the effects of erosion on crop yields were made by desurfacing land to simulate serious erosion. Results from major crop-producing areas throughout the USA (Bennett, 1939) showed that yields of major crops decreased an average of 77% where soil was removed to the subsoil level. The effect varied with different soil types and crops, but all yields were seriously reduced. Additions of fertilizer increased yields on the subsoil, but yields fell far short of those on fertilized topsoil.

Research in Arkansas during the 1930s and 1940s (Bartholomew et al., 1954 and 1955) showed that removal of about 15 cm of topsoil increased runoff and erosion and greatly reduced corn and cotton yields. Yields from the desurfaced soils could be increased, however, with careful selection of the cropping system and fertility practices.

In Ohio studies (Uhland, 1949; Stallings, 1950), yields of several crops were compared for (i) a virgin soil, (ii) the same soil with all topsoil removed, and (iii) the same soil with the removed topsoil from (ii) added to double the topsoil depth. Yields of corn, wheat, and oats on subsoil were less than half those on natural soil, and corn and wheat yields were considerably greater on double topsoil.

Similar Canadian studies near Ottawa (Ripley et al., 1961) showed that barley and alfalfa yields dropped progressively as the soil depth was varied in approximately 7.5-cm increments from 7.5 cm of soil added to greater than 15 cm removed. Fertilizers increased yields for all soil depths, but fertilized subsoil yields were much lower than even unfertilized topsoil yields.

These studies all showed that crop production on fertilized subsoil was much less than on surface soil. However, the crop yield was almost as great on subsoil of a well-fertilized deep-loess soil as on the topsoil (Englestad and Shrader, 1961b).

Studies of topsoil removal and addition to determine the effects of land leveling and terrace construction on crop production (Eck, 1969; Hauser, 1968; Heilman and Thomas, 1961; Olson, 1977; Lyles, 1977) have shown that the yield levels on many soils can be largely restored by fertilizers, but that some net yield loss may be expected as a result of the cutting and filling operations. However, because such operations are usually designed to improve water management or reduce erosion, the net long-term effect is often increased production.

Some early studies of the effect of erosion on production were made on farm fields with widely variable past erosion. Yields were measured for remaining topsoil thicknesses ranging from very thin to over 30 cm (Uhland, 1949). The yield reduction for corn was around 100 kg/ha per centimeter of decreased topsoil depth and was nearly linear for Corn Belt soils in Missouri, Indiana, and Iowa. Other studies in Georgia, Iowa, Ohio, Pennsylvania, Kansas, Washington, and New Jersey with corn and several other crops gave similar results (Stallings, 1950). Numerous other studies on natural soil conditions were cited in a recent overview article (National Soil Erosion-Soil Productivity Planning Committee, 1981).

Recent research has been conducted under modern farming conditions in Tennessee (Bell, 1977; Denton, 1978; Buntley and Bell, 1976), Maryland (Robinette, 1975), Kentucky (Frye et al., 1982), and Georgia (Langdale et al., 1979). The results generally show appreciable reductions in crop yields with increased erosion, especially on soils with restricting pans or on those with subsoils of considerably less yield potential.

Less rigorous and controlled experiments, such as field trials, demonstrations, and observations, can estimate crop yields for conditions where formal results are not available. Yield estimates for major U.S. soils at different levels of past erosion have been published in county soil surveys by the USDA–SCS in cooperation with state agricultural experiment stations. Some experiment stations have published similar detailed yield estimates (Buntley and Bell, 1976). The yields reported in these publications are regularly updated and provide reliable estimates of expected crop yields on farms with high levels of management.

Considerable research has evaluated erosion's effects on the physical, chemical, and biological properties of the soil that affect production. Other studies have tried to determine the production inputs needed to compensate for losses in production due to erosion. Many studies have evaluated erosion rates for a wide range of soil and cropping conditions.

13-3 FACTORS THAT AFFECT PRODUCTIVITY

The effect of removal or loss of surface soil on productivity depends on the soil profile (Baver, 1950) and on the chemical and physical characteristics of the soil horizons that consequently become part of the normal rooting depths of the crops to be grown. Occasionally, productivity is increased when surface soil is removed, especially if a buried profile of productive soil is exposed, but productivity usually is decreased by erosion.

Most research on productivity has attempted to determine nutrient inputs. The kinds of fertility problems in subsoils usually are the same as those in surface soils (Winters and Simonson, 1951), but levels of deficiency may differ among soil horizons. Fertility problems associated with subsoil but not with the corresponding surface soil material were reported by Carlson et al. (1961).

The most common intensification of nutrient deficiency with soil loss involves N and P. This is expected, since organic matter is virtually the only source of indigenous soil N and is the source of as much as 65% of the total P in the normal rooting-depth profile. Potassium deficiency has not been reported in yield-restoration studies in semiarid and subhumid regions. But K is deficient in the plow layer of soils in humid regions and hence is needed for yield restoration on eroded soils. Of the minor elements, zinc deficiency in zinc-sensitive crops such as corn (*Zea mays* L.) and potato (*Solanum tuberosum* L.) has been confirmed in subsoils (Carlson et al., 1961; Grunes et al., 1961). This deficiency is prevalent in corn grown on calcareous soils (Bauer, 1968; Gunderson et al., 1965). Iron deficiencies, frequently observed on soils calcareous to the subface (Reuss and Lindsay, 1963), likely would occur in associated subsoils.

Murray et al. (1939) found a positive correlation between corn yields and surface soil thickness up to 20 to 25 cm. However, yield responses to increasing thickness of surface soils varied among soils with differing subsoil permeabilities (Odell, 1950). Mixing various subsoil materials with surface soil increased yields over subsoil materials alone (Carlson and Grunes, 1958; Lutwick and Hobbs, 1964). Based on regression analysis of corn yields grown on subsoil and normal surface soil, N fertilizer substituted completely for loss of surface soil thickness in some years (Engelstad et al., 1961a).

The effects on crop yields of the physical properties of essentially undisturbed soils have received very little attention compared to the effects of nutrient and water properties. The same is true for studies of yield restoration. Haise et al. (1955) increased the percentage of aggregation, mean-weight diameter, and aeration porosity of subsoil developed on glacial till with applications of VAMA (copolymer of vinylacetate and partial methyl ester of maleic acid) and HPAN (hydrolyzed polacrylonitrile). Regardless of apparent improvement in subsoil structure, however, sugarbeet yields decreased with increased rates of these soil conditioners (linear organic polymers). Black and Greb (1968) measured several soil properties to show how they differed by soil genetic horizon, but they did not relate them directly to yield.

Water supply has an overriding influence on crop yield and on crop response to fertilization. Hence, to separate its effect from those of other soil parameters, precise information is required on water inputs and the profile distribution of soil water throughout the growing season.

Preparing a seedbed in exposed subsoils can be difficult. Seedling emergence can be inhibited by crusting or by improper planting depth. Lack of uniformity in plant population among treatments is a confounding factor because plant population affects yield.

13-4 CONSIDERATIONS WHEN DESIGNING EROSION/PRODUCTIVITY EXPERIMENTS

Since erosion of land is a progressive process, for which the appropriate time scale is decades and centuries, short-term effects of erosion usually are difficult to assess. Generally, a soil's productive potential, that is, its productivity, decreases as erosion progresses. This decrease usually accelerates with increased erosion: A similar amount of erosion during the second 10 years will reduce productive potential more than during the first 10 years, until production is seriously affected. Seldom, if ever, does the production trend for a specific crop decline to zero. Instead, after erosion proceeds to some level, production of that crop simply is no longer economical and it is not planted. However, the same land may be used subsequently for a less valuable crop, such as trees or pasture on former row-crop land. The researcher must determine where the present is on the time curve and what the trend is for the future.

Experimental research to quantify the effect of erosion on productivity is a complex undertaking. The range of necessary expertise may be rather broad. Seldom will a single researcher or even a small team have the knowledge required to design experiments that give definitive, broadly applicable results. Expertise is necessary in soils, crop production, water use, water management, erosion evaluations, statistics, and other disciplines.

Studies must be efficient in gleaning all possible information and must be broadly applicable and compatible with other research. Of particular importance is complete documentation of the soil, plant, topographic, and climatic characteristics of the experimental sites throughout the test period. Often in the past, too few variables were measured or were not measured uniformly; hence, rather diverse data were collected with limited use in large analytical efforts. For the same reasons, results of many experiments probably were never published, but were only reported locally or in theses without wide distribution. Key variables that should be characterized are listed in section 13-6.

Erosion/productivity research is constrained by the potentially massive cost required for comprehensive experimental efforts. For field studies, considerable land area is necessary. Numerous personnel may be needed, often when other responsibilities are also crucial. Studies usually require many years to sample typical climatic variabilities. The alternative is research very limited in scope. If a national commitment is made to quantify accurately the effect of soil erosion on productivity, it must be accompanied by adequate support for effective experimental research.

13-5 EXPERIMENTAL METHODS FOR EROSION/ PRODUCTIVITY RESEARCH

Most experimental research to quantify the effects of man-accelerated soil erosion on soil productivity has been conducted on land with varying past erosion or on land from which different amounts of surface soil were removed or added to simulate past soil erosion. Laboratory experiments have been limited. Related experiments have studied the indirect effects of erosion on soil properties, the increased technological inputs necessary to compensate for erosion, and those conditions identified as information gaps for making computer simulations. Following are discussions of the various experimental approaches, some of their advantages and disadvantages, and possible alternative procedures.

13–5.1 Experiments on Land with Variable Past Erosion

Erosion/productivity studies on farm fields with variable past erosion due to intensive cropping have the important advantages that the conditions are typical, not artificially induced, and that the test crop usually is grown on an entire field, not on plot-sized areas. But the researcher needs to resolve some important questions during the planning, conduct, and interpretation of these experiments. How can treatment sites be selected to ensure their comparability? Is the soil variability due to erosion or other causes? How can the extent of previous erosion be accurately determined? If erosion differences are studied at different locations on a slope, will runoff from upslope be permitted or restricted on lower sites? Should downslope areas, on which sediment from upslope erosion is deposited, be included in the study? How can unnatural differences in fertility, insects, and diseases be avoided?

How a researcher defines the term "productivity" can have a major bearing on how experiments are designed and conducted. A difference in crop yields among different degrees of erosion on the same soil individual[1] does not necessarily constitute a difference in productivity. A yield difference is a measure of production and can be a measure of productivity if the experiment is not confounded.

The current definition of soil productivity (Soil Conservation Society of America, 1976, p. 43-6; Soil Science Society of America, 1978, p. 13) is "The capacity of a soil in its normal environment for producing a specified plant or sequence of plants under a specified system of management." One interpretation of this definition is that uniformity of all manageable factors must be imposed on all differentially eroded plots. To fulfill this requirement, the researcher would establish uniformity in plant (seedling) population, soil water supply, fertility level, and other manageable factors. The resulting yields then would reflect the effect of nonmanageable factors (i.e., erosion) to the exclusion of manageable factors. The less uniformity imposed on the experiment, the greater the possibility of confounding, that is, the inability to establish cause-and-effect relationships. Imposing uniformity does not preclude using these manageable factors as variables in an experiment, but it requires that the experiment be enlarged to accommodate the additional controls.

Another interpretation of the definition is that only a uniform cultural system be used to establish the experiment. Any existing difference among differentially eroded plots at the initiation of the experiment, such as a differential soil water level within the rooting zone or resulting plant population, would be attributed to the consequences of past erosion on the soil properties causing the difference.

---

[1] A soil individual (polypedon) is a parcel of contiguous pedons, all of which have characteristics lying within the defined limits of a single soil series (Johnson, 1963).

With such an approach, yield measurements are made against a shifting, differential starting base that can vary among experiments and years, since the difference in any manageable factor among differentially eroded sites can vary from year to year. For example, although erosion can affect the infiltration rate, the difference in the amount of soil water present in otherwise identical rooting volumes can also vary because of rainfall characteristics and cultural practices. Similarly, seedling establishment can be affected by placement depth as a function of the planting equipment, of weather conditions after planting, and of seedbed conditions as determined by cultural practices and soil properties. Thus, if factors such as soil water supply and plant population are treated as dependent variables and differ annually among degrees of erosion of any soil, the probabilities of ascertaining cause-and-effect relationships are virtually nil.

One technique that integrates the impacts of the many contributing factors is the "sequential testing" approach, where studies are made on the sequence of two to four kinds of soils which typically occur on fields of sloping land (McCormack and Moorman, 1983). Yield data are collected for a sequence of soils within a field on which farming practices and weather are essentially uniform. Sites are carefully selected so that, insofar as possible, the eroded and uneroded phases of each selected soil series have the same slope position and steepness. McCormack and Moorman recommend that such studies be conducted on most major soil sequences, for various major crops, and for enough years to sample the range of typical climatic variations. Regression analyses of the data are suggested to extend the results to other soils.

When developing field studies on the effects of different amounts of past erosion, researchers must carefully define their goals. If all possible factors are controlled, more information will be gained about the effect of each factor, but some factors may be controlled that normally would vary. If, on the other hand, factors such as plant population and soil water supply are treated as dependent variables that may vary because of past erosion, the results may provide knowledge of the integrated effect of past erosion but will provide less understanding of the influence of each factor, thus limiting the application of the results for conditions where past erosion affects these factors differently.

13-5.2 Experiments on Land with Artificially Simulated Soil Erosion

Manual removal (scalping) of surface soil to simulate erosion has been used in numerous experiments. Yields after removal of different thicknesses of soil were taken as indicators of the effect of that much soil loss by erosion. However, normal removal by severe erosion is usually very nonuniform because of rilling and gullying, and both wind and water erosion may remove size fractions selectively. Also, in these simulations annual or multiannual mixing by tillage does not occur, with its attendant blending of surface with subsurface material and the extra aeration of subsurface soil. To illustrate, assume that erosion annually removes 1 cm from a soil with 15 cm of topsoil and that tillage for crop production is 15 cm deep. After the first year, tillage will mix 1 cm of the subsoil with the remaining 14 cm of topsoil, giving a mixture of 14/15 topsoil. During the second year, erosion will remove 1 cm of soil that is 14/15 topsoil and 1/15 subsoil. After the second year, another 1 cm of subsoil will be mixed with the 14:1 topsoil-subsoil mixture so that the surface 15 cm of soil will be 87% topsoil and 13% subsoil. Continuing for succeeding years, the mixture will be about 50:50 after 10 cm of erosion and will still be 25% topsoil after 20 cm of erosion.

This desurfacing approach has been very useful in evaluating the influence of cutting and filling for terrace construction and for land leveling to improve drainage or irrigation. However, when it is used to evaluate the effects of erosion on productivity, the results must be interpreted carefully.

Several modifications to mass removal of surface soil to simulate erosion merit consideration. One is to obtain the appropriate mixture of surface soil and subsoil that would result from severe erosion. This method avoids some unnatural characteristics, but not all. An approach for soils where the material below the tillage depth is similar to the surface material is to stockpile the scalped surface material, remove deeper layers to selected depths, and replace the stockpiled surface material. However, such methods have difficulty simulating the effects of long-term cropping on soil weathering, nutrient supply, and organic matter content. They also have difficulty accounting for changes in physical properties, such as soil structure and bulk density, that persist for considerable time.

A third approach is to add suitable soil to the surface of seriously eroded land, thus increasing the thickness of the surface layer in specific increments. This method has the advantage of selecting surface soil with specific characteristics and greater uniformity. Addition of soil on the surface of uneroded downslope sites may also be appropriate to evaluate the effects of deposited sediment on crop yields.

13-5.3 Experiments Using Rainfall/Runoff Simulators or Wind Erosion Simulators

An alternative to the manual removal of surface soil is to apply simulated rainstorms or windstorms, and thereby rapidly accelerate the rate of soil erosion. This method embodies more natural erosion processes and permits selective erosion of the soil, but it does not account for the long-term effects of crop growth and weathering, and it requires special equipment and considerable time and effort.

13-5.4 Laboratory Experiments

Experimental research under controlled conditions, as in a greenhouse or lysimeter, can broaden the range of treatments studied. Undisturbed soil cores or constructed soil profiles can be tested. Such controlled environments permit study of important cause-effect relationships. Disadvantages are the effort to transport the large volumes of soil, possible effects of soil disturbance and mixing, limited range of ambient conditions, and an unnatural environment. Laboratory findings must be verified by field studies.

13-6 STANDARDIZED MEASUREMENTS TO QUANTIFY FACTORS THAT AFFECT PRODUCTIVITY

Productivity is usually measured in terms of the yield of one or more marketable plant components. Yields result from the interaction of numerous physical, chemical, and biological factors. Although some of these factors cannot be measured satisfactorily, many can, which provides the challenge of selecting the proper parameters and measurement criteria. When selecting the measurements to be made, the primary research goal must be foremost, but measurements desired in national modeling efforts should also be considered.

A detailed description of the soil or soils at the research site must always be included. Information on slope steepness, length, aspect, and landscape configuration also should be obtained because of their effects on soil water, runoff, and erosion relations.

Factors external to the plant that determine its marketable yield are embodied in the soil and ambient environments. Within the soil environment, the factors affecting marketable yield can be categorized as physical, chemical, or biological.

13-6.1 Soil Environment

The soil environment includes many elements that affect crop yields. Listed below are data for physical, chemical, and biological properties that should be obtained during erosion/productivity experiments.

13-6.1.1 Physical Properties

1. Water-retention characteristics are needed to determine available water content. Minimum measurements should be field capacity and permanent wilting point.
2. Bulk density is needed to measure available water capacity and to evaluate restrictive horizons.
3. Primary particle-size distribution is needed to evaluate water-transmission characteristics and is desirable when the information is not available from a soil profile description or if a precise evaluation is needed.
4. Strength measurements are needed to evaluate crusts or restrictive layers that affect root penetration, including a laboratory characterization and periodic field measurements to assess long-term changes.

5. Aggregate stability measurements are needed when crusting (sealing of the surface) is a factor and are essential when wind erosion causes abrasion of plant seedling.
6. Measurements of infiltration capacity are needed to assess and predict infiltration, runoff, and sedimentation.
7. Measurements of shrink-swell potentials are needed when sodium concentration is a problem.

13-6.1.2 Chemical Properties

Measurements of nutrient supply or properties affecting the supply are considered essential for evaluation and interpretation of productivity. These include:
Organic carbon.
Total nitrogen.
Computed C:N ratio.
Nitrate and ammonium.
Total phosphorus
Organic phosphorus.
Inorganic phosphorus (by soil test procedure adapted for the region).
Cation exchange capacity.
Base saturation.
pH.
Measurements of other nutrients or chemical properties that affect production because of deficiency or excessive concentration should be made if the problem has been identified on a soil in the region, but these measurements are desirable for other locations as well.
Aluminum saturation.
Sulfur.
Micronutrients such as boron, copper, iron, molybdenum, manganese, and zinc.
Soluble salt concentration.
Sodium adsorption ratio (SAR).

13-6.1.3 Biological Properties

Measurements of these properties are desirable:
Heterotropic index.
Soil respiration.

13-6.1.4 Soil Landscape Characteristics

These relationships are needed to better understand erosional processes on a field or area basis:
Slope steepness and length.
Aspect.

13–6.2   Ambient Environment

Since the crop-production level of agronomically suitable soils is governed more often by water than by any other factor, a measure of the elements of the ambient environment that directly or indirectly affect the water supply is essential in all erosion/productivity experiments. The measurements include:

Precipitation (rain and snow)—date, amount, and intensity.
Irrigation—date, amount, and distribution system.
Total and net radiation.
Maximum-minimum temperature—wet and dry bulb.
Wind speed, direction, and unsheltered distance.
Pan evaporation or other estimate of potential evapotranspiration.
Canopy and soil temperature.

13–6.3   Plant Measurements

Crop yield is the standard for evaluating the effects of erosion on productivity. Measurements of the plant during the growing season often provide insights into causes of yield differences among experimental treatments. Plant measurements considered a minimum include:

Seedling (plant) population.
Leaf area index (LAI) or biomass, every two weeks.
Stage of plant development, weekly.
Yield components, e.g., number of spikes per unit area, number of seeds per spike, and seed weight.
Crop and residue yield.
Root distribution (qualitative).

13–6.4   Management Details

Since crop yields are influenced by management and cultural practices, details of crop establishment are a necessary part of the experimental record. As a minimum, these details should include:

Type and frequency of tillage.
Seeding rate, row spacing, and seeding depth.
Previous crop (cropping sequence).
Amount of crop residues present on the soil surface during the non-growing season and at seeding.
Rate, time, and placement of pesticides.
Rate, time, and placement of fertilizers.
Because the amount of residue influences the erosion rate, residue decomposition rate is a measurement desired by modelers. Suggested measurement frequency is about 6 weeks.

13-6.5 Other Important Factors for Erosion/Productivity Measurements

In any research, a carefully developed statistical plan is needed to select the proper experimental design and appropriate analysis of variance.

Research to measure the effect of erosion on productivity may best be targeted on benchmark soils of major land-resource areas. A list of these soils for the USA has been complied by the director of the Soil Survey Division, SCS. In most cases, such soils are the standards for comparing erosion effects and responses to management that can be expected on soils in the land-resource area. However, other soils may be preferable if the benchmark soil represents a limited area or is of lesser agronomic importance. Major soils with unique profile features and soils on which responses to important cropping and management practices have been determined should be considered.

In making a measurement, frequency, depth or height of the measurement, and increment length must be considered. Frequency is a factor to consider with measurement of ambient elements. Data-acquisition systems now available can log data at essentially any desired time interval. Hourly measurements, either as an integrated value or a point in time, of air and soil temperatures and wind speed will likely meet modeling needs.

Frequency is also important in measuring soil water and nutrients that are mobile in soil, specifically nitrate (NO_3-N). Changes in the position of mobile nutrients during the season can influence crop yield. The minimum frequency of measurement recommended for NO_3-N is at the beginning and end of the growing season. Models such as the Erosion-Productivity Impact Calculator (EPIC) (Williams and Renard, 1983) operate on a daily time step and thus use data obtained at frequent intervals for validation. Weekly measurements of soil water are recommended as a minimum.

Sensor location is a consideration in soil temperature and soil water measurements. Modelers using EPIC recommend temperature measurements during the growing season at depths of 5 and 15 cm and at 30-cm increments thereafter to about 200 cm. Additionally, measurements of soil temperature should be made at least monthly during the nongrowing season. The depth increment for soil-water measurements is usually 30 cm.

13-6.6 Uniformity of Measurements

Just as important as uniformity in the kinds of measurements is uniformity in the method and manner of making and expressing the measurement. For maximum utility, data collected for wide-area model development, model validation, and/or model testing must be in the same units or must be capable of being translated into the same units used in the model. The data must also be derived with the same, comparable, or correlatable techniques. For example, soil-water measurements can be made by gravimetric sampling and with various sensing devices, not all necessarily comparable in sensitivity. Depending on the method of measurement, expression of quantity can be limited to total water or expressed as both total and available. Also, scientists differ on standards acceptable as a measure of the upper and lower limits of "available" water. The method of measurement may affect the ease and, hence, frequency of measurement.

13-7   NEEDED EXPERIMENTAL RESEARCH

To state detailed recommendations for research that would provide needed knowledge on the effects of erosion on productivity would be presumptuous. Locally variable or unknown factors dictate the most needed research and the most appropriate techniques. However, some general suggestions and comments are appropriate.

Certainly, such research must focus primarily on how erosion will affect current and future productivity, not on what has happened in the past. Of greatest concern must be what will happen to the production of major crops as a result of different amounts and kinds of erosion, particularly for the major crop-producing areas of the USA, starting with soils in their current condition. Past yield information is useful only if it materially helps us understand what will happen. Furthermore, additional knowledge of erosion rates, sediment characteristics, the effects of erosion on soil properties, and other indirect effects of erosion on productivity are not useful in quantifying the effect of soil erosion on crop production unless these factors can be accurately correlated with consequent changes in crop yields. Our primary goal must be continued adequate crop production for the indefinite future, not just better evaluations of intermediate means toward that end.

Following are some of the major questions that experimental research needs to answer.

Which crops and crop cultivars are best adapted for different levels of past erosion for different soils and cropping conditions? Since commodity needs change, insects and diseases influence cropping patterns, and various other factors may emerge, the relative response of different crops and cultivars to different conditions must be established. Experience shows that crops and cropping patterns change as erosion progresses, but research is necessary for recommending changes rather than acknowledging them after they happen.

What fertility levels are optimal for different levels of past erosion? Fertility and crop-cultivar tests have been made almost exclusively on the least eroded, highly productive sites. Such tests need to be conducted on typical and subtypical situations as well.

How does past erosion affect crop insect, weed, and disease problems and the effectiveness of pesticides to control such problems? Since erosion usually varies widely within moderately to seriously eroded fields, each such field may have a wide variation in plant population, soil texture, organic matter content, and available water capacity. Such variability may foster problems on certain areas, and it certainly will influence the crop response to pesticides that are most conveniently applied at a constant rate over an entire field.

Will irrigation compensate for reduced water availability due to past erosion, and can it be used without seriously increasing future erosion? Wherever irrigation is feasible now or in the future, its potential for increased erosion as well as increased production should be evaluated. Reduction in plant-available water has been identified as a major, often dominant, consequence of soil erosion (National Soil Erosion-Soil Productivity Research Planning Committee, 1981), so methods to relieve this limiting factor are important research goals.

What tillage practices are most effective for different levels of past erosion on different soils for different crops? In addition to conservation-tillage practices, the applicability and effectiveness of other tillage methods are influenced by past erosion and can affect production. Tillage tools that prepare a good seedbed on well-structured soils may not be effective on some eroded soils. Fertilizer and pesticide placement on eroded soils may require different equipment. Such considerations may make a major difference in crop yield on eroded soils for certain cropping systems.

How rapidly does productive soil form from parent material for different profiles and cropping conditions? Rates of soil formation influence the long-term effect of erosion on productivity and, consequently, the tolerable rates of erosion. Research is needed to evaluate soil-formation rates under modern farming conditions for a wide range of soils, climates, and past erosion.

What means are required to restore productivity to seriously eroded land? Different methods will be effective for different conditions, and the economics of each situation will be crucial. Restoration of productivity may be impracticable or unwise for severe situations.

The researcher also must resolve other more general questions as the research plans are developed.

What experiments are needed to help define the economic relationships between soil erosion and crop production? Obviously, the economics of crop production are tremendously important in relation to all the preceding questions. However, since the costs of crop production and the value of different crops are not positively predictable, research must not be limited by the economic conditions of today or the foreseeable future. Studies are appropriate for conditions that are conceivably realistic, although currently impractical. And, where possible, an accurate evaluation of the economics of the tested treatments should be included.

What is the role of long-term studies of erosion and productivity? The need for, and problems with, long-term research must be fully explored, and, if such studies are merited, they should be initiated soon. The impact of current soil erosion on future productivity may be very important for hundreds or even thousands of years. Projects designed to continue for indefinitely long periods may be essential to resolve this issue, but they will require major commitments in funds, facilities, and personnel. They also pose the prospect of relatively infrequent and long-delayed results. We must decide whether long-term studies are the best approach or whether long-term information can be obtained by other means. The universal soil loss equation (Wischmeier and Smith, 1978) was the result of numerous relatively long-term studies throughout the USA, and it has had a tremendous impact on agricultural land management worldwide.

What questions can be effectively explored by laboratory research? Although field experiments likely will continue to dominate experimental research on this topic, laboratory, greenhouse, or lysimeter experiments have important advantages. The latter can be conducted under controlled environmental conditions, usually require less travel, and provide more precise evaluations. Furthermore, laboratory research may be helpful in making field research more effective and efficient by identifying the most important treatments for field testing.

What experiments are most needed to support mathematical modeling efforts? Recently developed physically and chemically based mathematical models such as EPIC (Williams et al., 1982; Williams and Renard, 1983; Dyke and Heady, 1983) can greatly expand the value and breadth of application of pertinent experimental data. Therefore, researchers should coordinate their experimental planning with appropriate modelers to develop experiments of common usefulness. Mathematical models can also be helpful in identifying major gaps in available experimental data and in selecting the most important research needs.

Obviously, procedural details for conducting erosion/productivity research can best be developed by each research team for its particular situation and goals. Various research approaches were discussed earlier, important data to acquire during the experiments are listed in section 13-6, and pertinent references provide further information. Consultations with other researchers are always wise in planning and conducting such research. A vehicle for national coordination of erosion/productivity research has been developed (see section 13-9).

13-8 AUTOMATED INFORMATION SYSTEMS

A frequent stumbling block to initiating research is lack of knowledge concerning what related research has been conducted, including general objectives, research locations, techniques used, and results. As funding sources become increasingly more selective, the researcher will be asked to demonstrate, quite specifically, the relation of proposed research to similar studies and to current trends in crop production and land management, the unique contributions of the proposed research, and the usefulness of existing methodologies. To ensure that the best decisions are made concerning the direction and coordination of research, the scientific community must fully use existing information. This information can only be obtained through knowledge of, and access to, published sources—journal articles, summaries, reports. In addition, unpublished information exists in theses, miscellaneous data files, and in-service documents; such information is of limited accessibility to all but parochial interests.

How can access to information specific to the effects of soil erosion on productivity be enhanced so that this information can be used more effectively? The answer is the use of automated information systems. Anderson (1982) defines an information system as an organization (of both human and mechanical elements) that maximizes the transformation of data into information. Numerous kinds of information systems, computer databases, and computer programs are available to various sectors of the agricultural scientific community. Current efforts to manage erosion/productivity information should focus on the development of two functionally different information systems. The first would catalog pertinent sources of data, and the second would serve as a repository for data.

A catalog of all sources of information relating to erosion/productivity, both published and unpublished, would greatly enhance the availability and use of previously conducted research. The widespread use of the system would encourage scientists to alert the user community to the existence of their unpublished works. Furthermore, graduate students and others who often do not formally publish their research results would be encouraged to do the same. The most valuable research contributes little to our pool of knowledge about a problem if critical summaries are not available to other scientists studying the same phenomena.

An information system that collects research data offers the potential for more complete analysis and use of the data by interested users than would otherwise be possible. A database management system allows manipulation and analysis of data concurrent with data collection and data loading. In addition, such an information system is cost effective. Experts contend that the cost of storing information electronically will soon be substantially lower than the cost of storing the same amount of information on paper (Office of Technological Assessment, 1981). Automated data systems have the potential of contributing greatly to progress in erosion/productivity research, and they merit a concerted effort toward development and use in this national research program.

13-9 NATIONAL COORDINATION OF RESEARCH TO QUANTIFY THE EFFECT OF SOIL EROSION ON CROPLAND PRODUCTIVITY

In 1980, the ARS–USDA established a National Soil Erosion-Soil Productivity Research Planning Committee (1981) to help foster research to meet the mandates of the national Soil and Water Resources Conservation Act (RCA) of 1977. A National Soil Erosion-Soil Productivity Planning Workshop was held in September 1981 at Lafayette, IN. Representatives of all known research efforts on this topic were invited. The attendees, including representatives of USDA–ARS, USDA–SCS, USDA–ERS, agricultural experiment stations, TVA, and Resources for the Future, were divided into four thrust groups: erosion mechanics, conservation tillage, mathematical modeling, and experimental research. Reports of the first three groups are included in other chapters of this volume, and those of the fourth are included throughout this paper but specifically as follows.

The objectives of the experimental research effort are (i) to describe techniques and experimental procedures to help determine how soil erosion affects productivity, (ii) to design and develop experiments to quantify the effect of soil erosion on productivity, (iii) to obtain experimental results that will be useful in improving the predictive capability of mathematical models for quantifying the effect of different amounts of erosion on productivity, and (iv) to develop practices to reduce the negative consequences of erosion on production and to increase production on seriously eroded land. Current (1983) and planned experimental research toward these objectives (with the appropriate contact persons and locations) include:

Evaluate the effect of loss of "A" horizon on productivity of Williams loam (Don L. Tanaka, USDA–ARS, Sidney, MT).

Effect of topsoil thickness on wheat production in a wheat-fallow system (Truman W. Massee, USDA–ARS, Kimberly, ID).

Evaluate the effect of soil erosion on productivity of soils developed on glacial till in the Northern Great Plains (Armand Bauer, USDA–ARS, Mandan, ND).

Evaluate the long-term effects of soil erosion on crop yields, inputs for crop production, and change in soil productivity with time (Tom McCarty, University of Missouri, Columbia).

Effects of topsoil thickness and soil horizonation (horizon sequence) on soybean yields of representative soils of the southeastern United States (D. E. Pettry, Mississippi State University, Mississippi State).

Quantify the effects of soil erosion on crop yields and elucidate mechanisms that control the effects of soil erosion on crop production (A. W. White, USDA–ARS, Watkinsville, GA).

Effects of long-term erosion on soil properties and productivity under natural conditions (L. R. Ahuja, USDA–ARS, Durant, Oklahoma).

Evaluate the effect of past erosion on major cultivated soils in Alabama (Ben F. Hajek, Auburn University, Auburn, AL).

Evaluate root growth in sub-horizons (Ben F. Hajek, Auburn University, Auburn, AL).

Effect of degree of erosion on soil productivity in North Carolina (J. W. Gilliam, North Carolina State University, Raleigh).

Develop a procedure for determining the effect of past erosion on present productivity (Dave Schertz, SCS, Purdue University, W. Lafayette, IN).

Quantify the influence of past erosion on soil productivity, Objective 4 of Southern Regional Project S-174 (Wendell Gilliam, S-174 Chm., North Carolina State University, Raleigh).

Determine effect of past soil erosion on soil productivity (Joe O. Sanford, USDA–ARS, Mississippi State University, Mississippi State).

Quantify the effect of soil depth above a fragipan on root development, water extraction, and yields (Donald D. Tyler, University of Tennessee-Jackson).

Long-term erosion effects on crop yields on irrigated lands (David L. Carter, USDA–ARS, Kimberly, ID).

How erosion affects the physical and chemical properties of soil and soil productivity of Minnesota soils (William E. Larson, University of Minnesota, St. Paul).

Determine the different soil physical, chemical and biological factors which are influenced by erosion and how changes in these factors influence the productivity of selected benchmark soils in the USA (Doug Karlen, ARS, Florence, SC).

Characterize physical and chemical properties of a limited number of severely eroded soils from several regions in the USA (Don McCormack, SCS, Washington, DC).

Soil erosion and soil productivity losses (Allan Olness, USDA–ARS, Morris, MN).

Evaluate and interpret data from 37-year standard runoff-erosion study conducted on a claypan soil (Robert Burwell, USDA–ARS, Columbia, MO).

Soil profile characteristics as a factor in determining the effect of soil erosion on productivity (Fred E. Rhoton, USDA–ARS, Oxford, MS).

Erosion-induced changes in soil respiration and soil organic matter and their effects on crop productivity (G. R. Benoit, USDA–ARS, Morris, MN).

The nutrient value of topsoil on furrow irrigated land (David L. Carter, USDA–ARS, Kimberly, ID).

The effects of production practices for soybeans on soil erosion, yield, and economics of production (Joe O. Sanford, USDA–ARS, Mississippi State, MS).

Economics of restoring productivity of severely eroded blackland soils (Joe O. Stanford, USDA–ARS, Mississippi State, MS).

Develop and evaluate individual practices and integrated systems for effective control of erosion and of sediment yield (Curtis Shelton, University of Tennessee, Knoxville).

Relationship of crop yields to erosion, soil moisture, and other physical and chemical properties on limestone valley soils (Ben F. Hajek, Auburn University, Auburn, AL).

FARMS Project: Field appraisal of resource management systems (Paul R. Hepler, University of Maine, Orono).

Crop systems, tillage, and irrigation to restore crop yields (George Langdale, USDA–ARS, Watkinsville, GA).

Estimates of long-term costs of soil erosion (Pierre Crosson, Resources for the Future, Washington, DC).

Determine relationships among soil erosion, potential productivity, soil characteristics, and costs of erosion (Lynn L. Reinschmiedt, Mississippi State University, Mississippi State).

Erosion losses and soil productivity changes in landscapes and major land resource areas in the Pacific Northwest (R. R. Allmaras, USDA–ARS, Pendleton, OR).

Evaluation of tillage systems for corn belt soils in Indiana (Don Griffith, Purdue University, W. Lafayette, IN).

Close-range photogrammetric techniques to measure soil erosion (A. W. Thomas, USDA–ARS, Watkinsville, GA).

Develop methods to extend conclusions from plot experiments to larger areas (A. S. Rogowski, USDA–ARS, University Park, PA).

Some of this research is ongoing, some is just getting started, and the rest is awaiting funding. We welcome information on other ongoing or planned experimental research. Further planning efforts and continued coordination of current research among projects and with closely related research in erosion mechanics, conservation tillage, and mathematical modeling is planned, but progress will be slow until more financial support is available.

13-10 ACKNOWLEDGMENTS

We thank those persons participating in the Experimental Research Thrust Group discussions and other Soil Erosion-Soil Productivity Planning Workshop participants for many of the items included in this paper. Special thanks to Caroline Yonker, Research Associate, Colorado State University, for her contributions in writing the section on Automated Information Systems.

REFERENCES

Anderson, D. L. 1982. Criterion for developing a comprehensive soil resource information system. M.S. thesis. Colorado State Univ., Ft. Collins.

Bartholomew, R. P., D. A. Hinkle, and K. Engler. 1954. Effect of rainfall characteristics and soil management practices on soil and water losses in northwest Arkansas. Arkansas Agric. Exp. Stn. Bull. 548.

----, ----, and ----. 1955. Soil and water losses in southwest Arkansas as influenced by rainfall characteristics and soil management practices. Arkansas Agric. Exp. Stn. Bull. 550.

Bauer, A. 1968. Effects of zinc fertilizer on corn yield in southeastern North Dakota. North Dakota Agric. Exp. Stn. Farm Res. 25(1):3-8.

Baver, L. D. 1950. How serious is soil erosion? Soil Sci. Soc. Am. Proc. 15:1-5.

Bell, F. F. 1977. Land use of agricultural production: Agricultural land use and soil erosion. p. 45-54. In L. F. Seatz (ed.) Ecology and agricultural production. University of Tennessee, Knoxville.

Bennett, H. H. 1939. Soil conservation. McGraw-Hill Book Co., New York.

Black, A. L., and B. W. Greb. 1968. Soil reflectance, temperature, and fallow water storage on exposed subsoils of a brown soil. Soil Sci. Soc. Am. Proc. 32:105-109.

Buntley, G. J., and F.F. Bell. 1976. Yield estimates for major crops grown on soils of west Tennessee. Tennessee Agric. Exp. Stn. Bull. 561.

Carlson, C. W., and D. L. Grunes. 1958. Effect of fertilization on yields and nutrient content of barley. Soil Sci. Soc. Am. Proc. 22:140-145.

----, ----, and J. Alessi. 1961. Fertilizer needs of subsoil areas exposed by land leveling operations. North Dakota Agric. Exp. Stn. Farm. Res. 21(9):12-15.

Denton, H. P. 1978. The effects of degree of erosion and slope characteristics on soybean yields on Memphis, Grenada, Lexington, and Loring soils. M.S. thesis, Univ. of Tennessee, Knoxville.

Deutsch, P. C., E. E. Jukkola, C. Weil, and C. M. Yonker. 1981. Catalog of resource information, project report 1.0. Cooperative project: USDI Bureau of Land Management, Denver, CO, and College of Agricultural Sciences, Colorado State University, Fort Collins.

Dyke, P., and E. O. Heady. 1983. Assessments of soil erosion and crop productivity with process and economic models—economic models. p. 105-117. In R. F. Follett (ed.) Soil erosion and crop productivity. American Society of Agronomy, Crop Science Society of America, and Soil Science Society of America, Madison, WI.

Eck, H. V. 1969. Restoring productivity on Pullman silty clay loam subsoil under limited moisture. Soil Sci. Soc. Am. Proc. 33(4):578-581.

Engelstad, O. P., W. D. Shrader, and L. E. Dumenil. 1961a. The effect of surface soil thickness on corn yields: I. As determined by a series of field experiments in farmer-operated fields. Soil Sci. Soc. Am. Proc. 25:494–497.

————, and ————. 1961b. The effect of surface soil thickness on corn yields: II. As determined by an experiment using normal surface soil and artifically-exposed subsoil. Soil Sci. Soc. Am. Proc. 25:497–499.

Frye, W. W., S. A. Ebelhar, L. W. Murdock, and R. L. Blevins. 1982. Soil erosion effects on properties and productivity of two Kentucky soils. Soil Sci. Soc. Am. J. 46(5):1051–1055.

Grunes, D. L., L. C. Boawn, C. W. Carlson, and F. G. Viets, Jr. 1961. Zinc deficiency of corn and potatoes as related to soil and plant analyses. Agron. J. 53:68–71.

Gunderson, O., D. Bezdicek, and J. MacGregor. 1965. Zinc deficiency on corn in Minnesota. Minnesota Agric. Ext. Ser. Bull. 322.

Haise, H. R., L. R. Jensen, and J. Alessi. 1955. The effect of synthetic soil conditioners on soil structure and production of sugar beets. Soil Sci. Soc. Am. Proc. 19:17–19.

Hauser, V. L. 1968. Conservation bench terraces in Texas. Trans. Am. Soc. Agric. Engr. 11(3):385–386, 392.

Heilman, M. D., and J. R. Thomas. 1961. Land leveling can adversely affect soil fertility. J. Soil Water Conserv. 16(2):71–72.

Johnson, W. M. 1963. The pedon and the polypedon. Soil Sci. Soc. Am. Proc. 27:212–215.

Langdale, G. W., J. E. Box, Jr., R. A. Leonard, A. P. Barnett, and W. G. Fleming. 1979. Corn yield reduction on eroded Southern Piedmont soils. J. Soil Water Conserv. 34(5): 226–228.

Lutwick, L. E., and E. H. Hobbs. 1964. Relative productivity of soil horizons, singly and in mixture. Can. J. Soil Sci. 44:145–150.

Lyles, L. 1977. Wind erosion: Processes and effect on soil productivity. Trans. Am. Soc. Agric. Engr. 20(5):880–884.

McCormack, D. E., and R. R. Moorman. 1983. Sequential testing—within-field data collection to determine erosion impacts. Proc. Intl. Conf. on Soil Erosion and Conservation, Honolulu, HI. 16–22 Jan. 1983.

Murray, W. G., A. J. Englehorn, and R. A. Griffin. 1939. Yield tests and land evaluation. Iowa Agric. Exp. Stn. Res. Bull. 252.

National Soil Erosion-Soil Productivity Planning Committee. 1981. Soil erosion effects on soil productivity: A research perspective. J. Soil Water Conserv. 36:82–90.

Odell, R. T. 1950. Measurement of the productivity of soils under various environmental conditions. Soil Sci. Soc. Am. Proc. 42:282–292.

Office of Technology Assessment. 1981. Computer based national information systems— technology and public policy issues. OTA 81-600144. Washington, DC.

Olson, T. C. 1977. Restoring the productivity of a glacial till soil after topsoil removal. J. Soil Water Conserv. 32(3):130–132.

Reuss, J. O., and W. L. Lindsay. 1963. Do Colorado crops need zinc and iron? Colo. State Univ. Coop. Ext. Ser. Pamphlet 59. Colorado State University, Fort Collins.

Ripley, P. O., W. Kalbfleisch, S. J. Bourget, and D. J. Cooper. 1961. Soil erosion by water. Canada Dep. of Agric. Pub. 1093, Ottawa.

Robinette, C. E. 1975. Corn yield study on selected Maryland soil series. M.S. thesis. Univ. of Maryland, College Park.

Soil Conservation Society of America. 1976. Resource conservation glossary. 2nd ed. Soil Conserv. Soc. Am., Ankeny, IA.

Soil Science Society of America. 1978. Glossary of soil science terms. Soil Science Society of America, Madison, WI.

Stallings, J. H. 1950. Erosion of topsoil reduces productivity. SCS-TP-98. USDA, Washington, DC.

Uhland, R. E. 1949. Crop yields lowered by erosion. SCS-TP-75. USDA, Washington, DC.

Williams, J. R., P. T. Dyke, and C. A. Jones. 1982. EPIC—A model for assessing the effects of erosion on soil productivity. Presented at the Third Intl. Conf. on State-of-the-Art in Ecological Modeling, Colorado State University, Fort Collins, CO, 24–28 May 1982.

----, and K. G. Renard. 1983. Assessments of soil erosion and crop productivity with process and economic models—process models (EPIC). p. 67–103. *In* R. F. Follett (ed.) Soil erosion and crop productivity. American Society of Agronomy, Crop Science Society of America, and Soil Science Society of America, Madison, WI.

Winters, E., and R. W. Simonson. 1951. The subsoil. *In* A. G. Norman (ed.) Adv. Agron. 3:1–92.

Wischmeier, W. H., and D. D. Smith. 1978. Predicting rainfall erosion losses—a guide to conservation planning. USDA Agric. Handb. 537. Washington, DC.

14

Regional Effects of Soil Erosion on Crop Productivity— Northeast

W. Shaw Reid
Cornell University
Ithaca, New York

14-1 SOILS OF THE NORTHEASTERN USA

Soils of the Northeast are extremely variable in properties that influence crop production. Major groups of soils are associated with the physiographic and topographic features of the landscape and are grouped into all or portions of five land-resource regions (USDA–SCS, 1981):

Lake States fruit, truck, and dairy region (eastern portion).

Northeastern forage and forest region.

East and Central general farming and forest region (northeast portions).

Northern Atlantic slope truck, fruit, and poultry region.

Atlantic and Gulf Coast lowlands, forest, and truck (northern portion).

The Northeast has four major topographic divisions—lowland plains, upland plateau, ridge and valley, and mountainous (Brady et al., 1957). The lowland plains is divided into three parts: the sea coast and coastal plain of New England and some Piedmont sections of Maryland, Delaware, and southern New Jersey; the Mohawk River Valley of New York and the lake plain area of western New York and northwestern Pennsylvania.

The ridge and valley division starts in south-central Pennsylvania and extends northeastward to include the Hudson Valley of New York.

The upland plateau includes the inland nonmountainous sections of New England and the northern Piedmont area of northwestern New Jersey, central Maryland, West Virginia, and southeastern Pennsylvania.

Published in R. F. Follett and B. A. Stewart, ed. 1985. *Soil Erosion and Crop Productivity*.
© ASA-CSSA-SSSA, 677 South Segoe Road, Madison, WI 53711, USA.

Mountainous areas with little agriculture include the rugged Adirondacks in New York, the Green Mountains in Vermont, the Berkshire Hills of Massachusetts, and the White Mountains in New Hampshire and Maine. The Allegheny, Catskill, and Taconic Mountain ranges have more gentle slopes with small, fairly level valleys and, therefore, can support some small farms.

Three general soil groups, based upon parent material, occur within the region: (i) soils developed on unconsolidated deposits of lacustrine and marine materials located in the Erie, Ontario, and Champlain Lake Plains, the Atlantic Coastal Plain, and scattered along the coast of New England; (ii) soils developed from glacial materials located in northern New Jersey, Pennsylvania, and New England; and (iii) soils developed from the underlying consolidated rock.

The areas of soils developed in the lacustrine and marine materials generally have nearly level to gently sloping landscapes and intensive agricultural production.

The glaciated landscapes usually have complex topography with narrow, nearly level valleys, steep sideslopes, and sloping to level hilltop plateaus. The soils commonly contain stones and/or gravel, and are not highly weathered. The glacial till is usually quite compact and tends to restrict water movement and root growth. Glacial till areas also tend to have a relatively high percentage of wet and/or shallow soils on the side slopes. Agriculture is usually concentrated either in the valleys or in a combination of valley and plateau.

Soils of the nonglaciated region have more profile development and less stones and gravel than the glaciated region. A wide range of slopes, from nearly level to very steep, occurs within the nonglaciated area. Agriculture is intensive only on the nearly level to rolling landscapes.

Most Northeast soils are in four orders: Spodosols, Inceptisols, Alfisols, and Ultisols. Entisols and Histosols also occur. There are five major suborders: Orthods, Ochrepts, Udalfs, Udults, and Aqualts. Most of the Orthods are in the great group of Haplorthods and are common in Maine, New Hampshire, Vermont, Massachusetts, and Connecticut, with scattered areas in New York. Much of the cation exchange capacity in these soils is the result of the organic matter that accumulates on or in the surface. Many of these soils tend to be droughty because they are shallow and sandy. They respond well to good management and to lime and fertilizers; therefore, they are good vegetable and potato (*Solanum tuberosum* L.) soils when the slopes permit mechanization and erosion protection.

The Ochrepts occur as two great groups, the Dystrochrepts and the Fragiochrepts. The Dystrochrepts in Pennsylvania and West Virginia are low in bases, are often on sloping topography, and are usually shallow to bedrock. The Fragiochrepts are mainly found in northern Pennsylvania and southern New York. These soils are low in bases and have a fragipan horizon, poor internal drainage, and gently to strongly sloping topography.

The dominant Udalfs of the region are Hapludalfs. These soils have a relatively thin subsurface clay horizon, are generally high in bases, occur on gentle slopes, and are considered the most productive of the region. The dominant Udults of the region are Hapludults. These soils have a relatively thin subsurface horizon of clay accumulation and are low in bases, but they respond well to management and support many crops.

The dominant Aquults are Ochraquults and are confined to the coastal areas of New Jersey, Maryland, and Delaware. They are nearly level to gently sloping, low in bases, and, without drainage, are seasonally wet. Where drained, a fairly wide range of crops can be grown.

Sandy loams, loams, and loamy sands are dominant in New England and along the coastal area of the rest of the region. Silt loams prevail throughout the remainder of the region. Localized areas of silty clay loams and clay loams occur in the lacustrine regions. Many of the soils are high in organic matter in the surface horizons, which contributes significantly to the cation exchange capacity (CEC) of the soils. The sandy soils have CECs from 5 to 10 c mol $(p+)$/kg, while the range for the silt loams and clay loams is between 10 and 20 c mol $(p+)$/kg.

The shallow and sloping soils of the Northeast have a very high potential for erosion damage. About 30% of the soils in the Northeast have bedrock at depths of 150 cm or less and a much larger percentage have bedrock at depths just greater than 150 cm. About 74% of the soils in the Northeast have T values of 6.7 t ha^{-1} yr^{-1} or less, compared to 40% nationally. Ninety percent have T values of 10 or less, compared to 48% nationally. An additional 5% of the soils have T values of 11 and erodibility values of 0.37 or more, making them susceptible to erosion. Therefore, 95% of the soils in the Northeast are susceptible to erosion damage.

Data in Table 14-1 from the National Resources Inventory (USDA–SCS, 1978) show that 24% of the cropland and 60% of the pastureland in the Northeast is eroding at 11.2 t ha^{-1} yr^{-1} or more. The erosion is less than 4.5 t ha^{-1} yr^{-1} on 55% of the cropland and 85% of the pasture. Thus,

Table 14-1. Sheet and rill erosion on cropland and pastureland in the northeastern USA (USDA–SCS, 1978).

	Erosion (t^{-1} yr^{-1})							
	Cropland				Pastureland			
State	<4.5	4.5–11.0	11.1–31.2	>31.2	<4.5	4.5–11.0	11.1–31.2	>31.2
	thousands of ha							
Connecticut	50	11	10	10	42	2	1	--
Delaware	89	100	25	6	8	--	--	--
Maine	270	30	39	28	99	--	1	--
Maryland	236	227	149	66	157	24	10	5
Massachusetts	80	21	11	2	34	1	1	--
New Hampshire	94	10	5	1	38	--	--	--
New Jersey	139	51	83	41	56	--	--	2
New York	1446	433	345	189	829	53	30	12
Pennsylvania	1142	484	429	224	560	83	48	35
Rhode Island	9	--	2	1	6	1	--	--
Vermont	187	21	11	2	34	14	1	--
Total	3742	1388	1119	570	1863	178	92	54
% of total	55	20	16	8	85	8	4	2

erosion exceeds the T-value on 24 to 45% of the cropland and 6 to 12% of the pasture in the Northeast.

These relatively low erosion rates compared to the potential of 95% of the soils are the result of corn-hay rotations on much of the sloping crop-lands and the occurrence of relatively large areas of nearly level to gently sloping land that can be used for continuous row crops such as corn (*Zea mays* L.). In most areas of the Northeast, there is more potential cropland than is being used and therefore little need to grow continuous row crops on the more sloping soils. Farmers of the Northeast are aware of erosion hazards, and most provide some type of soil protection. However, eco-nomic constraints often become more critical than erosion control, or are perceived to be so.

14–2 CLIMATE OF THE REGION

Precipitation within the Northeast varies between 60 and 125 cm (USDA, 1981). Most areas receive from 75 to 110 cm. Precipitation is dis-tributed relatively uniformly throughout the year, but the northern parts of the region get slightly less during the winter and all parts tend to get slightly more during the spring (April and May). Much of the December to March precipitation occurs as snow in the northern parts. Even though annual dis-tribution is relatively uniform, crops suffer from lack of water for short periods almost every year.

Rainfall intensity is maximum from June through September and water runoff and soil erosion are greatest during this period. In the northern part of the region, the soils are usually frozen from January through mid-March. Rainfall and snow melt during this period result in runoff but normally do not produce much erosion.

The average freeze-free season is from 110 to 200 days, but a 30-day variation is not unusual. The average annual air temperature ranges from 3 to 15°C. The colder areas are in Maine and the warmer areas are in the southern part of the region. Most of the region has an average annual air temperature in the 7 to 12°C range.

14–3 CROPS OF THE REGION

The 192 000 privately owned farms occupy only about 30% of the total land area of the Northeast (Table 14–2, U.S. Bureau of the Census, 1978). In 1978 just over 5.5 M ha were harvested cropland.

Much of the pastureland is not suited for cropland because of steep slopes, rock outcrops, or shall or wet soils. Within the past 20 years, farmers have tended to specialize in only one type of farm enterprise. For example, dairy farmers tend to grow only forage and grain crops to feed the cows, fruit growers grow only fruits, and cash grain farmers produce corn, small grains, and soybeans [*Glycine max* (L.) Merr.]. Specialization decreases the opportunities for crop selection and rotations as a means of erosion control.

Hectarage of hay exceeds that of any other crop except pasture in 9 of the 11 states in the Northeast (Table 14-3). Hay and pasture are important to the agriculture of the Northeast for the forage supplied to the dairy industry and for erosion control. The steeper slopes are usually in a rotation containing 5 or more years of hay with only 1 to 3 years of corn. Corn is important in the rotation for weed control and nitrogen use before reestablishment of the hay crop, but it increases the erosion potential because of lack of soil cover. Also, much of the corn grown on the dairy farm is for silage, thus reducing the organic residues left on the soil surface. Each dairy farm usually has some continuous corn fields. These fields tend to be on the nearly level slopes near the barn and receive much of the manure.

Vegetable crops occupy 258 000 ha (Table 14-3) and tend to be grown on the same areas year after year. Vegetables usually are grown on the sandy soils with nearly level to gentle slopes. Winter cover crops are usually

Table 14-2. Use of farmland in the northeastern USA (U.S. Bureau of the Census, 1978).

State	Number of farms	Area in farms	Cropland harvested	Cropland in pasture	Total pastureland
		%		ha	
Connecticut	4 560	16.1	72 420	22 470	51 030
Delaware	3 632	53.5	202 670	7 560	14 200
Maine	8 158	8.1	197 620	39 720	88 785
Maryland	18 727	42.9	612 425	100 335	188 145
Massachusetts	5 891	13.6	86 535	31 365	58 865
New Hampshire	3 288	9.4	55 405	18 845	42 970
New Jersey	9 895	21.8	249 205	33 590	59 475
New York	49 273	32.4	1 815 615	483 975	1 021 725
Pennsylvania	59 942	30.4	1 759 340	420 145	882 075
Rhode Island	866	11.1	10 125	3 140	5 655
Vermont	7 273	29.6	235 220	102 870	217 780
Total	192 137		5 541 260	1 593 910	3 355 835

Table 14-3. Hectareage of crops in the northeastern USA (U.S. Bureau of the Census, 1978).

State	All hay	Corn	Small grains	Soybeans	Potatoes	Vegetables	Fruit
				thousands of ha			
Connecticut	40	22			2	5	3
Delaware	9	68	9	107	2	17	--
Maine	102	17	17		48	5	4
Maryland	109	277	69	153	1	21	5
Massachusetts	54	18			2	6	4
New Hampshire	43	10				2	2
New Jersey	56	55	13	25		30	9
New York	1044	510	140	10	20	63	56
Pennsylvania	813	663	240	31	10	20	29
Rhode Island	5	1			2	1	--
Vermont	200	42	1			1	3
total	2475	1683	489	326	87	171	115

planted on vegetable lands but often do not provide adequate erosion control because a short growing season reduces fall growth and severe winter weather reduces stands.

Fruit crops (especially grapes) are grown in particular areas based upon the microclimate; therefore, more attention is given to climate than to soils or slopes. Most orchard-management schemes provide for a grass cover, especially between the rows. Some orchards that are clean-cultivated; therefore, on slope areas erosion is a problem.

14-4 EROSION AND CROP YIELDS

Experiments on the causes and effects of soil erosion were conducted by the SCS and others beginning in the mid-1930s. The two principal locations from which information was published on soil erosion and crop yields were the Arnot Experimental Station (Lamb et al., 1944) and the Marcellus Experiment Station (Free et al., 1946) in New York, with additional work located near Hammondsport and Geneva. The Arnot station was located in the upland-plateau region with acid, glaciated soils at an elevation of about 575 m. The valley floors of the region are about 300 m, and some of the hilltops exceed 600 m elevation. The Marcellus station was located in the high-lime region of glaciated soils and included work in the lake-plain region. This station was at an elevation of 300 m, but other experimental work was conducted at elevations from 100 to 500 m. The soils at the Arnot station had been cultivated for more than 60 years, while those at Marcellus had been cultivated for at least 150 years. Measurements were made of total rainfall, maximum intensity, runoff, water and soil losses for various slopes and slope lengths, amounts of crop cover, and contour planting, as well as other management factors. Crop yields were not always reported.

A survey of soil erosion conducted in New York and reported by Howe and Adams (1936) showed that entire farms were going out of production and overall crop yields were decreasing as a result of erosion. A similar survey on soil erosion was reported by A. L. Patrick (1938) for Pennsylvania. Therefore, no great need existed to document the effects of erosion on crop yields. Instead, the priorities were to determine how to recognize and control erosion. However, the influence of soil erosion on crop yields has been reported for a few studies in the Northeast.

A series of plots was established on a Bath silt loam (coarse-loamy, mixed, mesic Typic Fragiochrept) with a 20% slope to study the effect of fertilizer and crop rotation on soil loss, water loss, and corn yields (Table 14-4, Lamb et al., 1944). Applying fertilizer (although at low rates by today's standards) more than doubled corn yields and reduced soil and water losses by one-third. Growing the corn in rotation with clover more than doubled corn yields over those of fertilized plots and decreased soil erosion to less than 1 t/ha. Lamb et al. (1944) stated that erosion and continuous cultivation had depleted the soil of its organic matter, washed away much of the finer soil particles, and decreased aggregation. As a result, the soil had become compacted and the surface was sealed at the onset of rain, thus increasing runoff and soil loss. In addition, plant nutrients, especially nitrogen, were lost.

Potato yields were measured in a field in 1942 and 1943 (soil unknown, but probably a Fragiochrept). Yields were reduced by 50 to 100% from soil erosion, depending on the depth to what was reported as the C horizon (Table 14-5, Lamb et al., 1944). This C horizon could have been a fragipan, a slowly permeable subsoil, or compact glacial till. The quantity of fertilizer used was not reported, but the usual rate at that time for potatoes was 56, 49, and 47 kg/ha of N, P, and K, respectively.

In southern New York, Lamb et al. (1944) found that yields of oats (*Avena sativa* L.), beans (*Phaseolus vulgaris* L.), potatoes, and silage corn decreased as depth of topsoil decreased (Table 14-6). Soil organic matter content was also lower on soils with thinner topsoils.

On a Dunkirk silty clay loam, erosion reduced yields of sweet corn 3.3 t/ha (31%) and those of cabbage 6.5 t/ha (17%), even though 131 kg/ha of N was applied to the severely eroded plots and no N was applied to the non-eroded areas (Free et al., 1946). Roots, especially the smaller fibrous roots of the cabbage plants, were severely reduced in the eroded soil. This absence of fibrous roots evidently reduced N uptake.

On another field, the numbers and grade of nursery-rose stock were determined from severely eroded (mostly exposed subsoil) and moderately eroded areas (Table 14-7, Free et al., 1946). Only the number 1 and 1½ grades have commercial value; therefore, the yield of roses on the severely

Table 14-4. Effects of fertilizer and crop rotation on soil loss, water loss, and corn yields on Bath silt loam (coarse-loamy, mixed, mesic Typic Fragiochrept) from 1935 to 1943 at the Arnot Experimental Farm, NY (Lamb et al., 1944).

Cropping and fertilizer† practices	Water loss	Soil loss	Silage corn yield
	%	——— t/ha ———	
9 years corn + 0-0-0	9.5	6.7	4.0
9 years corn + 11-10-9	6.4	4.0	9.5
Corn-clover rotation + 0-24-0	2.3	0.8	22.1

† Fertilizer in kg/ha for N, P, and K, respectively.

Table 14-5. Effect of previous erosion on yields of potatoes in Steuben County, NY (Lamb et al., 1944).†

Depth to C horizon‡	Past erosion	Slope	Yield	
			1942	1943
cm		%	——— t/ha ———	
102	Some deposition	1	34.8	23.7
56	Moderate	9	29.7	--
46	Severe	8	27.8	16.9
36	Severe	10	23.3	12.6

† Data are averages from five plots.
‡ These figures probably indicate depth to a slowly permeable subsoil or fragipan rather than the C horizon as reported in 1944.

Table 14–6. Effect of previous erosion on crop yields and organic matter content
in southern New York in 1943 (Lamb et al., 1944).

Crop	Soil type	Slope	Depth of topsoil	Organic matter	Crop yield
		%	cm	%	kg/ha
Oats	Erie silt loam	12	15	2.5	820
	(coarse-loamy, mixed, mesic Aeric Fragiaqualf)	12	23	3.6	1 360
Beans	Darien silt loam	15	13	2.5	340
	(fine-loamy, mixed, mesic Aeric Ochraqualfs)	15	33	4.3	940
Potatoes	Mardin silt loam	6	33	2.9	940
	(coarse-loamy, mixed, mesic Typic Fragiochrept)	5	28	3.3	10 960
Silage corn	Canfield silt loam†	12	15	4.3	16 140
	(fine-loamy, mixed, mesic Aquic Fragiudalf)	10	33	5.5	25 460

† Currently correlated as Mardin silt loam.

Table 14–7. Grade and yield of rose plants on a severely and moderately eroded Camillus
silt loam (fine-loamy, mixed, mesic Dystric Eutrochrept) in Ontario County, NY,
during 1944 (Free et al., 1946).

Erosion	Depth of soil†	Organic matter	Rose plants per 100 m of row		
			Grade	Yield	Total
	cm	%			
Severe	36	0.9	No. 1	0	
			No. 1½	4	
			No. 2	37	41
Moderate	84	2.6	No. 1	42	
			No. 1½	20	
			No. 2	12	74

† Depth of soil over shale. Most, if not all, of the original topsoil was missing from the severely
eroded areas.

eroded soil was essentially nil. Under good management the moderately eroded land, however, continued to produce.

Erosion decreased yields of essentially all crops tested and on a number of soils. These experiments were conducted using little or no fertilizer nitrogen. Under these conditions the mineralization of nitrogen from the soil organic matter was probably one of the more important factors controlling yields. The percentage of organic matter always decreased as severity of erosion increased.

Since water runoff increased as the severity of erosion increased, the lower yields could have been the result of a lower quantity of available water.

14-5   CONTOUR PLANTING

Planting potatoes in rows on the contour rather than up and down the hill resulted in an annual average of 638 kg/ha more potatoes for 23 tests conducted between 1935 and 1943 (Lamb et al., 1944). The soils were Bath and Mardin silt loams with slopes of 10 to 20%. The advantage of contour planting ranged from −202 kg/ha for a high rainfall year to +2420 kg/ha in one year with moderate rainfall.

Crops planted on the contour on a Honeoye silt loam (fine-loamy, mixed, mesic Glossoboric Hapludalf) with a 10% slope produced 2697 kg/ha (43%) more grain in 1943 at Marcellus, NY, than up-and-down hill plantings (Free et al., 1946). On a Dunkirk soil with a 5% slope at Geneva, NY, the yields of sweet corn and cabbage were 35% and 5% greater when planted on the contour. Concord grape (*Vitis vinifera* L.) yields (shown in Table 14–8) were from adjacent vineyards on a well-drained Alton gravelly loam (loamy-skeletal, mixed, mesic Dystric Eutrochrept) with a slope of 8% or less (Shaulis and Carleton, 1947). Yields were related to position on the slope, contour, and organic matter. Where soil organic matter contents were similar, across-the-slope plantings produced 50% or more grapes per vine than up-and-down-the-slope plantings.

Thus, the collection of water by across-the-slope or contour rows was apparently important to yields of crops. The effect of additional water probably explains data of R. Musgrave (unpublished data, Cornell University, Ithaca, NY; Table 14–9) on a Honeoye silt loam soil near Aurora, NY. Corn yields increased an average of 1460 kg/ha/year on areas where topsoil had been added, even though adequate fertilizer N had been added to both soils. The increases in yields were large in the dry years, and no difference was obtained in the year of above-normal rainfall. Thus, the water-holding capacity of the soil is an important criterion in determining effects of erosion.

Table 14–8. Yield of Concord grapes and soil organic matter as influenced by slope position and direction of planting in Chautauqua County, NY, in 1946 (Shaulis and Carleton, 1947).

Position on slope	Soil organic matter	Yield
	%	kg/vine
Vineyard A: rows up and down slope		
Top	0.9	1.1
Middle	1.2	2.3
Bottom	2.1	3.0
Vineyard B: rows across slope		
Top	1.0	2.5
Middle	1.7	3.7
Bottom	2.1	4.7

Table 14-9. Effect of topsoil depth on the yield of corn at Aurora, NY, from 1962 to 1967
(R. Musgrave, Cornell University, Ithaca, NY, unpublished data).[†]

	Corn yields		Deviation from mean of May–Sept. precipitation
Year	Normal topsoil (20–25 cm)	Normal plus added topsoil (45–50 cm)	
	kg/ha		cm
1962	7970	8290	−3.0
1963	6280	7850	−4.0
1964	3450	6720	−13.7
1965	2320	4080	−12.4
1966	5650	7470	−2.8
1967	9790	9790	+10.3
Mean	5910	7370	−4.3

[†] Fertilizers applied yearly at 300, 90, and 160 kg/ha of N, P, and K, respectively.

14-6 EFFECTS OF GOOD MANAGEMENT OF ERODED SOILS

Most of the cultivated soils of the Northeast are sloping and are moderately to severely eroded. Lamb et al. (1950) conducted a series of experiments to determine if, under good management, soil productivity would recover from the effects of soil erosion. Good management was generally considered to include rotation of crops, adequate fertilization, and other erosion control measures such as cover crops and surface crop residues. Poor management, such as continuous cultivation of a grain or vegetable crop with no erosion control, was continued from previous studies to provide a control comparison.

The Bath flaggy silt loam at the Arnot Experimental Station was uniformly fertilized with 112, 49, and 93 kg/ha of N, phosphorus (P), and potassium (K), respectively, during both 1946 and 1947, plus 22 kg/ha of N in mid-July 1947. Corn was planted and then thinned to (44 000 plants per ha). The average corn grain yields are shown in Table 14-10. The fallow plots (Table 14-10) were from an experiment in which surface stones greater than 5 cm were removed to study their influence on soil loss and other factors (Lamb and Chapman, 1943). Early in the season, the corn plants on the eroded plots were stunted and showed deficiencies of phosphorus and possibly other nutrients, even though fairly high fertilizer rates were applied before planting. The root system on these plots was restricted to a small volume of fairly dry soil. The soil between the plants and rows was moist and contained few roots. On the less eroded plots the plants were large and green and had roots throughout the surface soil. Corn yields on the areas that had been in a sod or in a rotation containing a sod were twice those obtained from the continuous corn or fallowed plots. The organic matter and soil aggregation were also higher when a sod was included in the rotation. Previous fertilization of the corn-oats-clover rotation increased yields of the following corn crops more than continuous corn. Utilization of both the applied nitrogen and the soil water was very poor on the more eroded soils.

Table 14-10. Effect of past management and erosion on yield of fertilized† corn grown on Bath flaggy silt loam at the Arnot Experimental Station near Ithaca, NY in 1946-1947 (Lamb et al., 1950).

Soil management (1935-45)	Total soil loss (1935-45)	Corn grain yields (1946-47 avg.)	Soil organic matter	Soil aggregation
	t/ha	kg/ha	—————— % ——————	
Idle	4	4930	4.6	86
Meadow fertilized‡	0	4770	4.5	83
Rotation, potatoes-sweetclover	29	3770	4.1	75
Rotation, corn-oats-clover (fertilized)§	7	3490	4.7	79
Rotation, corn-oats-clover (unfertilized)	18	2510	3.8	80
Corn, continuous (fertilized)¶	76	1880	3.1	67
Corn, continuous (unfertilized)	108	1570	3.0	68
Fallow (stones in place)	166	2700	3.0	67
Fallow (stones removed)	307	1570	2.9	61

† Corn fertilized with 112, 49, and 9.3 kg/ha of N, P, and K, respectively, worked into the plow layer during 1946 and 1947. Lime was also mixed into the plow layer during 1946.

‡ 13.4 t/ha farm manure + 47 kg/ha P once every 3 years + lime.

§ 13.4 t/ha farm manure + 21 kg/ha P before corn + 21 kg/ha before oats + lime before clover.

¶ 11, 10, and 9 kg/ha of N, P, and K, respectively.

Studies by Lamb et al. (1950) indicated that the meadow area reached a bulk density of 1.5 g/cm³ at 26% moisture, whereas the fallow plot reached a bulk density of 1.6 g/cm³ at 22% moisture. The higher bulk density approaches the density at which corn roots will not penetrate the soil. A lowering of bulk density by including cover crops or by introducing organic matter (wood chips) into a Honeoye silt loam soil was reported by Ram and Zwerman (1960). Improved vegetable crop yields were also reported (Free, 1971).

Similar yield results were obtained by Lamb et al. (1950) on an Ontario sandy clay loam (Table 14-11), a Dunkirk silty clay loam, and a Honeoye silt loam (Table 14-12). Past erosion reduced yields under moderate fertilizer programs and led to a decrease in fertilizer efficiency.

In 1954 the erosion plots on Bath flaggy silt loam on the Arnot experimental area were regrouped according to their 1952 organic matter content. The groups are shown in Table 14-13 with the soil and erosion data. All these plots were under uniform treatment since 1946. Corn was grown in 1946, 1947, 1950, and 1954. Orchardgrass (*Dactylis glomerata* L.) and clover (*Trifolium pratense* L.) were grown in 1948-1949, oats were underseeded in 1951, and timothy (*Phleum pratense* L.) and clover were grown in 1952-1953. The hay crops were mowed, but the residue was not removed. No fertilizer was applied after seeding. In 1954 the plots were subdivided into quarters up and down the slope for a randomized Latin square with

Table 14–11. Effect of past management and erosion on yield of corn grain on an Ontario clay loam (fine-loamy, mixed, mesic Glossoboric Hapludalf) at Geneva, NY, in 1947–1948 (Lamb et al., 1950).

Past management 1936–46	Total soil loss (1936–46)	Corn grain[†] yield (1947–48 avg.)	Organic[‡] matter	Soil aggregation
	t/ha	kg/ha	%	
Bluegrass sod	trace	9361	2.4	55
Buckwheat sown in trash	20	9820	2.4	60
Soybeans fall-plowed	16	8540	2.1	40
Soybeans sown in trash	25	9210	2.1	54
Winter rye, summer fallow	173	6770	1.7	27
Continuous fallow	448	4960	1.3	16

† 90, 39, and 75 kg/ha of N, P, and K, respectively, was plowed under and 22, 10, and 16 kg/ha of N, P, and K, respectively, was applied in the band at planting.
‡ Original organic matter content was 2.0% in 1936.

Table 14–12. Effect of past management and erosion on crop yields grown on Honeoye gravelly silt loam at Marcellus, NY, from 1943 to 1948 (Lamb et al., 1950).

Past management 1939–42	Total soil loss 1939–42	Corn[‡] 1943	Oats 1944	Mixed hay 1945	1946	Corn grain 1947	1948	Soil organic matter 1943	1946
	t/ha	— kg/ha —		— t/ha —		— kg/ha —		—— % ——	
Meadow	3	6390	1110	3.3	5.1	4330	3890	2.6	2.6
Fallow	605	2990	750	3.0	4.4	3080	3260	2.0	2.2

† 18 t/ha manure was applied for corn in 1943 and 1947; 112, 49, 93 kg/ha of N, P, K, respectively was plowed under for corn in 1948; 15 kg/ha was applied annually for corn and 30 kg/ha for oats.
‡ Total dry-matter yield rather than grain because corn did not mature.

four rates of N, and were planted to corn. Nitrogen at rates of 0, 56, 112, and 224 kg/ha as ammonium nitrate (NH_4NO_3) was applied one-half at planting and one-half as a topdressing in late July. Phosphorus at 117 kg/ha and K at 224 kg/ha were applied before planting. Soil tests indicated that P was medium to high and K was high on all plots.

The 1954 grain yields are summarized in Table 14–14 (Free, 1957). Grain yields increased with increasing nitrogen up to 224 kg/ha for the plots with low organic matter (more highly eroded), but increased to only the 112-kg/ha N rate on the plots with high organic matter. Of special interest is that the maximum yield with adequate N (224 kg/ha) was the same for all plots. Thus, with sufficient N previous differences in erosion and/or organic matter were eliminated, at least for this location and year. Unfortunately, this experiment was not continued for a number of years to sample a wider range of growing-season rainfall. Rainfall for the 1954 growing season was normal and well distributed. This soil has a deep profile, unlike approximately 30% of the soils of the Northeast that are shallow to bedrock. The Bath soil has a fragipan at a greater depth than many associated soils.

Table 14-13. Soil and erosion data for groups of plots based on the 1952 organic matter content as a result of 1935-1945 management at the Arnot Experimental Station (Free, 1957).

1935-1945 Management	Organic matter	Soil aggregate stability	Soil pH	Soil loss 1934-1945
	—————%—————			t/ha
Group I				
Corn, continuous (fertilized)				
Corn, continuous (unfertilized)	3.4†	80	6.4	165
Fallow, stones removed				
Fallow, stones not removed				
Group II				
Meadow-buckwheat				
Meadow-buckwheat	3.8	84	6.5	63
Corn, continuous (fertilized)				
Corn, continuous (fertilized)				
Group III				
Oats-clover-corn (no lime)				
Oats-clover-corn (manure, no lime)	4.2	84	6.5	31
Meadow-buckwheat (unfertilized)				
Meadow-fallow-meadow				
Group IV				
Meadow (manure, lime)				
Idle (grass, weeds, brush)	4.7	87	6.6	3
Oats-clover-corn (manure, lime)				
Meadow, mulched				

† All data are means for each group of plots.

Table 14-14. Effect of soil organic matter and nitrogen fertilizer on 1954 corn grain yields at the Arnot Experimental Station (Free, 1957).

Nitrogen rate	Corn yields†				
	1952 soil organic matter level (%)				
	3.4	3.8	4.2	4.7	Mean
	————————— kg/ha —————————				
0	3520	4270	4900	4900	4390
56	4080	4830	5270	4830	4770
112	5020	4460	5520	5840	5210
224	5520	5520	5270	5400	5400
Mean	4520	4770	5270	5210	

† $LSD_{0.05}$ = 276 kg/ha corn grain at 15% moisture.

These data agree with those obtained for the forages in a corn-alfalfa (*Medicago sativa* L.) and timothy or corn-alfalfa, birdsfoot trefoil (*Lotus corniculatus* L.), and timothy rotation for three crop cycles over 20 years (Bowen, 1983). The crops were adequately fertilized annually but had differential lime rates applied over the years. There was no significant correlation between depth to soil mottling and yields of alfalfa grown on 160 plots of Bath, Mardin, and Volusia (fine-loamy, mixed, mesic Aeric Fragioquept) soils from 1972 to 1975 on the Mt. Pleasant Research Farm, Ithaca, NY. This farm, whose elevation is 550 m, is located only about 30 km from the Arnot station. The regression equation for yields of forage dry matter was

$$Y = 4903 + 13.9X$$

where Y is the yield of dry matter in kg/ha and X is the depth to mottling in cm (W. S. Reid and W. T. Bowen, 1983, unpublished data, Cornell University, Ithaca, NY). The regression line has a relatively small slope and a very small r^2, and the yields were not significantly correlated to soil depth. Yields normally were highest on the deepest soils, but some yields on the shallow, somewhat poorly drained, Volusia soils were also high. Persistence of the legumes was usually better on the deeper soils. The corn yields also agree with the forage data. Borst et al. (1945) also found that alfalfa established well and produced better yields than expected on severely eroded soils.

The data of Free (1957), Borst et al. (1945), and Reid and Bowen (1983, Cornell University, Ithaca, NY, unpublished data) tend to agree that near maximum yields can be produced on eroded soils, but they do not agree with my observations made in farmer's fields. Eroded fields consistently have poor early season corn growth and lower yields. The poor early season growth may be the result of poor air-water relations due to a decrease in aggregation and an increase in soil compaction.

14-7 SUMMARY AND CONCLUSIONS

Early research on the influence of erosion on soil productivity demonstrated drastic yield reductions when no or little fertilizer nitrogen was used. Most of the effects on yields could probably be atributed to the loss of organic matter and lower capacity for soil water. The lower organic matter contents resulted in less nitrogen available to the crop, deterioration in soil aggregation, poor root growth, and reduced crop yields.

Increasing the fertilizer, especially N, improved yields on eroded soils, but moderate fertilizer rates (higher rates than recommended at the time) did not result in yields as high as where little erosion had occurred. In one experiment in 1954, higher rates of fertilizer N did increase yields to those on the noneroded plots. However, the extra 100 kg/ha of N required for the eroded plots, at an annual cost of about $85/ha, would seldom be economical. Thus, in practice, production would be lower on the eroded soils.

The effects of poor soil structure and decreased root growth can be seen in many experiments and farmer's fields even today. These effects on yields have not been separated from other causes of poor growth and yield reductions, but they should be examined in more detail.

One serious limitation in the data on soil erosion for the Northeast results from the selection of soils used for the studies. Most of the information available is for deep, well-drained soils. Large areas of soils within the Northeast either have fragipans restricting root growth and water movement or are shallow to bedrock. As these soils erode, the water available to the plant is reduced. Also, the structure of the soil becomes less friable, and roots are not able to exploit as much soil for water and nutrients. Many of these more shallow soils are probably nearing their erosion limit for remaining productive.

Currently, most of the cultivated soils of the Northeast with slopes greater than 8% are in a corn-hay rotation, and erosion is minimized. If the corn in these rotations is increased, erosion will increase. Likewise, with the current trend to plant more hectares of corn, more sloping land must be planted and erosion will increase.

Erosion in the Northeast is still affecting crop yields. The shallow soils and soils with fragipans are probably the most immediately critical areas, but since most of the soils are not much deeper than 150 cm, almost all the cultivated soils of the Northeast are subject to losses in productivity by erosion.

REFERENCES

Borst, H. C., A. G. McCall, and F. G. Bell. 1945. Investigations in erosion control and reclamation of eroded land at the Northwest Appalachian Conservation Experiment Station, Zanesville, Ohio, 1934-42. USDA-SCS Tech. Bull. 888. U.S. Government Printing Office, Washington, DC.

Bowen, W. T. 1983. Yield response to limestone over three crop rotation cycles. M.S. thesis. Cornell University, Ithaca, NY.

Brady, N. C., R.A. Struchtemeyer, and R. B. Musgrave. 1957. The Northeast. p. 598-619. *In* Soils. Yearbook of Agriculture. USDA, U.S. Government Printing Office, Washington, DC.

Free, G. R. 1957. Effects of good management following soil erosion. Soil Sci. Soc. J. 21: 453-456.

----. 1971. Soil management for vegetable production on Honeoye soil with special reference to the use of hardwood chips. N.Y. Food and Life Science Bull. 2:1-20. Cornell University, Ithaca, NY.

----, E. A. Carleton, J. Lamb, Jr., and A. F. Gustafson. 1946. Experiments in the control of soil erosion in Central New York. Cornell Agric. Exp. Stn. Bull. 831. Ithaca, NY.

Howe, F. B., and H. R. Adams. 1936. Soil erosion in New York. Cornell Extension Bull. 347. Ithaca, NY.

Lamb, J., Jr., J. S. Andrews, and A. F. Gustafson. 1944. Experiments in the control of soil erosion in Southern New York. Cornell Agric. Exp. Stn. Bull. 811, Ithaca, NY.

----, and J. E. Chapman. 1943. Effect of surface stones on erosion, evaporation, soil temperature, and soil moisture. Agron. J. 35:567-578.

Patrick, A. L. 1938. Soil erosion survey of Pennsylvania. Penn. Agric. Exp. Stn. Bull. 354. State College, PA.

Ram, D. N., and P. J. Zwerman. 1960. Influence of management systems and cover crops on soil physical conditions. Agron. J. 52:473-476.

Shaulis, N., and E. A. Carleton. 1947. Higher grape yields from cross-slope plantings. Farm Research 13, #3. N.Y. Agric. Exp. Stn.

U.S. Bureau of the Census. 1978. 1978 Census of Agriculture. U.S. Government Printing Office, Washington, DC.

USDA. 1981. Soil, water, and related resources in the United States: Status, condition and trends. 1980 RCA Appraisal, Part I. USDA, Washington, DC.

USDA-SCS. 1978. 1977 National Resource Inventories. U.S. Government Printing Office, Washington, DC.

USDA-SCS. 1981. Land resource regions and major land resource areas of the United States. USDA-SCS Agric. Handb. 296. U.S. Government Printing Office, Washington, DC.

15 Effects of Soil Erosion on Crop Productivity of Southern Soils

G. W. Langdale
Agricultural Research Service
U.S. Department of Agriculture
Watkinsville, Georgia

H. P. Denton
North Carolina State University
Raleigh, North Carolina

A. W. White, Jr.
Agricultural Research Service
U.S. Department of Agriculture
Watkinsville, Georgia

J. W. Gilliam
North Carolina State University
Raleigh, North Carolina

W. W. Frye
University of Kentucky
Lexington, Kentucky

Accelerated soil erosion began on soils in the southern USA during the early 1800s with the increasing influx of European settlers (Dregne, 1982). Trimble (1974 and 1975) used principally archival facts to develop an erosive land use (ELU) component to fill data gaps and suggested that accelerated soil erosion did not exist prior to European settlement. Ruffin (1832) probably recorded the first crop-yield reductions caused by soil erosion in the South as well as the first restoration of soil productivity. A few decades later, Hilgard (1860) also warned that tillage practices without sound conservation principles were ruining the once-productive lands of the South. Like Ruffin, Hilgard (Jenny, 1961) recommended that farmers use marl and manure to maintain soil fertility as well as contour tillage to control soil erosion. Ruffin's efforts to restore soil productivity also included crop rota-

Published in R. F. Follett and B. A. Stewart, ed. 1985. *Soil Erosion and Crop Productivity.* © ASA-CSSA-SSSA, 677 South Segoe Road, Madison, WI 53711, USA.

tions with clover, but his tillage procedures were described as troublesome and imperfect.

In the 20th century, the Buchanan Amendment of 1928 (U.S. Congress, 1930) and Bennett's (1939) leadership inspired the first real soil erosion/soil productivity research in the South. Most of this research was summarized by Langdale and Shrader (1982) and Williams et al. (1981). However, research accomplished between 1930 and 1960 involved cropping-system technology that is not economically competitive for currently required crop yields on eroded southern lands. The objective herein is to assess impinging causes of soil erosion by water on southern soils and the erosion's effects on crop productivity. This objective deals with land resource areas east of the 96° west longitude (humid USA) and south of the Ohio River (Buol, 1973). Experimental sites of studies are all east of the Mississippi River, but results may be extrapolated farther west within several soil orders.

15-1 INHERENT ERODIBILITY OF SOUTHERN SOILS

Ultisols dominate the landscape of the humid-thermic, southern USA (Perkins et al., 1973). They occupy 872 000 km^2 in 13 Southern States and are associated with soils of all orders except Aridisols and Oxisols (Fig. 15–1). The high erodibility of Ultisols appears to be related to their low organic matter content, weak structure, and low permeability caused by intensive weathering. Only the Alfisols of the Mississippi-Tennessee loess belt have suffered greater rates of soil erosion than the Ultisols of the Piedmont. Alfisols have been subjected to severe soil erosion because of their high silt content and rolling topography, which affect their LS factors (slope length and steepness) in the universal soil loss equation (USLE) (Wischmeier and Smith, 1978). Alfisols with the greatest potential for soil erosion occur on lands bordering alluvial plains of the Mississippi and Ohio Rivers (Slusher and Lytle, 1973). Soils in "Pale-" and "Hapl-" great groups represent more than 50% (7.36 × 10^7 ha) of both the Ultisol and Alfisol orders and occupy most of the cultivated lands. Genesis of Ultisols and Alfisols is diverse because of varying climate, parent material, topography, runoff and water-table depths. Barnett (1976) experimentally determined K factors (soil erodibility in the USLE) on 20 southern soils, all Paleudult, Hapludult, and Hapludalf mapping units. These K factors averaged 0.44 (t•h)/(ha•N), ranged from 0.26 to 0.95 (t•h)/(ha•N), and were distributed similar to those of the highly erosive benchmark soils of the western Corn Belt, such as Shelby loam (Argiudolls), Marshall silt loam (Hapludolls), and Ida silt loam (Udorthents) (Wischmeier and Smith, 1978). Rainfall erosion indices (EI) and topography factors (LS) also vary widely to confound the inherent erodibility of both Ultisols and Alfisols in the region. Rainfall EI's range from about 260 N/h in Tennessee and Kentucky to 700 N/ha near Atlantic and Gulf Coastal areas (Wischmeier and Smith, 1978). Rainfall energy over most of the eroded southern landscapes equals or exceeds 400 N/h. Un-

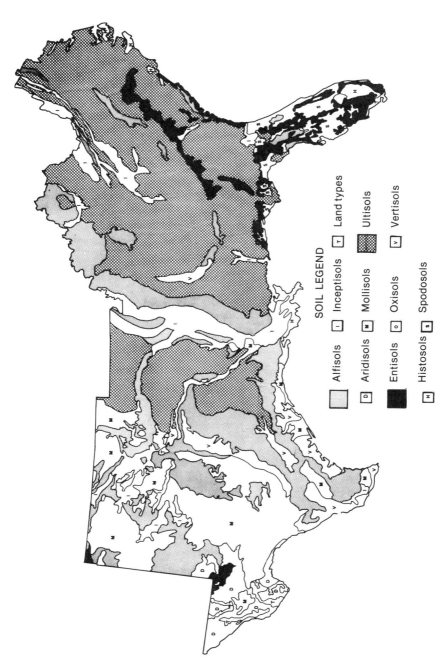

Fig. 15–1. Extent of Ultisols and Alfisols in the southern USA. Adapted from Buol (1973).

fortunately, most of this rainfall energy coincides with the conventional seasons for primary tillage in the south, late spring and early summer. Many southern landscapes are scarred with gullies and exhibit areas of spatially exposed subsoils. Only casual observations are required to conclude that conventional tillage practices during the past 150 years have not controlled management factors (C and P) of the USLE adequately.

15-2 PHYSIOGRAPHY AND CROP PRODUCTIVITY OF ERODED SOUTHERN SOILS

15-2.1 Background

The southern USA has a wide variety of soils with topography ranging from nearly level to very steep. The climate is humid and warm, and annual rainfall ranges from about 1140 mm in eastern Texas and Kentucky to about 1520 mm along the Gulf Coast. Most of the soils in the region are in udic, thermic families in *Soil Taxonomy* (Soil Survey Staff, 1975). For detailed delineation of soils, climate, and vegetation within the physiographic areas, see Buol (1973). Major physiographic divisions within the southern USA are shown in Fig. 15-2. Most research relating crop productivity to soil erosion is limited to the humid region east of the Mississippi River.

Ultisols are dominant throughout the Coastal Plain, Piedmont, and Interior Low Plateaus (Fig. 15-1 and 15-2). Significant areas of Alfisols also occur in the South, principally in the Southern Loess region along the Mississippi Delta and in the Interior Low Plateaus Province of Kentucky, Tennessee, and northwestern Alabama. Smaller areas of soils in other orders occur. However, research information on soil erosion in the South is confined to Ultisols and Alfisols.

In the southern USA many Ultisols have relatively thin sandy or loamy surface horizons (if only slightly eroded) overlying loamy or clayey subsurface horizons (Perkins et al., 1972). They have highly developed A and Bt horizons over partially weathered residuum from consolidated rock or unconsolidated sediments of different origins. The soils are generally infertile and acid with low base saturation in the argillic horizons.

Alfisols of the Southeast are soils with generally light-colored surface layers overlying more clayey subsoils (Slusher and Lytle, 1973). In some places, Alfisols have bedrock within a meter; in others, unconsolidated materials occur for greater depths, as in the silty soils developed in deep loess in a belt adjacent to the Mississippi Alluvial Valley. In the humid Southeast, Alfisols grade to Ultisols and are similar in some respects to Ultisols. They are typically acid in the upper part of the subsoil, but are not as highly weathered as the Ultisols, and have higher contents of Ca, Mg, K and other elements required for plant growth.

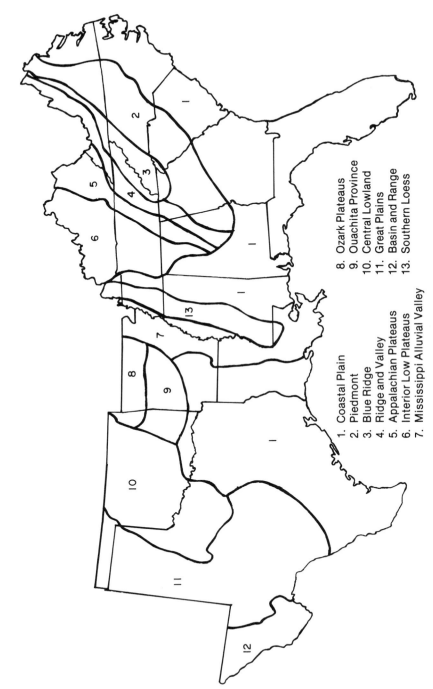

1. Coastal Plain
2. Piedmont
3. Blue Ridge
4. Ridge and Valley
5. Appalachian Plateaus
6. Interior Low Plateaus
7. Mississippi Alluvial Valley
8. Ozark Plateaus
9. Ouachita Province
10. Central Lowland
11. Great Plains
12. Basin and Range
13. Southern Loess

Fig. 15–2. Physiographic divisions of soils in the southern USA. Adapted from Buol (1973).

15–2.2   Southern Piedmont Region

Hapludults are the most extensive Ultisols on cultivated uplands in the Southern Piedmont (Perkins et al., 1973). These soils are part of an erosional landscape located in the foothills area southeast of the Appalachian and Blue Ridge mountains (Fig. 15–2). The Southern Piedmont extends about 1200 km from southern Virginia into Alabama and averages 250 to 300 km wide. The Southern Piedmont has a gently to strongly rolling topography with a few broad plateaus. The soils are developed in residuum from felsic and mafic igneous rocks, meta-volcanic, slates, sandstones, and Triassic sediments. Most of the weatherable minerals present in the rocks are absent in the soil and upper part of the saprolite. The original surface horizons of most Hapludults were loamy sands to sandy loams. They were relatively thin even under mature forests (Davis et al., 1931). Native fertility of the soils is low.

Soil-erosion problems resulting from agricultural activities on Southern Piedmont lands have long been recognized. The Southern Piedmont was one of ten original sites for national soil-erosion experiment stations in the early 1930s (Copley et al., 1944). Soil-erosion syntheses by Trimble (1974 and 1975) and R. B. Daniels (personal communication) suggest that the Southern Piedmont is one of the most severely eroded agricultural regions in the USA. Continuous, conventional cotton cultivation associated with high soil erodibility ($K = 0.40$ $(t \cdot h)/(ha \cdot N)$) and intense rainfall ($EI = 400$ N/h) contributed to this severe erosion. Carreker et al. (1977) and Langdale et al. (1979a) showed that, soil erosion on Southern Piedmont soils remains a problem without application of sound conservation practices.

Sporadic attempts have been made during the past 50 years to determine the effects of soil erosion on crop yields in the Southern Piedmont. Collins (1935), Copley et al. (1944), and Latham (1940) made the first attempts to determine the effects of Piedmont soil erosion on crop yields. They reported large yield reductions on severely eroded or desurfaced plots. Since little fertilizer or lime was used in that era, any yield reductions were due in part to nutrient deficiencies on acid subsoils. Unfortunately, this archaic data contributes little to understanding the effects of erosion under modern technological systems.

Effects of soil erosion on corn and soybean yields in the Southern Piedmont from several studies conducted since 1945 are reported in Table 15–1 and Table 15–2. Grain yields presented are those of the least and most severely eroded treatments used by Adams (1949), Langdale et al. (1979b), Stone et al. (1982), and White et al. (1983). Langdale et al. (1979b) concluded that, although corn yields on Southern Piedmont soils have increased more than 100% between 1949 and 1979 on both the least eroded and eroded treatments, relative yield difference remained about the same. The yield reductions on eroded sites were at least 40% unless corrective treatments other than fertilization were imposed.

Stone et al. (1982) demonstrated that landscape position significantly affects corn yields on eroded Southern Piedmont soils in North Carolina (Table 15-3). They attribute their highest yields to positions associated with converging rather than diverging waterflow on concave and convex slope forms, respectively. Munsell hues were also related to the extent of soil erosion in this study. The Munsell hue is closely related to the amount of B horizon material incorporated into the plow layer. Some of the yield differences, varying from 3.6 to 6.0 Mg/ha were explained by landscape position (Table 15-3). Some of the least eroded sites produced crop yields lower than the more intensely eroded sites in some fields. Most severely eroded areas, indicated by redder Munsell hues (5 YR), tended to occur on landscape positions with diverging flow, such as ridge crests and shoulder slopes. Thus, the effect of erosion on corn yields was confounded with the effect of landscape position. This confoundment commonly occurs in Piedmont landscapes and has not usually been considered in studies of erosion and crop yields.

In Georgia, White et al. (1983) and Bruce et al. (1983) showed that soybean yields on 24 farm fields averaged 2.77, 1.89, and 1.33 Mg/ha for slightly, moderately, and severely eroded sites, respectively. Yield reductions on the severely eroded sites averaged 52% in 1982 (Table 15-1). For all 24 fields, yields related positively to depth of the Ap horizon, depth to the top of the Bt horizon, and solum depth, and negatively with clay content in the surface horizon. Severely eroded sites had markedly lower levels of extractable P in the plow layer than slightly eroded soils. Moderately and severely eroded sites showed less favorable soil water regimes than the slightly eroded sites during the critical fruiting period of the soybean crop. The researchers concluded that ineffective root-zone recharge by rainfall on the eroded sites was largely responsible for the unfavorable soil water regime and was a major contributing factor to yield reductions.

15-2.3 Coastal Plain Region

The Coastal Plain is a large and important agricultural region of the South, extending in a belt bordering the Atlantic Ocean and the Gulf of Mexico from Virginia to Texas (Fig. 15-2). Much of the row-crop production in the South occurs in this region. From Virginia to Alabama the Coastal Plain and Piedmont regions join in an irregular zone called the fall line. Topography varies from nearly level to steep, but plateaus and swampy flats are common and abundant (Buol, 1973). Soils are derived from deposits of marine and fluvial origin. Ultisols, principally Paleudults and Ochraquults, dominate the region. The Ultisols commonly have sandy or loamy surface horizons and loamy to clayey subsurface horizons. Aquults are commonly found on broad, flat interfluves with little erosion hazard, while Udults are found on level to strongly sloping areas with erosion hazard.

Barnett (1976), Thomas et al. (1969), DelVecchio and Knisel (1982), and Sheridan et al. (1982) suggested that many of these soils possess an

Table 15-1. Crop-yield estimates associated with various levels of soil erosion on Ultisols of the southern USA.

Study	Family classification	Corn yields			Soybean yields		
		Least eroded	Eroded	Reduction	Least eroded	Eroded	Reduction
		Mg/ha		%	Mg/ha		%
Adams (1949)	Clayey, kaolinitic, thermic, Typic Hapludults	2.70	1.60	41	--	--	--
Langdale et al. (1979)	Clayey, kaolinitic, thermic, Typic Hapludults	4.67	2.23	52	--	--	--
Stone et al. (1982)	Clayey, kaolinitic, thermic, Typic Hapludults	4.40	3.65	17	--	--	--
White et al. (1983)	Clayey, kaolinitic, thermic, Typic Hapludults	--	--	--	2.78	1.33	52
Simpson (1974)	Clayey, kaolinitic, thermic, Typic Paleudults	--	--	--	1.68	1.32	21
Rhoton (1975)	Clayey, kaolinitic, thermic, Typic Paleudults	--	--	--	1.73	1.32	24
Batchelder and Jones (1972)†	Clayey, mixed, mesic, Typic Hapludults	4.82 6.32	0.95 5.07	80 (yr 1) 20 (yr 2-4)	-- --	-- --	-- --
McDaniel and Hajek	Fine-loamy, siliceous, thermic, Typic Paleudults	3.20	2.26	29	2.69	1.68	38
McDaniel and Hajek (1982)	Fine-loamy, siliceous, thermic, Plinthic Paleudults	5.74	4.42	23	2.59	1.48	48
Ultisol average		4.55	2.88	37	2.29	1.43	38

† Desurfaced comparison.

Table 15–2. Crop-yield estimates associated with various levels of soil erosion on Alfisols of the southern USA.

Study	Family classification	Corn yields			Soybean yields		
		Least eroded	Eroded	Reduction	Least eroded	Eroded	Reduction
		——— Mg/ha ———		%	——— Mg/ha ———		%
Denton (1978)	Fine-silty, mixed, thermic, Typic Hapludalfs	--	--	--	2.36	2.34†	1
					--	2.07‡	12
Denton (1978)	Fine-silty, mixed, thermic, Glossic Fragiudalfs	--	--	--	2.04	1.67†	18
					--	1.30‡	36
Denton (1978)	Fine-silty, mixed, thermic, Typic Paleudalfs	--	--	--	2.26	2.01†	11
					--	1.75‡	23
Frye et al. (1982)	Fine, mixed, mesic, Typic Paleudalfs	5.97	5.25	12	--	--	--
Frye et al. (1982)	Fine-silty, mixed, mesic, Typic Paleudalf	7.90	6.27	21	--	--	--
Alfisol average		6.94	5.76	17	2.22	1.86	16

† 2 to 5% slopes.
‡ 8 to 12% slopes.

Table 15-3. The effect of soil erosion and landscape position on corn yields
(Stone et al., 1982).

Degree of soil erosion	Munsell hues	Landscape position	Corn yields
			Mg/ha
Slight	10 YR	Head	4.4
Moderate	7.5 YR	Valley	6.0
	7.5 YR	Head	5.8
Severe	5 YR	Crest	3.6
	5 YR	Shoulder	4.4
	5 YR	Valley	4.4

erosive potential nearly equal to those of the Southern Piedmont. Experimentally determined K values range from 0.19 to 0.83 $(t \cdot h)/(ha \cdot N)$ (Barnett, 1976). Erosion index (EI) values usually range from 395 to 724 N/h. Most of this energy occurs during primary and secondary tillage of annual summer crops. Thomas et al. (1969) and DelVecchio and Knisel (1982) reported sediment yields greater than 30 Mg/ha with conventional till-monocrop management on soils in the middle Atlantic Coastal Plain. Recently estimated rates of soil erosion by Larson (1981) and USDA (1980) showed that soil losses exceeded the T-value (11.2 Mg ha^{-1} yr^{-1}) without adequate crop residue management (Campbell et al., 1979) for all row crops in Alabama, Georgia, Florida, and South Carolina. Most of these row crops are grown in the middle and upper Coastal Plains of these states.

The effects of soil erosion on crop productivity in the Atlantic and Gulf Coastal Plains appear to be little recognized. The Coastal Plain land-resource area was apparently not considered severely erosive when the 10 original soil-erosion experiment stations of the USA were funded in 1930 (U.S. Congress, 1930). Thus, relatively little research information is available from this area that has direct application to erosion and crop productivity. Winters and Simonson (1951) noted the scarcity of precise data on the relationships of exposed subsoil properties to plant growth.

Several researchers (Heilman and Thomas, 1961; Phillips and Kamprath, 1973) recognized soil fertility problems associated with land forming. Thomas and Cassel (1979) experienced significant yield decreases on land cuts in the Atlantic Coastal Plain immediately following land-forming procedures. They suggested that the A horizon thickness of the soil influenced corn yields more than plant available P, bulk density, available water-holding capacity, organic matter, and plant available K. They felt that the depth to toxic levels of exchangeable Al was in part a function of topsoil depth. This study supported Foy's (1974) and Rios and Pearson's (1964) conclusions that high exchangeable Al may be the most limiting root-growth factor.

Research initiated in 1981 provides the only available information with direct application to productivity of eroded soils in the Coastal Plain. McDaniel and Hajek (1982) conducted a survey of crop yields on slight and moderately eroded Paleudults in Alabama. They measured soybean and

Table 15-4. Crop yield and critical properties of slight and moderately eroded Coastal Plain soils† of Alabama (McDaniel and Hajek, 1982).

Crop	Soil erosion	Crop yield	Soil properties				
			Surface thickness	Clay content	Subsoil mixing	Fe₂O₃ content	P
		Mg/ha	cm	%			kg/ha
Soybeans	Slight	2.62	21	13	10	2.8	24
	Moderate	1.68	13	18	27	5.1	13
Corn	Slight	4.83	25	9	11	2.3	23
	Moderate	3.64	21	19	28	5.1	15

† Dothan, Malbis, Bama, Orangeburg, and Red Bay series.

corn yields on 30 and 22 fields, respectively, with both soil-erosion classes. Both corn and soybean yields were significantly reduced on 46% of the sites with moderate soil erosion. Average corn and soybean yields were reduced 26% and 43%, respectively (Table 15-1). Plow-layer soil properties often correlated with crop yields (Table 15-4). Although no subsoil property was correlated with crop yields, corn yield reductions were frequently observed on soils with yellowish-brown subsoils with plinthite. On some of their fields, crop yields on moderately eroded sites equaled or exceeded those of slightly eroded sites.

15-2.4 Southern Loess Region

Soils in the Southern Loess region are predominately Alfisols. Most were formed in deep loess deposits and are in fine-silty families (Slusher and Lytle, 1973; Springer and Elder, 1980). Common occurrences of continuous row cropping on sloping, silty soils where rainfall energy ranges up to 850 N/ha have resulted in soil-erosion rates among the nation's highest (Larson et al., 1983; Wischmeier and Smith, 1978; USDA–SCS, 1981). In 1975, erosion rates from cropland were estimated to average from 34 to over 56 Mg ha⁻¹ yr⁻¹ throughout the region (USDA–SCS, 1977).

Results of some studies of the effects of erosion on crop yields are given in Table 15-1. A study of the effects of past erosion and slope gradient on soybean yields on soils in the deep loess region of western Tennessee showed no significant difference in yields due to erosion or slope on Typic Hapludalfs (Denton, 1978). On Glossic Fragiudalfs and Typic Paleudalfs, yields were significantly lower on severely eroded areas on 8 to 12% slopes than on slightly eroded areas with 2 to 5% slopes. Reductions in yield were 36% and 23% for Fragiudalfs and Paleudalfs, respectively (Table 15-2). Yield reductions were attributed to lower moisture supplies on the steeper, more eroded areas, but the effects of increased erosion and of increased slope could not be separated. Within the 2 to 5% slope range, yields were not significantly different between erosion classes.

Overton and Bell (1974) studied corn yields over a period of years on a number of soils formed in loess and alluvium in western Tennessee. Their yields on moderately eroded and severely eroded areas were 7% and 32% lower, respectively, than yields on uneroded areas. Eroded areas were in general on steeper slopes, so part of the yield reductions may have been due to the drier environment associated with increased slope gradient rather than to increased erosion.

15-2.5 Interior Low Plateaus Region

The Interior Low Plateaus region of the southern USA includes central and much of western Kentucky, central Tennessee, and northwestern Alabama. Soils in northern and western Kentucky are primarily Fragiudalfs, Hapludalfs, and Paleudalfs formed in loess or in loess over sandstone, siltstone, or limestone (Slusher and Lytle, 1973). In south central Kentucky, central Tennessee, and northwestern Alabama, soils are predominately Paleudults, Fragiudults, Hapludalfs, and Paleudalfs, formed in residuum from limestone or cherty limestone or in a thin layer of loess overlying these materials (Perkins et al., 1973; Springer and Elder, 1980). Soils formed in loess generally are in fine-silty families, while those formed in limestone residuum are generally clayey. In the southern and western parts of this region, rates of soil erosion on cropland are almost as high as those in the Southern Loess region because of similar erosion hazards. In the northeastern portion of the region, estimates are lower, 11 to 34 Mg/ha annually (USDA–SCS, 1977).

Frye et al. (1982) found that on a fine, mixed, mesic, Typic Paleudalf (Maury series), formed from limestone in central Kentucky, corn yields were 12% lower on eroded than on uneroded areas (Table 15–2). On a fine-silty, mixed, mesic, Typic Paleudalf (Crider series), formed in loess over limestone in western Kentucky, yields were 21% lower on eroded areas. In the latter case, the yield reduction was partially due to lower fertility of the eroded areas. In both cases, the authors felt that lower moisture supply on eroded areas was the most important factor in reducing yields.

15-2.6 Southern Appalachian Ridges and Valleys Region

The Southern Appalachian Ridges and Valleys region includes portions of western Virginia, eastern Tennessee, northwestern Georgia, and northeastern Alabama. Soils on the ridges trending from southwest to northeast are largely Dystrochrepts and Eutrochrepts formed in residuum from sandstone and shale, with a few areas of Hapludults and Paleudults formed in residuum from shale and limestone. Soils in the intervening valleys are predominately Paleudults and Hapludults formed in residuum from limestone or occasionally shale (Springer and Elder, 1980). Soils in the valleys are generally in clayey families, except in alluvial or colluvial areas where they are usually loamy. These soils are less erosive than those in the other regions, and there is less row-crop cultivation and lower rainfall energy (ca.

200 to 400 N/h). However, slopes of cropland are often steeper. Erosion rates from row-cropped areas were estimated to be 34 to 56 Mg/ha annually in 1975 in most of the region (USDA–SCS, 1977).

Studies have been conducted in Virginia and Tennessee on soils derived from limestone. Soybean yields on gently sloping, uneroded areas were compared to those on sloping, severely eroded areas on clayey, kaolinitic, thermic Typic Paleudults (Dewey series) in eastern Tennessee. Yields were at least 20% lower on severely eroded areas (Table 15–1). The effect was attributed to lower moisture supply, resulting from both erosion and steeper slope. The effects of soil erosion could not be separated from those of slope. In western Virginia, Batchelder and Jones (1972) grew corn on a clayey, mixed, mesic, Typic Hapludult (Groseclose series) from which the surface soil had been removed, and then replaced over half the area. Yields in the first year from the area of exposed subsoil were only about 20% of those from the areas on which topsoil was replaced. After the soil-fertility problems were corrected, yields on the subsoil areas averaged 80% of the topsoil-covered areas in the second through the fourth year (Table 15–1).

15–2.7 Implications of Results from Soil Erosion-Crop Productivity Studies in the South and Problems Interpreting Existing Data

Limited studies conducted throughout the South indicate that soil erosion is more severe in some areas than in others and that effects on crop productivity may vary considerably with different soils. Overall results show that reductions in soybean and corn yields on severely eroded Ultisols averaged near 38 and 37%, respectively (Table 15–1). On Alfisols, yield reductions resulting from severe erosion were 16 and 17%, for soybeans and corn, respectively (Table 15–2). The Alfisols in these studies were generally in fine-silty families, while the Ultisols were generally in clayey families. The more favorable physical and chemical properties of the Alfisol argillic horizons apparently resulted in less yield reduction from erosion than on the Ultisols. On one study in thick loess no yield reductions occurred due to erosion of a Typic Hapludalf (Denton, 1978). Evidence from the entire region indicates that soil erosion is a much more severe problem when the subsoils contain fragipans, shallow rock, clayey argillic horizons, or strongly acid argillic horizons. Generally, the deep medium-textured soils show less crop-yield reduction from soil erosion than those with a medium-to-coarse-textured surface over clayey subsoils. Because of their infertility and the toxic Al concentrations associated with high acidity, subsoils of Ultisols and Ultic subgroups of Alfisols exposed by severe erosion exhibit many adverse chemical properties that can affect soil management and limit crop productivity if not corrected. Langdale and Shrader (1982) suggested that crop-yield reductions from erosion on Ultisols appear difficult to alleviate.

Some new studies on Ultisols in Alabama, Georgia, and North Carolina show that in some cases crop yields measured on moderately eroded sites equaled or exceeded those of slightly eroded sites (McDaniel and

Hajek, 1982; Stone et al., 1982; White et al., 1983). Considerably more research is needed to explain the crop-yield anomalies among soil erosion levels.

Although the present data base supports some statements about the effect of soil erosion on productivity of soils of the southern USA, there are some serious deficiencies. In general, studies of the effects of erosion on productivity have been of two types, those involving comparisons of normal and mechanically desurfaced soils, and those comparing soil areas thought to have been severely eroded in the past to areas thought to have undergone significantly less erosion. Studies on desurfaced soils do not simulate the effects of erosion on agricultural land except in the case of severe, rapid gullying. The erosion process involves the slow removal of materials over a period of time. During this process, some material is preferentially removed, organic matter and nutrients are mixed throughout the plow layer each season, new material is incorporated into the plow layer from below, structure formation occurs, and other processes of soil formation such as animal activity, mineral transformations, and translocations of materials take place. This is not simulated by the sudden removal of the top 250 mm of the soils. These studies are relevant only because they set an absolute upper bound on the damage that could be done by removal of the surface soil. They reveal little about what effect erosion is actually apt to have over time.

In most comparisons of sites with different degrees of erosion, the sites compared do not have the same microenvironments. Frequently, they were not the same soil with differences in surface properties as a result of erosion. Eroded sites, as defined by topsoil thickness, texture, or color, almost always occur on different areas of the landscape than uneroded sites. The severely eroded sites tend to be on narrow ridge crests, spurs, and side slopes, areas of diverging or linear overland flow and rapid water runoff. These areas are, in many seasons, more droughty than the broad ridge tops, footslopes, depressions, and head slopes on which uneroded sites most often occur. The uneroded sites are in positions of converging flow and less rapid runoff, or even of deposition of materials. These factors tend to result in larger productivity differences between areas or positions, which should not be attributed entirely to soil erosion. In fact, at least part of the productivity difference is due to differences in soil-moisture regime (Bruce et al., 1983), which would have occurred regardless of surface properties. Moreover, differences in surface properties may be largely a function of soil-formation factors other than erosion.

Most studies of soil erosion and productivity have not considered the effects of differences in landscape position and microenvironment, even when the landscape positions of the erosion classes were obviously quite different. This omission has generally resulted in overestimations of the productivity loss due to erosion.

Another problem with this type of study is that severely eroded and slightly eroded areas in the same field are usually managed alike, when in fact the achievement of optimum yields would probably call for different management strategies for the two conditions. The effect of this situation on productivity differences will depend on which erosion condition is dominant in a given field. This may partially explain why studies like those of Denton (1978), Stone et al. (1982), and McDaniel and Hajek (1982) have sometimes shown higher yields for eroded than noneroded sites in some fields. If the eroded sites represent the dominant condition, management strategies may be more suitable for those soil conditions than for those of the less eroded sites. The opposite effect of optimum management for the less eroded sites also occurs, resulting in an exaggerated estimate of the effect of erosion on productivity.

The problems identified here are not new and are not easily overcome. Past research has given some idea of the likely magnitude of soil erosion effects on soil productivity in the southern USA, but much more research is needed if quantitative predictions are to be made. Future study needs to address two issues. The first is the effect of changes in surface-soil properties on soil productivity. The second is the likely effect of further soil erosion on surface-soil properties. If these relationships can be quantified, then the truly important question of the effect of future soil erosion, as distinct from presumed past soil erosion, on crop productivity can be addressed quantiatively. Data from only a few studies on a few soils are now available, and these limited data are inadequate for clearly defining the effects of erosion on soils throughout the region.

15-3 IMPACT OF CONSERVATION PRACTICES ON SOIL PRODUCTIVITY

Lowdermilk (1953) suggested about 30 years ago that tillage procedures that permitted crop residues to remain at the ground surface were among the most significant contributions to American agriculture. This statement came near the end of the great era of cotton production in the South. Cotton cultivation probably was responsible for more soil erosion between 1850 and 1950 than all other combined agricultural endeavors (Trimble, 1974; Hendrickson et al., 1963; Copley et al., 1944; Carreker et al., 1977; Pieters et al., 1950). Rates of soil erosion near the end of this period often exceeded 50 Mg/ha on 7 to 11% sloping lands.

Ruffin (1832) probably used the first mulch tillage to control soil erosion and restore soil productivity on eroded soil in Virginia during the early 1800s. The next recorded event is cited by Lowdermilk (1953) in northern Georgia during the mid-1900s. He describes the conservation practices used by a farmer for approximately 20 years on land with slopes of up to 17%. His principal tillage implement was a 100-mm-wide, bull-tongue plow used to chisel his topsoil rather than incorporate crop residues. After 20 years, this farmer was still growing crops on topsoil of nearly original depth, while his ridiculing neighbors were plowing subsoil.

Research in the South during the 1940s and 1950s confirmed the utility of efforts to restore soil productivity as described early by Ruffin (1832) and

Hilgard (1860). Researchers (Adams, 1949; Beale et al., 1955; Hendrickson, 1963; Baver, 1950) used cool-season legumes and small grains to provide mulches for summer annual crops. Tillage procedures were crude because of poor conservation-tillage technology. Conservation-tillage procedures evolved from the mulch-bulk method to the wheel-track method and reduced soil erosion on runoff plots as much as 90% (Beale et al., 1955). On severely eroded landscapes, continuous cool- and warm-season legumes were used for several years to slowly improve soil productivity (Pieters et al., 1950; Baver, 1950). In less than 10 years, runoff was reduced to 1 to 3% of former levels and forage yields were raised to economically competitive levels. During this period of tillage and cropping innovation, soil erosion increased in cotton and corn production unless these crops were rotated with sod crops. Reduced till-rotation systems on eroded lands gave positive yield responses. These increases were attributed to higher soil contents of organic carbon and nitrogen and to changes in water stable aggregation (Beale et al., 1955; Carreker et al., 1977). Soil fertility, germ plasm, pesticides, and tillage implements were inferior at this time. When the technology of soil fertility improved in the 1960s, Batchelder and Jones (1972) restored corn yields on mulched, desurfaced sites in western Virginia after four years of cropping.

In the late 1970s and early 1980s, conservation technology began to emerge that could potentially restore productivity of southern soils. Although many studies are in planning or initiation stages, assuming that this technology will be adapted in double crop systems, up to 12.0 Mg/ha of crop residues may be returned to some eroded soil surfaces (Langdale et al., 1984). With these multicrop conservation tillage systems, soil erosion rates will be reduced (Langdale et al., 1979a and 1983; McGregor et al., 1975) to levels equal to or less than the soil formation rates discussed by McCormack et al. (1982). However, extra cost will be necessary to improve soil productivity of spatially eroded landscapes. Until the resource-product price ratios become more favorable to the producer, the cost of reclamation of most severely eroded soils for row-crop or optimum forage production may be prohibitive. Soil productivity of these severely eroded lands may be more economically improved with a long-term soil bank program for perennial legumes, such as occurred during the late 1950s and early 1960s (Pieters et al., 1950). Surface mulches for both short- and long-term soil-productivity reclamation processes appear essential.

15-4 SUMMARY

Trimble's (1974 and 1975) historical synthesis suggests that man-induced soil erosion in the South began about 1830. Ruffin (1832), Hilgard (1860), Bennett (1939), and Lowdermilk (1953) attested to the devastating effects of soil erosion on southern crop production. Approximately 1.5 centuries of extensive cotton cultivation caused most of the soil erosion in the South, particularly in the Southern Piedmont (Trimble, 1974). Trimble's synthesis also shows that accelerated soil erosion subsided in the Southern Piedmont about 1967. These positive results coincide with the application of soil-erosion research reported by Copley (1944), Adams (1949), Baver (1950), Pieters et al. (1950), and Beale et al. (1955). Soybean cultivation appears to have replaced cotton cultivation within the past two decades and now serves as the greatest man-induced soil erosion stimulus in the South (USDA, 1980; Larson, 1981).

Ultisols and Alfisols dominate the farm landscape in the South. Conventional cultivation of these soils remains the greatest contributor to accelerated soil erosion. Buol (1973), Barnett (1976), and Wischmeier and Smith (1978) have documented the inherent erosive nature of both Ultisols and Alfisols. Adverse subsoil properties cause severe soil erosion on the Ultisols to appear more damaging because of low soil fertility and because many Ultisols have clayey or acid subsoils with high Al concentrations (Winters and Simonson, 1951; Langdale and Shrader, 1982). Limited research during the last 40 years indicates that yield reductions on severely eroded Ultisols average near 38% for soybeans and 37% for corn. On Alfisols, the same yield reductions resulting from severe soil erosion average 16% for soybeans and 17% for corn.

Quantification of the effects of erosion on productivity from previous studies is very difficult because of confounding factors of landscape position and nonoptimum levels of management for eroded conditions. In comparisons of eroded versus noneroded sites in fields, the noneroded sites have often been located on landscape positions of inherently higher productivity than those of the eroded sites.

Some recent evidence suggests that soil erosion in the South has intensified again during the past decade (Larson, 1981; Larson et al., 1983). Research on soil erosion in the South from 1930 to 1960 provided considerable information related to lower crop-productivity systems. However, the limited research currently available is inadequate for clearly defining the effects of soil erosion on highly productive cropping systems.

REFERENCES

Adams, W. E. 1949. Loss of topsoil reduces crop yields. J. Soil Water Conserv. 4:130.

Barnett, A. P. 1976. A decade of K-factor evaluation in the Southeast. p. 97–104. *In* Soil erosion: Prediction and control. Soil Conserv. Sco. Am., Ankeny, IA.

Batchelder, A. R., and J. N. Jones, Jr. 1972. Soil management factors and growth of *Zea mays* L. on topsoil and exposed subsoil. Agron. J. 64:648–652.

Baver, A. L. 1950. How serious is soil erosion. Soil Sci. Soc. Am. Proc. 42:1–5.

Beale, O. W., G. W. Nutt, and T. C. Peele. 1955. The effects of mulch tillage on runoff, erosion, soil properties and crop yield. Soil Sci. Soc. Am. Proc. 19:244–247.

Bennett, H. H. 1939. Soil conservation. McGraw-Hill Book Co., New York.

Bruce, R. R., A. W. White, A. W. Thomas, and G. W. Langdale. 1983. Effect of water erosion upon physical character of rooting volume of Typic Hapludults. Agron. Abstr. American Society of Agronomy, Madison, WI, p. 196.

Buol, S. W. (ed.). 1973. Soils of the southern states and Puerto Rico. Southern Cooperative Series Bull. No. 174. North Carolina State Univ., Raleigh.

Campbell, R. B., T. A. Matheny, P. G. Hunt, and S. C. Gupta. 1979. Crop residue requirements for water erosion control in six southern states. J. Soil Water Conserv. 34(2):83–85.

Carreker, J. R., S. R. Wilkinson, A. P. Barnett, and J. E. Box, Jr. 1977. Soil and water management systems for sloping lands. USDA ARS-S-160.

Collins, W. O. 1935. Soil erosion experiments. College Agric. Bull. 35(10b): Serial No. 613. Univ. of Georgia, Athens.

Copley, T. L., L. A. Forest, A. G. McCall, and F. C. Bell. 1944. Investigations in erosion, control and reclamation of eroded land at the Central Piedmont Conservation Experiment Station, Statesville, NC. USDA Tech. Bull. 873.

Davis, W. A., E. F. Goldston, and C. H. Warner. 1931. Soil Survey of Franklin County, North Carolina. USDA, Bureau of Chemistry and Soils, Series 1931 Number 21.

DelVecchio, J. R., and W. G. Knisel. 1982. Application of a field scale nonpoint pollution model. p. 227–236. In E. G. Kruse et al. (ed.) Environmentally sound water and soil management. Proc. Specialty Conf., Orlando, FL. 20–23 July 1982. American Society of Civil Engineers.

Denton, H. P. 1978. The effects of degree of erosion and slope characteristics on soybeans yields on Memphis, Grenada, Lexington and Loring soils. M.S. thesis. University of Tennessee, Knoxville.

Dregne, H. E. 1982. Historical perspective of accelerated erosion, and effect on world civilization. p. 1–14. In B. L. Schmidt et al. (ed.) Determinants of soil loss tolerance. Spec. Publ. 45. American Society of Agronomy, Madison, WI.

Foy, C. D. 1974. Effects of aluminum on plant growth. p. 601–642. In E. W. Carson (ed.) Plant root and its environment. University Press of Virginia, Charlottesville.

Frye, W. W., S. A. Ebelhar, L. W. Murdock, and R. L. Blevins. 1982. Soil erosion effects on properties and productivity of two Kentucky soils. Soil Sci. Soc. Am. J. 46:1051–1055.

Heilman, M. D., and J. R. Thomas. 1961. Land leveling can adversely affect soil fertility. J. Soil Water Conserv. 16:71–72.

Hendrickson, B. H., A. P. Barnett, and O. W. Beale. 1963. Conservation methods for soil of the Southern Piedmont. USDA Info. Bull. 269.

Hilgard, E. W. 1860. Report on the geology and agriculture of the state of Mississippi. Jackson, MS.

Jenny, Hans. 1961. E. W. Hilgard and the birth of modern soil science. Collana Delta Rivista Agrochimica No. 3. Simposio Internazionale di Agrochimica, Pisa, Italy.

Langdale, G. W., A. P. Barnett, R. A. Leonard, and W. G. Fleming. 1979a. Reduction of soil erosion by the no-till system in the Southern Piedmont. Am. Soc. Agric. Engr. Trans. 22(1):83–86, 92.

----, J. E. Box, R. A. Leonard, A. P. Barnett, and W. G. Fleming. 1979b. Corn yield reduction on eroded Southern Piedmont soils. J. Soil Water Conserv. 34(5):226–228.

----, W. L. Hargrove, and J. E. Giddens. 1984. Residue management in double crop conservation tillage systems. Agron. J. 76(4):689–694.

----, H. F. Perkins, A. P. Barnett, J. C. Reardon, and R. L. Wilson, Jr. 1983. Reduced soil and nutrient runoff losses associated with in-row chisel planted soybeans. J. Soil Water Conserv. 28(3):297–301.

----, and W. D. Shrader. 1982. Soil erosion effects on soil productivity of cultivated cropland. p. 41–51. In B. L. Schmidt et al. (eds.) Determinants of soil loss tolerance. Spec. Publ. No. 45. ASA and Soil Science Society of America, Madison, WI.

Larson, W. E. 1981. Protecting the soil resource base. J. Soil Water Conserv. 36(1):13–16.

----, F. J. Pierce, and R. H. Dowdy. 1983. The threat of soil erosion to long-term crop production. Science 219(4584):458–465.

Latham, E. E. 1940. Relative productivity of the A horizon of the acid sandy loam and the B & C horizons exposed by erosion. J. Am. Soc. Agron. 32:950–954.

Lowdermilk, W. C. 1953. Conquest of the land through seven thousand years. USDA Info. Bull. 99.

McCormack, D. E., K. K. Young, and L. W. Kimberlin. 1982. Current criteria for determining soil loss tolerance. p. 95–111. *In* B. L. Schmidt et al. (eds.) Determinants of soil loss tolerance. Spec. Publ. 95. American Society of Agronomy, Madison, WI.

McDaniel, T., and B. F. Hajek. 1982. Past erosion reduces current yields. Alabama Agric. Exp. Stn. Highlights of Agric. Res. 29(3):10.

McGregor, K. C., J. D. Greer, and G.E. Gurley. 1975. Erosion control with no-till cropping practices. Am. Soc. Agric. Engr. Trans. 18(5):918–920.

Overton, J. R., and F. B. Bell. 1974. Productivity of soils on the West Tennessee Experiment Station for corn. Tenn. Farm Home Sci. 90:35–38.

Perkins, H. P., H. J. Byrd, and F. T. Ritchie, Jr. 1973. Ultisols-light-colored soils of the warm temperate forest lands. p. 73–86. *In* S. W. Buol (ed.) Soils of the southern states and Puerto Rico. Southern Cooperative Series Bull. 174. North Carolina State Univ., Raleigh.

Phillips, J. A., and E. J. Kamprath. 1973. Soil fertility problems associated with land-forming in the Coastal Plain. J. Soil Water Conserv. 28:69–73.

Pieters, A. J., P. R. Henson, W. E. Adams, and A. P. Barnett. 1950. Sericea and other perennial lespedezas for forage and soil conservation. USDA Circ. 863.

Rhoton, W. E. 1975. Productivity of Emory and Dewey soils for soybeans. M.S. thesis. University of Tennessee, Knoxville.

Rios, M. A., and R. N. Pearson. 1964. The effects of some chemical environmental factors on cotton root behavior. Soil Sci. Soc. Am. Proc. 29:232–235.

Ruffin, E. 1832 (1961 reprint, ed. J. C. Sitterson) An essay on calcareous manures. The Belknap Press of Harvard University, Cambridge, MA.

Sheridan, J. M., C. V. Booram, Jr., and L. E. Asmussen. 1982. Sediment delivery ratios for a small Coastal plain agricultural watershed. Am. Soc. Agric. Engr. 25(3):610–615, 622.

Simpson, D. V. 1974. Productivity of Dewey and Emory soils for soybeans. M.S. thesis. University of Tennessee, Knoxville.

Slusher, D. F., and S. A. Lytle. 1973. Alfisols-light-colored soils of the humid temperate areas. p. 61–72. *In* S. W. Buol (ed.) Soils of the southern states and Puerto Rico. Southern Cooperative Series Bull. 174. North Carolina State University, Raleigh.

Soil Survey Staff. 1975. Soil taxonomy: A basic system of soil classification for making and interpreting soil surveys. Agric. Handb. 436. U.S. Government Printing Office, Washington, DC.

Springer, M. E., and J. A. Elder. 1980. Soils of Tennessee. Univ. of Tennessee Agric. Exp. Stn. Bull. 596.

Stone, J., R. Daniels, J. Gilliam, J. Kleins, and K. Cassel. 1982. Relationship among corn yields, surface horizon color and slope form in some clayey North Carolina Piedmont soils. Agron. Abstr. American Society of Agronomy, Madison, WI, p. 257.

Thomas, A. W., J. R. Carreker, and R. L. Carter. 1969. Water, soil and nutrient losses on Tifton loamy sand. Georgia Agric. Exp. Stn. Bull. 491.

Thomas, D. J., and D. K. Cassel. 1979. Land forming Atlantic Coastal Plain soils: Crop yield relationships to soil physical and chemical properties. J. Soil Water Conserv. 30:20–24.

Trimble, S. W. 1974. Man-induced soil erosion on the Southern Piedmont, 1700–1970. Soil Cons. Soc. Am., Ankeny, IA.

----. 1975. A volumetric estimate of man-induced erosion on the Southern Piedmont. USDA–ARS-S-40.

United States Congress. 1930. House Committee on Appropriations, Subcommittee of 1928. Hearing of the Agricultural Appropriations Bill for 1930. U.S. (70th) Congress, 2nd Sess. p. 310–330.

USDA. 1980. Soil and Water Resources Conservation Act: Summary of Appraisal, Part I and II, and Program Report. USDA, Washington, DC.

USDA–SCS. 1977. Cropland erosion. U.S. Government Printing Office, Washington, DC.

USDA–SCS. 1981. America's soil and water: Condition and trends. U.S. Government Printing Office, Washington, DC.

White, A. W., Jr., R. R. Bruce, A. W. Thomas, G. W. Langdale, H. F. Perkins, and R. L. Wilson. 1983. Effect of soil erosion on soybean production in the Southern Piedmont of Georgia. Agron. Abstr. 75:205.

Williams, J. R., R. R. Allmaras, K. G. Renard, L. Lyles, W. C. Moldenhauer, G. W. Langdale, L. D. Meyer, W. J. Rawls, G. Darby, R. Daniels, and R. Magleby. 1981. Soil erosion effects on soil productivity: A research prospective. J. Soil Water Conserv. 36(2): 82-90.

Winters, E., and R. W. Simonson. 1951. The subsoil. Adv. Agron. 3:1-92.

Wischmeier, W. H., and D. D. Smith. 1978. Predicting rainfall erosion losses—a guide to conservation planning. Agric. Handb. 537. USDA. Washington, DC.

16 Regional Effects of Soil Erosion on Crop Productivity—Midwest

J. V. Mannering and D. P. Franzmeier
Purdue University
West Lafayette, Indiana

D. L. Schertz
Soil Conservation Service
U.S. Department of Agriculture
Washington, DC

W. C. Moldenhauer and L. D. Norton
Agricultural Research Service
U.S. Department of Agriculture
West Lafayette, Indiana

The Midwest is one of the most productive agricultural areas of the world. But the combination of climate, slope, and intensive cultivation has resulted in serious soil erosion. Recent USDA (1981) estimates show that average annual soil losses on a high percentage of sloping cropland exceed soil loss tolerance values (T values) in every state in the Midwest.

The effect of accelerated soil erosion on crop productivity varies widely depending on climatic conditions and a combination of soil properties such as depth, texture, structure, and nutrient-supplying power of both surface and subsoils (Langdale and Shrader, 1982; Williams et al., 1981; and Young, 1980). Extreme conditions range from almost total inability to produce crops (for example, with severe erosion on a shallow soil) to little, if any, yield reduction (for example, with severe erosion on heavily fertilized, deep silt soil).

Early research to show the effects of soil erosion on crop yields in the Midwest is well represented by Uhland (1949). He showed a definite and steady reduction in crop yields and organic matter as soil erosion continued to reduce the thickness of the topsoil. However, since 1949, large applications of inorganic fertilizer, especially nitrogen, have tended to mask the loss of productivity due to excessive erosion (Englestad and Shrader, 1961;

Published in R. F. Follett and B. A. Stewart, ed. 1985. *Soil Erosion and Crop Productivity.*
© ASA-CSSA-SSSA, 677 South Segoe Road, Madison, WI 53711, USA.

Englestad et al., 1961; Williams et al., 1981; and Young, 1980). This confounding effect has greatly decreased the usefulness of much earlier research (Williams et al., 1981; Young, 1980).

Recent research continues to show that accelerated soil erosion adversely affects yield even with improved technology. For example, in Minnesota, Cardwell (1982) reported that soil erosion reduced yield potentials 8% over 50 years. Kentucky researchers (Frye et al., 1982) reported that even moderate erosion reduced grain yield from corn (*Zea mays* L.) 12 and 21%, respectively, on Maury and Crider soils. Although the addition of high rates of nitrogen and zinc partially restored yield loss due to severe erosion in eastern South Dakota, the difficulty in preparing a seedbed in exposed subsoils as well as increased crusting still resulted in yield reductions (Olson, 1977). A recent economic analysis in Iowa (Rosenberry et al., 1980) showed that even with higher rates of fertilizer to offset erosion losses, yield generally declines as soil shifts from one erosion phase to another. Hagen and Dyke (1980), using yield estimates from published soil surveys, estimated the average corn yield loss in the Corn Belt was about 188 kg ha^{-1} cm^{-1} of soil loss. Benbrook (1981) points out that, even where excessive erosion does not change crop yields, farmers are required to rely more on nonsoil production inputs. He asks (p. 119) "How long will this substitution of energy-intensive farm inputs for soil remain affordable?" Certainly, recent evidence indicates that erosion can significantly reduce crop yields. However, since this research is limited spatially and temporally (Williams et al., 1981), additional research is needed.

The objectives of this paper are (i) to review presently used estimates for crop yield reduction due to erosion in the Midwest, (ii) to document how these estimates were developed, (iii) to report new research efforts in the Midwest that are designed to provide quantitative field data relating soil erosion to productivity, and (iv) to discuss the strengths and weaknesses of past and present research and develop recommendations for future research.

16–1 PROGRAMS IN USE AND THEIR DEVELOPMENT

Yield estimates for soil series and phases of soil series, such as slope and erosion phases, have been used to evaluate the effect of erosion on crop yields in several states. These estimates are published in bulletins from the state universities and in the Soil Interpretations Record of the National Cooperative Soil Survey. An example of an entry in the Soil Interpretations Record for the Miami series is shown in Table 1. According to these estimates, corn yield for the Miami series decreases from 6900 kg ha^{-1} (110 bu/acre) for the "0–2% slope, noneroded" phase to 5018 kg ha^{-1} (80 bu/acre) for the "12 to 18% slope, eroded" phase. Separate interpretation records are issued for other variations of the Miami series, such as the moderately permeable substratum phase. A Soil Interpretations Record is

Table 16–1. A sample entry in the Soil Interpretations Record for the Miami series (5-81).

Class-determining phase	Land capability class	Corn		Soybeans	
%		kg/ha	bu/ac	kg/ha	bu/ac
0–2	I	6899	(110)	2554	(38)
2–6	II	6899	(110)	2554	(38)
2–6 eroded	II	6586	(105)	2486	(37)
2–6 severely eroded	III	6272	(100)	2352	(35)
6–12 eroded	III	5958	(95)	2218	(33)
6–12 severely eroded	IV	5645	(90)	2150	(32)
12–18	IV	5331	(85)	2016	(30)
12–18 eroded	IV	5018	(80)	1882	(28)
12–18 severely eroded	VI	--	--	--	--

prepared initially by a field soil scientist working in an area where the series occurs. It is then reviewed in the state SCS offices and by cooperators of the soil survey, such as experiment stations and the U.S. Forest Service, in the states in which the series is found. Once issued, the records are revised from time to time, but not on a prescribed schedule.

This procedure does have limitations. For example, crop yields for map units, by their very nature, are difficult to characterize. Because yields depend so greatly on weather, long-term yield measurements are necessary. Over time, however, yields tend to increase because of new technology. Yields also depend greatly on the soil-management skills of the farmer.

In addition, few, if any, long-term plots have been established specifically to characterize yields of a certain soil. Instead, plots established for some other purpose are used, but these often do not provide optimum information. Such plots, commonly designed to test the yield of crop varieties or the response to fertilizer applications, are placed on uniform soil areas. They are seldom placed on eroding hillslopes. Furthermore, if drought or excessive rainfall occurs, yields are so greatly diminished that they are often not reported. Yet, these extremes greatly affect long-term yields. Under ideal conditions, a group of soils may have similar yields, but under stress some yield considerably more than others. These relationships are often difficult to characterize using data from plots designed for other purposes.

16–1.1 Procedures Used in Midwestern States

In most states, yield estimates are derived from a system, or model, in which the yields of all soils of the state are considered relative to each other. In this way, the relative values are kept in better perspective than if only one series is considered at a time. The nature of the evaluation and the amount of data considered vary from state to state. We will consider a few examples.

16-1.1.1   Iowa

In the Midwest, Iowa has the best database for estimating the effect of erosion on corn yield and soil productivity. In the 1950s, L. C. Dumenil (Engelstad et al., 1961), established a large set of plots around the state to characterize the effect of management practices on corn yield. He selected certain counties to represent the major soil associations of Iowa. Within these counties, he selected fields from the Conservation Needs Inventory tracts, a 2% randomly selected sample of 65-ha (160 acres) quarter sections. In these fields, he selected his plots. Yields were measured, by hand-picking, whenever corn was grown on these plots between 1955 and 1967. Records were kept of management practices and soil test values. Many of the plots were on sloping soils. Some of these had slight or moderate erosion, but few had severe erosion. The yield measurements were recorded in theses and unpublished records.

Soil scientists at Iowa State University (Fenton et al., 1971) and SCS examined these data to establish trends in the effects of slope, erosion, and other soil properties on corn yield, and extrapolated these trends to soils on steeper slopes and to those with more erosion. The effects of slope and erosion varied with the characteristics of the subsoil and soil depth.

These trends were projected to other soil map units through the use of a Corn Suitability Rating (CSR). The CSR considers the corn yield potential of the soil and allows for a rotation with sod crops, which produces less income, to control erosion. The highest yielding soil of the state, a Muscatine soil (fine-silty, mixed, mesic, Aquic Hapludoll), was assigned a CSR of 100. Then, CSR points were deducted for soil properties associated with lower yields (Table 16-2).

As an eample, consider the CSR computation for the Fayette soil (fine-silty, mixed, mesic, Typic Hapludalf):

2-5% slope, moderately eroded:

Index	100
Well-drained	0
Forest vegetation	− 10
2-5% slope	− 5
Moderate erosion	− 2
	83

9-14% slope, severely eroded:

Index	100
Well-drained	0
Forest vegetation	− 10
9-14% slope	− 30
Severe erosion	− 5
	55

Table 16-2. Some of the guidelines used in establishing Corn Suitability Ratings in Iowa for deep, moderately fine-textured soils (Fenton et al., 1971).

Soil property	Adjustment factor
Slope	
A. 0–2%	Index soil
B. 2–5%	−5
C. 5–9%	−20
D. 9–14%	−30
Erosion	
1. none to slight	Index soil
2. moderate	−2
3. severe	−5
Native vegetation	
prairie	Index soil
prairie/forest	−5
forest	−10

Table 16-3. Comparison of estimated yield and Corn Suitability Rating (CSR) for some slope and erosion phases of Fayette silt loam.

Slope gradient	Erosion class								
	Slight			Moderate			Severe		
%	kg/ha	bu/ac	CSR	kg/ha	bu/ac	CSR	kg/ha	bu/ac	CSR
0–2	7212	(115)	90	--†	--	88	--	--	--
2–5	7087	(113)	85	--	--	83	--	--	80
5–9	6773	(108)	70	6586	(105)	68	6271	(100)	65
9–14	6209	(99)	60	6021	(96)	58	5645	(90)	55

† Dashes denote phases not commonly mapped.

Calculated CSR values for several possible slope and erosion conditions and yields for the more extensive combinations are given in Table 16-3. The CSR values decrease more than yields with increasing slope gradient and degree of erosion. For example, for the 9 to 14% slope in the severely eroded phase, the CSR is 55, but the yield is 80% of the 2 to 5% slope in the slightly eroded phase. This means that when corn is grown on steeper, more eroded soils, relatively high yields are obtained, but corn cannot be grown continuously on these steeper slopes if erosion is to be controlled.

16–1.1.2 Indiana

In 1968, Indiana adopted the Iowa evaluation system. Plot data were available for several nearly level, uneroded soils, but few data were available for sloping and eroded soils, so the Iowa experience was largely relied on to estimate the effects of slope and erosion on yield. Some adjustments were made, however, based on local data and experience. For example, in a natural drainage sequence of soils, the poorly drained soil commonly has the lowest yield in Iowa, but, when drained, the poorly drained soil has the highest yield in Indiana.

The system in Indiana was further developed by Walker (1976). Walker used corn yield directly instead of CSR and used five index soils instead of one, but the system of assuming yields for index soils and subtracting penalty points was similar. The system was tested and refined in 11 counties by comparing corn yields calculated using the model and averaged for the corn-producing soils with yields reported for the counties by the USDA Statistical Reporting Service from 1964 to 1974. These yield estimates were reviewed by county extension agents and SCS district conservationists who evaluated them against their general knowledge of soil productivity according to kinds of soils. No attempt was made to include the influence of crop rotation on these yield estimates. Galloway and Steinhardt (1981) reviewed the model and presented its results.

16-1.1.3   Illinois

Somewhat different models were used in other states. In Illinois, Fehrenbacher et al. (1978) estimated yields for the 0–2%, uneroded phase of each soil series and adjusted downward for increasing slope and erosion using one schedule for soils with favorable subsoils and another schedule for those with unfavorable subsoils.

16-1.1.4   Ohio

In Ohio, crop yields for 22 major soils occurring in different major land-use areas were summarized from research results and farm demonstrations. Yield estimates were made for other 0–2% uneroded soil series of the state by ranking them relative to the reference soils based on the experiences of extension personnel and state and federal soil scientists (Bone and Norton, 1981).

Yield values for corn, soybeans, wheat, oats, and hay were estimated for each soil, and the ratio of biomass production from these crops relative to that of a reference soil (Ross sil) was used to further refine the rankings. This ratio times 100 was the productivity index.

Yields for sloping ($<$ 18%) and eroded soils were then adjusted downward from the yields for the 0 to 2%, uneroded phase, similar to the Illinois system, according to the following guidelines:

Slope (%)	Yield change (%)
0–2 (A)	0
2–6 (B)	4
6–12 (C)	8
12–18 (D)	16

Erosion	Yield change (%)
Slight	0
Moderate	4
Severe	12

Recently, a model similar to Walker's (1976) was developed and tested on some Ohio soils, but this approach has not yet been adopted statewide (Derringer, 1982).

16-1.1.5 Minnesota

In Minnesota, observations from more than 600 farmers from 1956 to 1972 were used to estimate yields, mainly for uneroded soils (Rust and Hanson, 1975).

Several states use some type of index system to allow for the contribution of the various crops to the total productivity of the soil. On moderately sloping soils, rotation of sod crops must often be used with row crops to control erosion. Other soils, such as very steep or very coarse-textured soils, are better suited for continuous pasture or forest than for row crops. To integrate the contribution of various crops into one index for all soils, yield or production levels must be established for the different crops and the portion of time that a soil is used for the various crops must be determined. The distribution of crops may be according to recommended rotation (Iowa), total production assuming all crops were grown (Ohio), or the actual distribution of crops, from census data, for various kinds of soils (Minnesota) or sections of the state (Illinois).

16-1.2 Examples of Yield Estimates from Illinois, Indiana, Iowa, Minnesota and Ohio

The data in Table 16-4 were derived from bulletins from the respective state universities. They represent the Alford and Fayette series or a class of soils that includes these series. Both Alford and Fayette soils formed in thick loess deposits under forest vegetation and are classified fine-silty, mixed, mesic, Typic Hapludalfs. Relative yields were calculated using the 2 to 6% slope uneroded phase as a base because this is a commonly occurring phase and bulletins from several states list yields for it.

In the Illinois bulletin (Fehrenbacher et al., 1978), yields are listed as 4955 kg/ha (79 bu/acre) for basic management and 8090 kg/ha (129 bu/acre) for high-level management for Fayette silt loam, 0–2% slopes, uneroded. Graphs are given to show the decrease in yield, under both management levels, due to increasing slope and erosion for soils with unfavorable subsoil characteristics and those with favorable subsoils (which includes Fayette). Entries in Table 16-4 were calculated from the graph for basic management.

In Indiana (Walker, 1976; Galloway and Steinhardt, 1981), yields were tabulated for all the phases of listed soils for high-level and average management. Entries in Table 16-4 are based on average management of Alford silt loam.

Yields for Iowa are listed for the phases of Fayette silt loam given by Fenton et al. (1971).

Table 16–4. Estimated corn yields for various slope and erosion classes of deep, well-drained soils with light-colored silt loam surface horizons.

Slope gradient	Base yield†		Slight	Moderate	Severe
			Erosion class		
%	kg/ha	bu/ac	——— % of base yield ———		
Illinois (Basic management)‡					
1 (0–2)	4955	(79)	102	97	
4 (2–6)			100	95	85
9 (6–12)			96	91	81
15 (12–18)			90	84	74
Indiana (average management)‡					
0–2	6272	(100)	100	96	
2–6	6272	(100)	100	96	92
6–12	5770	(92)	92	88	84
12–18	5018	(80)	80	76	72
Iowa (high-level management)‡					
0–2					
2–5	7087	(113)	100		
5–9			96	93	88
9–14			88	85	80
Minnesota (moderate level management)‡					
0–2					
2–6	6586	(105)	100		
6–12			86		
12–18			71		
Ohio (average management)‡					
0–2			104	100	
2–6	6899	(110)	100	96	88
6–12			96	92	84
12–18			88	84	77

† Base yields represent the Alford and Fayette series or a class of soils that includes these series. Relative yields were calculated using the 2–6% slope, none to slight erosion phase. Sources of were derived from bulletins from the respective state universities (see text).

‡ Basic management includes the mimimum input considered necessary for crop production to be feasible. High level management includes inputs that are near those required for maximum profit with current technology. Average management and moderate level management reflects what the majority of farmers are using.

Yields for Minnesota are listed for the phases of Fayette silt loam listed by Rust and Hanson (1975), who include only uneroded phases.

Values for Ohio were calculated from the yield given by Bone and Norton (1981) for Alford silt loam, 2 to 6% uneroded phase, and the percentage reduction for slope and erosion that applies to all soils.

16–1.3   Effect of Slope and Gradient

Erosion of cropland usually increases with increasing slope gradient. Table 16–5 shows the combined effect of slope gradient and erosion class

Table 16-5. Comparison of relative yields in same slope and erosion classes
for the soils represented in Table 16-4.

	Slope and erosion class		
	4%—slight	9%—moderate	15%—severe
		% of 4%—slight	
Illinois	100	91	74
Indiana	100	88	72
Iowa	100	89	<80†
Minnesota	100	<86‡	<71‡
Ohio	100	92	77

† For 9–14% slope.
‡ For slight erosion.

for the combination for which information is generally available in the state bulletins. Tables 16-4 and 16-5 show that estimated trends of the effects of slope and erosion on corn yield are similar in the five states. To what extent this means that independent investigators have discovered a universal truth or that people in the separate states cooperated in making the estimates is an unanswered question.

16-2 LIMITATIONS OF PRESENT PROCEDURES

Measurement of the yield potential of soil depends so greatly on weather, management skills, available technology, and other factors, that a test designed to account for all these variables would be too massive to carry out. Moreover, the prevailing conditions at the end of such an experiment would probably be so much different from those incorporated in the test that the results would have little relevance to actual conditions. For example, those who work closely with tillage research have observed that corn yields on eroded soils is depressed less with no-till methods than with conventional tillage methods. Thus, research based on a conventional tillage system might have limited applicability to a no-till system which could be the prevailing practice when the study is completed. Also, measuring the effects of some closely controlled variable could have little useful application if that variable could be closely controlled in the field.

Thus, the judgment of knowledgeable people with the assistance of some experiments is probably the only practical way to estimate the productivity of soils. We presently have too few of these experiments and are forced to extrapolate from a small amount of data.

16-3 NEW RESEARCH EFFORTS

Efforts have been initiated in several states to provide additional field data relating soil loss to crop productivity. One such effort, a cooperative study between USDA–ARS and SCS and Purdue University, began in fall 1981. The 3-year study in Benton, Montgomery, and Whitley Counties in

Indiana is evaluating there major soil series: Corwin, Miami, and Morley (fine-loamy, mixed, mesic, Typic Argiudoll; fine-loamy, mixed, mesic, Typic Hapludalf; and fine, mixed, mesic, Typic Hapludalf, respectively). One soil type is being evaluated per county. In each county, erosion classes of none to slight, moderate, and severe are being evaluated in five fields, three times in each field. Fields and erosion classes within fields were selected by soil scientists using detailed soil surveys and on-site verification. Sites within fields were marked through use of a compass and a measuring wheel and are used in each year of the study. All fields were in corn in 1981, but some sites were in soybeans in 1982.

16-3.1 Test Procedures

Each corn plot is three rows wide and 6.10 m long. In 1981, corn yields were determined for each row by harvesting every fifth ear. The sample was weighed to obtain an average ear weight and then multiplied by the total number of ears in each row. The percentage of moisture was determined using a composite sample of shelled corn from the picked ear sample. Corn yields for each row were then adjusted to 15.5% moisture and averaged over the three rows to obtain the plot yield. Corn yields for each erosion class were determined by averaging the three replications in each field. County averages were determined by averaging yields of the five different farms, each having the same soil type.

In 1982, a more elaborate corn-yield analysis was made to develop a statistically reliable sampling procedure. Each ear in each row was numbered and then harvested and placed in a bag labeled for that particular row. In the laboratory, each ear was weighed and then shelled. A shelled corn sample for each ear was weighed and placed in an oven at 48°C for 48 h and reweighed to determine moisture content. Corn yields were then determined based on the weight and percentage of moisture of each individual ear. Overall average yields were determined as in 1981.

In 1982, one plot each of none to slight, moderate, and severe erosion was used in each county to monitor soil water through the growing season. Replicated soil water samples were taken at planting, 2 weeks before silking, and every 2 weeks thereafter until plots were harvested. After harvest, soil pits were dug and three soil cores were taken by a hand coring device for each horizon. Bulk samples were also taken from each horizon and used to determine particle size. Water content was then determined at -0.1, -0.33, -1, -2, and -15 bars of pressure. Bulk densities of each horizon were calculated and used to convert the gravimetric soil water samples to volumetric measurements and then to compare with water content under various pressures. This comparison will provide some indication of the soil water status of the sites during the growing season, which can be related to periods of plant stress.

In 1982, soybean yields were determined by harvesting three rows, each 3 m long, in the center of each plot. For drilled beans, yields were determined by harvesting a strip 1.5 m wide and 3 m long in the center of each plot. The cut soybeans were allowed to dry, then threshed, and the percentage of moisture was determined. Yields were then calculated based on 13.0% moisture. Yield averages by erosion class were computed as for corn.

Amounts of fertilizer and pesticides were controlled only in the sense that in a particular field the same amounts were applied to noneroded and eroded sites. Variety of seed was also constant on individual fields. All applications of fertilizer and pesticides and planting of seed were done by the individual farmers using their normal operations. No attempt was made to select any specific level of management.

Duncan's Multiple Range test was used to determine statistical differences of crop yields between erosion classes.

16–3.2   1981 Results and Discussion

Particle-size analyses were determined for the Ap horizon at each site (Table 16–6). As a rule, severely eroded sites contained a significantly higher percentage of sand and clay than did the none to slightly eroded sites.

The relationship of erosion class to soil-available phosphorus and to organic matter content is shown in Table 16–7. In Benton County, the Corwin prairie soil, which is inherently higher in organic matter than the Miami or Morley forested soils, lost significant amounts of organic matter as erosion increased from slight to moderate to severe. The Miami and Morley forested soils were reduced much less in organic matter as erosion increased; the only significant reduction was between the none to slightly eroded site and the severely eroded site in Whitley County.

Generally, available phosphorus levels were quite high in the Ap horizon but decreased significantly as soil erosion increased from none to slightly eroded to severely eroded.

Table 16–6. Effect of the degree of soil erosion on particle-size distribution for the Ap horizon. Values shown are the means of three sites.

County	Erosion class	Sand	Silt	Clay
		%		
Benton	None to slight	21.6 a*	57.3 a	20.9 ab
	Moderate	26.5 a	53.8 a	19.6 a
	Severe	33.6 b	43.5 b	23.1 b
Montgomery	None to slight	16.7 a	67.9 a	15.5 a
	Moderate	24.5 ab	57.4 b	18.1 b
	Severe	31.3 b	46.5 c	22.1 c
Whitley	None to slight	36.6 a	44.8 a	18.5 a
	Moderate	33.2 ab	43.8 a	23.0 b
	Severe	29.6 b	42.1 a	28.4 c

* Numbers for a given county and particle size not followed by the same letter are significantly different at the 5% level.

Table 16–7. Effect of the degree of erosion on phosphorus and organic matter soil test levels of the Ap horizon. Values shown are the means of three sites.

County	Erosion class	Phosphorus		Organic matter
		kg/ha	lb/ac	%
Benton	None to slight	69.0 a*	(61.6)	3.03 a
	Moderate	68.1 a	(60.8)	2.52 b
	Severe	45.6 b	(40.7)	1.86 c
Montgomery	None to slight	106.4 a	(95.0)	1.83 a
	Moderate	96.5 ab	(86.2)	1.64 a
	Severe	76.4 b	(68.2)	1.51 a
Whitley	None to slight	89.0 a	(79.5)	1.91 a
	Moderate	74.4 b	(66.4)	1.76 ab
	Severe	55.8 c	(49.8)	1.60 b

* Numbers for a given county not followed by the same letter are significantly different at the 5% level.

Table 16–8. Effect of the degree of soil erosion on corn yields. Values shown are means of three sites.

County	Erosion class	Corn yield	
		kg/ha	bu/ac
Benton	None to slight	8216 a*	131
	Moderate	7777 ab	124
	Severe	7526 b	120
Montgomery	None to slight	9094 a	145
	Moderate	8781 a	140
	Severe	7213 b	115
Whitley	None to slight	7338 a	117
	Moderate	7213 a	115
	Severe	6648 a	106

* Numbers for a given county not followed by the same letter are significantly different at the 5% level.

Corn yields were significantly lower on severely eroded sites than on none to slightly eroded sites in two of the three counties (Table 16–8). Differences between erosion classes were greatest where yields were highest. Where yields were lowest (Whitley County), no significant differences were found.

The 1981 cropping season was unusually wet, with heavy spring rains delaying planting for as much as 4 weeks on the sites in the study. The abundant rainfall throughout the growing season was beneficial to the severely eroded sites by preventing water stress, but was most likely detrimental to the none to slightly eroded sites by keeping them wetter for longer periods. In many instances, the excessive spring rains resulted in poor weed control by herbicides, especially on the none to slightly eroded sites. Also, due to wetter conditions, the none to slightly eroded sites may have lost more nitrogen by leaching and denitrification, resulting in lower yields than normal. Rainfall data for the 1981 growing season are presented in Table 16–9.

Table 16-9. Total monthly precipitation and departure from normal for locations near the test sites for 1981 growing season.†

Month	Fowler		Crawfordsville		Columbia City	
	Precip.	Depart.	Precip.	Depart.	Precip.	Depart.
			cm			
Apr.	14.96	4.37	13.59	2.77	16.89	7.57
May	14.63	3.94	20.70	9.27	11.23	1.80
June	8.86	−1.98	6.76	−5.13	21.92	12.37
July	14.96	4.88	20.22	10.52	11.00	0.64
Aug.	22.15	14.55	12.98	4.83	7.64	−0.66
Sept.	7.77	0.61	10.72	2.69	12.12	5.10
Total	83.33	26.37	84.97	24.95	80.80	26.82

† Rainfall data from National Oceanic and Atmospheric Administration, Environmental Data and Information Service, National Climatic Center, Asheville, NC, Vol. 86, 4–9.

16-3.3 1982 Results

Due to more elaborate analysis, the 1982 crop-yield data have not been finalized. However, preliminary analysis and visual evaluation at harvest indicate considerably wider yield differences between none to slightly eroded sites and eroded sites.

16-4 DISCUSSION AND RECOMMENDATIONS

Developing a meaningful and usable relationship between soil erosion and crop productivity is a complex and frustrating task. A review of research in the Midwest is certainly inadequate to supply the needed answers. Reasons for the lack of information have been well documented, most realistically and precisely in Office of Technology Assessment (1980). Because of the complexity of the problem, some would suggest that we dismiss it and move on to less complex problems. However, scientists have not yet given this problem our best effort, and with increased emphasis and coorindation we can develop much more precise predictions of the effects of soil erosion on crop productivity.

More specifically, we need models that are site specific and that will predict with some degree of confidence the effect of a level of soil erosion on crop productivity. Before we can gain confidence in any model, we need to develop a sufficient database so that the model can be critically tested. To provide the database, we need to increase the number of well-organized and well-conducted field experiments on benchmark soils. There is some encouraging news in this area in the Midwest, since new field studies have been or soon will be initiated in several states under the coordination of a new regional project, NC 174 (Soil Productivity and Erosion). Skilled modelers are ready and waiting to test their models once these data are available. Even though this type research is slow and difficult, we must direct more effort to establish the relationship between soil erosion and crop productivity to adequately preserve the productivity of this most important natural resource.

REFERENCES

Benbrook, C. 1981. Erosion vs. soil productivity. J. Soil Water Conserv. 36:118–119.

Bone, S. D., and L. D. Norton. 1981. Ohio soils with yield data and productivity index. Coop. Ext. Serv. Bull. 685, AGDEX 524. The Ohio State University, Columbus.

Cardwell, V. B. 1982. Fifty years of Minnesota corn production: Sources of yield increase. Agron. J. 74:984–990.

Derringer, G. D. 1982. The relationship of soil properties to corn productivity in Ohio. M.S. thesis. The Ohio State Univ., Columbus.

Englestad, O. P., and W. D. Shrader. 1961. The effect of surface soil thickness on corn yields: II. As determined by an experiment using normal surface soil and artificially expoed subsoil. Soil Sci. Soc. Am. Proc. 25:497–499.

––––, ––––, and L. C. Dumenil. 1961. The effect of surface soil thickness on corn yields: I. As determined by a series of field experiments in farmer-operated fields. Soil Sci. Soc. Am. Proc. 25:494–497.

Fehrenbacher, J. B., R. A. Pope, I. J. Jansen, J. D. Alexander, and B. W. Ray. 1978. Soil productivity in Illinois. Coop. Ext. Serv. Circ. 1156. University of Illinois, Champaign-Urbana.

Fenton, T. E., E. R. Duncan, W. D. Shrader, and L. C. Dumenil. 1971. Productivity levels of some Iowa soils. Iowa Agric. Exp. Stn. Spec. Rept. No. 66.

Frye, W. W., S. A. Ebelhor, L. W. Murdock, and R. L. Blevins. 1982. Soil erosion effects on properties and productivity of two Kentucky soils. Soil Sci. Soc. Am. J. 46:1051–1055.

Galloway, H. M., and G. C. Steinhardt. 1981. Indiana's soil series and their properties. Coop. Ext. Serv. AY 212, Purdue University, West Lafayette, IN.

Hagen, L. L., and P. T. Dyke. 1980. Yield-soil loss relationship. Paper presented at Workshop on Influence of Soil Erosion on Soil Productivity, Washington, DC. March. USDA–ARS.

Langdale, G. W., and W. D. Shrader. 1982. Soil erosion effects on soil productivity of cultivated cropland. In B. L. Schmidt et al. (ed.) Determinants of soil loss tolerance. Spec. Pub. 45. American Society of Agronomy, Madison, WI.

Office of Technology Assessment. 1980. Impacts of technology on U.S. cropland and rangeland productivity. Congressional Board of the 97th Congress. Library of Congress Catalog Card No. 82-600596. Washington, DC.

Olson, T. C. 1977. Restoring the productivity of a glacial till soil after topsoil removed. J. Soil Water Conserv. 32:130–132.

Rosenberry, P., R. Knutson, and L. Harman. 1980. Predicting the effects of soil depletion from erosion. J. Soil Water Conserv. 35:131–134.

Rust, R. H., and L. D. Hanson. 1975. Crop equivalent rating guide for soils of Minnesota. Minnesota Agric. Exp. Stn. Misc. Report 132. University of Minnesota, St. Paul.

Uhland, R. E. 1949. Crop yields lowered by erosion. USDA–SCS Tech. Publ. 75.

USDA. 1981. Soil, water, and related resources in the United States: Status, condition, and trends. 1980 RCA Appraisal, Part I. Washington, DC.

Walker, C. F. 1976. A model to estimate corn yields for Indiana soils. M.S. thesis. Purdue Univ., West Lafayette, IN.

Williams, J. R., R. R. Allmaras, K. G. Renard, L. Lyles, W. C. Moldenhauer, G. W. Langdale, L. D. Meyer, W. J. Rauls, G. Darby, R. Daniels, and R. Magleby. 1981. Soil erosion effects on soil productivity: A research perspective. J. Soil Water Conserv. 36(2):82–90.

Young, K. K. 1980. The impact of erosion on the productivity of soils in the United States. p. 295–303. In M. DeBoodt and D. Gabriels (ed.) Assessment of erosion. John Wiley and Sons, Ltd., Chichester, NY.

17 Regional Effects of Soil Erosion on Crop Productivity— Great Plains

Earl Burnett
Agricultural Research Service
U.S. Department of Agriculture
Temple, Texas

B. A. Stewart
Agricultural Research Service
U.S. Department of Agriculture
Bushland, Texas

A. L. Black
Agricultural Research Service
U.S. Department of Agriculture
Mandan, North Dakota

17–1 HISTORICAL PERSPECTIVE

As settlers pushed west from the subhumid eastern edge of the Great Plains, they found that crop production became more erratic and precarious. Early-day conservationists warned of the erosion that would take place in many parts of the Great Plains if the land were cultivated. As long as those soils most susceptible to erosion remained primarily in grass, erosion by wind and water was not serious. As agriculture expanded, however, the semiarid Plains region was opened to homesteading, land was broken from sod, and erosion problems—particularly wind erosion—became increasingly serious.

Following World War I, high wheat prices, coupled with the development of power machinery, led to the rapid expansion in cultivated land and the large-scale production of wheat and other crops. Even marginal soils that should have been left in grass were plowed and seeded to wheat. This expansion in cultivated land took place primarily in the decade of 1915–25.

Published in R. F. Follett and B. A. Stewart, ed. 1985. *Soil Erosion and Crop Productivity.*
© ASA-CSSA-SSSA, 677 South Segoe Road, Madison, WI 53711, USA.

While the expansion of cultivated lands would likely have occurred in any event, the rate of expansion was probably enhanced as the USDA established a series of dryland experiment stations throughout the Great Plains, some as early as 1903. Substations of the state agricultural experiment stations were also established in several Great Plains States shortly after the first USDA stations. The principal thrust of all these stations was crop adaptation and cultural practices for the soils and climate of the region. In 1914 the USDA Division of Dryland Agriculture established field stations at 22 locations in the Great Plains (Fig. 17-1). Initially, these dryland stations, like their predecessors, focused their major efforts on evaluating crops and crop varieties for the local area. However, since the vagaries of climate and the susceptibility of many soils to erosion had been recognized, research was designed to develop crop rotations and management practices to maximize dryland crop production and control erosion.

While various parts of the Great Plains experienced drought during the expansion of cultivated land, the Plains as a whole received average or better rainfall before 1930. The widespread drought of the 1930s and the resulting wind erosion focused national attention on the Great Plains. Severe wind erosion in much of the Great Plains, particularly in southeast Colorado, southwest Kansas, western Oklahoma, and northwest Texas, resulted in the infamous Dust Bowl.

Surveys made during and shortly after the drought years of the 1930s in the Southern Great Plains wind-erosion area (Joel, 1937) suggested that 43% of the area had serious erosion damage. Finnell (1948a) estimated that about 2.6 million ha in the Southern Plains were removed from cultivation because of erosion but that about 10.3 million ha of less erodible soils made it through the drought cycle with no significant erosion damage and were

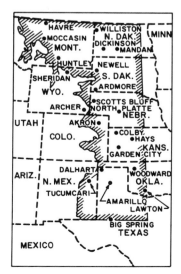

Fig. 17-1. Locations where dryland cropping experiments were conducted from 1903 to 1938 by USDA and SAES. Adapted from Cole and Mathews (1940).

put back into cultivation in the 1940s. He noted, however, that some lands remained idle or were returned to grass because erosion had made them unproductive. Finnell (1948b) pointed out that 59% of the serious erosion had occurred on poorer lands that probably should never have been cultivated.

With increased rainfall and higher crop prices in the early 1940s, the tendency to plow grasslands resumed. This grasslands area of less than 43 to 66 cm of annual rainfall lies along the western fringe of the dry farming region (Fig. 17–2). During the late 1940s and 1950s, the area experienced dust storms comparable to those of the 1930s. Some cultivated lands were returned to grass during the soil-bank period but again put to the plow to produce grain for export in the early 1970s.

Dust storms account for a significant amount of the total erosion in the Great Plains. However, the subhumid eastern portion of the Great Plains is frequently subjected to water erosion that has seriously damaged some erodible soils situated on rolling terrain (Enlow, 1939; Stallings, 1950; Adams et al., 1959; Greiner, 1982; Hill et al., 1944; Smith et al., 1954). While some of these erodible lands have been returned to grass or been placed in a good conservation cropping system, others still have unacceptable erosion rates (USDA, 1982).

17–2 CHANGES IN PRODUCTIVITY

17–2.1 Dryland Studies

Soil erosion is generally considered detrimental to crop productivity because the sorting action of wind or water tends to remove the finer soil fractions containing most of the nutrients available to crops. In addition, erosion may reduce soil depth, which may adversely affect rooting volume and the water-storage capacity of the soil. In many cases, erosion reduces water infiltration, thereby increasing runoff and reducing water available

Fig. 17–2. Fragile lands (dashed lines) on western edge of Great Plains. Solid lines are 43- and 63-cm annual precipitation isohyets. Adapted from Finnell (1948b).

for plant growth. Subsoils exposed by erosion may have poorer tilth and therefore reduced productivity.

Crop productivity can decrease in the absence of erosion, as cropping depletes soil organic matter and plant nutrients. The rate of decline in productivity is related to kind of crop, disposition of crop residues, inclusion of soil-building crops in the rotation, fertilizer use, soil texture, and depth of soil. Erosion is assumed to accelerate the rate of decline in crop productivity, but erosion effects are difficult to separate from cropping effects.

Great Plains soils, particularly the moderately-fine to fine-textured soils in the subhumid eastern side, were high to very high in N and organic matter before cultivation. But, as with soils in all parts of the world, the organic matter content of the soil declined with cultivation and cropping. The rate of decline was related to many factors, including kind of crop, tillage, erosion, and soil characteristics. Haas et al. (1957) measured changes in total soil N and organic C at many dryland experiment stations. Over a cropping period of 30 to 43 years, average organic C decreased 42% and total N decreased 39% compared to virgin grasslands. Losses of both N and C were greater with row crops (corn, *Zea mays* L., grain sorghum, *Sorghum bicolor* L.) than with small grains, and alternate sequences of small grains and fallow crops lost more N and C than continuous small grain. An attempt to relate the loss of N to crop yields at six locations was confounded by extreme weather variations between years and locations. Yields of row crops declined significantly in some instances while wheat (*Triticum aestivum* L.) yields did not. Wheat yields did tend to be lower with continuous cropping than with alternate cropping and fallow, even though losses in soil N were greater under alternate cropping and fallow. Changes in soil N were not significantly related to changes in yield. In some cases, yield changes were greater than N changes, while in others the reverse was true. No attempt was made to assess whether erosion caused some of the losses of N, C, or crop yield.

One of the earliest assessments of cropping effects on crop productivity in the Great Plains is that of Throckmorton and Duley (1932) who reported the decline in average corn yields in Kansas by 10-year periods between 1867 and 1924 (Fig. 17-3). They stated that yields declined, even though corn cultivars grown in the last half of the study period were better adapted than those grown during the first half. The yield decline was more rapid in the subhumid, eastern part of Kansas than in the west. They noted that Kansas soils cultivated for about 30 years had lost 25-30% of the total nitrogen and 30-35% of the original organic matter. Since the crop yields they report are averaged across the state, the yield decline must be attributed to a combination of cropping and erosion.

Finnell (cited by Baver, 1951) compared the average yield of winter wheat in 14 southwest Kansas counties with average rainfall by 3-year periods from 1928 to 1948 (Fig. 17-4). Average yields declined with decreasing rainfall during the drought years and increased with increasing rainfall following the drought. These counties were in the heart of the Dust Bowl, and the soils must have experienced serious erosion, but the average yield in the last period was more than 200 kg/ha higher than during the first period

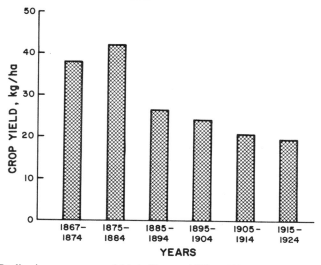

Fig. 17-3. Decline in average corn yields in Kansas, 1867 to 1924 (Throckmorton and Duley, 1932).

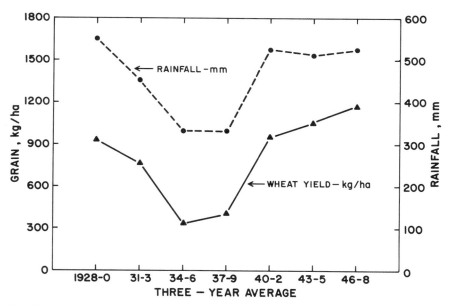

Fig. 17-4. Average wheat yield by 3-year periods in 14 southwest Kansas counties (Baver, 1951).

even though rainfall was comparable. Granted that erosion could have reduced yields on the more erosive soils, the higher yield potential on the deeper soils is evident in the increased yields in the last year. Improved varieties, improved cultural practices, and perhaps the use of fertilizer probably contributed to the higher yields in the late 1940s.

In a separate evaluation of cropping and erosion effects on crop productivity in the Southern Great Plains, Finnell (cited by Baver, 1951) stated that the 5-year average wheat yield on new land was 1750 kg/ha. Wheat yields on land cultivated 6 and 27 years averaged 1565 and 1285 kg/ha, respectively. About 3% of the decline was attributed to nutrient losses by cropping and about 4% to erosion effects. No record is available to show how the 3 and 4% values were derived.

At Colby, KS, Kuska and Mathews (1956) reported that even though organic C and total N had declined 30–35% after cropping began, crop yields had not declined proportionately. They showed average precipitation and crop yields by 5-year periods over a 36-year period. Average yields of most crops for the last 5 years of the experiments were very near the average for the first 5 years, when precipitation averaged about the same. However, crop varieties were changed during the experiment, and the higher yielding, improved varieties could have masked productivity declines from cropping and erosion. The authors indicated that the land had been cropped since 1905 and that wind erosion occurred before and during these studies. They stated that crop yields on the eroded land were "not appreciably different" from yields where little erosion had occurred. This suggests that so many production variables confounded the crop yields that conclusions relating to productivity decline are not warranted.

Much of the other published data on crop yields in the Great Plains also appears to be confounded with effects of variable erosion, precipitation, soil types, and changes in crop varieties. Thus, the actual productivity declines due to either cropping or erosion cannot be quantified. This is also true for reported declines in organic matter and N. For exmaple, Moomaw (1925) reported extremely variable crop yields at Dickinson, ND, over a 17-year period. Crop yields the 16th year were as high as or higher than the first few years of the study. In the Southern Great Plains, Moldenhauer and Keating (1958) reported variable crop yields on a sandy loam soil at Big Spring, TX, primarily due to variable precipitation. Harvey et al. (1961) reported similar crop yield variability at Lubbock, TX. Although crop yields varied considerably, Mathews and Barnes (1940) reported serious declines in milo (*Sorghum bicolor* L.) yields at Dalhart, TX, as a result of wind erosion. Dalhart is located in part of the Dust Bowl that experienced the most severe wind erosion during the 1930s. The soils are loamy sands to sands and are highly erosive. Mathews and Barnes (1940) concluded that crop yields fluctuated more on normal (noneroded) than on eroded lands. When water supplies were favorable, yields were usually much higher on the normal lands. When water supplies were severely limiting, yields were lower on the normal lands, but the magnitude of the yield difference was not as great as during the more favorable years. Mathews and Barnes attributed the differences to impaired fertility on the eroded lands which severely restricted yields in years with favorable rainfall. In low rainfall years, increased early vegetative growth of milo on normal lands resulted in greater drought stress and crop injury later in the season.

Fryrear (1981) used multiple linear regression techniques to relate the crop-yield data from Big Spring, Lubbock, and Dalhart, TX, to years of cropping. He estimated that grain sorghum (milo) yields declined only 3% at Big Spring, 17% at Lubbock, and 35% at Dalhart. Kafir (*Sorghum bicolor* L.) (forage) yields were reduced almost 40% at both Dalhart and Big Spring. Kafir (for forage) probably uses more water and nutrients than grain sorghum because of greater production of dry matter over a longer growing season and may be a better indicator of changes in crop productivity. Fryrear concluded that improvements in crop varieties and cultural practices should have increased yield potential but that the improvements did not keep pace with the factors that were reducing crop production.

Few research results have been published on changes in crop yields due to water erosion and cropping in the Great Plains. One such study, reported by Baird and Knisel (1971), was conducted in the Texas Blackland Prairie on Houston Black clay soil (fine, montmorillonitic, thermic, Udic Pellustert). For 5 years (1938–1942), two adjoining watersheds of about 120 ha each were farmed similarly with crops and cultural practices common at that time and place. This included cropping about 80% of one watershed in a rotation of cotton-corn-cotton-oats (*Avena sativa*) with cotton on one-half and corn and oats each on one-fourth of the cultivated land each year. This sequence was continued on one watershed through 1968. On the other watershed, a conservation program was established in 1942–1943 that included increasing the area of permanent grasses, terracing cultivated fields, establishing grass waterways, and changing from the 4-year to a 3-year rotation with fall-seeded oats and sweetclover-cotton-corn. Corn was gradually replaced with grain sorghum on both watersheds. Crop varieties were changed on both watersheds as varieties with greater yield potential became available. Fertilizers were first used on the conservation watershed in 1949 and on the nonconservation watershed in 1962.

Sediment yield from the conservation watershed was reduced to about 12% of that from the nonconservation watershed (Baird, 1964). Terraces with the 3-year crop rotation reduced storm runoff about 23%, but terraces alone had little effect on runoff (Baird et al., 1970).

Yield data for cotton lint for the 1938–1968 period are shown in Fig. 17–5. Since treatments were the same from 1938–1942 and potential benefits from sweetclover could not be fully expressed through the first cycle, there was little difference in yields through 1949. Beginning in 1950, phosphorus was applied to the sweetclover in the rotation on the conservation watershed, and from 1950 to 1962 cotton yields on the conservation watershed increased 62% more than on the nonconservation watershed. Fertilizer was used on cotton on the nonconservation watershed during 1963–1965 but not on the conservation watershed, where fertilizer was applied to sweetclover the year before cotton. During 1963–1965, cotton yields on the nonconservation watershed were only 59% of those on the conservation watershed, suggesting that production factors other than fertility were limiting yields.

Corn, although not well adapted to the area, was grown on both watersheds during 1938–1966. Yield differences (Fig. 17–6) were not significant from 1938 to 1948. Corn yields increased on the conservation watershed during 1950–1961 probably due to the legume crop in the rotation. From 1963 to 1966 fertilizer applied to the nonconservation watershed resulted in

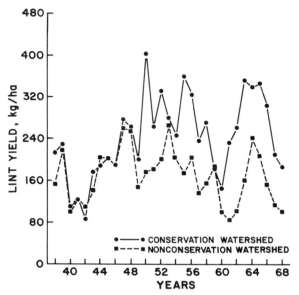

Fig. 17-5. Annual yields of lint cotton from two watersheds, Riesel, TX, 1938-1968 (Baird and Knisel, 1971).

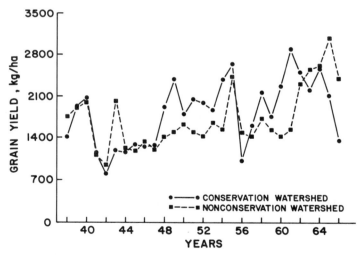

Fig. 17-6. Annual yields of corn from two watersheds, Riesel, TX, 1938-1966 (Baird and Knisel, 1971).

equal or better yields on the nonconservation watershed. Apparently, the nutrients lost to erosion were adequately replaced by fertilizer.

Yield data for grain sorghum for the period 1957-1968 are shown in Fig. 17-7. During 1957-1962, 60% more grain was produced on the conservation watershed than on the nonconservation watershed. However, with the application of fertilizer to the nonconservation watershed beginning in 1963, this yield difference was no longer apparent. As with corn, the use of

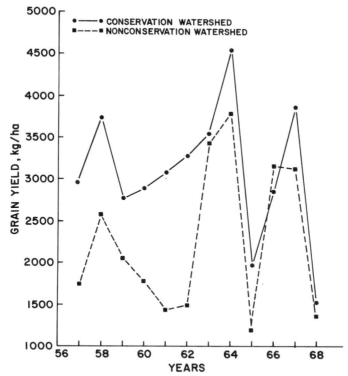

Fig. 17-7. Annual yields of grain sorghum from two watersheds, Riesel, TX, 1957–1968 (Baird and Knisel, 1971).

fertilizer on grain sorghum masked any detrimental effect soil erosion may have had on production. Considering production of all crops on deep Houston Black clay soils, adequate fertilizer can apparently offset any detrimental effects of erosion on corn and grain sorghum yields.

17-2.2 Simulated Erosion and Land Leveling Studies

In several studies in the Great Plains, variable thicknesses of soil have been removed to simulate loss by soil erosion. Other studies, generally in irrigated areas, have addressed the problem of restoring productivity to subsoils exposed by land leveling. In neither type of study, however, can the effects of removal of topsoil on crop productivity be directly compared with those caused by erosion because the time scale of removal is so different. The erosion process is slow, and weathering of newly exposed soil accompanies erosion. The effect of surface soil removal on crop productivity depends on soil profile characteristics. Simulated erosion is nonselective, but actual erosion removes the finer particles. In most instances, instantaneous removal of topsoil would likely affect productivity more drastically than would slow removal by erosion. However, removal of surface soil has, in some instances, increased productivity because the subsoil was more fertile than the topsoil. This may be true of soils with buried surface horizons and soils with a shallow sand mantle over a sandy, clay loam subsoil (Chepil et al., 1962).

One of the early simulated-erosion experiments was reported by Smith et al. (1967) in the subhumid Texas Blackland Prairie at Temple, TX. The area had been cropped for a number of years and considerable natural erosion had occurred. Two plots of Austin clay (fine-silty, carbonatic, thermic, Entic Haplustoll) on a 4% slope were desurfaced to a depth of 38 cm in 1931. One plot was immediately reseeded to native grasses and was used to measure organic matter accumulation. The other plot was cropped, and crop yields were compared with those on similar plots that had not been desurfaced. The crop sequence from 1931 to 1944 was corn-oats-cotton, without fertilizers. Then, sweetclover fertilized with phosphorus was grown in rotation with cotton from 1945-1949. Crop yields, which were low throughout the period, were markedly lower on the desurfaced plot, but yields did not change with time.

The desurfaced plot was idle from 1952-1960, with enough volunteer grasses and weeds to prevent serious erosion. In 1961 a third plot was desurfaced to a depth of 38 cm. All plots were cropped to grain sorghum and fertilized at a rate of 54-23-0 kg/ha in 1961-1965. Grain-yield data are shown in Fig. 17-8. In 1961 and 1962 grain yields on the newly desurfaced plot were much lower than on the 1931 desurfaced plot but were not different in 1963-1965. Average yields on plots with normal erosion (about 21 t ha^{-1} yr^{-1}) were higher than either of the desurfaced plots, except in 1962 when the 1931 desurfaced plot produced almost as much grain as the plots

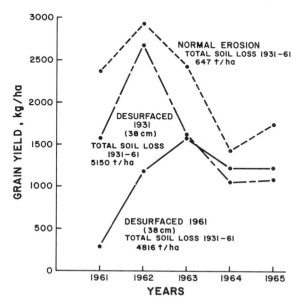

Fig. 17-8. Effect of topsoil removal (38 cm) on sorghum grain yield, Temple, TX, 1961-1965 (Smith et al., 1967).

with normal erosion. Even so, average yield was 30% less 30 years after desurfacing. The between-year variability, relatively low yields, and lack of replication prohibit definitive conclusions, but it appears that with adequate fertilization, crop yields on severely eroded Austin clay can approach yields on soils with normal erosion.

In a comprehensive soil-removal study begun in 1955 at Akron, CO, subsoil layers of a Weld silt loam, shallow phase, a member of the fine, montmorillonitic, mesic, Aridic Paleustolls, were exposed to depths ranging from 0 to 38 cm by excavation over a horizontal distance of 45.7 m (B. W. Greb, personal communication). The plots were subdivided, with sub-blocks cropped to wheat-fallow from 1956–1962. Wheat yields were reduced appreciably only after the maximum soil removal. The addition of N alone did not affect the wheat yield on plots with any soil-removal treatment. Since only extreme soil removal affected wheat yield, we would not expect normal soil erosion to decrease wheat yields on the Weld soil.

Olson (1977) studied the effect of topsoil removal on corn production on a Beadle silty clay loam (fine, montmorillonitic, mesic, Typic Argiustoll) on the eastern edge of the Great Plains. The Beadle soil is on glacial till and has a genetic sodic claypan at a depth of 15–30 cm. In 1966 topsoil was removed to depths of 0, 30, and 45 cm, and fertility subtreatments were applied as follows (kg/ha): (i) 56–22–0, (ii) 56–22–0 + 9 kg/ha Zn, (iii) 168–44–437 + 9 kg/ha Zn, and (iv) 168–44–0. In 1969 in a separate experiment topsoil was removed to depths of 0, 30, and 45 cm, but 15 cm of topsoil were replaced. Fertilizer treatments were the same as treatments (i), (ii), and (iv) in the first experiment. The average corn grain yields are shown in Table 17–1. Removing 30 cm of topsoil decreased yields, but removing an additional 15 cm did not further decrease yields. The addition of Zn increased yields on the plots with topsoil removed. Removing a portion of the soil profile had little effect on grain yield when the topsoil was replaced. Olson reported, however, that seedbed preparation of the exposed subsoils was difficult and that the resulting poor tilth caused poor seed germination and emergence. Corn plants on the subsoil plots showed nutrient deficiency symptoms, particularly in the early season. In 1982 Olson (personal com-

Table 17-1. Effects of topsoil removal with and without replacement of 15 cm of topsoil and with fertilizer treatments on corn grain yields on Beadle silty clay loam cut 0, 30, and 45 cm, Madison, SD (Olson, 1977).

	Grain yield for three levels of topsoil removed (cm)		
Fertility treatment	0	30	45
N–P–K + Zn (kg/ha)		kg/ha	
No topsoil replaced			
56–22–0 + 9	3625	2988	2827
168–44–37 + 9	3724	3092	3016
15 cm topsoil replaced†			
56–22–0	4308	4595	3933
56–22–0 + 9	4689	4741	4136
168–44–0	4579	4686	4568

† Topsoil removed to indicated thickness in 1969 but with 15 cm replaced on all plots.

munication) stated that the subsoil plots were still in poor physical condition and that it was still often difficult to establish a stand. Corn grown on these plots still exhibits early nutrient deficiency symptoms, is smaller throughout the growing season, and yields less than surrounding areas. On the glacial till soil on which these experiments were conducted, both poor soil tilth and nutrient deficiency contribute to yield reduction. Continued soil erosion would be disastrous to the long-term productivity of such soils.

Eck et al. (1965) and Eck (1968 and 1969) evaluated crop growth response to fertilization on Pullman silty clay loam soil (fine, mixed, thermic, Torrertic Paleustoll) near Amarillo, TX, cut to depths of 0, 10, 20, 30, and 41 cm under both limited water and fully irrigated conditions. Pullman is a deep soil with a dense subsoil at depths of 20–50 cm that restricts water movement and root growth. The topsoil was removed from the experimental area in 1960, and grain sorghum was grown through 1962 under full irrigation. Fertilizer rates and grain yields are shown in Table 17–2. Without N fertilizer, grain sorghum yields were reduced at all levels of topsoil removal. The heavily fertilized (N_2P_2) 10-, 20-, 30-, and 41-cm cut treatments produced yields approximately 107%, 100%, 100%, and 80% of those obtained on the undisturbed soil, respectively.

During 1963–1965 the study was continued under limited water conditions consisting of preplanting irrigation only. Seeding rates and fertilizer treatments were reduced to levels suitable for use with limited water (Table

Table 17–2. Effects of depth of topsoil removal and fertilizer treatments on sorghum grain yields, fully irrigated (1960–1962) and pre-irrigated only (1963–1965), Bushland, TX (Eck, 1968).

Cut	Check	N_0P_2†	N_1P_2	N_2P_0	N_2P_1	N_2P_2
cm			kg/ha			
Fully irrigated experiment: Average grain yield 1960–1962						
41	1040 c‡	1010 c	4870 b	4720 b	5140 ab	6220 a
30	1400 d	1380 d	5920 bc	5130 c	6640 ab	7310 a
20	1870 d	2090 d	6180 b	4760 c	7320 a	7350 a
10	2330 c	2480 c	7730 a	5540 b	7350 a	7900 a
0	3720 b	3600 b	6960 a	7310 a	7130 a	7340 a
Limited moisture experiment: Average grain yield 1963–1965						
41	1400 b	1410 b	2870 a	3290 a	3090 a	3190 a
30	1540 b	1520 b	3370 a	3300 a	3390 a	3180 a
20	2030 c	1850 c	3360 ab	2970 b	3420 a	3180 ab
10	2580 b	2600 b	3420 a	3530 a	3420 a	3420 a
0	3020 c	3340 bc	3830 a	3280 bc	3610 ab	3680 ab

† Fertilizer treatments were
  N_1—156 kg/ha N during 1960–1962 and 78 kg/ha N during 1963–1965 (Soil NO_3-N plus fertilizer N)
  N_2—224 kg/ha N during 1960–1962 and 112 kg/ha N during 1963–1965 (Soil NO_3-N plus fertilizer N)
  P_1—25 kg/ha P during 1960–1962 (No P applications made in 1963–1965)
  P_2—39 kg/ha P during 1960–1962 (No P applications made in 1963–1965)
‡ Means in the same row followed by the same letter are not significantly different at 5% level.

17-2). Grain yields and fertilizer treatments are also shown in Table 17-2. The yield data indicate that the N rate of 78 kg (N₁) restored soil productivity on all topsoil-removal plots. The data show that under limited water conditions fertilizer did not completely restore sorghum grain yields where topsoil had been removed. Sorghum grain yields on fertilized, cut areas were limited by water, not by plant nutrients. Water-holding capacities were reduced by topsoil removal, and, although the decrease was small, it reduced the amount of water available to the crop and, consequently lowered yields.

Land-leveling experiments in North Dakota on Gardena loam soil (coarse-silty, mixed, Pachic Udic Haploboroll) were reported by Carlson et al. (1961). Topsoil was removed to a depth of 30 cm on one area. Fertilizer treatments were applied to the cut area and to an adjacent undisturbed area. These treatments and 1954 and 1956 corn grain yields are shown in Table 17-3. Yields in 1954 were very low on the unfertilized cut area, were improved somewhat with P + Zn, but required 200 kg/ha of N in addition to P + Zn to surpass the unfertilized, undisturbed area. In 1956, yields were still greater on the undisturbed area than on the cut area, probably because the residual N from 1954 and 1955 applications was inadequate to maintain yields. No soil physical problems were encountered in this study. Fertility deficiencies can be corrected by fertilizer; however, for maximum production, the cut area of the Gardena soil generally needed twice as much N and P as undisturbed soil, plus about 17 kg/ha of Zn.

Similar land-leveling experiments with comparable results have been reported in Montana (Reuss and Campbell, 1961; Black, 1968), in Colorado (Robertson and Gardner, 1946), in Oklahoma (Gingrich and Oswalt, 1965), and in other states (Mickelson, 1968; Haas and Willis, 1968; Hauser and Cox, 1962). Generally, if subsoil horizons are markedly different from topsoil in either chemical or physical characteristics, crop yields are reduced. If nutrient deficiencies are the only problem, yields can be increased comparable to those new topsoils, but if poor physical conditions exist in the subsoil, it is very difficult to restore productivity.

Table 17-3. Effect of topsoil removal and fertilizer treatments on corn grain yield on Gardena fine sandy loam soil at Upham, ND, 1954 and 1956 (Carlson et al., 1961).

			Grain yield			
1954 fertilizer treatment			1954		1956	
N	P	Zn	C†	U‡	C	U
			kg/ha			
0	0	0	280	2885	1280	3945
0	50	17	480	3375	1345	4065
200	50	0	2020		2150	
200	0	17	1680	3535	1505	3625
200	50	17	3825	3615	1625	3765

† C = Area from which 30 cm surface soil was removed.
‡ U = Area from which no surface soil was removed.

17-2.3   Relation of Topsoil Thickness to Yield

Lyles (1975 and 1977) proposed a procedure for evaluating wind erosion effects on soil loss and subsequent crop yields. He related topsoil thickness or topsoil removal (excluding fertilizer effects) to crop yield, computed the potential average soil loss using the wind erosion equation, and, by converting annual soil loss to thickness of soil removed, estimated the corresponding loss in crop yield. He used published data to estimate the effect of topsoil thickness on crop yields (Table 17-4). A linear relationship was assumed between soil thickness and yield, although soil may be more productive in the top 10-15 cm than at lower depths. Lyles used a modifed wind erosion equation to compute the annual soil loss rate, which was converted to thickness of soil loss per year at several sites in the Great Plains (Table 17-5). These data and the data reflecting crop yield reductions per centimeter of topsoil removed are then combined to estimate the annual reductions in crop yield. The estimated annual yield reductions in wheat and grain sorghum yields due to wind erosion under various cropping systems as calculated by Lyles (1977) are shown in Table 17-6.

Although this procedure for estimating erosion effects on crop productivity appears to have merit, it probably tends to underestimate the effects because of the assumption that yield decreases are linearly related to thickness of topsoil removed. That assumption is unlikely to be true, particularly for developed soils in which profile characteristics change significantly with depth. Yields tend to decrease more rapidly with the loss of the upper part of the topsoil than with the lower part. The average data may be useful in looking at estimated yield reductions over large areas. To be site-specific, a factor for soil depth should be incorporated into the procedure, because yield reductions due to erosion are likely to be much greater on thin soils than on deep soils.

In an on-going study, Mielke and Schepers (personal communication) applied 0, 10, and 20 cm of topsoil to an eroded soil in eastern Nebraska. The experimental site was on a Crofton-Nora soil association (fine-silty, mixed, mesic, Typic Ustorthent) that had been in corn for 5 years and in a

Table 17-4. Estimated crop yield reduction per centimeter of topsoil loss at several Great Plains locations (Lyles, 1975).

Location	Wheat yield reduction per cm of topsoil loss		Grain sorghum reduction per cm of topsoil loss	
	kg/ha	%	kg/ha	%
Geary Co., KS	0.3	6.2		
Manhattan, KS	0.3	4.3		
Akron, CO	1.3	2.0		
Bushland, TX (irrigated)			0.7	5.2
Bushland, TX (preseeding irrigation only)			0.5	4.1
Temple, TX (nonirrigated)			0.5	5.7

Table 17–5. Average potential soil loss for different wind erodibility groups and crop rotations at several locations in the Great Plains (Lyles, 1977).

Location	Wind erodibility groups					
	1	2	3–4L	5	6	7
	cm/yr					
Wheat-fallow rotation						
Northern Plains†						
Wheat	0.74	0.25	0.13	0.08	0.05	0.03
Fallow	0.01	0	0	0	0	0
W. Kans.-E. Colo.						
Wheat	2.84	0.97	0.53	0.30	0.23	0.18
Fallow	0.20	0.01	0	0	0	0
W. Texas						
Wheat	4.01	1.42	0.79	0.43	0.36	0.25
Fallow	1.94	0.23	0.07	0.02	0.01	0.01
Wheat-sorghum-fallow rotation						
Nebr.-S. Dak.						
Wheat	1.40	0.53	0.30	0.18	0.13	0.10
Sorghum	0.01	0	0	0	0	0
Fallow	1.31	0.49	0.27	0.16	0.12	0.09
W. Kans.-E. Colo.						
Wheat	3.40	1.35	0.79	0.46	0.38	0.28
Sorghum	0.20	0.01	0	0	0	0
Fallow	3.36	1.28	0.73	0.43	0.34	0.26
W. Texas						
Wheat	4.27	1.63	0.94	0.56	0.46	0.36
Sorghum	1.94	0.23	0.07	0.02	0.01	0.01
Fallow	4.10	1.47	0.82	0.47	0.37	0.28

† N Dak., S. Dak., Nebr., Mont., Wyo.

Table 17–6. Estimated annual reduction in wheat and grain sorghum yields resulting from wind erosion under various crop rotations in the Great Plains (Lyles, 1977).

Location	Wind erodibility groups					
	1	2	3–4L	5	6	7
	kg/ha					
Wheat-fallow rotation						
Northern Plains†	15.1	5.2	2.8	1.6	1.2	0.8
W. Kans.-E. Colo.	60.3	19.5	10.7	6.0	4.8	3.6
W. Texas	118.3	33.0	17.1	9.1	7.5	5.2
Wheat-sorghum-fallow rotation (wheat)						
Nebr.-S. Dak.	36.1	13.5	7.5	4.4	3.2	2.4
W. Kans.-E. Colo.	92.1	34.9	20.2	11.9	9.5	7.1
W. Texas	136.6	44.1	24.2	13.9	11.1	8.7
Wheat-sorghum-fallow rotation (sorghum)						
Nebr-S. Dak.	54.0	20.2	11.3	6.5	4.7	3.6
W. Kans.-E. Colo.	137.6	52.2	30.2	17.8	14.2	10.7
W. Texas	204.0	65.8	36.2	20.8	16.6	13.0

† N. Dak., S. Dak., Nebr., Mont., and Wyo.

corn-sorghum sequence for the 10 previous years. The site was badly eroded, with the structureless subsoil exposed. In December 1980, topsoil that had eroded from the field and been deposited at the bottom of the slope was placed on the eroded slope. Corn yields in 1981 were significantly higher for the 10- and 20-cm topsoil thickness compared to the no-topsoil treatments. Although plant populations were not significantly different, the 10% increase in plant population for the 20-cm treatment suggests that soil physical factors as well as fertility may affect the results of this continuing experiment.

17-3 CURRENT SITUATION

17-3.1 Crop Yield Trends

Although dryland crop yields in the Great Plains are very erratic from year to year because of the vagaries of climate, average annual crop yields have been increased over time by changes in technology and increased inputs. This trend continues in many parts of the Great Plains today.

Fertilizer use is also increasing. For example, in Oklahoma the average of 21 kg of N/ha applied to wheat in 1965 had increased to 65 kg of N/ha in 1980. Similarly, P use rose from 13 to 24 kg of P/ha in the same period. Total plant nutrient consumption in Oklahoma from 1955-1980 is shown in Fig. 17-9. We do not know how much of the fertilizer is being used to offset the effects of erosion. Possibly, most of it is being used on land that is not experiencing excessive erosion (Willis, personal communication).

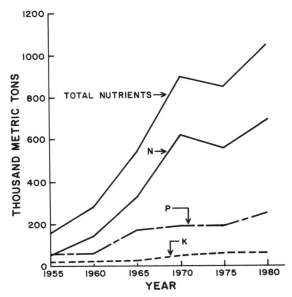

Fig. 17-9. Consumption of plant nutrients in Oklahoma, 1955-1980. Source: Oklahoma Fertilizer Statistics, 1981.

Similar trends are evident in Montana. From 1976 to 1982, nitrogen fertilizer use has increased from 49 000 to 78 000 t. In the same period, available P consumption increased from 31 000 to 35 000 t. Wheat yields were 33 kg/ha in 1983 compared to 22 kg/ha in 1950, and corn yields increased from 28 to 84 kg/ha in that period (Haderlie, personal communication).

17–3.2 Erosion Situation

Sheet and rill erosion and wind erosion estimates for the 10 Great Plains states are shown in Table 17–7 (USDA, 1982). The values for sheet and rill erosion are affected by adjacent areas of higher rainfall outside the Great Plains, but the wind erosion values are representative of Great Plains conditions. Erosion was still severe in parts of the Great Plains states in 1977. In Montana, out of a total of 6 million ha of cropland, almost 2 million are eroding at rates greater than 11 t ha^{-1} yr^{-1}, and over 400 000 ha are eroding at a rate greater than 31 t ha^{-1} yr^{-1} (Haderlie, personal communication). In New Mexico 44% of the total cropland is eroding at a rate greater than 11 t ha^{-1} yr^{-1}, and much of the dryfarmed cropland is eroding at rates of 40 t ha^{-1} yr^{-1} (Margo, personal communication). Similar severe wind erosion occurs on the sandy erosive soils of the Texas South Plains and in southeastern Colorado. Wind erosion is less severe in the Northern Plains than in the Southern Plains except in localized areas with few soil conservation practices.

The Soil Conservation Service, the Agricultural Research Service, and the State Agricultural Experiment Stations are trying to ascertain effects of current erosion on crop yields. Only limited information is available. For example, SCS is collecting crop yield data in Nebraska from soils with different erosion rates. A Uly silt loam eroding at 9 t ha^{-1} yr^{-1} produced 2700 kg/ha of wheat compared to 2440 kg/ha on the same soil eroding at 20 t ha^{-1} yr^{-1} (Sullivan, personal communication).

Table 17–7. Estimated average annual water erosion (sheet and rill) and wind erosion from cropland in the Great Plains states, 1977 (USDA, 1982).

State	Erosion	
	Sheet and rill	Wind
	——— t ha^{-1} yr^{-1} ———	
Colorado	5.6	20.0
Kansas	8.5	6.6
Montana	2.4	8.5
Nebraska	12.9	2.8
New Mexico	4.5	25.7
North Dakota	4.5	3.9
Oklahoma	8.2	6.7
South Dakota	5.5	6.6
Texas	7.8	33.4
Wyoming	3.5	5.4

Although rates of wind and water erosion are excessive for a considerable part of the Plains, this does not mean that Federal-state conservation programs are ineffective. Estimates of water erosion rates of 8 t/ha on Class I–IV cropland in Oklahoma under existing cropping systems and applied conservation practices would rise to 32 t/ha without any applied conservation practices (Willis, personal communication). Wind erosion rates would increase more spectacularly—from 8 to 66 t/ha. This estimate suggests that present conservation programs are effective but need to be extended to marginal lands that are still being cropped. Additional technology is needed to extend improved practices to these marginal lands, or the lands should be returned to permanent grass. Some croplands have had conservation practices removed, such as terraces, because of economic pressures and larger farm equipment.

17–4 CONCLUSIONS

Erosion, particularly by wind, has been and continues to be severe in many parts of the Great Plains. Erosion adversely affects crop productivity over the long term by removing plant nutrients, reducing infiltration, and reducing topsoil thickness. Water erosion can reduce productivity directly by reducing plant stand due to washing or sediment deposition and wind erosion can reduce productivity directly by sandblasting small plants.

Separating the effects of erosion on crop yields from the natural decline due to cropping and from increases due to technological advances is not possible quantitatively. The preponderance of evidence from the literature, however, supports the position that erosion has reduced yields in the Great Plains. Improved technology (improved crop cultivars, increased use of fertilizers, improved pesticides, and improved machinery) have masked the effect of erosion. Without the improved technology, crop yields likely would have declined.

The seriousness of the decline in productivity due to erosion is related to the degree of erosion, to soil profile characteristics, and to climate. The sorting action of erosion tends to remove the fine fraction, which contains most of the soil organic matter and plant nutrients, and thereby reduces fertility. Soil profile depth determines the crop rooting volume and plant-available water. Truncation of the topsoil on deep soils is less serious in reducing plant rooting depth than the same amount of truncation on shallow soils. Thus, the time scale for serious depletion of productivity is much shorter on shallow soils than on deep soils. The physical characteristics of the exposed subsoil after erosion removes the surface soil can decrease crop productivity. If it is clayey, seedbed preparation can be difficult, adversely affecting plant-stand establishment and yield. In most cases, deep soils of uniform silt loam, loam, or sandy loam texture continue to be productive with good management in spite of moderate erosion. On the other hand, shallow soils or soils with favorable surface soil characteristics, but underlain by a dense subsoil, require progressively larger inputs to maintain crop productivity as erosion proceeds.

The detrimental effects of erosion on crop productivity are accentuated under the harsh climate of the Great Plains, particularly on the western side where rainfall is lower and more erratic. Thus, the fragile soils on the western edge of the Plains are marginal for crop production, and every effort should be made to keep them protected with grass. On the less erosive soils in the Great Plains, proper management of crop residue through the use of conservation tillage systems can maintain soil productivity for future generations.

REFERENCES

Adams, J. E., R. C. Henderson, and R. M. Smith. 1959. Interpretations of runoff and erosion from field scale plots on Texas Blackland soil. Soil Sci. 57:232–238.

Baird, R. W. 1964. Sediment yields from Blackland watersheds. Trans. Am. Soc. Agric. Engr. 7:454–456.

––––, and W. G. Knisel. 1971. Soil conservation practices and crop production in the Blacklands of Texas. USDA Conserv. Res. Rep. 15.

––––, C. W. Richardson, and W. G. Knisel. 1970. Effects of conservation practices on storm runoff in the Texas Blackland Prairie. USDA Tech. Bull. 1406.

Baver, L. D. 1951. How serious is soil erosion? Soil Sci. Soc. Am. Proc. 15:1–5.

Black, A. L. 1968. Conservation bench terraces in Montana. Trans. Am. Soc. Agric. Engr. 11:393–395.

Carlson, C. W., D. L. Grunes, J. Alessi, and G. A. Reichman. 1961. Corn growth on gardena surface and subsoil as affected by applications of fertilizer and manure. Soil Sci. Soc. Am. Proc. 25:44–47.

Chepil, W. S., W. C. Moldenhauer, and H. M. Taylor. 1962. Deep plowing of sandy soil. USDA–ARS Prod. Res. Rep. 64.

Cole, J. S., and O. R. Mathews. 1940. Relation of the depth to which the soil is wet at seeding time to the yield of spring wheat on the Great Plains. USDA Circ. 563.

Eck, H. V. 1968. Effect of topsoil removal on nitrogen-supplying ability of Pullman silty clay loam. Soil Sci. Soc. Am. Proc. 32:686–691.

Eck, H. V. 1969. Restoring productivity of Pullman silty clay loam subsoil under limited moisture. Soil Sci. Soc. Am. Proc. 33:578–581.

––––, V. L. Hauser, and R. H. Ford. 1965. Fertilizer needs for restoring productivity on Pullman silty clay loam after various degrees of soil removal. Soil Sci. Soc. Am. Proc. 29:209–213.

Enlow, C. R. 1939. Review and discussion of literature pertinent to crop rotations for erodible soils. USDA Circ. 559.

Finnell, H. H. 1948a. The dust storms of 1948. Sci. Am. 179:7–11.

––––. 1948b. Dust storms come from the poorer lands. USDA Leafl. 260.

Fryrear, D. W. 1981. Long-term effect of erosion and cropping on soil productivity. p. 253–259. In Special Paper 186, Geol. Soc. Am., Boulder CO.

Gingrich, J. R., and E. S. Oswalt. 1965. Soil management practices for cotton production on leveled irrigated land. Oklahoma Agric. Exp. Stn. Bull. B-636.

Greiner, J. H., Jr. 1982. Erosion and sedimentation by water in Texas: Average annual rates estimated in 1979. Texas Dep. Water Resour. Rep. 268. Austin, TX.

Haas, H. J., C. E. Evans, and E. F. Miles. 1957. Nitrogen and carbon changes in Great Plains soils as influenced by cropping and soil treatments. USDA Tech. Bull. 1164.

––––, and W. O. Willis. 1968. Conservation bench terraces in North Dakota. Am. Soc. Agric. Engr. Trans. 11:396–398.

Harvey, C., D. L. Jones, and C. E. Fisher. 1961. Dryland crop rotations on the Southern High Plains of Texas. Texas Agric. Exp. Stn. MP-544.

Hauser, V. L., and M. B. Cox. 1962. Evaluation of Zingg conservation bench terraces. Agric. Engr. 43:462–464, 467.

Hill, H. O., W. J. Peevy, A. G. McCall, and F. G. Bell. 1944. Investigations in erosion control and reclamation of eroded land at the Blackland Conservation Experiment Station, Temple, TX, 1931–1941. USDA Tech. Bull. 859.

Joel, A. H. 1937. Soil conservation reconnaissance survey of the Southern Great Plains wind-erosion area. USDA Tech. Bull. 556.

Kuska, J. B., and O. R. Mathews. 1956. Dryland crop-rotation and tillage experiments at the Colby (Kans.) Branch Experiment Station. USDA Circ. 979.

Lyles, L. 1975. Possible effects of wind erosion on soil productivity. J. Soil Water Conserv. 30: 279–283.

----. 1977. Wind erosion: Processes and effect on soil productivity. Trans. Am. Soc. Agric. Engr. 20:880–884.

Mathews, O. R., and B. F. Barnes. 1940. Dryland crops at the Dalhart (Texas) Field Station. USDA Circ. 564.

Mickelson, R. H. 1968. Conservation bench terraces in eastern Colorado. Trans. Am. Soc. Agric. Engr. 11:389–392.

Moldenhauer, W. C., and E. E. Keating. 1958. Relationships between climatic factors and yields of cotton, milo, and Kafir on sandy soils in the Southern High Plains. USDA Prod. Res. Rept. 19.

Moomaw, L. 1925. Tillage and rotation experiments at Dickinson, Hettinger, and Williston, ND. USDA Bull. 1293.

Olson, T. C. 1977. Restoring the productivity of a glacial till soil after topsoil removal. J. Soil Water Conserv. 32:130–132.

Robertson, D. W., and R. Gardner. 1946. Restoring fertility to land where levelling operations have removed all the topsoil and left raw subsoil exposed. p. 33–35. *In* Proc. Am. Soc. Sugarbeet Technology.

Ruess, J. O., and R. E. Campbell. 1961. Restoring productivity to leveled land. Soil Sci. Soc. Am. Proc. 24:302–304.

Smith, R. M., R. C. Henderson, E. D. Cook, J. E. Adams, and D. O. Thompson. 1967. Renewals of desurfaced Austin clay. Soil Sci. 103:126–130.

----, ----, and O. J. Tippit. 1954. Summary of soil and water conservation research from the Blackland Experiment Station, Temple, Texas, 1942–53. Texas Agric. Exp. Stn. Bull. 781.

Stallings, J. H. 1950. Erosion of topsoil reduces productivity. USDA Rep. SCS-TP 98.

Throckmorton, R. I., and F. L. Duley. 1932. Soil fertility. Kansas Agric. Exp. Stn. Bull. 260.

USDA. 1982. Basic Statistics: 1977 National Resources Inventory. Stat. Bull. 686. Washington, DC.

18 Regional Effects of Soil Erosion on Crop Productivity— The Palouse Area of the Pacific Northwest

R. I. Papendick
Agricultural Research Service
U.S. Department of Agriculture
Pullman, Washington

D. L. Young
Washington State University
Pullman, Washington

D. K. McCool
Agricultural Research Service
U.S. Department of Agriculture
Pullman, Washington

H. A. Krauss
Soil Conservation Service
U.S. Department of Agriculture
Spokane, Washington

Soil erosion has taken a heavy toll of topsoil[1] from large areas of prime croplands in the Pacific Northwest since the region came under cultivation about 100 years ago. One such area is the Palouse Basin in eastern Washington and the western side of the Idaho panhandle, which is one of the most productive, nonirrigated, wheat-growing areas in the world. It also is one of the most erodible areas in the nation with erosion rates on some slopes of 200 to 450 t/ha (90 to 200 tons/acre) in a single winter season (USDA, 1978). Soil surveys show that since cultivation began, all the original topsoil has been lost from 10% of the cropland, and from one-fourth to three-

---

[1] In this paper, the term topsoil refers to the surface layers of Mollisols, rich in organic matter, which make up most of the soils of the study area.

Published in R. F. Follett and B. A. Stewart, ed. 1985. *Soil Erosion and Crop Productivity.*
© ASA-CSSA-SSSA, 677 South Segoe Road, Madison, WI 53711, USA.

fourths of the original topsoil has been lost from another 60% of the culti-vated area in the Basin (USDA, 1978). Other dryfarmed croplands of the Pacific Northwest experiencing serious erosion include parts of the Columbia Plateau of Oregon and Washington, and the intermountain areas of southeastern Idaho.

Much of the erosion is caused by rain and melting snow on saturated or frozen soil and is accelerated by steep topography and by farming practices that leave the soils unprotected over the winter. Snowmelt from deep ac-cumulations common on steep north and east slopes can produce heavy erosion on downslope positions. Both wind erosion, which is prevalent in low-rainfall areas, and tillage erosion, the downhill movement of soil caused by moldboard plowing on steep slopes, can move large amounts of topsoil from specific areas, but neither is as extensive as erosion caused by water runoff.

The loss of topsoil has exposed the subsoil on many ridges and slopes, making them less productive, and soil movement has covered many fertile bottomlands with silt deposits up to 2 m deep. Erosion has also left some land surfaces irregular and has already removed small portions of some fields from cultivation. Both effects increase the cost of farming and reduce farm income. Cultivating the exposed subsoils requires greater power re-quirements for tillage and creates problems with machine operations, seed-bed preparation, planting, and soil crusting during seedling emergence. All these factors will affect the productivity of the soil.

The objective of this paper is to develop erosion-soil productivity rela-tionships for a study area within the Palouse Basin. Our analysis takes into account only the direct effect of topsoil loss on crop yields. Other specific effects which can affect yields, such as organic matter decline and changes in soil physical properties, are not considered.

18-1 THE STUDY AREA

Our study concentrated on the eastern part of Whitman County, WA, which lies adjacent to the western border of the Idaho panhandle. This area lies within the heart of the Palouse, a land feature characterized by steep, rolling loessial hills and with an agriculture specializing in the production of winter wheat (*Triticum aestivum* L.) with secondary crops of spring barley (*Hordeum vulgare*), dry peas (*Pisum sativum*), and lentils (*Lens culinaris*). The soils are silt loams derived mainly from loess, most of which have formed under a native bunchgrass cover. Most soils are permeable, well drained, and deep enough to store the winter precipitation. The depth of loess in much of the cropland areas is 30 m or more over bedrock with mini-mum depths of 3 m on some parts of the landscape.

Whitman County constitutes about 72% of the Palouse River Basin, which has been targeted by the USDA as a critical erosion area. Average annual erosion in the study area is estimated at about 27 t ha^{-1} yr^{-1}, but varies considerably with land capability class.

Pawson et al. (1961) present data for a typical Palouse hill showing division according to land capability class, and for each class the percentage of field surface that it occupies and the average percent slope and topsoil depth (Fig. 18-1). These data are taken from measurements made by the SCS in eastern Washington in about 1950. Figure 18-1 shows that the soil on the lower part of the slope generally has a thicker A horizon than on the upper slopes, which are either more eroded or had less topsoil before cultivation. For example, the depth of topsoil on the lower slopes ranges from 50 to 60 cm compared with less than 15 cm on the hilltops.

18-2 WHEAT YIELDS AND TECHNOLOGY ADVANCES

Average wheat yields for Whitman County have been summarized in the Palouse Cooperative River Basin Study (USDA, 1978) for 1934–1976. These yield records are extended to 1982 and presented graphically in Fig. 18-2. Since 1934, the average wheat yield has increased from about 1750 kg/ha to over 3400 kg/ha. These yields are an average based on actual grower yields over a range of managements and soils and show the net effects of technological improvements which override any obvious decline in soil productivity from erosion.

Trends in wheat yields are steadily upward from the beginning of the period until the early 1970s where there is greater annual fluctuation and an indication of leveling off (Fig. 18-2). The increase in yield beginning in the mid-1930s is attributed to adoption of legume green manures in the crop

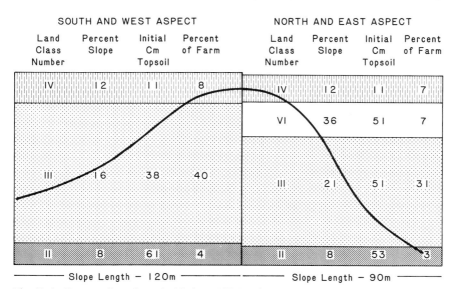

Fig. 18-1. Cross section of a typical Palouse hill showing the percent slope, slope length, depth of topsoil, and percent of land area occupied for the different land capability classes. Based on Pawson et al. (1961).

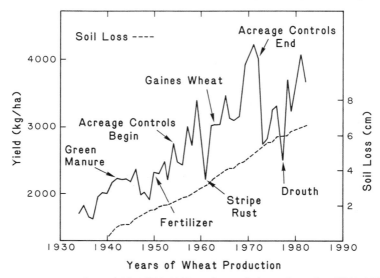

Fig. 18-2. Average wheat yields (1934–1982) and cumulative soil erosion (1940–1982) for Whitman County, WA. The wheat yield graph was reproduced from the USDA (1978) study and updated to 1982. The erosion data were obtain from records of the SCS, Spokane, WA. Some major technology inputs and other factors that affected production are also indicated on the time scale.

rotation to increase the supplies of N and organic matter. This practice gave way in the late 1940s and early 1950s to commercial N fertilizer and sharper yield increases. The yield increases were sustained during the 1950s as N fertilizer rates were increased and more farmers applied fertilizer, and in the 1960s with the introduction of semi-dwarf wheats which could be fertilized more heavily with nitrogen for increased production without lodging. Also, introduction of herbicides during the 1950s improved weed control in wheat. During 1954–1973, acreage allotments were in effect, and the idled land was fallowed to increase crop yields the following year. Yields declined when acreage controls were lifted in 1973, which reduced fallowing and encouraged more annual cropping and farming of more marginal lands.

Figure 18-2 shows that over 6 cm of topsoil were lost from croplands during the period but that, despite this erosion, wheat yields have continued to increase. Furthermore, these advances in yield, which average about 1.5% per year in Fig. 18-2, will probably continue in the foreseeable future. This progress masks the adverse effects of erosion on soil productivity and must be considered in any meaningful assessment of erosion impact on crop yields. The questions are, What would wheat yields be if erosion had been controlled during these years of record? and how will erosion, if allowed to continue, affect future yields?

18-3 RELATIONSHIP OF YIELDS TO TOPSOIL DEPTH

It is difficult to separate soil erosion's effects on crop yields from other factors that contribute to loss of soil productivity. For example, in the

Pacific Northwest, soil productivity has been decreased through depletion of soil organic matter, loss of plant nutrients, and deterioration of soil structure either from breakdown of soil aggregates or soil compaction, or both. All these factors may be enhanced by erosion, but any can occur without loss of topsoil and can result in a net loss of soil productivity.

Some data are available to analyze erosion-productivity relationships for the deep loessial soils of the Palouse. In 1952–1953 Pawson et al. (1961) collected more than 800 samples of winter wheat yields and made soil measurements from typical Palouse hills to study the relationship of yield to depth of topsoil and organic matter content. They defined topsoil as the dark layer showing discoloration due to enrichment with organic matter, which is fairly distinct for Mollisols. In all cases, the wheat followed dry peas and was fertilized with 90 kg/ha of N, which they presumed was adequate for maximum yield. They concluded, using curvilinear regression techniques, that the deeper the topsoil, the less effect an additional increment of topsoil had on yield of wheat, other things being equal. For example, a site with 20 cm of topsoil yielded 87 kg more wheat per ha than a site with 18 cm of topsoil, whereas the difference in yield was only 17 kg/ha where there was 61 vs. 58 cm of topsoil depth (Pawson, 1961). They also showed that the effect of topsoil depth on yield was dependent on the organic matter content of the topsoil. Where the topsoil was thin, the organic matter content was low and vice versa. Hence, part of the apparent effect of topsoil depth on yields is likely due to variation in organic matter content.

Wetter (1977) reported a linear relationship between thickness of the mollic epipedon and yield of winter wheat, which he determined from 90 paired measurements made on different slope positions in the Palouse during 1970–1975. The epipedon thickness was taken as the depth at which the brown organic discoloration was not evident or the structure or texture changed markedly. This corresponds closely with the depth of topsoil measurements of Pawson et al. (1961). Yield measurements were of wheat following either fallow, dry peas, or lentils in a 41- to 57-cm average annual precipitation zone. Using a linear response curve, decreases in wheat yield averaged 54 kg/ha per cm decrease of topsoil thickness over a range of 0 to 61 cm of epipedon depth.

Frymire (1980) and Hipple (1981) collected field data on topsoil depth and associated winter wheat yields during 1974–1979 for 11 soil series across four vegetational zones in Latah County, ID, which is sometimes called the Idaho Palouse. Wheat yield and topsoil depth were regressed for the different soil series. In some instances, the effects of slope steepness and aspect, crop variety, erosion rate, weed density, seed rate, fertilizer, farmer cooperator, and position on slope were also studied. Both studies showed that there was usually a significant positive correlation between topsoil depth and wheat yield and that the relationship was influenced by soil series, which in turn are largely associated with slope position on the landscape. Topsoil depth generally affected yield the most on soils having a shallow A horizon or a fragipan that restricted the soil volume available for rooting. The effect was least in a dry year and for thick topsoils located on warmer parts of the landscape (south and west exposures). In 1978, a year when water was limiting, the average yield for eight soil series was increased by 13 kg/ha per cm of topsoil thickness, whereas in 1979, a year with adequate moisture, average yield for six soil series was increased by 66 kg/ha for each cm increment of topsoil (Hipple, 1981). In Hipple's (1981) study, the average wheat yield for three soil series was increased by 98 kg/ha in 1974 and by 90 kg/ha in 1976 per cm thickness of topsoil.

Soil erosion has been a major factor in soil development and, along with differences in microclimate, contributes considerably to the characteristic soil variability in the steeper areas of the Palouse. Many soil series in these areas are classified in part according to the amount of erosion that has occurred, and erosion continues to introduce variation in these soils, often over short distances.

Differences in topsoil depth on different parts of the Palouse landscape are not always due solely to man-induced erosion but also to differences in combined geologic and accelerated erosion and other soil-forming processes. Moreover, many soils in proximity with each other may have markedly different subsoil properties which may influence yield and, thus, bias any correlations between topsoil depth and yield. Erosion is also insidiously causing a shift to soil series where erosional processes are dominant in soil development at the expense of other series where erosion has played a lesser role in soil formation. Thus, a simple relationship between yield and topsoil depth across a range of soil series is highly unlikely. Nevertheless, the studies show that although considerable scatter exists, the effect of topsoil depth in relation to yield of winter wheat can be approximated and used to assess how erosion effects crop productivity.

18-4 TECHNOLOGICAL IMPACTS ON PRODUCTIVITY

If other factors remain constant, crop yields will generally decline as topsoil is lost. However, as Fig. 18-2 shows, crop yields have been increasing in the Palouse. Through the years, improved technology has raised crop yields on both deep and eroded topsoils, thereby masking the true yield decline from erosion. However, studies indicate that *potential* yields have been reduced by erosion. For example, the Palouse Cooperative River Basin Study (USDA, 1978) estimated that without soil erosion, improved technology by the mid 1970s would have increased average wheat yields in the Basin by 30 to 40% above the average yield of 3400 kg/ha. Moreover, most of the present yield increase from technological advances comes from parts of the fields with the least erosion damage (USDA, 1978).

Krauss and Allmaras (1982) used data from the Palouse Cooperative River Basin Study (USDA, 1978) to separate the erosion-caused decreases in soil productivity from technology impacts. The effects on wheat yields of technology and soil erosion were separated using long-term changes in wheat yields, paired measurements of winter wheat yield, and associated topsoil thickness using the Wetter (1977) data, historical erosion rates, and landscape-distributed erosion rates. They showed that an average topsoil loss of 13.4 cm over a 90-year period in Whitman County, WA, decreased wheat yields by 725 kg/ha. However, technology increased wheat yields by 1446 kg/ha over the past 40 years, which gave a net increase in productivity of 721 kg/ha (Krauss and Allmaras, 1982). Krauss and Allmaras (1982) noted that productivity gains from technology on land classes less vulnerable to erosion may mask net yield declines on steeply sloped land classes that have erosion rates much higher than average rates. They also speculate that it will be much more difficult for technology to mask future yield declines from erosion because of a slowing in the rate of technological progress.

The future effects of soil erosion on productivity, or, conversely, the economic payoff from soil conservation, will depend not only on the present topsoil depth and future erosion rates, but also on the rate and nature of future technical progress. According to Young (1982), technical progress can shift the yield-topsoil function in either a multiplicative or uniform additive manner as illustrated in Fig. 18-3. A shift in the response function by the same amount at all topsoil depths is an example of uniform additive technical progress. On the other hand, a fixed percentage increase in the response function at all depths is an example of a multiplicative interaction. With multiplicative technical progress, the absolute payoff from technical progress is larger on deep than on shallow topsoils (Young, 1984).

Figure 18-4 shows response curves for wheat yields and topsoil depths for the Wetter (1977) data from the early to mid-1970s and the Pawson et al. (1961) data collected about 20 years earlier. Both curves are Mitscherlich-Spillman functions fit to sample observations. The upper curve was developed by Taylor (1982) from 89 measurements made on farmers' fields by Wetter (1977) in the eastern Palouse in 1970–1974. The lower curve was developed by Young et al. (1982) from the relationships reported in Pawson et al. (1961). The mathematical form of the Pawson et al. (1961) function was

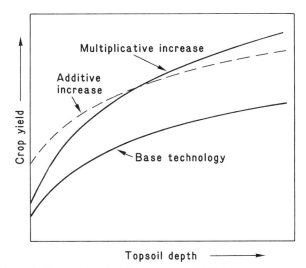

Fig. 18-3. Schematic illustration of additive versus multiplicative technology impact on crop yield for different topsoil depths.

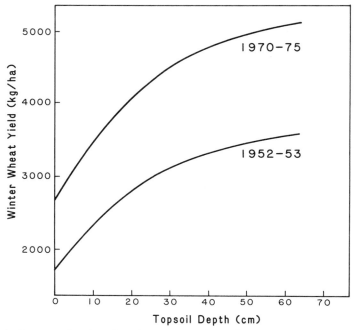

Fig. 18-4. Response functions for wheat yields and topsoil depth fit to sample observations made in the early 1950s (Pawson, 1961) and again about 20 years later (Wetter, 1977). Both curves are Mitscherlich-Spillman functions. Young et al. (1982) provides details of the mathematical derivation and statistical significance of the functions presented in this figure. The slope of the 1970–1975 curve was shown to be greater than that of the 1952–1953 curve at the 0.10 level of significance.

adjusted to arrive at a relationship directly comparable to Taylor's (1982) response function. A comparison of the two curves show greater yield growth on the deeper topsoils in the Palouse, which supports the multiplicative technology impact. Over the 20-year interval, technology gains increased yields by 970 kg/ha for zero topsoil depth and 1500 kg/ha for 50 cm of topsoil depth, or 55% more for the deeper topsoil.

Kaiser (1967) cites unpublished data obtained from studies by the SCS, which also indicate that wheat yields in the Palouse have tended to increase more on the deeper topsoils. These increases may, however, be smaller on the soils of the steep north and east slopes which have more problems with wetness, disease, and weed infestations.

A survey of 272 farmers in southeastern Washington and northern Idaho provided further support for a multiplicative technology impact. On the average, the farmers expected future technology to boost yields more on deeper topsoils. They expected that winter wheat yield would increase by about 800 kg ha^{-1} yr^{-1} over the next 50 years on typical slopes but only about one-third this on hilltops that have very shallow topsoils (Young, 1984).

Whether or not the impact of future technical progress will have a multiplicative or additive effect on yields is crucial in forecasting the impact of erosion on crop productivity. As Young (1984) shows, ignoring reasonably expected technological advances will consistently underestimate the yield loss from soil erosion if technical progress shifts the yield-topsoil function multiplicatively. With a multiplicative shift due to technology, the impact of technical progress will be greatest on the deeper topsoils. In this case, erosion reduces both current yields and the payoff from future technical progress. On the other hand, if technology shifts the topsoil function uniformly at all topsoil depths, ignoring technology will not bias estimates of erosion-induced losses in crop productivity. Instead, future gains due to technology are equal on both eroded and deep topsoils, so erosion does not inflict a "double" penalty as it does with a multiplicative effect (Young, 1984).

18-5 RELATIONSHIPS FOR ASSESSING THE IMPACT OF EROSION ON CROP PRODUCTIVITY

The effect of soil erosion on crop yield can be assessed from the response functions in Fig. 18-4 developed by Taylor (1982) and Young et al. (1982). These functions expressed algebraically are

$$y_D = a + b(1 - R^D) \text{ for the base year} \qquad [1]$$

and

$$y_D = A + B(1 - R^D) \text{ for T years later} \qquad [2]$$

where
y_D = wheat yield in kg/ha for topsoil depth D in a given period
a,b,A,B,R = statistically estimated response function parameters (a, A) ≥ 0 and $0 < R < 1$
D = topsoil depth in cm.
Since Eqs. [1] and [2] have the same mathematical form, they can be combined, which gives

$$y_t = (a + a't) + (b + b't)(1 - R^D) \qquad [3]$$

where
y_t = wheat yield in kg/ha in year t at topsoil depth D
t = annual time index from base year, t = 0, 1, . . ., T
$a' = (A - a)/T$
$b'(B - b)/T$.
The constants a' and b' in Eq. [3] account for the yield change due to technological progress. At zero topsoil depth, yield is increased at the rate of a' kg ha^{-1} yr^{-1} over the time period T, and, at infinite topsoil depth, yield is increased at the rate of $(a' + b')$ kg ha^{-1} yr^{-1}.

Where average annual erosion rates can be specified, the topsoil depth, D, in Eq. [3] can be expressed as a function of time. Krauss and Allmaras (1982) computed annual soil erosion rates by land capability subclass for Whitman County, WA, using the land area in each subclass and the percentage of the average annual erosion associated with each subclass. These values for the different capability subclasses, which are commonly associated with different parts of the Palouse landscape (see Fig. 18-1), are presented in Table 18-1. Also included in the table are topsoil depths taken from data by Pawson et al. (1961), which were estimates made by SCS in about 1950. Since the average erosion rates are taken as constant, the topsoil depth, D, in Eq. [3] can be expressed as

$$D = D_o - A_s t \qquad [4]$$

where D_o is the topsoil depth in cm at the base year, and A_s is the average annual erosion rate in cm/year. Substituting in Eq. [3] the expression for D in Eq. [4] gives

$$y_t = (a + a't) + (b + b't) [1 - R^{(D_o - A_s t)}] \qquad [5]$$

Equation [5] gives the combined effect of technical progress and erosion rate on wheat yields for different topsoil depths as a function of time.

18-6 IMPACT OF EROSION ON WHEAT YIELDS

We used Eq. [5] to make projections of wheat yields for the four major land classes on the highly erosive Palouse landscape as depicted in Fig. 18-1. Class VI land was included in the analysis, recognizing that wheat yields and yield increases from technology on these steep north and northeast slopes may be less than those on deep topsoils on other exposures. For the different land classes, we used initial topsoil depths and long-term average annual erosion rates as reported in Table 18-1. The values of the parameters of Eq. [5] that were used for the projections are listed in Table 18-2. These values are based on Pawson et al. (1961) and Taylor's (1982) original regression estimates from field observations in the eastern Palouse.

Table 18-1. Initial topsoil depth and long-term erosion rates in Whitman County, WA.

Land class	1950 topsoil depth, D_o†	Erosion rate, A‡	
	cm	mt ha^{-1} yr^{-1}	cm/yr§
IIe	57	11.2	0.086
IIIe	43	15.7	0.120
IVe	11	44.8	0.343
VIe	51	96.3	0.754
County average	40	21.1	0.161

† From SCS estimates made in about 1950 and published by Pawson et al. (1961).
‡ From Krauss and Allmaras (1982).
§ Based on a soil bulk density of 1.31 g/cm³.

Use of 1953–1973 data, the period bracketed by Pawson et al. (1961) and Taylor's (1982) response functions, gave estimates of a′ and b′ equal to 48.5 and 29.8, respectively. However, Palouse wheat yields grew considerably faster during 1953–1973 due to improved technology than they did over longer periods within the past 50 years. Consequently, we adjusted a′ and b′ downward to 32.4 and 19.9, as listed in Table 18-2, which brings them closer to the more typical rates of wheat yield growth experienced during the longer period, 1950–1980. These slower, longer-term rates were judged to be better estimates of future rates of wheat yield growth in the Palouse.

Figure 18-5 shows that increases in wheat yields projected to the year 2000 remain nearly linear for the class II (deep topsoil) and class III (moderately deep topsoil) lands which have the least erosion. However, yield reaches a maximum and eventually declines later in the 50-year period for the class VI (deep topsoil) land that has excessive erosion. In sharp contrast

Table 18-2. Parameters used in Eq. [1], [2], and [5].†

Parameter	Equation	Value
a	1, 5	1644 kg ha⁻¹
b	1, 5	2126 kg ha⁻¹
A	2	2615 kg ha⁻¹
B	2	2722 kg ha⁻¹
a′	5	32.4 kg ha⁻¹ yr⁻¹
b′	5	19.9 kg ha⁻¹ yr⁻¹
R	1, 2, 5	0.96

† All values except a′ and b′ are derived from Young et al. (1982).

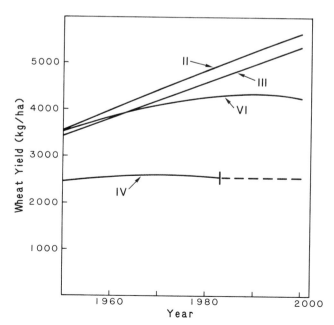

Fig. 18-5. Wheat yields for land capability classes of the Palouse projected for 1950–2000 using Eq. [5].

with the deeper topsoils, wheat yields on the highly erodible class IV land with a shallow topsoil are much lower, and the yield levels out and subsequently declines much earlier than on the class VI land. For the conditions established, all the topsoil on the class IV land is eroded by the early 1980s, and we have arbitrarily chosen to project a constant yield for the deep subsoils of the Palouse. Figure 18–5 also illustrates that the differences in yield between the land classes widen with time. This is a direct result of differences in erosion and in topsoil depths; that is, (i) with less erosion the yield enhancement from technology is greater (the effect of multiplicative technology), and (ii) yields are less affected by erosion on the deeper topsoils (the effect of the curvilinear response function between topsoil depth and yield).

Figure 18–6 presents yield reductions from erosion projected to the year 2000 in comparison with maximum potential yields without erosion, assuming past rates of technological progress are maintained. The area between the two yield curves with and without erosion for each land class gives the cumulative yield loss in kg/ha due to erosion. These losses are summed for two 25-year periods in Table 18–3. Reductions are small and would be masked by technological gains for the slowly eroding land classes II and III. According to the projections in Table 18–3, the annual yield losses over the 50-year period average 23 kg/ha and 61 kg/ha for the class II and III lands, respectively. Note, however, that 79% of the total projected yield loss for

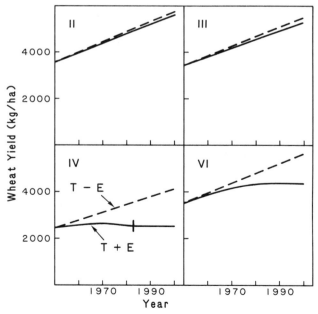

Fig. 18–6. Maximum potential wheat yields assuming historic technology and no erosion (T − E) compared with yields assuming historic technology and erosion (T + E) for land capability classes of the Palouse projected for 1950–2000 using Eq. [5].

both classes over the 50-year period occurs during the last 25 years. This reflects the accelerating impact of erosion on wheat yields.

Erosion causes much larger yield reductions in the highly erodible land classes IV and VI, where average annual yield reductions of 729 and 489 kg/ha are projected. Considering both accumulated erosion and the impact of technology, projected yields by the year 2000 (calculated from Eq. [5]) are 54, 141, 1573, and 1364 kg ha^{-1} yr^{-1} less than maximum potential yields without erosion for land classes II, III, IV, and VI, respectively. This represents an average of 435 kg ha^{-1} yr^{-1} when weighted according to the distribution of the four land classes reported in Fig. 18–1. In effect, this yield reduction from erosion amounts to a total crop loss once every 11 to 13 years across the Palouse sample area.

Erosion control can be achieved, within limits, a number of ways. A practice holding considerable promise for effective control of soil loss is no-till farming. With no-till, crops are seeded directly into residue-covered surfaces without seedbed preparation. Results of research on field runoff plots (D. K. McCool, unpublished data) and field studies by the SCS in the Palouse Basin show that no-till farming can reduce erosion rates in the Palouse to 5 mt/ha or less, even on the more erodible parts of the landscape (USDA, 1978). Figure 18–7 compares yields projected from 1980 to the year 2000 for no-till farming with erosion held to 5 mt/ha with yields for conventional tillage on the highly erodible class VI land, assuming yields for a given topsoil depth and the impact of technology are the same for the two

Table 18–3. Projected wheat yield loss from soil erosion for two 25-year periods.

Land class	Period		Total (50 years)
	1950–1975	1975–2000	
	kg/ha		
II	241	931	1 172
III	649	2 402	3 051
IV	7 812	28 644	36 456
VI	4 094	20 379	24 473
Weighted avg. all classes	1 936	7 494	9 430

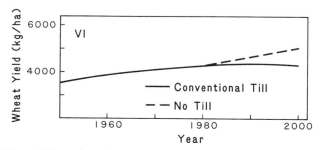

Fig. 18–7. Wheat yields projected for 1950–2000 with conventional tillage and historic erosion (solid line) compared with projected yields where no-till farming to control erosion is introduced in 1980 and carried to the year 2000. The projections were made with Eq. [5].

practices. As a result of the soil saved, the cumulative yield advantage with no-till (graphically approximated from the area between the two curves in Fig. 18–7) in 10 years (1980–1990) is 1567 kg/ha, and in 20 years (1980–2000) is 6932 kg/ha. Thus, in this hypothetical case, a farmer using no-till would harvest an equivalent of about 1.5 additional crops over 20 years compared with conventional practices because of the soil saved. The return would of course be less on the less erodible parts of the landscape. These estimates assume that yields are equal with no-till and conventional tillage on topsoils of equal depth, but, no-till maintains deeper topsoil through time by reducing erosion rates.

As erosion proceeds, the decline in soil productivity accelerates as more soil is lost and the topsoil becomes shallower. Thus, the impact of erosion on productivity is most severe on thinner topsoils. Table 18–4 shows the value of a centimeter of topsoil in terms of wheat yield for the various land classes at three different times, assuming the same erosion rates as in the other yield projections. The yield values were obtained by differentiating Eq. [5] with respect to topsoil depth. With the exception of Class IV land, the value of a centimeter of topsoil doubles or almost doubles in terms of yield from 1950 to 2000. As would be expected, topsoil value is less and increases more slowly over time with deeper topsoils and low erosion rates (classes II and III) compared with a much higher initial value and a greater increase over time for Class IV land, with its shallow topsoil and high erodibility (Table 18–4). The percentage value of topsoil for Class VI land, which has an initial topsoil depth comparable to classes II and III, increases much more over time (from 11 to 74 kg/ha over 50 years) than the other classes because of its excessive erosion rate. The value of topsoil increases by roughly 100% during the 50-year period when averaged (weighted basis) over all land classes.

18–7 CONCLUSIONS

Most of the cropland soils in the Pacific Northwest have suffered some productivity loss from erosion since they were first cultivated. Moreover, the rate of loss will accelerate in the years ahead if erosion goes unchecked. Clearly, productivity loss is most severe on the steep, highly erodible land

Table 18–4. Projected reduction in wheat yield for each centimeter of topsoil lost.

Land class	1950 topsoil depth	Year		
		1950	1975	2000
	cm		kg/ha	
II	57	8	11	15
III	43	15	21	28
IV	11	55	97	113
VI	51	11	29	74
Weighted avg. all classes	40	20	32	43

classes and where topsoils have become shallower. Land classes having moderately low rates of erosion ($<$ 15 mt ha^{-1} yr^{-1}) and relatively thick topsoils ($>$ 30 cm) have lost little productivity so far. However, slow erosion over future years has the potential of degrading the productivity of even some of the prime land classes of the region.

Wheat yields are projected to increase from technological progress in the foreseeable future. This yield growth will essentially mask the adverse effect of erosion on productivity. However, on land classes where the topsoil was originally thin or lost by accelerated erosion, projected yields will level off or even decline with continuing erosion because successively shallower topsoils become less productive and respond less to technological advances. Thus, future wheat yields and yield growth in the Palouse and other areas of the Northwest wheatlands that are subject to erosion will depend more and more on how well erosion is controlled. Major changes in tillage and cropping practices will be required to improve or even sustain productivity on land classes subject to heavy erosion or where past erosion has taken a heavy tool of topsoil. For example, greater use of sod and green-manure crops in the rotation on erosion-prone areas may be necessary to reduce soil loss and to increase the soil organic matter content. In conjunction with changes in cropping systems, increased attention should be given to continuous no-till planting, a practice still largely in the research stage in the wheatland region. Greater use of spring crops and retention of surface residues over the winter are also effective ways to reduce erosion. Research on erosion principles and management concepts to control soil loss, combined with application of sound management systems, appears to hold the key for reducing future yield losses due to soil erosion.

REFERENCES

Frymire, W. L. 1980. Topsoil depth (mollic epipedon) and its effect on crop productivity in Latah County, Idaho. M.S. thesis. Univ. of Idaho. (Washington Library Network No. 80-73791).

Hipple, K. W. 1981. Genesis, classification, and economics of deep loessial agricultural soils in Latah County, Idaho. Ph.D. diss. Univ. of Idaho, Moscow. (Diss. Abstr. 82-24568).

Kaiser, V. G. 1967. Soil erosion and wheat yields in Whitman County, Washington. Northwest Sci. 41(2):86–91.

Krauss, H. A., and R. R. Allmaras. 1982. Technology masks the effects of soil erosion on wheat yields—A case study in Whitman County, Washington. p. 75–86. *In* B. L. Schmidt et al. (ed.) Determinants of soil loss tolerance. Spec. Pub. 45. American Society of Agronomy and Soil Science Society of America, Madison, WI.

Pawson, W. W., O. L. Brough, Jr., J. P. Swanson, and G. M. Horner. 1961. Economics of cropping systems and soil conservation in the Palouse. Agric. Exp. Stn. of Idaho, Oregon, and Washington, and ARS, USDA Bull. 2.

Taylor, D. B. 1982. Evaluating the long-run impacts of soil erosion on crop yields and net farm income in the Palouse annual cropping region of the Pacific Northwest. Ph.D. diss. Washington State Univ., Pullman. (Diss. Abstr. 82-15171).

USDA. 1978. Palouse cooperative river basin study. SCS; Forest Service; and Economics, Statistics, and Cooperative Service. U.S. Government Printing Office, Washington, DC.

Wetter, Fred. 1977. The influence of topsoil depth on yield. Tech. Note 10. USDA–SCS, Spokane, WA.

Young, D. L. 1984. Modeling agricultural productivity impacts of soil erosion and future technology. p. 60–85. *In* B. C. English, J. A. Maetzold, B. R. Holding, and E. O. Heady (ed.) Future agriculture technology and resource conservation. Iowa State Univ. Press, Ames.

Young, D. L., D. Hoag, and D. Taylor. 1982. An empirical test of the multiplicative impact of technical progress on the topsoil-yield response function for winter wheat in the eastern Palouse. STEEP Agric. Econ. W.P. 82-3, Dep. of Agric. Economics, Washington State Univ., Pullman.

19 Effects of Soil Erosion on Productivity in the Southwest

Kenneth G. Renard and Jerry R. Cox
Agricultural Research Service
U.S. Department of Agriculture
Tucson, Arizona

Donald F. Post
University of Arizona
Tucson, Arizona

Any discussion of the soil erosion problem in the Southwestern USA must consider the two distinctly different agricultural systems in the area: (i) the irrigated agronomic crop system found on the deep, nearly level, alluvial soils in the valleys, and (ii) the rangeland system found on the gently sloping to steep alluvial fans and mountain slopes. The rangeland soils have a wide range of characteristics but are generally more gravelly, more cobbly, and shallower than the irrigated alluvial soils. Generally, productivity problems resulting from erosion are minimal on the irrigated agronomic crop system and severe on the rangeland vegetation system.

Our objectives are to present historical evidence of soil erosion and its effect on land productivity, based primarily on experience in southern Arizona, and to explain new modeling techniques that we believe should aid in predicting soil erosion and plant productivity on these lands.

19-1 RANGELAND AND IRRIGATED CROPLAND PRODUCTIVITY (PRIOR TO 1900)

The Spanish established agricultural and livestock industries in the Southwest at about the same time that other European nations began colonization along the east coast of the USA. Father Kino, in 1687, travelled the northern frontier of Mexico, which currently includes southern Arizona, southern New Mexico, and west Texas. He later introduced grazing animals and established missions throughout the area. Livestock pro-

Published in R. F. Follett and B. A. Stewart, ed. 1985. *Soil Erosion and Crop Productivity.*
© ASA-CSSA-SSSA, 677 South Segoe Road, Madison, WI 53711, USA.

spered on the desert grasslands, and extensive cattle herds grazed the area by 1840.

The Santa Cruz Valley, between Nogales and Tucson, AZ, was described by Bartlett in 1854 (Humphrey, 1958): "We were off this morning (from Tucson). . . and soon entered a thickly wooded valley of mesquite. A ride of nine miles brought us to San Xavier de Bac. . . . A mile farther we stopped in a fine grove of large mesquite trees near the river, where there was plenty of grass. The bottoms (between San Xavier and Tubac) in places were several miles wide . . . and covered with tall, golden colored grass (big sacation grass) . . . divided by a meandering stream a dozen yards wide and as many inches deep, this shaded by cottonwood, willow and mesquite trees" (p. 203) (Fig. 19–1).

Today, the Santa Cruz Valley, between Nogales and Tucson, described so elegantly by Bartlett in 1854, is typified by a channel 95 km long and about 9.5 m wide and 6.1 m deep (Fig. 19–2). If soil weight is assumed to be 1450 kg/m^3, then 8 million t of soil have been removed in the past 100 years. The distance between these Arizona cities represents only a small portion of lowland channel erosion within the Upper Santa Cruz Basin and a minor portion of lowland channel erosion that has occurred in southeastern Arizona.

Historians have assumed that livestock were equally dispersed over the entire area. However, upland water development and fencing began after 1930 (Wagner, 1952). Before 1930, livestock grazing and irrigated agriculture were more likely confined to riparian lowland areas where surface water supplies were available. If this assumption is correct, grazing and irrigated agriculture were limited to about 20% of the land area, or an estimated 1.5 million ha.

Southeastern Arizona is an arid or semiarid region. Could vegetation covering 20% of the area support 1.5 million cattle or 1 animal ha^{-1} yr^{-1} in 1891? Alkali sacation [*Sporobolus airoides* (Torr.)] and big sacaton (*Sporobolus wrightii* Munro ex Scribn.) are coarse perennial bunchgrasses that were widely distributed in alluvial floodplains in the Southwest before 1900 (Hubbell and Gardner, 1950). The soils on these flood plains are enriched alluvial sediments derived from the surrounding mountains. Their textures are usually medium to moderately fine, and they have high organic matter content. Initially, the soils were classified as Haplustolls, but their taxonomic classification was changed to Torrifluvents, because the solum remains dry more than 90 days during the year (Richardson et al., 1979). These soils receive large amounts of floodwaters during the summer, which, combined with their high available water-holding capacity, makes them extremely productive.

The annual, aboveground, net production of alkali and big sacaton ranges from 6000 kg/ha in dry years to 10 000 kg/ha in wet years at a riparian sites in southeastern Arizona and southwest Texas (J. R. Cox, 1982, unpublished data). In wet years, these areas could easily support 1.5 million cattle, but in dry years the same areas could support less than 1 million head.

Fig. 19-1. The Santa Cruz River, near Tucson, AZ, before 1900 (Photograph provided by the Arizona Historical Society).

Fig. 19-2. The same area of the Santa Cruz River, near Tucson, AZ, in 1980.

Griffiths (1901), Thornebar (1905), Wooten (1916), and Hubbell and Gardner (1950) concluded that dense sacaton grasslands in lowland riparian areas slowed floodwaters, trapped sediments, and enhanced soil fertility. When sacaton grasslands were plowed, irrigation systems transported water directly from rivers and streams to cropland (Cooke and Reeves, 1976). Following the removal of the sacaton grasslands, either by grazing or farming, there were no barriers to reduce water velocity. Runoff from storms, between 1893 and 1900, entered the lowlands and eroded the soils that produced the forage for the livestock industry. Channel trenching resulted in the lowering of shallow water tables that had irrigated croplands (Griffiths, 1901; Cooke and Reeves, 1976).

Annual precipitation variability is greater in southeastern Arizona than at any other location in the contiguous USA (Hershfield, 1962). Precipitation measurements made 16 km apart also showed wide aereal variability in the same year (Renard and Brakensiek, 1976). Precipitation variability within short distances directly affects upland forage productivity, but, because of runoff accumulation from upland areas in lowland areas (Osborn and Renard, 1973), sacaton grassland riparian areas are expected to have a more stable forage productivity.

19-2 RANGELAND PRODUCTIVITY (1900-1980)

Major soil erosion and vegetation changes occurred in lowland riparian areas between 1893 and 1900. These land changes have had a major effect on land productivity and have necessitated a need for new methods for assessing soil erosion.

The development of railway systems in the Southwest in the latter part of the 19th century allowed stockmen to move large herds of cattle and sheep into the area and provided for rapid distribution of agricultural products (Griffiths, 1901; Bahre, 1977). Humphrey (1958) estimated that 1.5 million cattle grazed in southeastern Arizona by 1891.

Passage of the National Recovery Act, implementation of the Work Progress Administration, and creation of the Civilian Conservation Corps in the 1930s contributed to upland water development and provided fencing to separate grazing units (Cox et al., 1982; Johnsen and Elson, 1979). Livestock, previously concentrated in lowlands, were then redistributed over new grazing lands covering the remaining land area.

Populations of range cattle were relatively stable between 1920 and 1970 and correspond with upland water developments and the continuing processes of providing new grazing areas (Table 19-1). Total cattle in 1980 are generally equivalent to 1910 populations, but more than 50% of the cattle in 1980 were maintained in feedlots, while 99% of the cattle were supported on rangelands in 1910.

In the Southwest, the cyclic wet periods were followed by overstocking, and dry periods were followed by livestock die-offs (Wagner, 1952). With each successive cycle, perennial grass productivity has decreased (Fig. 19-3), shrub densities have increased (Hastings and Turner, 1965) (Fig. 19-4), and cattle populations on rangelands have decreased 87% in 90 years (Table 19-1).

In summary, many factors have contributed to the decrease in rangeland productivity. These factors are conversion of land to agronomic crops, invasion of brush species, grazing practices, channelization, and soil erosion. It is difficult to quantify the magnitude and the interactions of each factor in the overall assessment.

Table 19-1. Cattle populations in southeastern Arizona counties between 1890 and 1980. Populations of range cattle were determined by using published estimates or by subtracting estimated dairy and feedlot cattle from estimated county populations.

| Year | Counties | | | | | Cattle populations | |
	Cochise	Graham	Pima	Pinal	Santa Cruz	Total	Range
				1000 head			
1890						1500†	1500
1900	172	85	98	42	43	400	438
1910	150	98	43	42	44	377	375
1920	84	47	64	45	27	267	263
1930	91	42	88	21	30	272	268
1940	91	33	58	53	26	261	250
1950	65	51	41	38	27	222	210
1960	71	74	83	64	33	325	250
1970	68	60	72	221	24	445	240
1980	67	35	40	207	16	365	188

† Estimate from Humphrey (1958).

19-3 CROPLAND PRODUCTIVITY (1900–1980)

Irrigated agriculture expanded between 1900 and 1960 in southeastern Arizona (Table 19-2), and most of the remaining sacaton grasslands were plowed by 1940. Between 1940 and 1960, irrigated agriculture rapidly expanded to all areas of southern Arizona. Irrigated land use decreased 88% in southeastern Arizona between 1960 and 1980 (primarily due to lowering water table), with an additional 50% decline projected by the year 2020 for the entire state (Arizona Water Commission, 1977). This change has had a major impact on production of cotton (*Gossypium hirsutum* L.) and alfalfa (*Medicago sativa* L.). Differences between 1960 and 1980 show that irrigated farmland has decreased about 1.0 million ha (Table 19-2).

Vegetation on rangeland and abandoned farmland currently consists of widely spaced half-shrubs and shrubs (Cox et al., 1982). Raindrop impact on bare areas between shrubs reduces infiltration and enhances runoff from the shrubland. Runoff, which also comes from roofs and pavement (Fig. 5 and 6), causes downstream flooding (Schulz and Lopez, 1974).

Abandoned farmland has also created a serious wind erosion problem. Dust storms evolving from these lands have caused several highway accidents. Productivity is affected, but at the present the most serious problem is human safety.

19-4 ADVANCES IN TECHNOLOGY FOR PREDICTING EROSION AND PRODUCTION ON RANGELANDS AND CROPLANDS

The rate of soil erosion from upland Arizona rangelands is believed to be significantly less in recent decades than in the early part of this century.

Fig. 19-3. Upland range in the Santa Rita Experimental Range in 1920 (Photograph provided by the Arizona Historical Society).

Fig. 19-4. The same location in the Santa Rita Experimental Range in 1980.

For example, many upland soils had a thin, 3 to 6 cm (or deeper), loamy surface horizon before they eroded. This loamy horizon creates a favorable rooting medium for establishment of plants, particularly desert grasses. During erosion, coarse particles accumulate on the surface. Shaw (1927) identified the coarse particles as erosion pavement, a surface covering of stone, gravel, or coarse soil particles that accumulated as sheet or rill erosion removed the finer soil particles. Lowdermilk and Sundling (1950) suggested that an accumulation of rock fragments on the soil surface was equivalent to soil at similar depths and to layers of uneroded soil found to contain similar amounts of rock fragments. Figure 19-7 is a picture of a typical rangeland soil profile (Hathaway soil) showing the distribution of coarse fragments throughout the profile. Figure 19-8 illustrates the surface condition of a typical rangeland soil with its present erosion pavement.

Table 19-2. Irrigated agriculture in southeastern Arizona between 1900 and 1980, and estimates of abandoned farmland in 1980.

| Year | Counties | | | | | |
	Cochise	Graham	Pima	Pinal	Santa Cruz	Total
	1000 hectares					
1900	2	7	3	5	1	18
1910	2	16	4	10	2	34
1920	5	13	7	12	1	38
1930	153	55	114	31	2	355
1940	368	85	119	240	68	880
1950	145	164	143	353	51	856
1960	258	199	126	300	75	958
1970	37	21	20	105	1	184
1980	36	19	19	90	2	166
Abandoned farmland†	332	180	124	263	73	972

† Abandoned farmland figures were obtained by subtracting 1980 estimates from peak production years.

We suggest that in many rangeland areas the rate of erosion has been stabilized because of the erosion pavement and that a modified range ecosystem now exists. Coarse particles on the surface absorb the impact of raindrops and reduce runoff velocity on the land surface, and thus reduce erosion. Evidence of the effect of erosion pavement on infiltration, as obtained with rainfall simulators, has been mixed. Renard (1970) showed that infiltration increased as the combined cover of shrub, grass, litter, and erosion pavement increased. Tromble et al. (1974) showed an increase in infiltration with rock and gravel on the plot surface. Dadkhah and Gifford (1980) conducted rainfall simulator experiments and simulated compaction effects. As trampling and compaction increased, there was little relationship between rock cover and infiltration rate when erosion pavement ranged from 5 to 10%. Noncompacted or ungrazed soils had increased rock cover, which was associated with increased infiltration. They also reported that rock cover did not have a significant effect on sediment production. However, plot size may have influenced their results (Foster et al., 1981).

Rainfall simulators are not ideal for measuring infiltration and erosion, or for comparing results of different studies, because plot sizes, simulation durations, and simulated rainfall characteristics differ. Erosion pavement may be related to infiltration, as shown in the schematic diagram of Fig. 19-9. Thus, the schematic relationship for a specific soil would be adjusted for the effect of other factors known to affect infiltration and erosion, such as compaction and antecedent moisture.

19-4.1 Rangeland Erosion Pavement and Soil Moisture

Southwestern rangelands are characterized by extreme climatic variability. Certainly, the variability associated with annual, seasonal, spatial, and temporal precipitation is well documented by the work of Renard and

Fig. 5. The village of Tucson, in the Arizona Territory, before 1900 (Photograph provided by the Arizona Historical Society).

Fig. 19-6. The city of Tucson, AZ, in 1980. Note increases in housing density in foreground.

Brakensiek (1976), Osborn (1968), and Osborn et al. (1979). Simulation models have been used to assist with quantification of rainfall variability (Osborn et al., 1979; Osborn et al., 1980a and 1980b; Fogel and Duckstein, 1969; Fogel et al., 1971; Gifford et al., 1967; Smith, 1974; and Smith and Schreiber, 1973 and 1974).

Erosion pavement may also reduce the amount of evaporation from bare soils. Kimball (1973) found that mulches retarded water-vapor movement at the soil-air interface.

Jury and Bellanticoni (1976a and 1976b) found that surface rocks had pronounced effects on both the temperature and water flow. The net vertical heat flow would be either upward or downward in a soil with a rock cover (erosion pavement), depending upon prior conditions. During dry periods a slightly greater amount of moisture was always stored under the

Fig. 19-7. The soil profile for a Hathaway soil containing large amounts of coarse material. Following erosion of the finer particles, the coarse material becomes the erosion pavement.

Fig. 19-8. A typical surface view of the vegetation and erosion pavement on an Arizona rangeland soil.

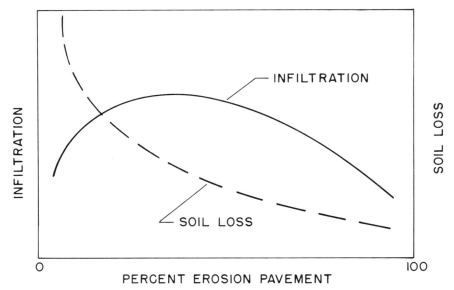

Fig. 19-9. A schematic diagram of the interaction of infiltration and erosion as a function of erosion pavement.

rock compared to adjacent bare soil. Furthermore, the additional soil moisture under the surface rock cover persisted to a depth of 15 cm.

Experimental data to illustrate how soil erosion might affect the productivity of the soil pedon are essentially nonexistent for rangeland areas of the Southwest. Wight and Siddoway (1982) applied the concept of soil loss tolerance (T value), as developed for cropland, to rangelands. They stated that the fragility of rangelands, the irreversibility of erosion damage, and the large margins of error associated with soil loss estimates make it difficult to develop meaningful T values for rangeland. However, a number of recently developed models contribute significantly to our understanding.

19-4.2 Rangeland Forage Production Models

The development of analytical modeling principles associated with digital computers has changed the way much research is conducted. A series of one or more known physically based principles can now be used in a complex problem to conduct a series of numerical experiments (generally a series of mathematical expressions which we call a model), and then to design field experiments with measurements to verify, improve, or calibrate the model (or model coefficients). Thus, a complicated problem can be simplified, eliminating the need for field experiments over all possible conditions. Although quantitative data to substantiate how erosion affects soil productivity are not available for the Southwest, we can use some of the analytical models to arrive at some inferences and to design some experiments to quantify the problem.

Wight and Hanks (1981) predict herbage production using a relationship between vegetation production and precipitation, soil moisture, and climatic variables. The Wight and Hanks yield equation is

$$Y = Y_p (T_a/T_p) \qquad [1]$$

where Y is the actual site yield (kg/ha), Y_p is the potential site yield (kg/ha), T_a is the actual transpiration (mm), and T_p is the potential transpiration (mm). An alternative to this equation is

$$Y = K_e T \qquad [2]$$

where Y is the actual site yield (kg/ha), K_e is the water-use efficiency factor expressed as kg of dry matter produced per kg of water used, and T is the actual transpiration (kg/ha). Lane et al. (1983) discussed the advantages and disadvantages of the method and pointed out that the problems in estimating Y_p and K_e. Values for K_e have often been determined in the greenhouse, so it is not known how this factor applies to the field where water stress, competition among species, spatial variability of soil characteristics, and relative amounts of soil water loss by bare soil evaporation and transpiration are important factors.

An alternative approach to the problem of modeling soil productivity involves more comprehensive water balance models such as the Chemical, Runoff, and Erosion from Agricultural Management Systems (CREAMS) (Knisel, 1980), the Erosion Productivity Impact Calculator (EPIC) (Williams et al., 1983), or the Simulation of Production and Utilization of Rangelands (SPUR) (Wight, 1983). These models are intended for different uses, but they all contain the same algorithms for infiltration (USDA-SCS, 1972) and evapotranspiration (Ritchie, 1972). Evapotranspiration calculations are based on mean daily temperature and solar radiation, soil evaporation based on soil physical properties, and plant transpiration based on a seasonal leaf area index. The evapotranspiration model includes a procedure to reduce computed evaporation and transpiration when soil moisture is limiting, a situation common to arid and semiarid rangeland. Application of algorithms, specifically in the CREAMS model, to arid and semiarid rangeland conditions has been attempted by Lane and Nyhan (1981), Hakonson et al. (1982), and Lane et al. (1983) with considerable success.

Lane et al. (1983) showed prediction accuracy and precision as a function of model complexity (Table 19-3). The simplest model, mean annual net production, obviously does not reflect the annual variability in production, and the confidence interval ranged from 147 to 455 kg/ha. Annual precipitation alone explains 51% of the variance and reduced the width of the confidence interval by 19%. Seasonal estimates of transpiration (estimated by CREAMS) explain 90% of the variance and reduce the confidence interval by 63%. The most significant point with this illustration is that, to reflect production losses due to erosion, detailed measurements

Table 19-3. Summary of regression analysis of predictor variables (x) with standing above
ground net biomass of perennial shrubs and grasses (y) at Rock Valley, NV,
1968 and 1971-1976 (Lane et al., 1983).

				Summary of predictions		
	Regression equation $y = a + bx$			% Explained variance[†]	95% CI width[‡]	% Reduction in 95% CI
Predictor	a	b	R^2	$100\,R^2$	(kg/ha)	width[§]
$x = \bar{y}$ = mean	0	1.0	0.0	0	147-455	0
x = annual precip.	-21	2.21	0.51	51	177-425	19
x = seasonal precip.¶	136	2.40	0.74	74	211-391	42
x = annual trans.	27	6.94	0.84	84	229-373	53
x = seasonal trans.¶	40	9.33	0.90	90	244-358	63

† Percent explained variance, or relative improvement over using the mean annual net produc-
tion as a predictor.
‡ Width of the 95% confidence interval about the mean annual net production.
§ Percent reduction in the width of the 95% confidence interval about the mean annual net
production.
¶ Seasonal precipitation and transpiration from January through May.

are required to reflect soil physical and chemical properties as input to a
physically based model such as CREAMS.

19-5 CONCLUSION

Past soil losses over the Southwest cannot be estimated accurately.
However, upland and lowland arroyo development reductions in livestock
population, abandoned farmland acreages, and shrub invasions indicate
that land abuse has had a major effect on rangeland and cropland productivity, especially in southeastern Arizona.

Physically based models that describe the important processes known
to affect soil productivity have considerable promise for quantifying how
erosion affects soil productivity. Furthermore, research planned and
conducted in concert with such models can greatly reduce the number of
sites necessary to quantify the spatial variability encountered in the
rangeland areas of the region.

REFERENCES

Arizona Water Commission. 1977. Phase II, Arizona state water plan: alternative futures.
Arizona Water Commission.

Bahre, C. J. 1977. Land-use history of the research ranch, Elgin, Arizona. J. Ariz. Acad. Sci.
12 (Supplement 2):3-32.

Cooke, R. U., and R. W. Reeves. 1976. Arroyos and the environmental change in the American Southwest. Oxford University Press, London.

Cox, J. R., H. L. Morton, T. N. Johnsen, Jr., G. L. Jordan, S. C. Martin, and L. C. Fierro.
1982. Vegetation restoration in the Chihuahuan and Sonoran Deserts of North America.
USDA-ARS Rep. ARM-W-28.

Dadkhah, M., and G. F. Gifford. 1980. Influence of vegetation, rock cover, and trampling on
infiltration rates and sediment production. Water Resour. Bull. 16(6):979-986.

Fogel, M. M., and L. Duckstein. 1969. Point rainfall frequencies in convective storms. Water Resour. Res. 5(6):1229-1237.

----, ----, and J. L. Sanders. 1971. An event-based stochastic model of areal rainfall and runoff. p. 247-261. *In* USDA-ARS Misc. Pub. 1275.

Foster, G. R., J. R. Simanton, K. G. Renard, L. J. Lane, and H. B. Osborn. 1981. Discussion of: "Application of the universal soil loss equation to rangelands on a per-storm basis." J. Range Manage. 34(2):161-166.

Gifford, R. O., G. L. Ashcroft, and M. D. Magnuson. 1967. Probability of selected precipitation amounts in the Western Region of the United States. Western Region Res. Pub. T-8. Univ. of Nevada, Reno.

Griffiths, D. 1901. Range improvement in Arizona. USDA, Bureau of Plant Industry Bull. 4.

Hakonson, T. E., L. J. Lane, J. G. Steger, and G. L. DePoorter. 1982. Some interactive factors affecting trench cover integrity on low-level waste sites. p. 377-399. *In* Proc., Nuclear Regulatory Comm. Symp. on Low-Level Waste Disposal: Site Characterization and Monitoring, Arlington, VA. 16-17 June. Office of Nuclear Material Safety and Safeguards.

Hastings, J. R., and R. M. Turner. 1965. The changing mile. Univ. of Arizon Press, Tucson.

Hershfield, D. M. 1962. A note on the variability of annual precipitation. J. Appl. Meteorol. 6:575-578.

Hubbell, D. S., and J. L. Gardner. 1950. Effects of diverting sediment-laden runoff to range and cropland. USDA Tech. Bull. 1012.

Humphrey, R. R. 1958. The desert grassland. Botanical Rev. 24:193-252.

Johnsen, T. N., Jr., and J. W. Elson. 1979. Sixty years of change on a central Arizona grassland—juniper woodland ecotone. USDA-ARS Rep. ARM-W-7.

Jury, W. A., and B. Bellanticoni. 1976a. Heat and water movement under surface rocks in a field soil: I. Thermal effects. J. Soil Sci. Soc. Amer. 40:505-509.

----, and ----. 1976b. Heat and water movement under surface rocks in a field soil: II. Moisture effects. J. Soil Sci. Soc. Amer. 40:509-513.

Kimball, B. A. 1973. Water vapor movement through mulches under field conditions. Soil Sci. Soc. Amer., Proc. 37(6):813-818.

Knisel, W. G., Jr. (ed.) 1980. CREAMS: A field-scale model for chemicals, runoff, and erosion from agricultural management systems. USDA Conserv. Res. Rep. 26. Washington, DC.

Lane, L. J., and J. W. Nyhan. 1981. Cover integrity in shallow land burial of low-level wastes: Hydrology and erosion. p. 419-430. *In* ORNL/NFW-81/34. DOE Low-Level Waste Mgt. Program, Proc. 3rd, New Orleans, LA.

----, E. M. Romney, and T. E. Hakonson. 1984. Water balance calculations and net production of perennial vegetation in the Northern Mojave Desert. J. Range Mgt. 37(1):12-18.

Lowdermilk, W. C., and H. L. Sundling. 1950. Erosion pavement, its formation and significance. Trans. Am. Geophys. Union 31(1):96-100.

Osborn, H. B. 1968. Persistence of summer rainy and drought periods on a semiarid rangeland watershed. Int. Ass. Sci. Hydr. Bull. 13(1):14-19.

----, and K. G. Renard. 1973. Management of ephemeral stream channels. J. Irrig. Drain. Div., ASCE 99(IR3):207-214.

----, ----, and J. R. Simanton. 1979. Dense networks to measure convective rainfall in the Southwestern United States. Water Resour. Res. 15(6):1701-1711.

----, L. J. Lane, and V. A. Myers. 1980a. Two useful rainfall/watershed relationships for southwestern thunderstorms. Trans. Am. Soc. Agric. Engr. 23(1):82-87, 91.

----, E. D. Shirley, D. R. Davis, and R. B. Koehler. 1980b. Model of time and space distribution of rainfall in Arizona and New Mexico. USDA-SEA Rep. ARM-W-14.

Renard, K. G. 1970. The hydrology of semiarid rangeland watersheds. USDA-ARS Rep. 41-162.

----, and D. L. Brakensiek. 1976. Precipitation on intermountain rangeland in the Western United States. p. 39-59. *In* Workshop, US/Australia Rangeland Panel, Proc. 5th, Boise, ID. Utah Water Res. Lab., Logan.

Richardson, M. L., G. D. Clemens, and J. C. Walker. 1979. Soil survey of Santa Cruz and parts of Cochise and Pima Counties, Arizona. USDA–SCS, USDA–FS, and Arizona Agric. Exp. Stn.

Ritchie, J. T. 1972. A model for predicting evaporation from a row crop with incomplete cover. Water Resour. Res. 8(5):1204–1213.

Schulz, E. F., and O. G. Lopez. 1974. Determination of urban watershed response time. Hydrology Paper 71. Colorado State Univ., Ft. Collins.

Shaw, C. F. 1927. Erosion pavement. The Geogr. Review 19:638–644.

Smith, R. E. 1974. Point processes of seasonal thunderstorm rainfall. 3: Relation of point rainfall to storm and areal properties. Water Resour. Res. 10(3):424–426.

––––, and H. A. Schreiber. 1973. Point process of seasonal thunderstorm rainfall. 1: Distribution of rainfall events. Water Resour. Res. 9(4):871–884.

––––, and ––––. 1974. Point process of seasonal thunderstorm rainfall. 2: Rainfall depth probabilities. Water Resour. Res. 10(3):418–423.

Thornber, J. J. 1905. Range improvement. Univ. of Arizona Agric. Exp. Stn. Annu. Rep. 16: 17–22.

Tromble, J. M., K. G. Renard, and A. P. Thatcher. 1974. Infiltration for three rangeland soil vegetation complexes. J. Range Manage. 27(4):318–321.

USDA–SCS. 1972. National Engineering Handbook, Hydrology: Section 4, Chapters 4–10. U.S. Government Printing Office, Washington, DC.

Wagner, J. J. 1952. History of the cattle industry in Southern Arizona, 1540–1940. Univ. Arizona Social Sci. Bull. 20. Univ. of Arizona Press, Tucson.

Wight, J. R. (ed.). 1983. SPUR—Simulation of production and utilization of rangelands: A rangeland model for management and research. USDA–ARS Misc. Pub. 1431.

––––, and R. J. Hanks. 1981. A water-balance climate model for herbage production. J. Range Manage. 34(3):307–311.

––––, and F. G. Siddoway. 1982. Determinants of soil loss tolerance for rangelands. p. 67–74. *In* Determinate of Soil loss tolerance. Spec. Pub. 45. American Society of Agronomy.

Williams, J. R., P. T. Dyke, and C. A. Jones. 1983. EPIC—A model for assessing the effects of erosion on soil productivity. p. 553–572. *In* Proc. Third Int. Conf. on State-of-the-Art in Ecological Modeling. 24–28 May, Colorado State Univ., Fort Collins.

Wooten, E. O. 1916. Carrying capacity of grazing ranges in Southeastern Arizona. USDA, Bureau of Plant Industry Bull. 367.

20 Restoration of Crop Productivity on Eroded or Degraded Soils

W. W. Frye
University of Kentucky
Lexington, Kentucky

O. L. Bennett
Agricultural Research Service
U.S. Department of Agriculture
Beckley, West Virginia

G. J. Buntley
University of Tennessee
Knoxville, Tennessee

"Erosion in Nature is a beneficent process without which the world would have died long ago. The same process, accelerated by human mismanagement, has become one of the most vicious and destructive forces that have ever been released by man" (Jacks and Whyte, 1939, p. 3). Damage done by accelerated erosion may be loss of fertility, loss of organic matter, deterioration of soil structure, decreased infiltration rate, diminished workability, decreased available water-holding capacity, or, more frequently, a combination of these (National Soil Erosion-Soil Productivity Research Planning Committee, USDA, 1981; Larson et al., 1981; Frye et al., 1982). The usual result is lower productivity of the soil.

For each soil, there is a maximum rate at which erosion on an annual basis can be allowed and productivity of the soil still maintained over a long period (Wischmeier and Smith, 1978). This permissible or tolerable level of annual soil erosion is defined as T in the universal soil loss equation (USLE). T values for most cropland soils range from 6.7 to 11.2 Mg ha^{-1} yr^{-1}. Where the rate of soil loss exceeds T, soil productivity may be permanently diminished. The higher the rate of excessive erosion or the longer it occurs, the more detrimental to soil quality it is likely to be.

Published in R. F. Follett and B. A. Stewart, ed. 1985. *Soil Erosion and Crop Productivity.*
© ASA-CSSA-SSSA, 677 South Segoe Road, Madison, WI 53711, USA.

The USDA estimated the average annual soil loss by the water-erosion (USLE) and wind-erosion equations at each of the sample sites used in the 1977 National Resource Inventories (USDA, 1981). Table 20-1 shows a nationwide summary from the inventories of estimated sheet, rill, and wind erosion by soil-loss categories on nonfederal cropland, pastureland, forest land, and rangeland. If erosion-damaged soils can be equated to soils that are losing approximately 11 Mg ha^{-1} yr^{-1} or more soil, the values in Table 20-1 indicate that at least 98 million ha of soils on nonfederal lands in the USA, about 18% of those lands, are damaged by erosion.

Erosion is more damaging to the quality and productivity of some soils than others. On some soils, it may cause little or no permanent reduction in productivity, while on others reduction may occur so slowly as to be unnoticed. If the rate of reduction is equal to or less than the rate of increase in crop production due to technological inputs, such as fertilizer management, irrigation, improved varieties, and more effective pest control, crop yields may be maintained or increased during and after erosion (Fig. 20-1). The farmer, observing no decrease or perhaps an increase in crop yields, may assume, usually incorrectly, that soil erosion is not diminishing the productivity of the soil and that he is not experiencing a financial loss (Murdock et al., 1980).

The effects of soil erosion on productivity depend largely on the thickness and quality of the topsoil and on the nature of the subsoil. Productivity of deep soils with thick topsoil and excellent subsoil properties may be virtually unaffected by erosion, if the fertility lost by erosion is replaced. However, most soils are shallow or have some undesirable properties in the subsoil that adversely affect yields. In either case, productivity will decrease as the topsoil gets thinner and undesirable subsoil is mixed into the Ap horizon by tillage, or as water-storage capacity and effective rooting depth are decreased. Thus, in general, crop yields decrease in proportion to the degree of past erosion.

Shrader et al. (1960) used primarily the texture of the subsoil to place several soils of Iowa into three groups according to the reduction in yield capacity and decrease in workability resulting from loss of topsoil by erosion. The finer the texture of the subsoil, the more damaging was the erosion. In Georgia, Langdale et al. (1979) found a close relationship between corn (*Zea mays* L.) yields and the depth to the Bt horizon in a Typic Hapludult. Depth to the clayey subsoil affected grain yields more than

Table 20-1. Erosion on nonfederal land in the USA (USDA, 1981).

Land use	Sheet, rill, and wind erosion (Mg/ha)				Total area in each use
	<11	>11‡	11–31	>31	
	Million ha†				
Cropland	110		38	19	167
Pastureland	48		4	9	54
Forest land	143		5	2	150
Rangeland	137	28			165

† Values are rounded for convenience.
‡ Category applied only to rangeland.

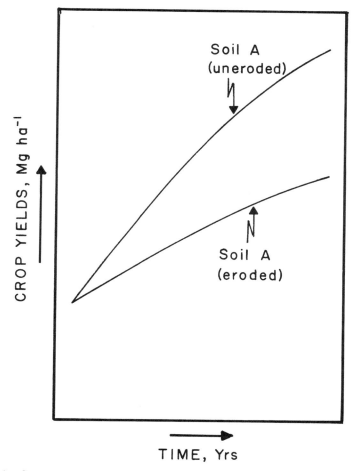

Fig. 20–1. Conceptualized effect of erosion on soil productivity over time with increased technological inputs (Murdock et al., 1980).

stover yields. The 3-year average grain yields were 2.2 and 4.7 Mg/ha on severely eroded and moderately eroded soils, respectively, compared to 6.4 Mg/ha on an uneroded alluvial soil. As the depth to the Bt horizon decreased from 41 to 25 cm, yield of corn grain decreased almost 0.15 Mg ha^{-1} yr^{-1} cm^{-1} of soil lost. Beasley (1972) reported yield reductions of 15% and 30%, respectively, where 5 and 15 cm of topsoil had been eroded.

Batchelder and Jones (1972) drastically decreased the yields of corn grain in Virginia by artificially removing the topsoil of a Typic Hapludult, exposing a subsoil with more than 50% clay content. Restoring soil fertility alone was not enough to restore the productivity of the soil. However, yields were restored to almost that of the unaltered soil by adding fertilizer and irrigation water. Ripley et al. (1961) reported that artificial removal of 7.5 cm of topsoil caused a 21% decrease in yield of barley (*Hordeum vulgare*) averaged over a 10-year period. Removal of 15 cm of topsoil resulted in a 58% decrease in yields. Where small amounts (by present-day standards) of fertilizers were applied, yield reductions were less, 19% and 46%, respectively, with 7.5 and 15 cm of topsoil removed.

Evidence suggests that the productivity of some soils may never be restored completely following excessive erosion (Frye et al., 1982), or at least that the process of restoration would be slow or might require extraordinary and expensive treatment. Frye et al. (1982) reported research on a moderately eroded Maury soil (Typic Paleudalf) which had been in permanent bluegrass (*Poa pratensis* L.) pasture for about 60 years. Averaged over 4 years, grain yields of no-tillage corn grown in mulches of different winter-cover crops were 12% less on eroded soil than on uneroded soil at that site. Dry matter yields of four cover crops were 28% lower on the eroded soil.

In contrast, Engelstad and Schrader (1961) artificially removed the topsoil of a Marshall soil in Iowa, which resulted in decreasing the corn yields by more than one-half, but application of extra N fertilizer to replace the N lost by removal of the soil organic matter restored the productivity to that of the unaltered soil (Fig. 20-2).

Larson et al. (1981, p. 31) stated that we know little about the relationships between soil productivity and the properties of eroded soils, prompt-

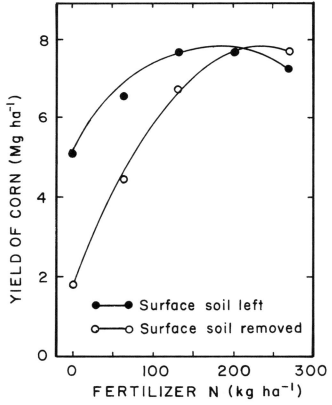

Fig. 20-2. Effect of N fertilizer on yield of corn grain on artificially exposed Marshall subsoil (Englestad and Shrader, 1961).

ing them to ask, "Are yields actually reduced (by erosion), or can input of chemical fertilizers, etc., sustain production?" Available data appears to be conflicting, further emphasizing the need to define relationships between erosion and productivity. An understanding of these relationships is necessary in determining effective practices to restore the productivity of erosion-damaged soils.

The purpose of this chapter is to improve the understanding of soil productivity in relation to the properties of eroded soils and how reduced productivity can be restored to erosion-damaged soils. Specific objectives are:

1. To discuss the effects of excessive soil erosion on chemical and physical properties of soils and the relationship of the effects to soil productivity.
2. To examine the feasibility of using soil management practices, chemical fertilizers and other amendments, vegetative cover, or mechanical means to either maintain or enhance current productivity or to restore diminished productivity of soils damaged or degraded by erosion.

In fulfilling the first objective, we cite evidence that erosion diminishes the crop-production potential of soils mainly by modifying certain soil properties. Major emphasis is on the second objective. We discuss the feasibility of using several means to either maintain or enhance current productivity of soils or to restore diminished productivity of soils damaged or degraded by erosion. Several soil management practices are discussed, including conservation tillage; chemical fertilizers, lime, manures, and organic wastes; vegetative cover from cover crops, crop residues, or mulches; irrigation; and earth moving, including both beneficial and detrimental effects. Included is a discussion of why it is difficult to restore the productivity of some soils.

20-1 EFFECTS OF EROSION ON CHEMICAL PROPERTIES

Soil erosion affects the chemical properties of soils mainly in three important ways: (i) loss of soil organic matter, (ii) loss of soil minerals containing plant nutrients, and (iii) exposure of subsoil material with low fertility or high acidity.

20-1.1 Soil Organic Matter

20-1.1.1 Loss of Organic Matter

Erosion decreases the organic matter content of the soil in two ways. Loss of topsoil from the surface decreases the total amount of organic matter by making the topsoil layer thinner. As the topsoil layer becomes thinner, tillage mixes subsoil material with topsoil. The subsoil of most soils, especially those formed under forest vegetation, is very low in organic matter content and dilutes the organic matter content of the topsoil when mixed. The effect is a lower organic matter concentration in the Ap horizon. For example, a soil with an Ap horizon 20 cm thick with 4%

organic matter and a bulk density of 1.25 Mg/m³ would lose 5 Mg/ha of organic matter for each centimeter of topsoil removed by erosion. Ignoring the usual low content of organic matter in the subsoil and the change in bulk density caused by erosion, 1 cm/ha of soil lost by erosion decreases the percentage of organic matter by about 1/20th of the initial percentage.

The above calculations probably underestimate the amount of organic matter lost from a soil as a result of erosion because of the tendency for erosion to selectively remove organic matter and finer soil particles. Stoltenberg and White (1953) clearly showed that eroded material was higher in organic matter and plant nutrients than was the soil from which the material eroded (Table 20–2). Most nutrients lost in eroded material are probably contained in the organic matter.

A Crider soil studied by Frye et al. (1982) had significantly lower organic matter in the Ap horizon where eroded than did the uneroded soil of the same series at the same site. However, there was no significant difference in the organic matter content of eroded and uneroded Maury soil at another site that had been in permanent bluegrass pasture for several decades, although corn grain yields and yields of winter cover crops were significantly less where the Maury soil was eroded than where it was uneroded. This finding suggests that restoring the organic matter of an eroded soil does not necessarily restore the productivity of the soil. In the case of the Maury soil, the reduced productivity was thought to result from the decreased available water holding capacity due to the higher clay content of the Ap horizon of the eroded soils.

20–1.1.2 Direct Effects of Loss of Organic Matter

Barrows and Kilmer (1963) concluded that significant losses of soil organic matter were accompanied by removal of N and P by water erosion. In terms of soil N, each Mg/ha of organic matter lost by erosion represents approximately 60 kg of N. Therefore, in the previous example, 300 kg/ha total soil N would be lost with the removal of 1 cm of soil.

Loss of organic matter also affects the use and effectiveness of herbicides. Herbicides often injure crops on severely eroded soils that are low in organic matter, even when the herbicide is used at a recommended rate. Increasingly, herbicide labels are specifying a minimum soil organic matter content for their safe and effective use. The dependency of today's farmers on chemical weed control makes this consequence of soil erosion a problem of concern.

Table 20–2. Mean composition of surface soil and eroded material from a group of six watersheds in Indiana with an average soil loss of 19.7 Mg ha⁻¹ yr⁻¹ (Stoltenberg and White, 1953).

	Organic matter	N	P	K	CaCO₃ and MgCO₃
			% by weight		
Surface horizon	3.3	0.16	0.02	0.01	1.0
Eroded material	4.1	0.28	0.04	0.06	1.5

20-1.2   Soil Fertility Status

20-1.2.1   Soil Nitrogen, Phosphorus, and Potassium

Soil erosion significantly decreases the fertility of soils. Beasley (1980) and Taylor (1967) estimated that the approximately 2.7 billion Mg of soil eroded from U.S. farms and forest lands each year have an average analysis of 0.10% N, 0.07% P, and 1.25% K for a total of over 38 million Mg of plant nutrients lost by erosion. Placing an accurate economic value on plant nutrients and soil organic matter lost by soil erosion would be difficult if not impossible, but, according to Beasley (1980), the value would be billions of dollars.

Frye et al. (1982) studied the effects of erosion on the fertility of two Kentucky soils. Erosion reduced soil test P and K levels for a Crider soil but not for a Maury soil, which is naturally high in P and had been heavily fertilized with K. Burwell et al. (1975) found that sediment transport accounted for more than 95% of N and P lost and most of the total K lost from fallow, continuous corn, and rotation corn treatments. Miller and Krusekopf (1932) compared the removal of plant nutrients by erosion and by crops from a Shelby silt loam soil in Missouri where corn was grown continuously. They found that the amounts removed by erosion expressed as a percentage of the crop removal were as follows: N, 55%; P, 90%; K, 605%; Ca, 550%; Mg, 290%; S, 85%. A rotational cropping system of corn-wheat-clover markedly decreased the erosion losses of the nutrients to 22%, 36%, 214%, 212%, 97%, and 30%, respectively, of that removed by the crop. Workers in Georgia (Thomas et al., 1968) and Minnesota (Timmons et al., 1968) found that cropping systems that reduced soil erosion also reduced N and P losses.

20-1.2.2   Secondary and Micronutrients

In humid regions where soils tend to be mature and strongly weathered, certain secondary and micronutrients associated with mineral matter are removed by weathering and soil formation. In such cases, the soil organic matter may be the most abundant indigenous source of the nutrients. For example, much of the total S in humid-region soils is in the organic matter, as is the case to a lesser degree for B, Cu, Mo, and Zn.

When the soil's supply of a nutrient depends on the soil's organic matter content, erosion will decrease the supply of that nutrient at an abnormally high rate. In addition, the concentration of plant nutrients is usually higher in the material eroded than in the soil from which the material eroded (Table 20-2).

20-1.2.3   Soil pH and Lime Requirement

Erosion tends to increase the soil acidity and lime requirement of humid-region soils, particularly in the Southeast, in at least three important ways: (i) replacement of topsoil with the more acid subsoil as erosion occurs, (ii) selective removal of the base-forming elements (K, Ca, and Mg) from the topsoil, and (iii) removal of applied lime before it reacts to neutralize soil acidity.

In such soils, the natural processes of soil formation and nutrient cycling are responsible for the tendency of the subsoils to be more acid than the topsoils. Movement of Al from A to B horizons results in accumulation of Al compounds which are mainly responsible for strongly acid subsoils. Base nutrients recycled by plants are returned to the soil mainly at the soil surface in organic matter. Therefore, eroded soils usually are more acid and have higher lime requirements than uneroded soils (Frye et al., 1982). An additional reason that eroded soils may have higher lime requirements is the higher buffering capacity associated with the clay mixed into the plow layer as a result of erosion.

20-2 EFFECTS OF EROSION ON PHYSICAL PROPERTIES

Soil erosion causes changes in several of the physical properties of soils. Since individual soil properties, such as structure, texture, bulk density, infiltration rate, depth favorable for root development, and available water-holding capacity, are so closely interrelated, a change in one property results in a change in the interaction among all the properties. Because of this interrelationship, the consequences of erosion damage to one physical property are difficult to isolate. Erosion damage to the physical properties involved is probably best expressed in terms of the consequences of the combined damage. Although several consequences can and do occur, the most serious is reduction in the water-supplying capacity of the soil and the associated reduction in production potential (Batchelder and Jones, 1972; Frye et al., 1982). We will discuss erosion damage to the physical properties of the soil in this context.

20-2.1 Soil Water in Relation to Texture and Structure

Subsoil material exposed by erosion is lower in organic matter and often higher in clay, or sometimes sand, than the original surface horizon. Baver et al. (1972) pointed out the importance of organic matter in both the development of structural aggregates and their stability. They also stated that structure degradation by raindrop impact, the major cause of aggregate deterioration, is less in soils in which aggregates are stabilized by humus than by clay. These changes resulting from erosion of the Ap horizon, although they may not seem great, significantly affect water relationships of the soil. When the exposed subsoil material is high in clay, total pore space per unit volume increases, pore size decreases, and total water-holding capacity increases, but available water-holding capacity decreases. The infiltration rate decreases and runoff rate increases. In addition, since the structure aggregates are usually less stable than those in the original surface horizon, they are more easily broken down by the impact of raindrops (Baver et al., 1972), and surface crusts may form to further reduce the infiltration rate and increase the rate of runoff. Thus, the recharge potential of the soil water is reduced. In total, these changes decrease the available water-supplying capacity of the soil and lower its production potential.

As more of the clayey subsoil material is incorporated into the plow layer, the moisture range within which the soil can be easily and safely worked becomes narrower and more limiting. If worked too wet, soil structure tends to break down even further, and the soil may be left puddled. This further decreases pore space, aeration, infiltration and percolation, and increases bulk density. Soil compaction often becomes a problem. If tilled when too dry, the clayey subsoil material in the plow layer becomes cloddy and difficult to work into a suitable seed bed without a number of additional energy-consuming trips across the field. The increased clay content, deteriorated structure, and lower organic matter in the surface horizons of many erosion-damaged soils also result in increased power demands for normal tillage operations. In addition, the puddling and crusting associated with eroded soils often interfere with germination and retard or restrict seedling emergence. This is especially critical for those plants having large cotyledons that must move through the soil to emerge.

Where the exposed subsoil material is higher in sand than the original topsoil, the total pore space per unit volume decreases, both total and available water-holding capacities decrease, infiltration rate increases, and runoff rate decreases. The benefits of the increased infiltration and decreased runoff resulting from the sandier textured plow layer are more than offset by the decreased total and available water-holding capacities. This combination of changes reduces the water-supplying capacity of the soil and lowers its production potential.

Eroded soils, especially those developed in sandy material, are more susceptible to soil compaction in the form of plow pans and traffic pans. These compacted zones drastically decrease the effective rooting depth and seriously reduce the water-supplying capacity of the soil.

20–2.2 Soil Water in Relation to Soil Depth

Soil erosion also decreases the effective rooting zone of many soils in ways unrelated to compaction. The water-storage capacity of soils that are shallow to root-restricting horizons or materials can be reduced significantly by soil erosion. For example, the Grenada soil, a common fragipan soil in the southern loess region of the USA, holds about 0.25 cm of available water per centimeter of soil above the fragipan. Thus, each 4 cm of soil above the fragipan can store 1 ha-cm of available water. This means that a relatively uneroded Grenada soil 60 cm thick above the fragipan can store 1500 m³/ha of available water, whereas a severely eroded Grenada with only 20 cm of soil above the fragipan can store only 500 m³/ha of available water (Table 20–3).

Table 20–3. Relationship between depth of soil to a fragipan and available water storage.

Depth to fragipan	Available water storage	
cm	cm	Mg/ha
60	15	1500
50	12.5	1250
40	10	1000
30	7.5	750
20	5	500
10	2.5	250

Frye et al. (1983) studied corn grain yields on a Zanesville soil (fine-silty, mixed, mesic, Typic Fragiudalf) in which the depth to a fragipan ranged from 0.3 to 0.6 m. In 3 of the 4 years, susceptibility to water stress increased and grain yields decreased with decreasing depth to the fragipan horizon. Only during a season in which rainfall was moderately high and abnormally well-distributed was this effect not observed. They concluded that for shallow fragipan soils productivity is unpredictable, yield uncertainty is high, and yields are highly dependent on rainfall distribution during certain critical times.

As important as water-storage capacity is, it is not possible, even on deep, uneroded soils, to produce high crop yields using only the water stored in the soil at planting time. The soil water must be replenished by rainfall periodically during the growing season. Ironically, as explained earlier, erosion damage to the physical properties of the soil usually reduces the recharge potential of soil water as well as decreasing the infiltration capacity.

20–3 MAINTAINING AND RESTORING SOIL PRODUCTIVITY

20–3.1 Soil Organic Matter

As stated previously, restoring the organic matter content of an eroded soil does not necessarily restore the soil's productivity to the level of the uneroded soil. Nevertheless, increasing the organic matter content is one of the most effective and practical ways to help restore the productivity of eroded soils. Volk and Loeppert (1982) reported that yield potential increased by an average of 21% for each 1% increase in soil organic C.

The organic matter content can be increased by several soil-management practices. Beale et al. (1955) studied the effects of mulch tillage on runoff, erosion, soil properties, and crop yields over a 10-year period in South Carolina. Corn grown by no-tillage in a vetch and rye mulch increased the soil organic matter in the Ap horizon from 1.5% to 2.6% during the 10 years, while the organic matter under the plowed check plots decreased to 1.2% from the initial value of 1.5% over the same period. Their study also showed that the degree of soil aggregation and the stability of the soil structure increased during the 10 years under the mulched treatments, but was reduced considerably under the plowed check treatment. Results of an Ohio study on a conventionally tilled cropland soil, reported by Salter et al. (1941), showed a strong relationship between cropping systems and soil organic matter content (Table 20-4). All cropping systems decreased the organic matter relative to the content at the beginning of the experiment, but continuous corn was the most detrimental. A legume crop in the rotation every third year resulted in the least decrease in soil organic matter.

The potential for a winter legume cover crop to increase the productivity of a soil over time was shown by Ebelhar (1981). With a hairy vetch (*Vicia villosa*) cover crop and 100 kg/ha fertilizer N each year, grain yields of no-tillage corn increased from 8.1 Mg/ha in 1977 to 10.1 Mg/ha in 1981, presumably as a result of the hairy vetch cover crop. The increase was 0.51 Mg/ha per year relative to the corn residue and 100 kg/ha fertilizer-N treatment.

Research over a 7-year period on a moderately eroded Cecil soil in Georgia by Adams et al. (1970) showed the importance of cover crops in improving the productivity of eroded soils. Winter cover crops of vetch or rye (*Secale cereale*) as green manure significantly (5% level) increased corn grain yields over those of continuous corn without a green manure crop, except at the highest N fertilizer level (Table 20-5). Whether corn stalks were removed or left did not appear to affect grain yields. Without N fertilizer, vetch resulted in higher grain yields than did rye. Grain yields were not as high without a cover crop and with the highest rate of fertilizer N (180 kg/ha) as with vetch and no N fertilizer or with rye and 45 kg/ha of N added (Table 20-5).

Table 20-4. Effect of 32 years of various cropping systems on soil organic matter content (Salter et al., 1941).

Cropping system	Organic matter	Percent of originals[†]
	Mg/ha	%
Continuous corn	14.3	37
Continuous oats	25.5	65
Continuous wheat	24.6	63
Rotation: corn-oats-wheat-clover-timothy	30.0	77
Rotation: corn-wheat-clover	33.2	85

† Soil organic matter was 39.2 Mg/ha under cropland conditions at start of study.

Table 20-5. Effect of vetch or rye as green manure crops on corn grain yields on a moderately eroded Cecil soil (clayey, kaolinitic, thermic Typic Hapludult), 1958–1964 (Adams et al., 1970).

Cropping treatment[†]	Fertilizer N (kg/ha)			
	0	45	90	180
	Grain yield, Mg/ha			
Corn	1.6 c[‡]	4.0 b	4.7 b	4.8 a
Corn-vetch	5.3 a	5.3 a	5.6 a	5.5 a
Corn-rye	3.9 b	5.5 a	5.7 a	5.9 a

† Corn stalks left in all treatments shown.
‡ Values in same column followed by same letter are not significantly different at 5% level.

Restoring organic matter to the surface horizon of erosion-damaged soils over time may eventually return most of the physical properties to near their original condition, with the exception of soil texture. As the organic matter content of the surface horizon increases, the degree of aggregation and the stability of the aggregates also increases. This in turn increases porosity, lowers bulk density, increases infiltration, decreases runoff, and increases water-recharge potential. Increasing organic matter generally increases the total water-holding capacity as well as the water-supplying capacity of the soil, but apparently does not increase available water-holding capacity except in sandy soils. Organic matter holds much of its water with tension too great to be available to plants. This increase in water-supplying capacity is a major factor in restoring the production potential to erosion-damaged soils, especially under rain-fed water management.

Restoring organic matter also has beneficial effects on the chemical properties of soils. The C:N ratio of organic matter added to soils varies widely (80:1 to 12:1). The ratio narrows to about 10:1 after the materials have been incorporated in the soil and humification has occurred. Most of the soil N is in the organic form. The slow decomposition of humus makes possible the storage of N in the soil and its gradual release for plant use. Organic matter also is an important mobilizer of plant nutrients. Acids produced during the decomposition of organic matter facilitate the release of plant nutrients from soil minerals, and, as the organic matter decomposes, the plant nutrients contained in it are mineralized and available for plant use.

Humus, or highly decomposed organic matter, usually is colloidal. It has a much higher cation-exchange capacity per unit weight than does colloidal-sized mineral matter such as clay. As a consequence, humus functions as a strong buffer in the soil, slowing changes in soil reaction and minimizing the risks of herbicide injury to crops. In addition, humus adsorbs fertilizer nutrients and mineralized plant nutrients that might otherwise be lost by leaching.

Biologically, organic matter supplies the food for the microorganisms in the soil. These microorganisms are important in the decomposition of that organic matter, formation of humus, and transformation of plant nutrients into forms that are available to the higher plants.

20–3.2 Soil Amendments

20–3.2.1 Lime and Fertilizers

Research in Iowa by Engelstad et al. (1961) and Engelstad and Shrader (1961) showed that adding N fertilizer restored productivity to an artificially eroded Marshall soil, a deep loess soil (Fig. 20–2), but at greater expense and lower production efficiency. Frye et al. (1982) concluded from erosion studies with two soils in Kentucky that neither low-intensity use, such as permanent pasture, nor optimum fertilizer amendments can restore a strongly developed soil to its original production potential after severe damage by erosion. However, they acknowledged that productivity of eroded soils can be increased by management. Batchelder and Jones (1972) found that 2 years of fertilization and liming improved corn yields on exposed subsoils commensurate with those on surface soils when the water supply was increased by mulching and limited irrigation. After 3 years, yields of corn on the mulched, unirrigated subsoil were higher than those on the irrigated topsoil, probably because the mulch increased available soil water, reduced soil temperatures, and provided a source of organic matter and plant nutrients.

20-3.2.2 Manures and Organic Wastes

Animal manure and other organic wastes have long been recognized as one of the best methods of maintaining fertility, productivity, and soil organic matter. On eroded soils, the value of manures and organic waste goes far beyond their value as sources of plant nutrients. They serve as a mulch, reducing runoff and further erosion. They also increase soil aggregation, promote infiltration, and are good sources of organic matter, microbes for the rhizosphere, and organic ligands for chelation and movement of major plant nutrients into the soil profile (Tan et al., 1985). Organic wastes have been successfully used for soil stabilization on disturbed areas with extremely low pH to promote rapid growth and root development of plants. Excellent results have been obtained on acid mine spoils using manures, composted garbage, sewage sludge (Stout et al., 1982; Mathias et al., 1979), tannery waste, hardwood bark (Sarles and Emanuel, 1977), and other organic residues. In many cases, the use of organic wastes provides the only means of establishing a suitable plant cover in hostile soil environments such as severely eroded or denuded areas (Bennett, 1978; Bennett et al., 1976; Stout et al., 1982).

20-3.3 Earth-Moving Restoration

Filling and leveling the land may be necessary before severely eroded and gullied soil can be used for crop production. Land leveling often creates a soil condition similar to growing plants on exposed subsoils or on reclaimed surface-mine soils. Reuss and Campbell (1961) indicated that these soils are usually deficient in N and P, but they obtained good yields when both elements were supplied in adequate amounts or when a heavy manure application was plowed down. Gardner (1941), Robertson and Gardner (1946), and Whitney et al. (1950) indicated that the major factor limiting crop production on desurfaced soils in Colorado was the lack of adequate N and P. Carlson et al. (1961) also reported deficiencies of N and P plus some micronutrients in infertile subsoils. Bennett (1978) and Bennett et al. (1976) found that both chemical and physical properties of regraded mine soils must be considered before satisfactory reclamation and revegetation could be obtained. Heavy applications of lime, fertilizer, and organic waste usually were necessary for satisfactory restoration.

Reshaping fields severely damaged by erosion may decrease the productivity over a large area of the field by "borrowing" soil from less eroded or uneroded areas to fill in gullied areas. The exposed subsoil may be ripped or scarified to increase infiltration. Diversion terraces may need to be installed to protect an eroded area from run-on water before reclamation is started. This is especially true where run-on water accumulates.

Transporting topsoil sediments from depositional areas within a field to eroded areas of the field has been suggested as a means of restoring the productivity of eroded soils. The economic feasibility would depend on whether the increase in productivity offsets the cost of the operation.

20-3.4 Cropping and Tillage Systems

For restoration efforts to be beneficial, present and future soil erosion must be controlled on the site. Jacks and Whyte (1939) stated that one of the outstanding features of the modern view of soil conservation is the emphasis on the superior value of biological erosion control in contrast to mechanical methods. They likened biological control by means of plant cover to treating a disease by dieting or to maintaining good health by temperate living. They compared mechanical control to surgery. The vegetative cover needed to control soil erosion can be provided by crop residue, by cover crops grown during the intercrop period specifically for erosion control, and by crop rotations that include forage, pasture, and small grain crops that provide sufficient and timely cover to control erosion. The effectiveness of live vegetative cover is controlled mainly by the cropping system, and the effectiveness of crop residue is controlled mostly by the tillage system.

20-3.4.1 Cropping System

Controlling further erosion is an extremely important part of the restoration process, and the cropping system is probably the most important factor in controlling erosion. Generally, the effectiveness of the cropping system in decreasing surface runoff and controlling soil erosion is determined by the amount of cover provided during critical periods of the year, as suggested by Table 20-6. The influences of the cropping system on

Table 20-6. Effect of cropping system on runoff and soil erosion
(Miller and Krusekopf, 1932).

Treatment	Runoff as percentage of rainfall	Soil loss
	%	Mg/ha
No crop, plowed 10 cm, cultivated regularly	31	94
Continuous corn	29	45
Continuous wheat	23	22
Corn-wheat-clover rotation	14	7
Bluegrass sod	12	0.7

soil organic matter, soil structure, surface water-storage capacity, and infiltration rate are also important. Table 20–6 shows that the cropping system has proportionally greater effect on soil erosion than on runoff. This is best illustrated by the ten-fold decrease in soil loss with only a 2% decrease in runoff with bluegrass sod compared to the corn-wheat-clover rotation. However, decreasing surface runoff and increasing infiltration is probably the major way in which the cropping system improves the productivity of soils where the water-supplying capacity has been diminished by erosion.

Plant cover decreases runoff and soil erosion by slowing the flow of water over the soil surface and protecting the soil from the erosive forces of water or wind. Straw mulch applied at rates of 1 and 2 Mg/ha reduced runoff velocities to about 50% and 33%, respectively, of velocities with no mulch (Mannering and Meyer, 1963). In the same study, infiltration increased by 2.5 and 7.5 cm, respectively. Less well known is the effect of the cropping system on soil porosity and organic matter content. Wischmeier and Mannering (1965) showed that residue reduced runoff even when turned under (Fig. 20–3). Runoff averaged 40% less where residues were turned under each year rather than removed at harvest. They concluded that soil loss decreases by about 12% for each 2.24 Mg/ha of cornstalks mixed into the soil. However, continuous cultivation, especially with conventional tillage, tends to decrease the porosity and organic matter content of soils. Continuous corn for 10 years, even with a small grain cover crop each year, resulted in a sizeable decrease in the organic matter content when compared to a 60-year-old bluegrass sod in adjacent plots (Blevins et al., 1983). Organic C levels in the first 5 cm of the soil were about 3.8%, 2.8%, and 1.4% for bluegrass sod, continuous no-tillage corn, and continuous conventional tillage corn, respectively.

Certain crop rotations can decrease, or in some cases reverse, the tendency for organic matter to decline with row cropping. For example, Long (1959) found that double-cropping increased the amount of roots added to the soil regardless of how much of the above-ground yield was removed. Although the differences were not statistically significant, double-cropping with corn and small grain silage tended to build up the organic matter content more rapidly than single-cropped corn silage in a 4-year study of no-tillage in Kentucky (Table 20–7). Single cropping with conventional tillage in that same study resulted in less soil organic matter than single cropping with no-tillage.

The importance of crop residues in increasing the organic matter content was also shown by Barber (1979). Where residues from continuous corn were removed from the soil for 5 years, the organic matter content was significantly lower ($P < 0.05$) than where the residues were returned—2.8% compared to 3.0%. Cropping systems that include legumes increase organic matter more rapidly and maintain it at a higher level than nonleguminous row crops. Larson et al. (1978) reported a linear relationship between the annual rates of crop residues and organic content over an 11-year period (Fig. 20–4). Approximately 5 Mg ha^{-1} yr^{-1} residue was required to maintain the organic C at its original level.

Table 20–7. Effect of cropping systems on soil organic matter at 0–15 cm depth in Pope and Huntington soils (Wells et al., 1983).

Cropping system	Soil organic matter	
	Pope soil	Huntington soil
	%	
Beginning of study	4.88	2.06
End of study:		
Double-crop (no-till)	5.59	2.20
Single-crop (no-till)	5.23	2.18
Single-crop (conv. till)	4.83	1.97
LSD (0.05)	N.S.	N.S.

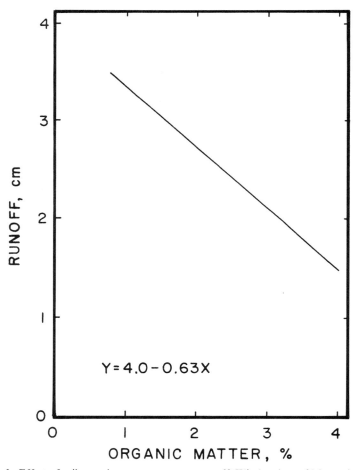

Fig. 20–3. Effect of soil organic matter content on runoff (Wischmeier and Mannering, 1965).

An experiment with soil organic matter has been conducted on a loamy sand in Michigan since 1963 (Lucas, 1982). Soil analyses in 1968, 1976, and 1980 suggested that the organic matter content was being maintained where corn grain was being produced but was decreasing slightly where corn silage was removed each year, except on plots where manure was applied. Manure increased soil organic matter substantially under both harvest methods. Overall, silage plots averaged about 0.6% less organic matter than grain plots.

Lal (1977) showed the relationship between mulch rate and soil loss on different slopes (Fig. 20-5). Three points are obvious from this graph: (i) the greater the mulch rate, the lower the soil loss rate; (ii) the steeper the slope, the greater the mulch rate required to control erosion; and (iii) the effect of no-tillage on erosion control was equivalent to approximately 5 Mg/ha mulch with conventional tillage.

20-3.4.2   Tillage System

The tillage system used can enhance or destroy the effects of plant cover in protecting the soil from erosion. Table 20-8 shows the effect of the tillage system on the amount of the soil surface covered by crop residue in a

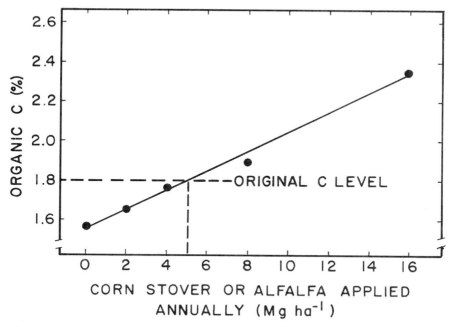

Fig. 20-4. Relationship between annual application rates of crop residues and organic C after 11 years (Larson et al., 1978).

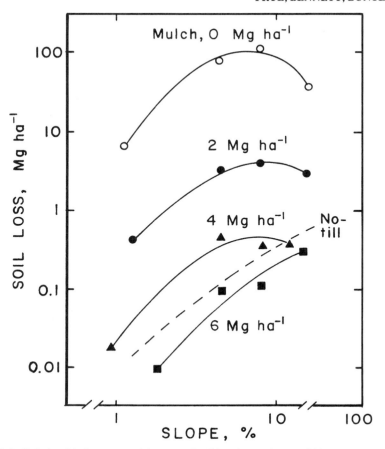

Fig. 20-5. Relationship between mulch rate and soil loss by erosion on different slopes (Lal, 1977).

continuous corn cropping system (Griffith et al., 1977). While tillage was once used mainly to prepare a seedbed and control weeds, the tillage system has become a very important soil conservation measure in recent years. Also, the tillage system may aid in restoring the productivity of erosion-damaged soils.

Conservation tillage systems which leave the soil protected by vegetative cover increase infiltration of water, reduce runoff, improve soil water-use efficiency, and increase the soil organic matter content. Crop residues conserve soil and water whether mixed into the soil or left on the surface, but they are far more effective when left on the surface (Mannering and Meyer, 1963; Wischmeir and Mannering, 1965).

Harrold and Edwards (1972) reported dramatic effects of surface cover on soil loss. A single July storm of 13 to 14 cm rainfall produced nearly equal amounts of runoff from watersheds with conventionally tilled corn and no-tillage corn, but soil loss from the plowed watershed was about 100 times greater with contoured rows and 700 times greater with straight rows.

Table 20–8. Effect of tillage system on amount of residue cover in continuous corn (Griffith et al., 1977).

Tillage system	Soil covered†
	%
Conventional, fall, moldboard plow	1
Till-plant	8
Disk twice	13
Chisel plow	19
Strip rotary	62
No-tillage	76

† Average of four measurements taken immediately after planting corn.

With no-tillage corn on a 21% slope, runoff was 6.4 cm (approximately 50% of the rainfall), and soil loss was about 0.07 Mg/ha. Where the soil had been plowed on a 6% slope, runoff was 5.8 cm and soil loss was about 7 Mg/ha. With conventional tillage and straight rows on a 7% slope, the storm was devastating, causing 11.2 cm of runoff and almost 50 Mg/ha of soil loss. Undoubtedly, some of the greater soil loss was due to the more erodible condition of the conventionally tilled soil resulting from the "fluffing-up" by plowing and seedbed preparation.

On the surface of the soil, crop residue absorbs the impact of raindrops, helping to prevent disintegration of soil aggregates and keeping pores open into the soil. It slows surface runoff and decreases wind velocity at the surface, decreasing the erosiveness of those agents. Finally, it filters out soil particles, preventing their transportation over more than short distances.

In addition to leaving more residue on the surface, conservation tillage usually increases the soil organic matter content, as shown by Blevins et al. (1983) with no-tillage and conventional tillage. The general relationship seems to be that, when tillage is reduced, organic matter content tends to increase, especially in the top few centimeters of soil. Probable reasons for this include (i) a less oxidative microbial environment with reduced tillage resulting from greater water content slowing organic matter decomposition (Doran, 1980), (ii) organic matter being added mostly at the soil surface and gradually becoming mixed into the soil, and (iii) less removal of the organic matter by erosion.

20–4 SUMMARY AND CONCLUSIONS

According to the 1977 National Resource Inventories, an estimated 98 million ha of soil in the USA has more than 11 Mg ha⁻¹ yr⁻¹ erosion, a rate considered excessive for most soils based on commonly used equations for estimating soil loss due to wind and water erosion (USDA, 1981). Many studies have shown that excessive erosion decreases the productivity of soils. The amount of decrease in soil productivity resulting from a given amount of erosion depends on the thickness and quality of the topsoil and on the nature of the subsoil. Loss of plant-available water-supplying capacity associated with erosion seems to be the most important damage to shallow soils or soils with heavy clay subsoils. In other soils, loss of fertility may be the major production-limiting effect of soil erosion.

Before attempting to restore productivity of eroded soils, the production limitations must be identified. Some soils can be restored simply by replenishing lost fertility, other soils require more drastic restorative measures, and some soils may never be fully restored.

For most severely damaged soils, efforts to restore productivity should be aimed at increasing the water-supplying capacity. This includes increasing the organic matter content, the infiltration rate, and the plant-available water-holding capacity through the use of crop residues, cover crops or green manure crops, or organic wastes such as animal manure and sewage sludge. More extreme efforts may include practices such as transporting soil from deposition areas to erosion-damaged areas.

Restoring crop productivity is often very expensive and requires long-term efforts. The most economically prudent alternative is to control erosion before the soil is damaged and its productivity diminished.

REFERENCES

Adams, W. E., H. D. Morris, and R. N. Dawson. 1970. Effects of cropping systems and nitrogen levels on corn (*Zea mays*) yields in the Southern Piedmont region. Agron. J. 62:655–659.

Barber, S. A. 1979. Corn residue management and soil organic matter. Agron. J. 71:625–627.

Barrows, H. L., and V. J. Kilmer. 1963. Plant nutrient losses from soils by water erosion. Adv. Agron. 15:303–316.

Batchelder, A. R., and J. N. Jones, Jr. 1972. Soil management factors and growth of *Zea mays* L. on topsoil and exposed subsoil. Agron. J. 64:648–652.

Baver, L. D., W. H. Gardner, and W. R. Gardner. 1972. Soil physics. 4th ed. John Wiley and Sons, New York.

Beale, O. W., G. B. Nutt, and T. C. Peele. 1955. The effects of mulch tillage on runoff, erosion, soil properties, and crop yields. Soil Sci. Soc. Am. Proc. 19:244–247.

Beasley, R. P. 1972. Erosion and sediment pollution control. Iowa State University Press, Ames.

––––. 1980. Erosion and sediment pollution control. 6th ed. The Iowa State University Press, Ames.

Bennett, O. L. 1978. Reclamation of lands disturbed by surface mining. Proc. Int. Soc. Soil Sci. 3:249–258.

––––, W. H. Arminger, and J. N. Jones, Jr. 1976. Revegetation and use of eastern surface mine spoils. p. 195–215. *In* Land application of waste materials. Soil Conserv. Soc. Am., Ankeny, IA.

Blevins, R. L., G. W. Thomas, M. S. Smith, W. W. Frye, and P. L. Cornelius. 1983. Changes in soil properties after 10 years continuous non-tilled and conventionally tilled corn. Soil and Tillage Res. 3:135–146.

Burwell, R. E., D. R. Timmons, and R. F. Holt. 1975. Nutrient transport in surface runoff as influenced by soil cover and seasonal periods. Soil Sci. Soc. Am. Proc. 39:523–528.

Carlson, C. W., D. I. Grunes, J. Alessi, and G. A. Reichman. 1961. Corn growth on Gardena surface and subsoil as affected by applications of fertilizer and manure. Soil Sci. Soc. Am. Proc. 25:44–47.

Doran, J. W. 1980. Soil microbial and biochemical changes associated with reduced tillage. Soil Sci. Soc. Am. J. 44:765–771.

Ebelhar, S. A. 1981. Nitrogen from winter-annual legume cover crops for no-tillage corn. M.S. thesis. Univ. of Kentucky, Lexington.

Englestad, O. P., and W. D. Shrader. 1961. The effect of surface soil thickness on corn yields: II. As determined by an experiment using normal surface soil and artificially exposed subsoil. Soil Sci. Soc. Am. Proc. 25:497–499.

----. ----, and L. C. Dumenil. 1961. The effect of surface soil thickness on corn yields: I. As determined by a series of field experiments in farmer-operated fields. Soil Sci. Soc. Am. Proc. 25:494–497.

Frye, W. W., S. A. Ebelhar, L. W. Murdock, and R. L. Blevins. 1982. Soil erosion effects on properties and productivity of two Kentucky soils. Soil Sci. Soc. Am. J. 46:1051–1055.

----, L. W. Murdock, and R. L. Blevins. 1983. Corn yield-fragipan depth relations on a Zanesville soil. Soil Sci. Soc. Am. J. 47:1043–1045.

Gardner, Robert. 1941. Why is subsoil unproductive? Colo. Agric. Exp. Stn. Bull. 464. Fort Collins.

Giffith, D. R., J. V. Mannering, and W. C. Moldenhauer. 1977. Conservation tillage in the eastern Corn Belt. J. Soil Water Conserv. 32:20–28.

Harrold, L. L., and W. M. Edwards. 1972. A severe rainstorm test of no-till corn. J. Soil Water Conserv. 27:30.

Jacks, G. V., and R. O. Whyte. 1939. Vanishing lands: A world survey of soil erosion. Doubleday, Doran and Co., New York.

Lal, R. 1977. Soil-conserving versus soil-degrading crops and soil management for erosion control. In D. J. Greenland and R. Lal (ed.) Soil conservation and management in the humid tropics. John Wiley and Sons, New York.

Langdale, G. W., J. E. Box, Jr., R. A. Leonard, A. P. Barnett, and W. G. Fleming. 1979. Corn yield reduction on eroded Southern Piedmont soils. J. Soil Water Conserv. 34:226–228.

Larson, W. E., R. F. Holt, and C. W. Carlson. 1978. Residues for soil conservation. In W. R. Oschwald (ed.) Crop residue management systems. Spec. Pub. 31. American Society of Agronomy, Madison, WI.

----, L. M. Walsh, B. A. Stewart, and D. H. Boelter (ed.). 1981. Soil and water resources: Research priorities for the nation. Soil Science Society of America, Madison, WI.

Long, O. H. 1959. Root studies on some farm crops in Tennessee. Bull. 301. Univ. of Tenn. Agric. Exp. Stn., Knoxville.

Lucus, R. E. 1982. Understanding soil organic matter changes. In Proc. Ninth Michigan seed, weed, and fertilizer school. 14–15 December 1982. Department of Crop and Soil Science. Michigan State University, East Lansing.

Mannering, J. V., and L. D. Meyer. 1963. The effects of various rates of surface mulch on infiltration and erosion. Soil Sci. Soc. Am. Proc. 27:84–86.

Mathias, E. L., O. L. Bennett, and P. E. Lundberg. 1979. Use of sewage sludge to establish tall fescue on strip mine spoils in West Virginia. p. 307–314. In W. E. Sopper and S. N. Kerr (ed.) Utilization of municipal sewage effluent and sludge on forest and disturbed land. Pennsylvania State University Press, State College.

Miller, M. F., and H. H. Krusekopf. 1932. The influence of systems of cropping and methods of culture on surface runoff and soil erosion. Res. Bull. 177. Mo. Agric. Exp. Stn., Columbia.

Murdock, L. W., W. W. Frye, and R. L. Blevins. 1980. Economic and production effects of soil erosion. p. 31–35. In Proc. Southeastern Soil Erosion Control and Water Quality Workshop, Nashville, Tenn. November 1980. Nat. Fert. Dev. Center, Muscle Shoals, AL.

National Soil Erosion-Soil Productivity Research Planning Committee, USDA. 1981. Soil erosion effects on soil productivity: A research perspective. J. Soil Water Conserv. 36:82–90.

Reuss, J. O., and R. E. Campbell. 1961. Restoring productivity to leveled land. Soil Sci. Soc. Am. Proc. 25:302–304.

Ripley, P. O., W. Kalbfleisch, S. J. Bourget, and D. J. Cooper. 1961. Soil erosion by water: Damage, prevention, control. Pub. 1083. Research Branch, Canada Dep. of Agric., Ottawa.

Robertson, D. W., and R. Gardner. 1946. Restoring fertility to land where leveling operations have removed all the topsoil and left raw subsoil exposed. p. 33–36. *In* Proc. Am. Soc. Sugar Beet Tech. Fort Collins, CO.

Salter, R. M., R. D. Lewis, and J. A. Slipher. 1941. Our heritage—The soil. Ext. Bull. 175. Ohio Agric. Exp. Stn., Columbus.

Sarles, R. L., and D. M. Emanuel. 1977. Hardwood bark mulch for revegetation and erosion control on drastically disturbed sites. J. Soil Water Conserv. 32:209–214.

Shrader, W. D., F. W. Shaller, J. T. Pesek, D. F. Slusher, and F. F. Riecken. 1960. Estimated crop yields on Iowa soils. Spec. Rep. 25. Agric. and Home Econ. Exp. Stn., Coop. Ext. Serv., Iowa State Univ., Ames.

Stoltenberg, N. L., and J. L. White. 1953. Selective loss of plant nutrients by erosion. Soil Sci. Soc. Am. Proc. 17:406–410.

Stout, W. L., H. A. Menser, O. L. Bennett, and W. M. Winant. 1982. Cover establishment on an acid mine soil using composted garbage mulch and fluidized bed combustion residue. Reclam. and Revegetation Res. 1:203–211.

Tan, K. H., J. H. Edwards, and O. L. Bennett. 1985. Effect of sewage sludge on mobilization of surface applied calcium. Hort. Sci. (In press).

Taylor, A. W. 1967. Phosphorus and water pollution. J. Soil Water Conserv. 22:228–231.

Thomas, A. W., R. L. Carter, and J. R. Carreker. 1968. Soil and water nutrient losses from Tifton loamy sand. Trans. Am. Soc. Agric. Engr. 11:677–679.

Timmons, D. R., R. E. Burwell, and R. F. Holt. 1968. Loss of crop nutrients through runoff. Minn. Sci. 24:16–18.

USDA. 1981. Soil, water, and related resources in the United States: Status, condition, and trends. 1980 RCA Appraisal, Part I. USDA. Washington, DC.

Volk, B. G., and R. H. Loeppert. 1982. Soil organic matter. p. 211–268. *In* V. J. Kilmer and A. A. Hanson (ed.) Handbook of soils and climate in agriculture. CRC Series in Agric. CRC Press, Boca Raton, FL.

Wells, K. L., L. W. Murdock, and W. W. Frye. 1983. Intensive cropping effects on physical and chemical conditions of two soils in Kentucky. Commun. in Soil Sci. and Plant Anal. 14:297–307.

Whitney, R. S., R. Gardner, and D. W. Robertson. 1950. The effectiveness of manure and commercial fertilizer in restoring the productivity of subsoils exposed by leveling. Agron. J. 42:239–245.

Wischmeier, W. H., and J. V. Mannering. 1965. Effect of organic matter content of the soil on infiltration. J. Soil Water Conserv. 20:150–152.

----, and D. D. Smith. 1978. Predicting rainfall erosion losses: A guide to conservation planning. Agric. Handb. 537. USDA, Washington, DC.

21 Conservation Tillage Systems and Soil Productivity

R. R. Allmaras
Agricultural Research Service
U.S. Department of Agriculture
St. Paul, Minnesota

P. W. Unger
Agricultural Research Service
U.S. Department of Agriculture
Bushland, Texas

D. W. Wilkins
Agricultural Research Service
U.S. Department of Agriculture
Pendleton, Oregon

Consistent with the Soil and Water Resources Conservation Act of 1977, current USDA policy is that conservation tillage (CT) is the "leading cost-effective practice" for soil erosion control (USDA, 1982). The goal is adoption of CT systems in nearly all 182 major land resource areas (MLRA) in the USA (USDA, 1981). Other recent actions, such as establishment of a Conservation Tillage Information Center by the National Association of Conservation Districts, reinforced the USDA decision. These actions and associated farmer enthusiasm recognize that much CT technology is available, but they also challenge public and private workers in research and extension to provide new technology.

A study by the Office of Technology Assessment (1982) indicated many perceived and real barriers to adoption, including: problems of pest control, especially weeds; unfavorable soil drainage and associated low soil temperature; poor information about the benefits of soil erosion control; and management difficulties. The mix of real and perceived barriers indicates much confusion about existing and undeveloped technology to make CT function in all areas of the USA. Perhaps this confusion should not be unexpected because CT will bring about a major technological change in agriculture.

Published in R. F. Follett and B. A. Stewart, ed. 1985. *Soil Erosion and Crop Productivity.*
© ASA-CSSA-SSSA, 677 South Segoe Road, Madison, WI 53711, USA.

Although barriers to adoption are cited frequently, CT is estimated as being accepted for 25% of U.S. cropland (Office of Technology Assessment, 1982), and estimates of the expected acceptance by the year 2010 range from 50 to 84% of U.S. cropland. This adoption rate will require much new tillage technology and a better understanding of the associated new field ecology. Consequently, this paper has the following objectives:

1. Review the benefits from CT systems, with emphasis on tillage goals and their consequent value to resource conservation.
2. Review existing schemes for guiding the use of CT technology as related to soils, climate, topography, and crop rotations. National soils and other data banks should be useful for expansion of these schemes.
3. Give a skeleton overview of crop yield trends for CT systems as compared to conventional tillage systems.
4. Suggest areas for research emphasis to assure that the national objective to achieve maximum adoption of CT systems is consistent with limiting natural and biological features.

21-1 ATTRIBUTES OF A CONSERVATION TILLAGE SYSTEM

A conservation production system is achieved when high current production and sustained long-term soil productivity are ensured. The latter can be that productivity consistent with the T-value definition given by Wischmeier and Smith (1978), in which T-value is "the maximum level of soil erosion that will permit a high-level of crop productivity to be sustained economically and indefinitely" (p. 2). The productivity index (PI) response to removal of soil (Pierce et al., 1983) and the projected yield decline in the Erosion Productivity Impact Calculator (EPIC) of Williams and Renard (1985) are newer and more quantitative methods for estimating soil-productivity responses to soil erosion.

If tillage is to be a subsystem of a conservation production system, it must be a conservation oriented subsystem. For brevity, we will use the term "conservation tillage system," which can consist of conservation tillage and conservation-neutral tillage components. Conservation tillage keeps crop residues on or near the surface, produces a rough soil microrelief, or both, to minimize soil erosion and improve soil-water relations. A conservation-neutral component (such as clean tillage during a period of low soil erosion hazard) can be used in a conservation tillage system because it may facilitate a subsequent CT. Operationally, CT is attained by general preference for a primary-tillage implement other than the moldboard plow and by using herbicides rather than shallow cultivation to control weeds. This definition of a CT is similar to that made independently by Mannering and Fenster (1983).

Some definitions of CT, such as that given by the Office of Technology Assessment (1982), differ from that given above because they exclude use of the moldboard plow, which always buries crop residues deeper than other implements. Our definition of CT does not exclude use of the moldboard plow when needed to produce surface roughness for soil erosion control under special conditions. Exclusive reliance upon chemical rather than chemical plus cultural means for weed control is also not suggested by our definition. Our definition, furthermore, concentrates on the use of such implements as chisel plows, sweeps, rod weeders, discs, mulch or skew treaders, and planters designed to move through residues and precisely place seed and fertilizers under conditions more adverse than on bare seedbeds.

A conservation tillage system is an organized integrated whole made up of diverse but interrelated and independent parts. Thus, tillage must be integrated with other parts (inputs) into a conservation tillage system. Moreover, some inputs can only be applied by integration with the conservation tillage system. The crop rotation (or sequence) is essential to the operation of a conservation tillage system because crop and pest interactions, fertilization, and timeliness of operations are just a few of the many factors to be considered during the planning needed for successful adoption of a new CT. As an example, planning begins with the distribution of crop residues as they emerge from harvesters. This distribution is critical if the subsequent operation is no-till planting. Another planning requirement is the control of herbicide residual, since such control can facilitate CT. With crop rotation as part of the conservation tillage system, moldboard plows and discs can be used when the danger of erosion is slight or nonexistent in some part of the rotation. Non-inversion or shank-type equipment can be used when necessary to meet conservation objectives in other parts of the rotation. Systematic and greater use of no-till methods is also facilitated in rotations.

New developments and accelerated adoption of CT systems have changed both conventional and conservation tillage. Evolutionary directions in conventional tillages in the absence of herbicides were toward soil fragmentation and complete burial of crop residues (Triplett and Van Doren, 1977), increased speed of field travel as tractor horsepower increased, and lower manpower per unit of production (Council for Agricultural Science and Technology, 1983). These evolutionary directions were inconsistent with soil conservation. However, with the recent increased emphasis on soil conservation, the new evolutionary direction toward reduced tillage (Collins, 1982) will diminish the current and dramatic differences between the two tillage systems. Associated with these trends are changes in shape and arrangement of the common tillage tools (Buckingham, 1976) and planters (Breece et al., 1981). This conservation evolution for all tillage tools is associated with substitution of herbicides for tillage to control weeds (Office of Technology Assessment, 1982). These evolutionary trends have also supplied us with many types and modifications of tillage tools.

This evolutionary trend in tillage systems makes it more difficult to compare the performance of tillage systems for crop production. Rather than comparing conventional vs. conservation tillage, one should compare stated or described tillage operations and alternatives. Perhaps even better are comparisons over time for a particular tillage alternative, such as no-till. Other conservation measures accomplishing some of the same objectives as CT are terraces, divided slopes, strip crops, and buffer strips. However, a

CT system is usually less costly and more flexible because tillage systems can evolve to accommodate other production inputs. Mitchell et al. (1980) found that only in very special cases were structures such as terraces cost effective if the long-term soil productivity was not considered.

21-2   RESOURCE CONSERVATION BENEFITS OF CONSERVATION TILLAGE SYSTEMS

Assessments regarding adoption of CT systems often presume incorrectly that the resource conservation benefits are clearly evident. An understanding of causes and effects is necessary to clarify the relations, first, by quantifying the most important ends or goals of tillage, then by developing theoretical relations between these goals and their effects on runoff, water erosion, wind erosion, and other resource-related damages or benefits. Finally, a correct assessment requires theoretical relations between these conservation effects and soil degradation.

Conservation tillage systems (previously called stubble mulching) were apparently first developed in the Great Plains to control major soil erosion by wind in the 1930s (McCalla and Army, 1961). Water erosion had earlier ravaged the humid or eastern portions of the USA, but control alternatives there were centered more on terraces, contouring, and strip cropping (USDA, 1938). The first national efforts to develop stubble mulching occurred in about 1943 to 1947 when researchers at six U.S. locations and one Canadian location began joint experiments, which provided observational information on water conservation and control of wind and water erosion. Much of this effort involved the development of tillage and planting machinery for handling surface residues (McCalla and Army, 1961).

Chepil and Woodruff (1959) provided early observational data on wind erosion and specified, for instance, that 2 t/ha of anchored straw was necessary in Kansas in 1954 to control wind erosion. Van Doren and Stauffer (1943) had reduced runoff and soil erosion in Illinois by using surface residues of soybean (*Glycine max* L.), corn (*Zea mays* L.), and wheat (*Triticum aestivum* L.). Later, Zingg and Whitfield (1957) compared the effectiveness of several tillage systems for controlling runoff and erosion in the West.

Early reports indicated no water conservation responses to tillage systems in the arid climate of the West and occasional net soil water storage under surface residues in the humid climate of the East (McCalla and Army, 1961).

21-2.1   Goals of a Tillage

In the 1960s and 1970s, two separate lines of investigation greatly clarified the benefits of CT systems. As combinations of tillage equipment and systems expanded, new methods were developed to characterize the tillage goals and, in turn, their relation to runoff, soil erosion, and water conservation. Meanwhile, research on soil erosion quantified climate, topography, soil, and cultural factors affecting the losses. This research further sought to use tillage for erosion control and to understand interactions of tillage in the fundamental mechanics of wind and water erosion.

A research philosophy for describing the goals of tillage and their subsequent effect on soil erosion was first presented by Larson (1964) when he discussed the use of soil parameters to evaluate tillage needs and operations. These soil parameters are the same as the process goals later discussed by Stickler (1982). Some of Larson's parameters for the zone between rows were microrelief, associated microdepressions, and plow-layer storage. For the row zone, he considered width and depth, geometric-mean diameter of secondary aggregates, and dry bulk density. Both Larson (1964) and Stickler (1982) recognized that this investigational philosophy would encourage greater flexibility in the choice of tillage equipment and operations and in the development of new machine systems and adjunct practices.

Field methods were developed to measure and characterize two parameters: the range of rnadom roughness and total porosity (Allmaras et al., 1966; Burwell et al., 1963; Currence and Lovely, 1970; Kuipers, 1957). These two parameters were sensitive to tillage implement, traffic, and soil type, roughly in decreasing order. They were later found to be sensitive to water content and dry bulk density of the soil to be tilled (Allmaras et al., 1967; Ojeniyi and Dexter, 1979). The range of these two parameters is affected by time of year for tillage and tillage implement (Table 21-1). Textures of the soil from which the data of Table 21-1 were derived ranged from loam to clay. For additional information on these parameters see Burwell and Larson (1969), Currence and Lovely (1970), Dexter (1977 and 1979), and Johnson et al. (1979).

Table 21-1. Random roughness and fractional total porosity of tilled layer as affected by tillage systems (adapted from Voorhees et al., 1981).

Tillage method		Random roughness measured after spring tillage		Total fractional porosity measured after spring tillage	
Fall	Spring	Range	Mean†	Range	Mean†
		——— cm ———			
Plow‡	--	1.96–7.26	3.24	0.54–0.96	0.72
Tandem disk‡	--	1.75–1.83	1.79	0.54–0.69	0.61
Chisel‡	--	1.35–1.68	1.52	0.68–0.69	0.68
Plow§	--	1.30–3.63	2.22	0.48–0.68	0.60
Plow	Disk and harrow	0.89–1.52	1.27	0.58–0.66	0.61
None	Plow	1.17–3.61	2.28	0.55–0.84	0.66
None	Plow, disk, and harrow	0.69–2.16	1.15	0.53–0.82	0.67
None	Plow and wheel tracked	0.93–1.15	1.06	0.52–0.56	0.54
None	Rotary till	1.55–1.75	1.65	0.64–0.67	0.66
None	None	0.46–0.86	0.59	0.52–0.53	0.52

† Average coefficient of variation for all treatments was 6 and 4% for random roughness and total fractional porosity, respectively.
‡ Measured after fall tillage, before overwintering.
§ Measured in spring before secondary tillage operations.

The smallest random roughness produced by fall moldboard plowing was greater than the maximum produced by discing or chiseling (Table 21-1). Moldboard plowing produced greater mean random roughness than any other spring tillage. Unfortunately, measurements have not been made extensively on soils other than Mollisols of the Corn Belt, which are known for their structural stability.

Voorhees et al. (1981) showed that dry-aggregate-diameter distribution and, hence, interaggregate space within a soil, is usually a response to tillage but that the porosity within aggregates is sensitive to long-term management of organic matter (Larson and Allmaras, 1971) or to compaction from wheel traffic (Voorhees et al., 1978). Thus, a significant body of technical information is available for characterizing the soil structural properties that are produced by tillage. For example, the particular soil structure desired can be determined based on its interaction with soil erosion or infiltration processes.

Placement of crop residue is another important tillage goal (Larson, 1964). Early descriptions of residue placement (Van Doren and Stauffer, 1943; McCalla and Army, 1961) focused on mass per unit area, which is still the most frequent characterization of residue placement (Gregory, 1982; Unger and McCalla, 1980). Other methods of determining surface cover include a point quadrant system (Mannering and Meyer, 1963; Van Doren and Triplett, 1973), line intercept, and photographic density (Sloneker and Moldenhauer, 1977).

Gregory (1982) developed theoretical relations between mass per unit area and fractional cover. This relation expresses the fractional surface cover:

$$F_c = 1 - e^{(-A_m \cdot M)} \qquad [1]$$

where M is the mass of mulch in kg/ha and A_m is a constant determined empirically. The values of A_m (multiplied by 10^4) for six crops were: soybean, 4 to 7; wheat, 5; corn, 2.5 to 4; sunflower (*Helianthus annuus* L.), 2; soybean stems, 2; and cotton (*Gossypium hirsutum* L.) stems, 1.

From the viewpoints of surface residue effectiveness, impediment to the passage of machinery, and plant damage, a description of standing vs. flat residue components should be emphasized. Conservation tillage equipment rarely operates in a field without significant variation in surface residue concentration and stubble height produced by harvest operations. Residue concentration within harvested wheat fields, for instance, could vary from 50 to 200% of the average concentration, and the on-the-ground component could vary from 10 to 300% of the average concentration. Cochran et al. (1977) demonstrated the consequences of such variation on stand and growth of no-till seeded wheat. Perhaps the average residue height, as considered in Woodruff et al.'s (1965) computation of a residue density in kg/m^3, would help to characterize surface residue configurations as they affect soil erosion or the ability to operate tillage and planting machinery.

Colvin et al. (1981) and Voorhees et al. (1981) summarized field observations of residue remaining on the soil surface after various tillage operations. Table 21-2 emphasizes the amount left on the surface after spring tillage, an amount that would be most critical for the wind or water erosive period in spring. (In a fall-seeded crop after fallow, the final and critical amount of residue on the surface before the winter erosive season would be the amount after fall seeding.) These measurements suggest about 30% decomposition of buried residue from late fall to early spring, which agrees with measurements made by Douglas et al. (1980). The measurements also suggest that soybean residue is more easily buried and decomposed than cereal residues. Corn and sorghum (*Sorghum bicolor* L.) residues react similarly, with less percentage burial than soybean residues and more than cereal residues. Colvin et al. (1981) suggested various tillage combinations, in both row and close-seeded crops, to minimize residue removal

Table 21-2. Plant residues remaining on soil surface after various combinations of fall and spring tillage operations (adapted from Voorhees et al., 1981).

Crop	Tillage implement used		Residues remaining†
	Fall	Spring	
			%
Oat and	None	None	85
wheat	None	Sweep	88
	None	Rodweeder	90
	None	Chisel plow	76
	None	Tandem disk	53
	None	One-way disk, 7.5 cm deep	60
	None	Moldboard plow and disk	12
	None	Moldboard plow	10
	Chisel plow	None	66
	Sweep	None	51
	Disk	Tandem disk	39
	Sweep	Tandem disk	32
	Chisel plow	Tandem disk	32
	Moldboard plow	Tandem disk	12
Sorghum	None	Sweep twice	52
	None	One-way disk and sweep	26
	None	Tandem disk and sweep	24
	None	Tandem disk twice	13
	Sweep	Sweep twice	43
	One-way disk	Sweep twice	36
	Tandem disk	Sweep twice	44
	Tandem disk	Tandem disk and sweep	9
Corn	None	No-till	84
	None	Tandem disk	45
	None	Moldboard plow and tandem disk	10
	Tandem disk	Tandem disk	41
	Chisel plow	Tandem disk	32
	Moldboard plow	Tandem disk	8
Soybean	None	No-till	63
	None	Moldboard plow and tandem disk	5
	None	Chisel plow	25
	Chisel plow	Field cultivator	25
	Chisel plow	None	20
	Moldboard plow	Disk	4

† Percent of original plant residues left on soil surface after spring tillage operation.

from the surface during an anticipated period of soil erosion. Results for both cropping systems showed a distinct trend for little loss of surface residue with chisels and sweeps and a nearly complete loss with the moldboard.

Little attention has been given to position and localization of buried residue. Although the decomposition rate of buried residues has received considerable emphasis (Jenkinson, 1971), the interest was not related to tillage. Data comparing decomposition rate for surface-placed vs. buried residues are also limited. Unger and Parker (1968) found similar decomposition rates for wheat straw mixed with the top 3 cm of soil or buried in a layer 3 cm deep. Greb et al. (1974) recovered from 1300 to 7500 kg/ha of buried wheat straw from the top 7.6 cm of soil under stubble-mulch systems in the semiarid Great Plains. Moldboard plowing was not used in these studies. Their recoveries, by a washing technique, ranged from 10 to 60% of precultivation residue amounts and were sensitive to duration of the wheat-fallow rotation and to type and time of primary tillage. More recently, Allmaras et al. (1982) used a combination of dry sieving and carbon dilution techniques to estimate the position of undecomposed cereal residues. Two tests with fall moldboard plowing (after wheat harvest), followed by preplant secondary spring cultivation, indicated only negligible wheat residues at depths less than 10 cm.

21–2.2   Water Erosion

Much data on water erosion as related to climate, topography, and cultural practices in the 1950 to 1970 era were incorporated in the universal soil loss equation (USLE) (Wischmeier and Smith, 1978). Tillage and residue placement practices were included in the C, or cover management, factor of the USLE. These practices were merely evaluated empirically as a soil-loss ratio by comparing soil losses with these practices to soil losses from a clean-tilled or fallow condition. The practices were usually described qualitatively, or by name of operation, rather than quantitatively, as related to the "goal of a tillage operation". The C value was computed from these soil-loss ratios and from the rainfall-energy intensity-duration (EI) accumulated for different periods of the crop rotation cycle.

Beginning in 1975 (Wischmeier, 1975), the soil-loss ratio was no longer merely measured empirically but was considered to involve three components: crop canopy to dissipate raindrop energy, surface mulches to dissipate raindrop energy and obstruct runoff flow, and a component consisting of "tillage and residue effects of the land use." Wischmeier (1975) included in this third component the residual effects of land use on soil structure, organic matter content, and soil density; the effects of tillage or its absence on surface roughness and porosity; and the effects of roots. He also stated that these effects are often substantial. Simultaneous studies with tillage systems have defined the last two components of the soil-loss ratio directly and in terms of tillage goals.

After field techniques became available to measure surface micro-relief (random roughness, ridging roughness, or depressional storage), soil porosity, and the fraction of surface covered by crop residue, numerous studies with rainfall simulations quantified their effects on energy to initiate runoff and soil erosion (Burwell and Larson, 1969; Cogo et al., 1984; Johnson et al., 1979; Meyer et al., 1970). For instance, Burwell and Larson (1969) found that the random roughness produced by moldboard plowing alone would withstand about three times as much rainfall energy (EI) before runoff began as that produced by moldboard plowing, discing, and harrowing. Later studies indicated that random roughness had even greater effects on soil erosion. Meyer et al. (1970) found that on 15% slopes, surface-residue covers of 34 to 49% reduced soil losses to less than 30% of that without mulch cover. Cogo et al. (1985) and Foster et al. (1982) demonstrated that residue attachment was critical for controlling soil erosion with residues on steeper slopes.

With further theoretical refinement of soil detachment and transport responses to the forces of raindrop impact and overland flow, as well as experimental refinements to simulate soil erosion on long slopes, soil erosion responses to random or ridge roughness, surface residue cover, and anchored residue (shallow incorporation) were quantified to more fully characterize the effects of tillage and crop residue management (Cogo et al., 1985 and 1984; Foster et al., 1982; Hussein and Laflen, 1982; Laflen and Colvin, 1981). These researchers jointly used field relationships to modify the C value for conservation tillage based on components describing the effects of random roughness, percentage of surface covered by residue, and anchor of residues in soil.

Laflen et al. (1985) have simply integrated these concepts into a compound C factor composed of four subfactors:

$$C = PLU \cdot CC \cdot RC \cdot SR \qquad [2]$$

where PLU is a prior land use subfactor, CC is a crop canopy subfactor (which here will be unity merely to focus on conditions at the soil surface), SR is a surface roughness subfactor, and RC is a residue cover subfactor. The subfactor PLU is simplified here to emphasize the impacts of no-till and incorporated residue:

$$PLU = CON \cdot e^{-0.012 \, RSDU}$$

where CON is 0.65 for no-till and 1.0 otherwise and RSDU is the average incorporated residue in kg/ha per mm of soil depth between the 10 and 100 mm deep. Here is where an improved measure of depth-of-residue incorporation by various implements is needed for different types of residue.

The surface roughness subfactor was estimated as:

$$SR = e^{-A} \qquad [3]$$

where $A = 0.026 \, [(10RB - 6)(1 - e^{-.35RS'})] \, e^{-0.18EC}$. In these equations, RB is the random roughness in cm, RS' is $10^{-3} \times$ tilled-zone residue mass (kg/ha), and EC is erosion (kg/ha) accumulated since the last tillage when random roughness was estimated. The tilled-zone residue mass includes the residue on the surface and that buried within the upper 100 mm. In Eq. [3], SR is expected to have a minimum value of about 0.3 for a rough surface and may increase to 1 for very smooth surfaces after excessive tillage, sufficient accumulated rainfall, or soil erosion. An interaction in Eq. [3] causes SR to decrease as the mass of tilled-zone residue increases.

The RC subfactor is derived as follows:

$$RC = e^{-B} \qquad [4]$$

where $B = 3.5\ M\ (6/RG)^{0.08}$. In Eq. [4], M is the fractional surface-residue cover, and RG is estimated as:

$$RG = 6 + [(10RB - 6)(1 - e^{-.35RS'})]\ e^{-0.18EC} \qquad [5]$$

The interaction term $(6/RG)^{0.08}$ in Eq. [4] accounts for the observation that residue is less effective for reducing runoff velocity and sediment transport when the soil surface is rough. When the surface is smooth, this term approaches 1.

Equations [3] and [5] both have a term, $e^{-0.18EC}$, which accounts for the reduction in random roughness (RB) effectiveness as erosion progresses. This term should be modified to express RB decline as a function of kinetic energy and soil resistance to the kinetic energy, as follows:

$$RB' = RB \cdot e^{-\alpha\beta(KE)} \qquad [6]$$

where KE is total kinetic energy of the rainfall in a storm or storm period. The term α should account for the protection afforded by surface residue, and the term β should account for inherent soil resistance. Kinetic energy of rainfall can be estimated by duration-intensity characteristics (Wischmeier and Smith, 1978). Justification for Eq. [6] is provided by the measurements of Van Doren and Allmaras (1978), Dexter (1979), and Johnson et al. (1979). Considerable field research is required to measure and model the soil factors, such as mechanical composition, organic matter, clod density, and wet aggregate stability, that affect the random roughness resistance to rainfall. In some MLRA's, freezing and thawing also will be a large part of the energy input (Voorhees et al., 1981; Zuzel et al., 1982).

Computations of the three subfactors of the C factor in the USLE are shown in Table 21-3 to illustrate the effects of tillage. The computations, based on Eqs. [1] through [5], assumed that 9000 kg/ha of wheat and corn residue was produced the previous year and that 2700 kg/ha of soybean residue was produced. The relative distribution of crop residue in each of the three soil layers is consistent with surface-residue reductions in Table 21-2 and the above discussion regarding residue burial with the three types of tillage. Random roughness estimates are consistent with values in Table 21-1. Even though there is no crop residue in the 1- to 10-cm layer for the no-till situation, the CON = 0.65 (used to compute PLU) accounts for undisturbed dead roots in this layer.

For wheat and corn, two distinct groupings of C values characterize the effectiveness of residue on the surface to control erosion and the reduced value of bare surfaces to control erosion. After soybeans, surfaces with the

Table 21-3. Components of the C factor as related to typical variations in full-width tillage and managements of three common types of crop residue.

Previous crop	Tillage†	Crop-residue mass‡			RB§	RSDU¶	M#	C factor and its subfactors††			
		Surface	1 to 10 cm	Below 10 cm				SR	RC	PLU	C factor
		kg/ha			cm						
Corn	none	9000	0	0	0.6	0	0.96	1.00	0.04	0.65	0.03
	chisel	6300	2700	0	1.5	30	0.89	0.79	0.06	0.70	0.04
	p-d-h	720	720	7560	1.2	8	0.22	0.94	0.47	0.91	0.40
	plow	720	720	7560	3.2	8	0.22	0.76	0.49	0.91	0.34
	disc	3600	5400	0	1.4	60	0.72	0.82	0.09	0.49	0.04
Wheat	none	9000	0	0	0.6	0	0.99	1.00	0.03	0.65	0.02
	chisel	6300	2700	0	1.5	30	0.96	0.79	0.04	0.70	0.02
	p-d-h	720	720	7560	1.2	8	0.30	0.94	0.36	0.91	0.31
	plow	720	720	7560	3.2	8	0.30	0.76	0.38	0.91	0.26
	disc	3600	5400	0	1.4	60	0.83	0.82	0.07	0.49	0.03
Soybean	none	2700	0	0	0.6	0	0.42	1.00	0.23	0.65	0.15
	chisel	1890	810	0	1.3	9	0.32	0.89	0.34	0.90	0.27
	p-d-h	150	180	1370	1.1	2	0.03	0.99	0.90	0.98	0.87
	plow	150	180	1370	3.0	2	0.03	0.93	0.90	0.98	0.82
	disc	1080	1620	0	1.2	18	0.19	0.91	0.53	0.81	0.39

† These are the full-width tillages performed before the planting operation; chisel is the use of a chisel plow; p-d-h is plow, disc, harrow; plow is moldboard plowing.

‡ 10³•RS is the sum of residue masses in the first two columns, which is the tilled zone residue mass.

§ Estimated from random roughness values in Table 21-1. Random roughness is estimated to be 10% less after soybeans than for similar tillage operations after corn and wheat.

¶ RSDU is the residue incorporated in the upper 1- to 10-cm depth expressed as kg/ha•mm.

M is the fractional cover of residue on the surface.

†† The SR, RC and PLU, respectively, describe the effects of random roughness, surface residue cover, and residue in the top 10 cm of soil. C is the product of these three components.

ıost residue cover have an erosion susceptibility in the range of those with
ıne least residue cover after corn and wheat. The computations apparently
do not accord sufficient erosion control to random roughness, as evidenced
by the small difference in C-factor estimates for the moldboard plow alone
vs. the moldboard, disc, harrow method of tillage.

After some soil erosion has occurred and the land is not tilled anew, the
values of SR and RC in Table 21-3 will be reduced in accordance with the
term $(-0.18EC)$ and associated interactions in Eqs. [3] and [5]. This
response to climatic factors can be improved by using relations involving
storm energy and soil resistance to energy forces of raindrop impact and
overland flow. Such a relation involving soil resistance can estimate more
accurately the SR subfactor in the C factor.

21-2.3 Wind Erosion

The wind erosion equation uses three factors related to the goals of
tillage—soil-erodibility index, soil-ridge roughness, and vegetative cover
(Woodruff and Siddoway, 1965). The soil-erodibility index is described as
the percentage of dry soil aggregates greater than 0.84 mm, soil-ridge
roughness as the height of soil ridges with a height/spacing ratio of 1:4 at
right angles to the wind, and vegetative cover as a mass per unit area located
above or on the soil surface. Very early in the development of the wind erosion equation, the last two factors were also depicted pictorially (Chepil and
Woodruff, 1963). In contrast to development patterns in the water erosion
equation, tillage-related parameters, rather than the tillage operation, were
early recognized as factors in wind erosion control. Soil management to
control wind erosion is directed at reducing windspeed to nonerosive
velocity at the air-soil interface or at reducing soil erodibility at this interface (Skidmore and Siddoway, 1978).

The effects of tillage tools on the soil-erodibility index may range from
10 to 75% (Siddoway, 1963). When residues were removed, moldboard
plowing produced a higher percentage of nonerodible aggregates than either
sweep or one-way disc tillage, but retention of cereal grain residues on the
surface by sweep or disc tillage as compared to burial by moldboard
plowing ultimately provided a higher percentage of nonerodible aggregates.
For the whole crop rotation, including all seasons, tillages that retained
surface residues gave the lowest soil-erodibility index. Recent refinements
of the wind erosion equation (Bondy et al., 1980), in which wind energy is
partitioned by seasons of the crop rotation, permit one to take advantage of
the low soil-erodibility index for limited periods and special conditions after
moldboard plowing. Black (1973) and Smika and Greb (1975) showed that
nonerodible aggregates increased 5 to 8% for each 1000 kg of wheat straw
per ha maintained near the surface by a tillage system excluding the moldboard plow. This effect of surface residue for producing a nonerodible soil
condition compares to a maximum difference of 13%, depending on tillage
tool, when crop residue was either burned or removed. (Differences in percentage of nonerodible aggregates produced by different tillage tools follow
differences in random roughness. Thus, when sieving equipment is unavail-

able, an operator might use pictorial random-roughness information to infer the fraction of nonerodible dry aggregates.)

Percentages of nonerodible aggregates are usually low in coarse-textured soils low in organic matter and in very-fine-textured clay soils slaked by freeze-thaw cycles. Medium-textured soils usually have the highest proportion of nonerodible aggregates. A simple method for estimating a nonerodible assembly of dry aggregates is when more than 50% of the soil surface is covered by aggregates greater than 1.0 mm in diameter.

Aerodynamic research conducted on ridge roughness indicates a ridge spacing of 5 to 10 cm is more effective when the height/spacing ratio is 0.25 and drag velocities are 100 cm/s. This ridge-roughness factor does not include the cloddiness factor, as expressed by the percentage of aggregates greater than 0.84 mm, but cloddiness on all parts of the ridges, and especially at the ridge peak, is necessary to retain ridge configuration and effect.

The vegetative factor for wind erosion control was originally specified as a mass of small-grain residue per unit area, lying on the ground with definite orientations related to wind direction (Lyles and Allison, 1981). Stalk orientation was not random. Yet the effectiveness of vegetation in controlling wind erosion depends on quantity and kind of residue as well as orientation with respect to the erodible soil on the surface. In CT systems, a flat orientation may be neither necessary nor feasible. The relative values of various crop residues and orientations given in Table 21–4 show that standing residue is about 75% more effective than a random and flat orientation for controlling wind erosion. Wheat straw in a random orientation on the surface is also more effective than the reference orientation described in the footnote of Table 21–4. Crop residues in flat-random orientation differed

Table 21–4. Effectiveness of crop-residue geometry for wind erosion control relative to the reference orientation (adapted from Lyles and Allison, 1981).†

Crop residue	Orientation‡	Row spacing	Height or length§	Equivalent residue mass¶
			cm	kg/ha
Winter wheat	S	25	25	280
	F-R	--	25	540
Sunflower	S	76	43	3060
	F-R	--	43	4210
Cotton	S	76	34	1790
	F-R	--	25	3330
Rape	S	25	25	700
	F-R	--	25	1740
Soybeans	F-R#	--	25	1230
Corn	S	76	16	1610

† Reference orientation of residue for the V (vegetation) factor is 25 cm long, dry small grain stalks lying flat on the soil surface in rows perpendicular to the wind direction. The planted row is parallel to the wind direction and rows are spaced 25 cm apart.
‡ S is standing, F-R is flat-random, in which case the residue stalks are overlapping with no row direction or cross-wind orientation.
§ Height in standing orientation or length of stalks in the flat-random orientation.
¶ Weight of residue required to give same wind erosion control as 1000 kg/ha of small grain stubble in reference orientation.
# 90% of stalks flat-random, 10% standing.

significantly, and most of the differences could be explained by stalk diameter and weight per unit volume. As either stalk characteristic increased, effectiveness for controlling wind erosion decreased (Lyles and Allison, 1981). The relative effectiveness of crop residue varied somewhat, depending on the mass of reference residue to which a comparison is made (Table 21-4, last column), so Lyles and Allison (1981) should be used for detailed evaluations.

Lyles and Allison (1976) experimentally determined factors of a standing stubble that afforded protection against wind erosion. This protection was defined in terms of a critical friction velocity ratio at the soil surface (CFVR):

$$(u^*/u^*_t)_s = 1.638 + 17.044 \, N \, A_s/A_t - 0.117 \, L_y/L_x + [(1.0236)^c - 1] \quad [7]$$

where
  u^* is the total friction velocity when a surface stabilizes at a given free stream velocity,
  u^*_t is the threshold friction velocity to initiate movement of erodible soil particles,
  N/A_t is the number of stalks per unit of ground area (cm^2),
  A_s is the area (cm^2) of a single stalk that faces the wind,
  L_y is the distance (cm) between stalks (center to center measured normal to the wind),
  L_x is the distance between stalks measured in the direction of the wind, and
  c is the percentage of soil aggregates greater than 1.0 mm.

The value of CFVR increases as standing stubble affords more protection. For a given standing stubble configuration, erosion starts when CFVR is exceeded. Some computed CFVR's to show the effects of standing stubble characteristics are given in Table 21-5. Where wheat stubble is short, a doubling of population increases CFVR only 0.59 compared to a 3.57 increase for the same change in tall stubble. At practical cutting heights, wheat stubble has a CFVR about 100% greater than corn or sorghum stubble. The orientation effect for corn stubble shows nearly the same CFVR for normal row orientation and uniform spacing but at least a 20% decrease for rows parallel to wind direction. When wheat stubble consists of 247 stalks/m^2 each 5 cm high, a well-managed surface of nonerodible aggregates (c = 50 in Eq. [7]) will increase CFVR from 2.12 (Table 21-5) to 2.33. The 5-cm-high wheat stubble configuration in Table 21-5 gives some indication of wind erosion protection when straw is partially incorporated so that straws extend 5 cm above the soil surface and the number of straws is not decreased significantly during incorporation.

21-2.4 Infiltration, Evaporation, and Soil Water Storage

A primary objective of CT is better soil water conditions. Infiltration, evaporation, and responses of soil water storage to tillage are not as well

Table 21-5. Effects of crop, stalk population, stalk height, and row direction of a standing stubble on the wind velocity required to initiate soil erosion (adapted from Lyles and Allison, 1976).

Crop	Stalk height	Population	CFVR for indicated orientation[†]		
			Normal	Uniform	Parallel
	cm	stalks/m²			
Wheat	5	247	--	2.12	--
	5	494	--	2.71	--
	30	247	--	5.09	--
	30	494	--	8.66	--
Corn	30	6	2.34	2.43	1.71[‡]
	61	6	--	3.15	--
Sorghum	23	11	--	2.26	--
	46	11	--	3.01	--

[†] Critical friction velocity ratio (CFVR) = $(u_*/u_{*t})_s$. The larger the CFVR, the greater the protection afforded against wind erosion. CFVR was computed assuming that all soil particles were erodible. If all soil particles are erodible, $c = 0$ in Eq. [7]. Computations assumed stalk diameters of 2.78, 17.7 and 25.4 mm for wheat, sorghum, and corn, respectively.

[‡] Corn in rows 102 cm apart.

understood as soil erosion responses. Greater water storage from surface residue was already observed before 1960 (McCalla and Army, 1961); moreover, net storage was more consistent and pronounced in humid than in arid or semiarid climates. Soil water storage has been field measured repeatedly, but models for water storage have met with limited success. Yet, several tillage goals have been identified as critical for a more moist environment and subsequent soil water storage. Soil water storage is the result of four processes: suppressed overland flow, infiltration, redistribution of water in soil, and evaporation. Soil water storage as related to tillage has been reviewed many times (Haas et al., 1974; Unger and McCalla, 1980; Unger and Phillips, 1973). We emphasize the relationship between tillage and the four processes involved in soil water storage.

Surface residues and rough surface microrelief are well recognized for suppressing overland flow and increasing infiltration. Moore and Larson (1979) demonstrated that surface microrelief, approximated by random roughness, provides an estimate of surface storage of water (when rainfall is greater than infiltration), and that runoff begins before the full potential of surface storage is met. After full surface storage is achieved, runoff and infiltration are the two active hydrological components, but full surface-storage potential declines as rainfall reduces surface roughness (Moore and Larson, 1979). (This surface storage does not include macrovoid storage in the tilled layer.) During this runoff period, infiltration may be decreasing due to development of a homogeneous surface seal on a smooth soil surface (Edwards and Larson, 1969) and a nonhomogeneous surface seal produced by shifting loci of low and high permeability in a randomly rough surface (Falayi and Bouma, 1975). In both cases, surface seal characteristics are highly transient. Overland flow rate is reduced as a consequence of surface storage on a randomly rough surface (Foster, 1982).

Burwell and Larson (1969) expressed infiltration responses to tillage-induced random roughness by observing the time to start of runoff from a set rate of simulated rainfall. Differences in infiltration for randomly rough surfaces, such as produced with moldboard plowing, and smoother surfaces, such as produced with plow-disc-harrow or no tillage, were as great as 10 cm of water before runoff occurred. Johnson et al. (1979) showed that surface roughness had a greater effect on soil erosion than on runoff, a finding confirmed by subsequent studies by Cogo et al. (1984 and 1985).

Many studies, typified by those of Mannering and Meyer (1963) and Meyer et al. (1970), indicate that surface residues significantly reduce the velocity of overland flow, so that infiltration is increased even if the infiltration rate is low. Meyer et al. (1970) showed that a wheat straw mulch of 0.56 t/ha (34% surface cover) reduced runoff velocity to less than half that on a bare surface on a moderately permeable Fox loam (Typic Hapludalf) on a 15% slope. A greater effect of surface residue mulch on soil erosion than on runoff (Meyer et al., 1970) is consistent with the hydrologic response to random roughness. Surface residues control runoff and erosion much longer than does random roughness because they are not destroyed by kinetic energy of the precipitation.

In some cases surface residues apparently do not reduce overland flow velocity and, thus, do not provide for high infiltration, especially if the soil is susceptible to crusting and seal development or has become compacted. Van Doren and Triplett (1973) and Whitaker et al. (1973) indicated that cultivation to break the crust obtained infiltration benefits in addition to those of the surface residue mulch. Soils high in silt or poorly drained exhibited this surface seal. Some indication of this same problem was given by Foster et al. (1982) who evaluated soil erosion control with unanchored surface residues. Lindstrom et al. (1981) showed that areas with traffic compaction and long-term no-till had much lower infiltration than tilled areas, even though surface residue was greater on the no-till areas. Gantzer and Blake (1978) found greater biochannel formation, but also higher bulk density, in the top 15 cm of these no-till plots on a Le Sueur clay loam (fine-loamy, mixed, mesic Aquic Argiudoll). Obviously, some measure of soil structural response to soil fauna activity and soil compaction is needed to interpret potential infiltration benefits from surface residues.

Numerous attempts have been made to simulate infiltration, but very few have considered tillage-induced soil conditions and/or the presence of residues on the surface. None of the simulations can account for all the tillage-induced properties and their dynamics during a rainstorm. Exclusive of residue disposition, the three most important tillage-induced properties related to infiltration are random roughness, interaggregate or macroporosity, and permeability of the pressure pan below the tilled layer. As water infiltration proceeds, a surface seal forms. The continuity and effect of this seal depend on surface roughness and destruction of structure. Meanwhile, water moves through the macrovoids, but these have reduced capacity to conduct water as they interact with the microvoid volume and become plugged. Tillage changes the macrovoid-volume component of tilled-layer porosity (Allmaras et al., 1977). Ultimately, infiltration progresses to a stage at which the pressure pan affects the flow system and, consequently, limits water conservation. If the pressure pan is near the surface (Lindstrom et al., 1981), the surface seal may form immediately and seal off the macrovoid space.

Linden (1979) viewed this complex system as a microporous flow subsystem, in which the tilled layer was sandwiched between a surface seal and the subsoil and was linked with a macroporous-flow subsystem which also conducted water and interacted with the microporous subsystem. Each system had a stability constant or function, that of the microporous subsystem being related to change of surface seal and that of the macroporous system being related to plugging. Van Doren and Allmaras (1978) suggested that surface seal conductivity should decline exponentially as the rainfall's kinetic energy accumulated and should have a "soil stability" constant related to soil properties such as texture (Moore, 1981) and to influences of organic matter on soil structure, (for example, the residues maintained near the surface by CT). A surface residue would influence infiltration primarily by reducing the fraction of the soil surface subject to raindrop impact. Linden (1979) suggested that macroporous conductivity should decline exponentially. The "macroporous conductivity stability" factor for a freshly tilled soil could be small to reflect a quick decline in the function of this system.

For a no-till soil with many worm and root channels, it could be larger to reflect continuing function during a greater accumulation of rainfall energy. The latter situation was modeled by Edwards et al. (1979) for biochannels produced in a soil that had been no-till planted for as many as 5 years.

During and after precipitation, infiltrated water redistributes downward within a soil profile, rapidly at first and then more slowly as redistribution continues. Simultaneous with water redistribution within a soil profile, evaporation proceeds, at first limited only by the energy available to evaporate water (Stage 1). Then, during Stage 2 evaporation, water movement to the evaporating surface from within the soil is the limiting factor. Gardner et al. (1970) showed that accelerated water redistribution rates may reduce evaporation significantly but that water redistribution is not influenced significantly by evaporation unless the quantity of infiltrated water is small. Again, management of soil compaction in a CT system to avoid impeded internal drainage can affect water storage. The time over which various evaporation stages operate depends on both time of year and evaporation rate within Stage 1 (Idso et al., 1974).

Evaporation from the soil surface should be considered generally as an unsaturated, nonsteady-state water movement (Linden, 1982). During Stage 1, surface residues can have three significant effects (Van Doren and Allmaras, 1978): increase resistance to water diffusion through a residue mat; increase reflection, which reduces radiation available for energy at the evaporating surface; and, when residue is standing, separate energy-exchange and evaporating surfaces. These effects also influence Stage 2 evaporation, but much less so. Partial surface coverage with crop residues is effective during Stage 1 evaporation, but more complete covers are needed in later stages. More research is required on both stages of these factors related to surface-residues.

Linden (1982) altered an evaporative model to make it sensitive to "tillage goals" and field observations. Potential evaporation may be increased because albedo is increased as random roughness increases. Soil-to-air contact area also increases when random roughness is increased, which increases actual, but not potential, evaporation. Actual evaporation was less for a uniformly dense soil than for a soil of a low density in the tilled layer. The latter configuration of soil density is typical of moderately deep tillage. However, a shallow tillage disturbance, which was simulated by a reduction in dry bulk density, reduced evaporation.

Field observations generally indicate that surface residues increase soil water storage, that storage occurs during seasons when precipitation is frequent and potential evaporation is low, and that tillage often is less effective than surface residues for increasing storage.

In the arid to semiarid Great Plains, Greb et al. (1967 and 1970) found that water storage during summerfallow increased as quantities of surface residue increased. Seventy percent of the net soil water gains occurred during spring months of the 14-month period beginning in the fall. Moreover, water storage was at soil depths greater than 60 cm. Other studies in this traditional summerfallow area indicated that tillage had no effect on long-term water storage as long as residue was not buried by moldboard plowing (Haas et al., 1974). Furthermore, a shallow tillage to break capillarity was required at the beginning of the warm summer season, even if weeds in grain stubble were controlled by herbicides.

In Texas, responses of soil water storage to surface residues were almost 100% greater in a winter fallow (July to May) than in a summerfallow (Unger, 1978). Unger and Wiese (1979) also obtained different water storages depending on how they managed the wheat residues, which ranged from 5000 to 9000 kg/ha at the beginning of the fallow season. No tillage during winter fallow conserved more soil water than sweeping, which in turn conserved more than discing. None of the tillages buried the straw deeper than 10 cm. These soil water storages were probably obtained under climatic conditions similar to the spring and early summer months in more northern latitudes.

Comparative profiles of soil water content under conventionally and no-till planted corn in Kentucky (Blevins et al., 1971; Unger and Phillips, 1973) showed consistently wetter profiles in the upper 60 cm early in the growing season with no-till. Undoubtedly, these effects related to influences of surface residue, first on infiltration and then on evaporation during the energy-limiting stage. Macroporous flow in cracks or biochannels was also likely in this no-till planted corn after grass sod. Soil water stored deeper than 80 cm was not measured to determine if macroporous flow was produced in the randomly rough surface. Macroporous flow can also occur following moldboard plowing with no secondary cultivation (Allmaras, 1967). An unusually large increase in water content was noted at depths ranging from 60 to 120 cm. Bruce et al. (1968) observed a similar effect in Mississippi on a Vertic Haplaquoll. Infiltration responses to tillage treatments in the studies of Burwell and Larson (1969) could not have occurred without macroporous flow.

21-2.5 Soil Environment Near the Soil Surface

Recent research indicates that shallow residue incorporation (top 10 cm of soil), as compared to deep incorporation with conventional tillage, can dominate the soil environment as a medium for crop production. Associated soil structure, residue incorporation, and chemical incorporation can drastically affect microbiological reactions. In addition to the effects on soil erosion and water conservation, this residue placement also affects soil temperature (Voorhees et al., 1981).

Numerous studies (Bauer and Black, 1981; Blevins et al., 1977; Dick, 1983; Doran, 1980; Unger, 1968) showed that crop residues on the surface or buried within 10 cm of the surface in CT increase organic matter concentrations in the top 10 cm of soil compared to conventional tillage in which the residues are buried deeper. Deeper burial was nearly always done by moldboard plowing in these studies. These differences were fully expressed only after 5 or more years, depending on the soil and climate. In double cropping of wheat and soybeans on acid soil (Hargrove et al., 1982), no-till produced only minor accumulations of organic matter in the upper 7.5 cm.

Edwards et al. (1979) show increased macropore space, possibly associated with the soil fauna activity resulting from greater amounts of surface residue and organic matter, less tillage disturbance, or both. Their observation and simulations of infiltration clearly established benefits of macropore space for enhancing infiltration. Field measurements by Triplett et al. (1968) on a Wooster silt loam after 12 years of CT for corn production indicated that a combination of no-till and residue cover (compared to moldboard plowing and associated secondary tillage) increased infiltration potential as much as 300%. Further evaluations indicated that infiltration was influenced more by soil structural stability than by the direct protection by residues against soil displacement and transport by rainfall. Studies by Gantzer and Blake (1978) and D. M. Van Doren (private communication, 1984, Ohio Agricultural Research and Development Center, Wooster), indicate that soil fauna activity and biochannel formation are usually encouraged in long-term no-till. The ecological requirements of soil fauna in cultivated crops must be understood to manage infiltration benefits from CT.

A soil crust can be contrasted to the layer produced by tillages that retain the crop residues on the surface or buried within the top 10 cm. A soil crust was characterized as a thin (several mm thick) soil layer with few large pores, a high bulk density, a platy or highly-oriented structure often stratified, and few or no organic matter fragments (Kemper and Koch, 1966). Crusting soils are often those with low wet strength of aggregates, low organic matter, high silt content, and severe compaction. Crusts become the seal against infiltration as modelled by Edwards and Larson (1969). This structure contrasts with aggregation resulting from residues incorporated at a shallow depth. Crop residues on the soil surface or buried no deeper than 10 cm increased the percentage of dry aggregates more than 0.84 mm in diameter (Black, 1973; Siddoway, 1963; and Smika and Greb, 1975). Smika and Greb (1975) also found that the percentage of dry nonerodible aggregates increased as the content of fats, oils, and waxes (organic matter soluble in ethyl ether) of the aggregates increased. Wet aggregate stability often increases dramatically with increased organic matter, but direct measurements of this trend have not been made in long-term comparisons of conventional tillage vs. CT. In some cases of no-till, the soil surface may be smooth enough (no microrelief) to permit soil crusting. Production of a large random roughness through tillage management has been demonstrated to provide non-uniform crusting and, therefore, to maintain simultaneous loci of low and high permeability (Falayi and Bouma, 1975). Consequently, high random roughness and high organic matter contents (crop residue) at or near the surface ensure maximum and prolonged infiltration.

The biological and biochemical properties of the top 10 cm of a soil change drastically when the soil is cultivated with implements other than a moldboard plow (Doran, 1980). This response is mainly due to returning the residue to the top 10 cm rather than burying it deeper. Measurements of microbial and biochemical components at various soil depths at several locations in the USA where long-term comparisons were made between conservation and moldboard-related tillage indicated greater microbial biomass with no-till, more potentially mineralizable N near the surface with no-till but greater overall N for conventional, a less oxidative environment with no-till, and increased potential for N immobilization coupled with less NO_3^- with no-till.

Pesticides, especially herbicides, are more likely to be used and to be located near the soil surface in CT systems. Sporadic field evidence is emerging that the soil-active persistence of some herbicides is decreasing in conventional tillage systems (Kaufman and Kearney, 1976; Hance, 1980). This unusual reduction in herbicide and other pesticide activity could be a serious challenge to continued CT because of the greater biomass available to degrade pesticides in the zone of soil application (Helling et al., 1971; Hance, 1980). Some fertilizers have chemical groups structurally similar to those in herbicides, which poses strong interactions of fertilizers and herbicides in these surface soil layers. This surface layer with high amounts of organic matter may also significantly change the hazards of plant disease (Cook et al., 1978; Cook and Baker, 1983). Interactions, especially of pesticides and soil-borne diseases (Hance, 1980), set the stage for even more difficulties in the management of tillage systems.

The soil environment near the surface could be discussed more thoroughly. However, enough research has been cited to indicate that major improvements in CT require that measurements be made to describe properties and activity in surface layers and not to continue to describe the Ap as merely a single layer.

21-2.6   Soil Compaction and/or Soil Illuviation Processes

Soil compaction may seriously reduce soil productivity even without soil truncation by erosion. Soil consolidation and shear planes near the surface or just below the Ap layer, as well as consolidation below the Ap layer and as deep as 120 cm (Erikkson et al., 1974; Gaultney et al., 1982), may reduce soil productivity because of both impeded water flow and mechanical impedance to plant rooting. Organic matter deterioration in the Ap horizon and illuvial processes (from the Ap horizon) with deposition in deeper horizons may also reduce long-term soil productivity. Acidification parallels this process, especially with the use of NH_4^+ fertilizers, and could be critical with CT (Blevins et al., 1977; Hargrove et al., 1982). Although not clearly defined, these processes were highlighted in a recent European symposium on soil degradation (Boels et al., 1982). Conservation tillage could affect these reactions differently than moldboard plowing, depending on fertilization and management of crop residues. Any heavy equipment loads (i.e., axle weights greater than 9 t), especially when soils are wet, could contribute to this soil deterioration with or without CT (Voorhees and Lindstrom, 1983).

21-3   CONSERVATION TILLAGE GUIDES TO SYSTEMATIZE TECHNOLOGY APPLICATION

The benefits of CT for soil and water conservation cannot be realized unless crop production in CT systems can provide a fair return for the investment. Tillage reductions usually associated with a CT system can reduce cost inputs to crop production, but unless yields can be sustained or increased, profits cannot be achieved. We analyze existing tillage guides with the assumption that yields can be sustained or even increased with an adapted CT system. From these existing guides, a similar approach should be possible for each MLRA in the USA.

In some landscapes, conventional tillage does not cause soil erosion even within a fraction of the tolerance value set by the USLE. However, the tillage goals and conservation benefits discussed above should encourage CT for higher and more profitable production, even though soil erosion is not serious. Many ecological factors affect performance of tillage operations, respond to tillage management, and control production. Some of these factors are discussed as related to tillage guides.

At least six tillage-planting guides have been developed in the Corn Belt (Table 21-6). Five were designed for use within a state. Of special interest is how each guide addressed the ecological factors affecting production. In these guides, soil and landscape factors emerged as dominant factors affecting production in CT systems.

The Indiana guide (Galloway et al., 1981) is the most comprehensive. Nine tillage-planting systems for corn were rated for their adaptability to 23

Table 21-6. Tillage-planting guides used for systematic application of conservation tillage in the Corn Belt states.

State	Crop	Number of tillage-planting systems	Application basis		Reference
			Class	Number of classes	
Indiana	Continuous corn	9	Tillage-soil group, and subgroup	$(10 \times y)$†	Galloway et al., 1981
	Rotation	--	Rating is a variation on continuous corn	--	Mannering and Griffith, 1981
Iowa	Rotation	3	Major soil area, principal soil association	$(10 \times \mu)$‡	Miller et al., 1982
Michigan	Continuous corn	1	Soil-management unit, soil-management group	6 (14×3)	Robertson et al., 1976
Minnesota	Rotation	5	Broad soil group	6	Swan and True, 1978
Ohio	Continuous corn	1	Tillage group	5	Triplett et al., 1973
Corn Belt	Corn	3§	Soil taxonomy	variable	Cosper, 1983

† y represents the subgroups within the tillage-management soil group. y is variable.
‡ μ is the principal soil association within the major soil area. μ is variable.
§ Three groups: all tillage systems, conventional and conservation tillage systems other than no-till, and only conventional systems.

groups of soil series. Within each of the 10 classes, internal soil-permeability categories were based mainly on the clay percentage, which characterized field drying time. Soil organic matter was also used to key on such tillage-related problems as soil structure and soil erodibility. Thus, these 23 groups considered natural drainage, landscape position, surface and subsoil character (of organic matter and texture), and surface slope. Each soil series in Indiana was classified into one of these 23 groups.

The nine tillage-planting systems were either shallow or deep tillage. Deep tillage included moldboard plowing or chiseling in fall or spring. The five shallow tillages were fall or spring discing followed by conventional planting, till-planting, ridge planting, and no-tilling. These tillage-planting systems along with characteristics for residue cover, cloddiness, and soil temperature were listed by Galloway and Griffith (1978). The basic criterion used in determining ratings given each tillage-planting system was the ability of tillage systems to maximize corn production while controlling runoff and erosion (both wind and water). Mannering and Griffith (1981) modified the rating to accommodate application of the tillage guide to a rotation of soybeans and corn, or soybeans-corn-meadow (or hay). The ratings and factor delineations in this guide were derived from evaluating many field experiments in the eastern Corn Belt together with specialist and grower experience. Mannering and Griffith (1981) noted that a soil temperature impediment to CT was more evident in northern Indiana and that it was an air temperature reaction independent of soil drainage.

Unlike Indiana, the Iowa tillage guide has not been extensively field tested. The Iowa tillage guide (Miller et al., 1982) has 10 major soil areas, each of which may consist of one or more of the 21 principal soil associations in Iowa. Soil and landscape factors of a suitable soil are soil texture, soil organic matter, natural drainage, and landscape position. Two related environmental factors are soil temperature and soil water content, factors which can be anti-productive because of delayed spring operations and early growth. Soil water content also affects soil compaction. Three tillage systems were considered: full width, strip tillage, and slot tillage. Full-width systems were suggested for soils somewhat poorly drained, poorly drained, and very poorly drained, because residues could be buried and densities of the plow layer could be decreased. Moreover, random roughness could be used to control erosion in situations where residues could not be tolerated on the soil surface. Strip tillage had greater advantages on moderately drained soils, whereas slot tillage was suited for excessively drained soils. As the hazard of soil erosion increased, preference would shift from full-width to strip to slot tillage. Finally, selected soil properties for predominant soil types (and dominant slope class) in each major soil area were rated for effect on tillage adaptability. Soil properties selected were soil wetness, surface-layer texture and temperature, subsurface fertility, and erosion potential. Each property was rated in relation to the three tillage systems.

Robertson et al. (1976) developed a no-till guide for continuous corn in Michigan. Soil adaptability was based on dominant profile (upper 150 cm), texture (14 conditions or levels), and natural drainage class (three classes). Within each of the 42 soil-management groups, six soil-management units corresponded to slope classes. Each major soil series in Michigan was classified into a soil-management group, and slope classes were obtained from soil maps. "Success opportunities" suggested no-till only on well-drained and moderately well-drained soils. These soil profiles have less than 40% clay. Because of natural soil compaction and/or cool and wet conditions produced by impaired drainage, no-till would not be suitable for soils with more than 50% clay and less than moderately well-drained. Herbicide efficacy was poor on soils with high organic matter and poor drainage in some soil-management groups.

The Minnesota tillage guide (Swan and True, 1978) approaches tillage adaptation by identifying six broad soil groups and associated problems to be managed by tillage in the southern half of Minnesota. These problems were soil crusting, low soil temperature, timeliness of tillage and planting, surface roughness, soil compaction, and soil erosion. Associated soil properties were drainage class, surface soil texture, and low organic matter content combined with medium soil texture. In contrast to other guides discussed above, water-supply deficiency was also considered. Five tillage practices and equipment were considered: conventional moldboard plowing with conventional secondary tillage, moldboard or non-moldboard primary tillage with reduced secondary tillage, spring moldboard plowing with secondary tillage only in the row, tillage in the row only, and no-tillage.

Based on the critical soil properties and the problems to be managed, one or more of the tillage combinations was recommended for major soils in each broad soil group.

The first tillage guide was developed for no-till in Ohio (Triplett et al., 1973). Five tillage groups were designated based on soil properties and their influence on corn response to no-till. In group 1, corn yields were better for no-till than for conventional tillage systems: the soils were moderately-well drained, had surface layer clay contents less than 40%, and were low in organic matter. They also had a glaciated, terrace, or residual origin. A mulch cover was required for high production, reduced evaporation and increased infiltration being the two responsible factors. Corn yields in group 2 were similar for no-till and conventional systems as long as surface or subsurface drainage was managed. Other than poorer drainage, soils in this group were similar to those in group 1. In groups 3 and 4, no-till corn yielded less than conventionally tilled corn because of poor to very poor drainage, surface soil with clay contents greater than 40%, and high enough organic matter contents to prevent the mulch response noted in group 1. Uncertain and unexplained no-till responses were included in group 5. In fact, fall tillage of most soils in these three tillage groups was preferred. No-till was suggested under restrictive conditions when fall tillage was missed.

A guide designated as "Corn Belt" in Table 21–6 recommended three groups of permissible tillage systems based on soil properties related to soil taxonomy (Cosper, 1983). The three groups were equal corn yields in all systems, equal yields in reduced tillage (CT excluding no-till) and conventional tillage, and higher yield in conventional than in CT tillage. The "all tillage" systems group included no-till. This guide was developed from many evaluations of the corn yield responses from tillages (conventional tillage, CT excluding no-till, and no-till), as related to taxonomic class of the soil series. Soil taxonomy includes climate and landscape factors in addition to soil. Soil moisture in the moisture-control sections of a soil (aquic, aridic, and torric; udic, ustic, and xeric), as designated in the subgroup category (Soil Survey Staff, 1975), was the strongest taxonomic character keyed upon. Included also were some expressions of combined soil moisture and temperature in the great soil group category as well as soil temperature expressions in the family category. By combining this system with SCS's Conservation Needs Inventories, Cosper (1983) summarized tillable land suitable in four Corn Belt states for inclusion in each tillage group. The Indiana tillage guide indicated that 38% of the tillable land was suitable for no-till, and the taxonomic system estimated 35% for Indiana. Proceeding west from Ohio to Iowa, the soil taxonomic guide indicated more potential for CT, especially no-till. However, acceptance of CT decreases from east to west across the Corn Belt, which suggests that something in addition to the factors encompassed in soil taxonomy is affecting CT suitability. Such factors as air temperature and length of growing season could be affecting corn adaptation to CT. About 33% of the tillable land in these four Corn Belt states was designated as unsuitable for CT.

Aside from the Corn Belt, tillage guides have not been used in the USA. Ketcheson and Stonehouse (1983) showed that corn yield responses to no-till relative to conventional tillage were explained by soil texture of Ontario soils and by previous crop (corn vs. grass-legume sod). This latter observation is similar to that of Mannering and Griffith (1981) in Indiana and could be related to friable surface conditions for good drill performance, good germination and emergence, and greater infiltration. In Great Britain, soil properties have been used to guide the adoption of direct drilling (Wilkinson, 1975). Direct drilling of cereals is similar to no-till except that residue from the previous crop is usually burned. Wilkinson examined soil and/or site factors responsible for success of seeding operations and then examined soil properties interacting with climatic and management factors during subsequent crop development. Soil/site characteristics used were spatial uniformity of microrelief and other soil properties, self-mulching property and soil resistance to compaction, natural porosity, infiltration and profile drainage, rate of biological activity and related susceptibility to special pests, range of friable consistence, slope steepness, and stoniness. Wilkinson (1975) used these characteristics in a limitations sense, so that success of direct drilling was reduced where more of these soil/site characteristics were limitations. This guide for direct drilling has been used for winter and spring cereals.

Success of these guides suggests development of guides for a region, such as the whole Corn Belt, and for MLRA's. Expansion into areas outside the Corn Belt will require guides for fall-sown as well as spring-sown crops and for significant shifts in climatic factors. Degree-day relations for crop development will be needed to forecast timeliness of tillage and seeding and penalty for timeliness failures. Many excellent data banks are available (such as those maintained by SCS and soil testing laboratories) that characterize physical, structural, biological, chemical, and taxonomic information by soil series. These data banks should be more extensively used, along with existing tillage, plant, and pest models to provide a functional interpretation of the impact of soil properties on tillage performance and effects.

Soil drainage class and permeability were recognized in all the guides. Soil drainage affects soil drying, timeliness of field operations, performance of tillage tools, tractive performance, early growth through both the direct and indirect effects of soil temperature, and soil compaction. It is appropriate, therefore, to quantify water flow associated with soil drainage class and internal soil properties. Much has been done merely by using textural information and compaction (Rawls et al., 1982). Soil drying and timeliness can be projected by jointly relating soil drainage properties and climate (Dyer and Baier, 1979). Poor drainage impacts on anoxia, damping-off diseases, and soil temperature were mentioned occasionally but were not quantified. More knowledge will be necessary to manage these problems in CT systems.

Self-mulching capability of soils, resistance to natural and implement-induced compaction, and random roughness produced by tillage were all discussed but not quantitatively. The first two are critical soil factors for machinery performance and timeliness, and the third is critical for conservation benefits when surface residues cannot be tolerated. Surface residues were encouraged in one tillage group in Ohio for inducing soil mulching in soils of medium texture and low organic matter. Only the guide to direct drilling (Wilkinson, 1975) discussed soil consistence properties as related to machine performance.

Soil consistence properties are measured on disturbed soil and, therefore, must be used along with in situ soil structure and other information to project self-mulching. Self-mulching properties are currently not identified taxonomically in categories above the soil series level. Suggested properties associated with self-mulching are clay content greater than 30%, wetting without saturation, and organic carbon/clay content values greater than 0.1 (R. Grossman, SCS, Lincoln, NE, private communication, 1983). Grieve (1980) showed that the structural stability of soil decreased as organic carbon decreased or as any of several indexes decreased; water content at the plastic limit, water drop stability, and pore volume stability determined by wetting against a small tension, against capillarity with no tension, or during flooding. Tillage, residues on or near the surface, soil fauna activity, and crop rotations should all be evaluated more thoroughly to guide management of self-mulching properties.

Frequently measured soil properties, such as organic matter, texture (grain-size distribution), and compression index, can be used to project soil impacts on natural and tillage-induced compaction (Gupta and Larson, 1982). Because self-mulching and low natural compaction are opposites of soil crusting, technology for describing and managing soil crusting could be assisted by evaluating natural compactability. Godwin and Spoor (1977) discussed the use of soil consistence, shear strength, and soil water content to project a work day in which the soil can withstand loading without shear and loss of intra-aggregate pores.

21-4 CROP YIELD RESPONSE TO CONSERVATION TILLAGE SYSTEMS

In this section, we assess the effects of CT systems on crop yields and identify factors that may have suppressed crop yields and acceptance of CT systems in tillage-management regions (TMR). These TMR are identified in Figure 21-1 as Pacific Northwest, Northern Great Plains, Southern Great Plains, Eastern Uplands, Corn Belt, Piedmont, and Coastal Plains. As more information is developed, these TMR can be subdivided or new ones formed. This identification of land masses into TMR corresponds to major land resource regions (MLRR) (USDA, 1981) in some instances and even an MLRA in others. The dominant physical characteristics of MLRR's and included MLRA's are described by land use, climate, soils, elevation and topography, water, and natural vegetation. Land use deals with principal crops and type of farming. The TMR description in Figure 21-1 disagrees also with the areal delineations of CT described in Soil Conservation Society of America (1977).

Climate and principal crops in each TMR, shown in Table 21-7, are related to expected tillage problems, but within these TMR's such informa-

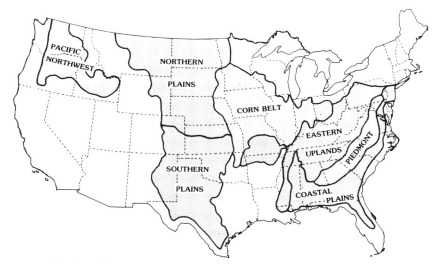

Fig. 21-1. Tillage-management regions (TMR) in the conterminous USA.

tion as soil type, weather variations, and crop rotation are needed to better understand the factors encouraging and discouraging CT. We emphasize development of CT in rainfed agriculture primarily because CT development in irrigated agriculture may have a different sensitivity to ecology in the TMR.

21-4.1   Corn Belt

Moderate precipitation (with about one-half during the growing season), low annual evaporation, and deep fertile soils are conducive to high crop yields in most years in the Corn Belt, but short-term drought (Amemiya, 1977) may reduce yields. Low temperatures, however, may shorten the growing season for warm-season crops, especially when CT retains large amounts of crop residue on the soil surface. Slopes range from very slight (often associated with poor drainage) to steep, where erosion may be severe. Major conservation needs for the Corn Belt, therefore, are water conservation, soil temperature management, drainage, and control of wind and water erosion. Soils vary widely in their physical characteristics, such as water regimes, soil texture, permeability, water-holding capacity, restrictive layers in the profile, and slope.

Studies of tillage-planting systems in Ohio, for example, documented the relationships between soils, tillage systems, crop yields, and soil erosion (Baxter et al., 1971; Harrold et al., 1970; Richey et al., 1977; Triplett, 1970; Triplett et al., 1973; Van Doren, 1965; Van Doren and Ryder, 1962). One Ohio study (Van Doren et al., 1976) showed that differences in corn yield between conventional and no-tillage systems were affected by soil type, immediate past cropping history (and tillage intensity), and time (Table 21-8).

Table 21-7. Climatic characteristics, major crops, and typical major land resource areas in seven tillage-management regions (TMR) of the USA.

Tillage management region	Typical major land resource areas†	Mean monthly temperature‡		Mean annual pan evaporation‡	Mean annual precipitation‡	Frost-free season‡	Growing season precipitation	Major crops
		Jan	July					
		°C		cm		days	% of annual	
Corn Belt	103, 104, 106, 107, 108, 110, 111, 114	−5, −10 to 0	22, 20 to 25	80, 70 to 100	80, 50 to 110	180, 130 to 230	50	corn, soybeans, wheat, hay, feed grains
Eastern Uplands	121, 134, 147	3, 0 to 10	22, 20 to 27	100, 70 to 110	120, 90 to 150	190, 120 to 280	<50 in south, >50 in north	cotton, corn, soybeans, small grains
Piedmont	136, 148	5, 0 to 10	24, 20 to 25	90, 80 to 100	120, 90 to 150	200, 160 to 240	<60	corn, wheat, grain, sorghum, soybeans, forages, cotton
Coastal Plains	133A	8, 5 to 10	25, 25 to 27	105, 100 to 120	130, 100 to 155	240, 200 to 280	>60 in east, <60 in west	corn, soybeans, grain sorghum, small grains, cotton
Southern Great Plains	73, 77, 78	5, −3 to 10	27, 26 to 28	150, 120 to 190	60, 40 to 100	170, 130 to 250	>70	wheat, corn, soybeans, grain sorghum, cotton, forages
Northern Great Plains	55B, 56, 72, 75	−10, −15 to 3	22, 20 to 25	90, 70 to 130	40, 25 to 60	130, 100 to 160	>70	wheat, corn, soybeans, grain sorghum, sunflowers, feed grains, forages
Pacific Northwest	7, 8, 9	0, −5 to 0	18, 15 to 20	90, 80 to 100	35, 15 to 60	160, 100 to 200	<45	wheat, peas, barley, lentils

† Major Land Resource Areas (USDA, 1981) are: (7) Columbia Basin, (8) Columbia Plateau, (9) Palouse Nez-Perce Prairie, (55B) Central Black Glaciated Plains, (56) Red River Valley, (72) Central High Tableland, (75) Central Loess Plains, (73) Rolling Plains and Breaks, (77) Southern High Plains, (78) Central Rolling Plains, (103) Central Iowa and Minnesota Till Prairies, (104) Eastern Iowa and Minnesota Till Prairies, (106) Nebraska and Kansas Loess—Drift Hills, (107) Iowa and Missouri Deep Loess Hills, (108) Illinois and Iowa Deep Loess Drift, (110) Northern Illinois and Indiana Heavy Till Plain, (111) Indiana and Ohio Till Plain, (114) Southern Illinois and Indiana Thin Loess and Till Plain, (121) Kentucky Bluegrass, (133A) Southern Coastal Plains, (134) Southern Mississippi Valley Silty Uplands, (136) Southern Piedmont, (147) Northern Appalachian Ridges and Valleys, and (148) Northern Piedmont.

‡ These observations are arranged to show characteristic value on first line and range on second line.

Table 21-8. Effects of soils rotation and duration of tillage management on corn response to no-till in Ohio (adapted from Van Doren et al., 1976).

Soil type and (taxonomic description)	Rotation‡	Corn yield†					
		Years 2-6			Years 7-11		
		Conv.	No-till	Increase	Conv.	No-till	Increase
		——kg/ha——		%	——kg/ha——		%
Wooster wilt loam	CC	6370	7060	9.7	8420	9400	11.6
(fine-loamy, mixed,	C-S	5900	6470	8.8	8720	9480	8.7
mesic, Typic	C-O-H	6850	6880	0.4	9720	10450	7.5
Fragiudalf)							
Hoytville silty clay loam	CC	6490	5720	−11.9	8000	6820	−14.0
(fine, illitic, mesic,	C-S	7000	6720	−4.0	8260	7920	−4.1
Mollic Ochraqualf)	C-O-H	6800	6830	0.4	8390	8180	−2.5

† Corn yields at equal plant density and weed control. Percentage increase in the yield quantity: 100 × [(no-till minus conv.)/conv.].
‡ CC is continuous corn, C-S is corn-soybeans, and C-O-H is corn, oats, hay.

These Ohio studies were based on equal plant densities for the different tillage systems, because, when planting systems and weed-control technology are developed, plant density will not be limiting and soil effects will then dominate. On Wooster silt loam, yields for no-till increased earlier for those rotations with the most intensive conventional tillage (Van Doren et al., 1976). The overall percentage increase on this soil for no-till was 7.8%. When Van Doren et al. (1976) adjusted for differences in percentage of surface residue cover in the two systems, they noted an overall yield increase of 9.9% for no-till. When no-till was used over an extended period, the yield advantage increased. Continuous improvement in the stability of soil structure and in soil-water relations were involved because, without surface residues, tillage intensity on the Wooster silt loam had to be increased to maintain corn yields. Moreover, when water was limited during the growing season, maximum tillage intensity was not sufficient to conserve water as effectively as did 75 to 100% soil cover. Other studies on Wooster silt loam (Triplett et al., 1968; Van Doren, 1965) concluded that the no-tillage system for corn production was superior to either conventional or other CT systems when normal or greater amounts of plant residue from the previous crop were allowed to remain on the soil surface.

Problem soils for CT in Ohio include the dark-colored and poorly drained soils, such as Brookston silty clay loam (fine-loamy, mixed, mesic, Typic Argiaquoll) and Hoytville silty clay loam. A tillage study on a Brookston soil (Baxter et al., 1971) showed the largest corn yield from tillage systems that provided the greatest amount of soil manipulation, that is, conventional tillage with fall or spring plowing. The smallest corn yields and the largest reduction (1140 kg/ha or 10.9%) were obtained with systems that provided the least soil manipulation, such as a no-till system. Table 21-8 shows that no-till was inferior on the Hoytville silty clay loam throughout the 10 years of record in continuous corn and corn-soybean rotation. However, nearly equal corn yields were obtained from the two tillages after a hay crop in the corn-oats-hay rotation. Van Doren et al. (1976) ascribed this adverse response under intensive cultivation to a *Pythium* type of root disease. Similar soil temperature problems were expected for no-till after the hay crop (based on the percentage of surface residue cover) as in continuous corn, yet the adverse yield was not produced.

Corn yield responses to tillage systems differ depending on soils in Indiana (Griffith et al., 1973; Mannering and Meyer, 1963; Mannering et al., 1966). Soils varied from a well-drained sandy loam, to a poorly drained silty clay loam, to a very poorly drained loam (Table 21-9). A conventional tillage system produced greater corn yields than the till-plant, strip tillage, and other reduced tillage systems on the fine-textured and poorly drained soils. Reduced, conservation, and strip tillage and planting performed better than the conventional system on well-drained soils. The amount of tillage and the percentage of soil cover associated with different tillage systems affected soil temperature and, consequently, plant growth, timely maturation, and potential corn yield. Unless specified otherwise, yield reductions from experimental CT in states other than Ohio often included the effects of poorer emergence and reduced stands. Weed-control failures might also have been involved.

Some Illinois soils also responded more favorably to no-tillage or reduced tillage than did others (University of Illinois, 1976). Corn yields obtained with five different tillage systems on five Illinois soils (Table 21-10) were reduced by reduced tillage only in the poorly drained soil. Water conservation and supply to the corn plant were a concern on the Cisne and Grantsburg soils, in which no-till gave significant yield increases. Tillage systems based on moldboard vs. chisel plow gave higher corn yields on one Illinois and three Nebraska soils (Cihacek et al., 1974). The Illinois soil and two of those in Nebraska were somewhat poorly drained.

Soils in Iowa showed less selectivity between soils and tillage systems than those in either Ohio or Indiana. On soils susceptible to compaction during tillage or a rainstorm, corn yields were higher with reduced secondary tillage (Larson, 1962). A surface mulch of plant residue on a poorly drained Grundy silty clay loam (fine, montmorillonitic, mesic, Aquic Argiudoll) in Iowa, reduced corn yields by 1140 kg/ha (Larson, 1962). Amemiya (1977) cited tests in Iowa showing that corn yields in CT were reduced because CT produced poorer stands, cooler soil temperatures in early spring, and restricted nutrient availability.

In southern Minnesota, conventional moldboard plowing in the fall, followed by preplant field cultivating, was compared to four CTs (no-till, fall chiseling followed by preplant discing and field cultivating, and ridge and flat surface configurations of till planting) for continuous corn production (Bauder and Randall, 1982) on a Webster clay loam (fine-loamy, mixed, mesic, Typic Haplaquoll). Corn yield reductions of 7 and 12%, respectively, were observed in the fall chisel and no-till treatments, whereas the till-plant and conventional systems gave the same yields. Reduced stand might have caused some of the yield reduction. Reduced soil temperatures, caused by weather and/or increased surface residue cover with CT after

Table 21-9. Corn yields with different tillage-planting systems on five Indiana soils (adapted from Griffith et al., 1973).

| Tillage-planting system | | Four-year (1967–1970) average corn yield | | | | |
Full-width primary tillage†	Secondary tillage-planting combination‡	Tracy sandy loam	Bedford silt loam	Blount silt loam	Pewamo silty clay loam	Runnymede loam
		kg/ha				
Sp. MB	Disc (2), plant	8400	6450	8670	7320	9140
Fall or sp. MB	(FC-planter)	8470	§	8870	6720	9480
Sp. MB	(WT-planter)	8670	6590	7860	6990	9070
Fall chisel	(chisel-planter)	8740	6720	7800	6450	8870
None	(sweep-planter)	9610	6790	7390	5710	9210
None	(rotary strip-planter)	8670	6720	5380	5580	8600
None	(coulter-planter)	8400	6590	4030	5380	7930
Modifier in soil drainage category		well	moderately well	somewhat poorly	poorly	very poorly
Taxonomic description		coarse loamy, mixed, mesic Ultic Hapludalf	fine-silty, mixed, mesic, Typic Fragiudult	fine, illitic mesic, Aeric Ochraqualf	fine, mixed mesic, Typic Argiaquoll	fine-loamy, mixed, mesic, Typic Argiaquoll

† MB is moldboard plow; sp. is spring; fall chisel depth is at least 20 cm. The system in the first row is conventional; all others reduce the number of operations but are not necessarily CT systems.

‡ When operations are separated by comma, secondary tillage operations are separate from planting; when enclosed in parentheses, the secondary tillage and planting operations are combined. FC-planter is a field cultivator with spring-loaded shanks combined with a planter; WT-planter is a wheel ahead of each planter opener; chisel-planter is a combination of chisel and planter using sweep shovels to a depth of 10 cm; sweep-planter is a till-planter combination of sweeps and seed openers operated in specially ridged fields; rotary strip-planter is a rotary tiller (20 cm wide and 10 cm deep) used with a planter; coulter-planter is a fluted coulter used ahead of a planter.

§ System not recommended for this soil.

Table 21–10. Corn yields from various preplant tillage combinations for continuous corn culture in five Illinois soils (adapted from University of Illinois, 1976).

Full-width primary, secondary tillage before planting	Mean corn yield				
	Proctor silt loam (1971–73)	Symerton silt loam (1968–71)	Drummer silty clay loam (1970–73)	Cisne silt loam (1971–73)	Grantsburg silt loam (1967–73)
	kg/ha				
Fall plow, disk	7800	9340	9140	--	--
Spring plow, disk	7800	--	--	6990	7460
Spring plow	--	--	--	6920	7800
Fall chisel	7930	9270	8870	--	--
None	7800	9540	6720	7190	8130
Soil drainage category	well-drained	well-drained	somewhat poorly-drained	well drained†	well drained†
Taxonomic description	fine-silty, mixed, mesic, Typic Argiudoll	fine-loamy, mixed, mesic, Typic Argiudoll	fine-silty, mixed, mesic, Typic Haploquoll	fine, montmorillonitic, mesic, Mollic Albaqualf	fine-silty, mixed, mesic, Typic Fragiudalf

† Water conservation is a significant production problem in these soils.

harvest of the previous crop, delayed planting or reduced early plant growth. Although the soil is poorly drained, periodic droughts from mid-June to mid-September cause greatest water stress (and lower yield) in plants whose growth is delayed by lower soil temperatures in the spring. Allmaras et al. (1972) observed similar corn responses to tillage treatments on a poorly drained, Nicollet clay loam (fine-loamy, mixed, mesic, Aquic Hapludoll) in southern Minnesota. They controlled soil temperature at planting time by managing corn-residue placement and random roughness from corn harvest until planting.

Soybean yields in a rotation with corn showed some of the same selectivity for soils, drainage, and tillage systems but were less sensitive than corn (Bone et al., 1976). Where no-till corn yields may decline 10% on poorly drained soils, soybean yields in no-till may decline 5% in wide row (76 cm) spacing and may not decline in row spacings approaching 18 cm (Griffith, 1982). These declines on poorly-drained soils are smaller for both crops in Indiana when planting can be delayed or when they are grown in the southern part of the state. In well-drained soils, both corn and soybeans showed a yield advantage of 10% for no-till.

Major deterrents to more widespread adoption of CT systems in the Corn Belt are delayed planting and lower yields on poorly drained soils, and lower soil temperatures at planting time when surface residues are present. The poor drainage problem is aggravated by CT because surface residues reduce runoff and evaporation and prolong soil drying. Strong evidence exists that yield losses from CT systems, especially no-till, are not as large in current experiments as in those conducted 10 to 15 years ago. In many cases, stand and weed control have been improved in CT systems. More recent experiments (Bauder and Randall, 1982) show that ridge-till planting systems are performing well in many instances but especially on poorly drained soils.

21–4.2 Eastern Uplands

Much of the CT research in the Eastern Uplands (Fig. 21–1) has been conducted in the Kentucky Bluegrass, Southern Mississippi Valley Silty Uplands, and Northern Appalachian Ridges and Valleys (Table 21–7). Precipitation in this TMR is generally favorable throughout the year, but short droughts may severely reduce crop yields in some years. Consequently, improved conservation and use of water along with improved control of water erosion are major conservation needs for the region.

The deep and well-drained Maury silt loam (fine, mixed, mesic, Typic Paleudalf) and similar soils in Kentucky are intensively no-till cropped (Blevins et al., 1977). Residues retained on the surface in no-till have reduced evaporation and increased soil water contents compared with conventional tillage (Blevins et al., 1971; Phillips et al., 1980). A 4-year study on Maury silt loam, for example, indicated that the soil water content in the upper 15 cm, during the growing season, averaged 29.5% under no-till compared to 24.4% under conventional tillage. Corn grain yields on Maury and other well-drained soils averaged 9240 kg/ha with no-tillage compared to 8780 kg/ha with conventional tillage (Phillips et al., 1980). Corn yields generally were lower with no-tillage than conventional tillage on poorly drained soils (Phillips et al., 1980), as in the Corn Belt.

No-tillage on the well-drained soils has caused soil management problems that could reduce crop yields unless corrected. Surface application of N fertilizer and an accumulation of decaying residues on the surface in no-till has rapidly decreased the pH of the surface soil (Blevins, 1982). This low pH decreases corn yields and impairs weed control with triazine (Kells et al., 1980a and 1980b; Slack et al., 1978). Surface applied lime corrected both problems (Blevins, 1982; Blevins et al., 1978). Nitrogen fertilization is more critical with no-tillage because the higher average soil water content provides a stronger leaching environment (Blevins, 1982; Phillips et al., 1980; Thomas et al., 1973). To minimize leaching losses of nitrates, Blevins (1982) recommended split applications of N fertilizer or N fertilization no sooner than 4 to 6 weeks after planting.

Conservation tillage in the Southern Mississippi Valley Silty Uplands is practiced on Lexington silt loam (fine-silty, mixed, thermic, Typic Paleudalf), in which soybean yields with no-tillage have equaled or exceeded those obtained with other types of tillage (Tyler and Overton, 1982a and 1982b; Tyler et al., 1983). Good results were obtained with no-tillage on fields that had been in continuous cultivation for at least 70 years with traditional annual moldboard plowing and discing and that have soil compaction as evidenced by a platy structure combined with high penetrometer and bulk density readings in the top 10 to 20 cm of the soil (Tyler and Overton, 1982b). Besides equal or higher soybean yields in the no-till, seed quality in a dry year was better because of greater availability of soil water and associated lower plant-water stress. Surface residues with no-till also reduced soil splash, which is required to infect plants with purple stain [*Cercospora kikuchii* (T. Matsn. and Tomoyaso) Chupp.] organisms (Tyler and Overton, 1982a). In another year, septoria brown spot (*Septoria glycines* Hemmi) infestation was lower, and stem canker [*Diaporthe phaseolorum* (Cke. and Ell.) Sacc. var. *caulivora* Athow and Caldwell] infestation was higher with no-tillage than with other types of tillage. Cyst nematode (*Heterodera glycines* Ichinohe) counts were also higher with no-tillage. However, seed yields were higher with no-tillage (Tyler et al., 1983).

An 8-year comparison of conventional tillage and planting (moldboard plowing of cover crop sod, discing, harrowing, and planting) of corn with no-till planting into a winter cover crop of winter rye (*Secale cereale* L.) in West Virginia indicated a distinct yield advantage for no-till (Moschler et al., 1973). Respective average corn yields were 8350 and 6960 kg/ha. The no-till treatment yielded less than the conventional treatment only once in 8 years. After some experience with no-till systems, lime applications were seen to increase the corn yield advantage of no-till. After corn harvest, both treatments were disced lightly before seeding winter rye. This tillage comparison on a Frederick silt loam (clayey, kaolinitic, mesic, Typic Paleudult) has given results similar to those in the Kentucky Bluegrass MLRA.

Although responses to CT (mainly no-tillage) have generally been favorable in the Eastern Uplands, there are some problems. These include: difficulty in control of weeds (Blevins et al., 1982); greater potential for insect, disease, and root damage (Phillips et al., 1980); need for more critical and timely management inputs (Blevins et al., 1983; Phillips et al., 1980); and lower temperatures with surface residues, which may delay planting in the spring (Phillips et al., 1980). However, this soil temperature problem is minor compared to that in the Corn Belt.

21-4.3 Piedmont

The Piedmont (Fig. 21-1) consists of the Southern Piedmont and Northern Piedmont MLRA's. Precipitation is nearly always adequate and temperatures are moderate throughout the year (Table 21-7). This climatic combination encourages double-cropping of summer and winter crops. Crop production, however, is mainly on sloping soils that are relatively infertile, highly erodible, low in organic matter, and easily compacted by rainfall and machinery. Summer drought occurs in the Northern Piedmont but more infrequently in the Southern Piedmont. The dominant conservation needs in the Southern Piedmont are erosion control on cropland and reestablishment or improvement of vegetative cover on pastures (Carreker et al., 1977). Cecil sandy loam (clayey, kaolinitic, thermic, Typic Hapludult) is a major soil throughout the Southern Piedmont.

Conventional clean tillage for corn normally involves moldboard plowing and harrowing each spring to prepare a seedbed following sod or corn. Corn is often grown continuously, either with or without a winter cover crop. Conservation tillage involves planting corn in the residues of annual cover crops or of 2-year-old perennial grasses or legumes, usually by no-tillage. Large fertilizer applications are usually required for maximum corn yields in the conventional system, except in some cases where legumes are grown in rotation with the corn (Carreker et al., 1977). Results with double-cropping (Table 21-11) show equal or superior yields of corn or grain sorghum for no-till seeding after forage harvest of wheat or barley (*Hordeum vulgare* L.) winter crops. Yields in double-cropping were low only when planting of warm season crops was excessively delayed by the wait for grain harvest. Alternative managements of a winter cover crop (Table 21-12) indicate the flexibility of double-cropping with no-till, but they also show the loss of corn yield when the fescue (*Festuca arundinacea* Schreb.) sod is not controlled full-width. Wheat also produced well when seeded into a cover crop or grass sod, provided adequate fertilizer was applied (Carreker et al., 1977).

In the Northern Piedmont, CT includes no-till corn production in double-cropping after grass or legume sod or small grain cover, no-till soybean cropping after small grain harvest, and various forms of reduced tillage in full-season corn (Bennett, 1977). Crop responses and problems are very similar to those noted for the Southern Piedmont.

Table 21–11. Corn or grain sorghum yield responses to conventional tillage and no-till in several double-cropping alternatives in the Southern Piedmont (adapted from Nelson et al., 1977).

Culture of warm-season crop				Four-year mean grain yield	
Tillage-planting method for warm-season crop	Seeding date	Treatment of winter crop†	Winter crop	Corn	Grain sorghum
				——— kg/ha ———	
Conventional‡	30 April	Incorporated	wheat	7500	4100
No-till	30 April	Forage harvest	wheat	8000	4200
	30 April	Forage harvest	barley	8300	4600
	4 June	Grain harvest	wheat, barley	6300	2500

† After removal of forage, the stubble height was 5 cm; after grain harvest, it was 25 cm.
‡ Conventional operations were moldboard plow, disk harrow (2), spiketooth harrow, and plant.

Table 21–12. Crop yield responses to the type of herbicide kill of winter crop in a double-cropping system with no-till planting of corn (adapted from Box et al., 1980).

	Corn and forage yield‡							
	Rainfed				Irrigated			
Herbicide treatment of winter crop†		Forage				Forage		
	Corn	Fescue (summer)	Winter rye	Fescue (winter)	Corn	Fescue (summer)	Winter rye	Fescue (winter)
	——————— kg/ha ———————							
20-cm strip kill	5400	3670	--	2390	8600	3100	--	1690
Full-width kill	8500	--	5960	--	10000	--	6150	--

† Corn rows were 102 cm apart; the same strip was killed before each planting; winter rye was killed and replanted again after corn harvest.
‡ Irrigated when soil water potential in 0 to 30-cm layer was less than −2 bar. Fescue yields are separated into those at the end of summer and of winter. All yields are the means of 3 years on the same site.

Conservation tillage, including no-tillage, has been widely accepted by producers in the Piedmont. Higher yielding systems and good erosion control have been achieved through use of CT in this TMR, but some of the disadvantages of CT are greater insect problems and the higher level of management ability required (Phillips, 1978).

21–4.4   Coastal Plains

The Coastal Plains in the Southeastern USA (Fig. 21–1), like the Southern Piedmont, has precipitation and temperatures (Table 21–7) generally favorable for year-round crop production. Although local relief in the Coastal Plains may be less than 10 m, soil erosion is severe because of high-intensity rainfall and shallow soils underlain by pressure pans or hardpans at depths of 20 to 30 cm. These pans restrict internal drainage and mechanically impede rooting. These factors, coupled with sandy soils having low water-holding capacities, often cause plant water stress within 3 to 7 days after rainfall. Consequently, frequent water replenishment during the growing season is required for maximum crop yields (Reicosky et al., 1977). Major conservation needs for the Coastal Plains, therefore, are erosion control, disruption of pans (pressure pans or hardpans), and water conservation. Surface residues not only protect against erosion in the Coastal Plains but also enhance water infiltration. However, short-term droughts can still reduce crop yields because of the limited water-storage capacity in the shallow, sandy soils where the pan restricts deep root penetration by most crops. Where the pan is ruptured, an increased water supply associated with root access to a larger soil volume minimizes the adverse effects of short-term droughts.

Typical cropping sequences include: continuous corn (with or without a winter cover crop); corn and soybeans in rotation; corn, soybeans, or grain sorghum double-cropped after wheat or barley for forage or grain; and corn or sorghum after winter legumes. Management practices in CT have a major influence on soil water contents at planting and water conservation from rainfall during the growing season (Campbell et al., 1984b). A delay in killing the winter cover crop or a mild kill can rapidly reduce soil water contents enough to reduce soybean yields (Campbell et al., 1984a) and corn yields (Table 21–13, Site 2). However, even clean tillage (no cover crop) provides soil recharge with water to only 73% of capacity measured in the top 60 cm. Where corn residues from the previous crops were main-

Table 21–13. Effect of rye cover crop and residue-management techniques on soil water of a Norfolk loamy sand (17 days after planting corn) and on corn grain yield (adapted from Campbell et al., 1984b).

Management of cover crop and planting	Water content at various soil depths (cm)				Water capacity	Corn yield	
	0–15	15–30	30–45	45–60		Absolute	Relative
	%, w/w				%	kg/ha	
Site 1							
Clean tillage-planting (no cover crop)	--	--	--	--	--	5540	100
No-till in corn stover (no cover crop)	--	--	--	--	--	7220	130
Site 2							
Clean tillage-planting (no cover)	8.9	9.9	19.4	21.4	73	6160	100
Disk rye cover (before planting)	5.7	6.6	15.7	19.5	41	5490	89
No-till rye cover (herbicide at planting)	5.4	5.6	13.7	18.5	29	4930	80
No-till 50% rye cover (herbicide at planting)	4.6	4.6	14.2	18.9	26	5040	82
No-till 50% rye cover (no herbicide)	2.4	3.9	14.1	18.2	17	3920	64

tained on the surface, yields were increased about 30% over those with clean tillage and planting (Table 21-13, Site 1). Although soil water contents were not measured, this was a likely response to water conservation by the corn residues on the surface.

Although random chiseling of the pans increases water infiltration, plant rooting depth, and subsequent water extraction by plants, these increases often are temporary in the Coastal Plains because of natural compaction of the soils and because of subsequent tractor and implement traffic, random with respect to the row. With deep chiseling and controlled traffic, cotton lint yields were 20% higher than where traffic was not controlled (Reicosky et al., 1977). A recent machinery development is an integral in-row deep chisel ahead of the seed opener to provide both minimum disturbance of surface residues (essentially no-tillage) and rupture of the pan (Harden et al., 1978). When corn was seeded with this equipment on Orangeburg sandy loam (fine-loamy, siliceous, thermic, Typic Paleudult), grain yields were 12 600 kg/ha compared with 9600 kg/ha without in-row chiseling (Langdale et al., 1981). For double-cropped soybeans and wheat in the Southern Piedmont, Touchton and Johnson (1982) found that subsoiling simultaneous with no-till planting of soybeans increased both the soybean yields and yields of the wheat seeded after a shallow (8-cm) discing of the soybean stubble.

Although CT systems reduce erosion, improve water conservation, and increase crop yields in the Coastal Plains, they have not been widely accepted. Major problems include inadequate information about weed control, cropping systems, species or cultivar selection, and insect and disease pests. Suitable equipment is not yet developed for trouble-free operation in heavy residue (Campbell et al., 1984b). Because in-row deep chiseling is energy intensive, its use might be increased if more were known about long-term effects of this practice on soil physical conditions, water relations, and crop responses.

21-4.5 Southern Great Plains

Climatic conditions vary widely among different areas within the Southern Great Plains (Fig. 21-1, Table 21-7). In addition, the deep soils range from sands to fine clays, and surface elevations range from near sea level to over 1300 m above sea level. Consequently, crop production practices and conservation needs vary widely. Annual cropping is widely practiced in the region, and double-cropping may be practiced, especially where water for irrigation is available. In the semiarid Southern High Plains, summer fallowing is used widely for more reliable crop production from the limited precipitation.

Major conservation needs include erosion control, water conservation, and improved cropping systems. A potential for water erosion exists throughout the region but is most common in the eastern part where precipitation is more abundant. Most wind erosion occurs in the semiarid western part, especially on sandy soils of the Southern High Plains and

Central Rolling Red Plains, where cotton, which produces very little residue, is extensively grown. Although annual cropping is widely practiced, crops throughout the TMR often experience drought some time during the growing season. Consequently, crops in most of the region, but especially the western semiarid portion, would benefit from the improved water conservation of CT. Conservation tillage systems are available for some crops in some areas, but research is needed to develop them for other crops throughout the region. Crops may be grown in monoculture or in rotations, such as wheat-sorghum-fallow, cotton-sorghum, and wheat-fallow.

Stubble-mulch tillage was developed mainly for controlling erosion (wind and water) by retaining crop residues on the soil surfaces. The practice was developed about 1940 at Bushland, TX, and other dryland research centers. Where sufficient residues are present (which often is not the case in the Southern Great Plains), stubble-mulch tillage controls erosion, conserves water, and increases crop yields (Johnson and Davis, 1972). Stubble-mulch tillage is used mainly for wheat and grain sorghum production under dryland conditions in the western semiarid portion of the region. It is used very little in more humid locations where a crop such as continuous wheat produces amounts of residue sufficient to interfere with tillage, fertilization, and seeding operations. It is not used under irrigated conditions.

When large amounts of residue are produced, conventional clean tillage systems are widely used. However, some progress has been made in developing and implementing CT systems under such conditions (Unger et al., 1977). Suitable, well-timed tillage operations that retain reduced amounts of surface residues on the surface are gaining acceptance in the more humid eastern part of the region and under irrigated conditions.

Research involving no-tillage has been conducted under dryland and/or irrigated conditions in the Southern Great Plains since the 1950s (Unger et al., 1977). Early work with wheat and grain sorghum on dryland showed no water conservation or yield benefits with no-tillage compared to sweep (stubble-mulch) tillage. New and improved herbicides and/or herbicide mixtures are improving the feasibility of no-till under rainfed conditions, such as for continuous wheat in Oklahoma (Fain, 1981) and double-cropping of cotton and corn in south Texas (N. Namken, personal communication, Weslaco, TX, 1983).

For a cropping sequence of irrigated winter wheat, fallow, and dryland grain sorghum, water storage during fallow averaged 15, 23, and 35% of precipitation with disc, sweep, and no-tillage treatments, respectively. Subsequent yields of dryland sorghum grain averaged 1930, 2500, and 3140 kg/ha, respectively (Unger and Wiese, 1979). Use of no-tillage rather than disc tillage during fallow also increased water storage and sorghum yields under irrigated conditions when wheat and sorghum were grown in rotation (Musick et al., 1977).

Although CT can enhance water conservation and increase crop yields in the Southern Great Plains, there are still impediments to widespread acceptance. Major problems include: poor weed control, inadequate amounts of residue, and poor equipment performance. Although weed control is a problem throughout, it is most severe in the more humid eastern part of the region, where weeds grow year-round. Biological similarities between weeds and crops, for example, cheat grass (*Bromus tectorum* L.) in wheat, restrain timely use of species-selective herbicides. Unpredictable shifts in weed types also complicate their control. At some locations where cotton is grown continuously in western Texas and eastern New Mexico, residue production is not adequate to control erosion or increase water conservation. Seeding equipment often performs poorly under high residue conditions, which causes poor seed placement and crop establishment. Under high soil water conditions on sticky clay soils, planters often clog. Other problems encountered with CT in the Southern Great Plains include reduced yields with continuous wheat in more humid areas where high residue yields increase potential for insects and diseases, interfere with pesticide application and effectiveness, and require a change in fertilizer application methods.

21-4.6 Northern Great Plains

The Northern Great Plains (Fig. 21-1) varies widely in pan evaporation relative to precipitation (Table 21-7) but has highly productive soils throughout. The short growing season in the northern portion restricts the choice of crops in a rotation. A combination of short growing season, drought periods, and cold winters limits opportunities for more than one crop per year. In most of the TMR, summerfallow is practiced to manage water conservation and production stability. The major conservation problems are wind and water erosion, water conservation, and the development of cropping systems to fully utilize precipitation within the constraints of low temperature and short growing season.

Since the pioneering development of machinery and soil management systems for stubble-mulch tillage in Nebraska in 1937, stubble mulch is practiced in the western and semiarid portions of the TMR where wheat-summerfallow rotations predominate. Even though stubble mulching has significant water-conservation benefits, wheat yields with this tillage system, compared to moldboard-plow systems, increased only marginally in the semiarid climates and decreased in the subhumid portions of the region (McCalla and Army, 1961). Investigations in Colorado, Nebraska, and Montana (Greb et al., 1970) indicated that crop-residue additions equivalent to those from wheat production in an existing stubble-mulch system increased water conservation from 1.5 to 7.0 cm and that higher rates of residue would lead to even more water conservation if they could be produced with improved wheat production. Wheat yields were improved by this stored water in years of normal to above-normal air temperatures and normal to below-normal spring precipitation.

In the subhumid portions of the TMR, rotations have a larger non-fallow component such as wheat-corn (sorghum)-summerfallow (Ramig and Smika, 1964) with stubble-mulch tillage in the summerfallow. In the northern part of the TMR, summerfallow is practiced once in 3 or 4 years in rotation with wheat, barley, corn, sugar beets (*B. vulgaris* L.), potatoes

(*Solanum tuberosum* L.), or flax (*Linum usitatissimum* L.) (Norum et al., 1957). Soybeans and sunflowers have recently become major crops in these rotations. These rotations are in a constant state of flux depending on many factors, one of which is tillage management.

Early attempts to grow continuous wheat with conventional tillage systems instead of wheat-summerfallow rotations tilled with stubble mulch were notoriously unsuccessful. For instance, in North Platte, NE, when average annual precipitation was 43 cm or less, crop failures were ≥ 30%, and average yields were about one-third of those in summerfallow (Smika, 1970). It was estimated that 58 cm or more of annual precipitation was needed for continuous cropping. Another early evaluation showed that herbicide-assisted or herbicide-only alterations to summerfallow were no better than stubble-mulch tillage for water storage and yield because weed control was unreliable and critical tillage inputs were not anticipated (Black and Power, 1965).

Significant advances in CT have improved water storage and crop yield in the Nebraska-Colorado area. Wicks and Smika (1973) used combinations of stubble-mulch tillage, soil active herbicides, and contact herbicides to improve production of winter wheat in a winter wheat-summerfallow rotation (Table 21–14). Soil water gains were larger in the fall and spring when only herbicides were used for weed control before late spring of the summerfallow season. These gains were produced by reduced transpiration (by weeds) and reduced soil water losses in an erect stubble. Greb and Zimdahl (1980) observed similar water conservation gains with weed control by herbicides and no tillage until the spring after wheat harvest on a Rago silt loam (fine, montmorillonitic, mesic, Pachic Argiustoll) in Colorado. Soil water gains from this practice (10-year average) were 3.8 cm, which improved wheat yields 20%. In a semiarid climate such as this, the 800 kg/ha of added wheat straw will improve water conservation and soil erosion control. Comparative tests in a wheat-summerfallow rotation in northwestern

Table 21–14. Water conservation and winter wheat yield in a winter wheat-summer fallow rotation as affected by five forms of fallow management of a Holdrege silt loam (adapted from Wicks and Smika, 1973).†

	Soil water gain				Surface residue at planting	Wheat yield
Fallow treatment‡	Harvest to 4 Nov	4 Nov to 14 Apr	14 Apr to 3 July	3 July to seeding		
	cm				kg/ha	
Moldboard plow	−4.4	7.5	10.3	1.2	0	2690
SM	0.4	7.1	11.3	1.4	1390	2880
SM + SAH	1.2	6.8	13.7	−0.2	1390	2910
SAH + CH + SM	1.1	8.0	15.5	−0.9	1650	3040
SAH + CH	3.4	9.2	16.6	−1.8	3040	3170

† Holdrege silt loam is a fine-silty, mixed mesic, Typic Argiustoll. All observations are six-year average.
‡ SM is stubble mulch; SAH is a soil-active herbicide; CH is a contact herbicide. In SM alone and SM + SAH, the stubble-mulch tillage began by sweeping after harvest; in SAH + CH + SM, the stubble-mulch tillage began in spring.

North Dakota did not show a response to weed control in the fall after wheat harvest (French and Riveland, 1980).

While use of herbicides in a stubble-mulch tillage system has made the wheat-summerfallow and small grain (wheat)-summerfallow-corn (sorghum) rotation more productive, another change is being researched to utilize no-till planting for soybean production instead of summerfallow after small grain (Burnside et al., 1980). The possibility of a rotation such as wheat (small grain)-soybeans-corn (sorghum)-summerfallow to replace the original wheat-summerfallow of the 1940s, illustrates that CT may facilitate new, more intensive crop rotations. Residual herbicides in these studies have nearly always been more effective if applied without tillage of the small-grain stubble after harvest.

In subhumid and northern parts of the Northern Great Plains, advances in CT and associated greater productivity will depend on inclusion of crops other than small grain in the rotation. Yields of annually cropped spring wheat were highest with moldboard plowing (Table 21–15). Three distinct yield levels distinguish no-till, preplant tillage of the non-moldboard type, and preplant moldboard tillage. Tillages after harvest and before planting that produced the highest yields were, in descending order, fall plowing, spring plowing, shallow fall tillage followed by spring plowing, and fall plowing followed by field cultivating. Yield losses when moldboard plowing was not used were caused by erratic plant populations and weed infestation even though herbicides were used as recommended. Studies in the Red River Valley on a Fargo clay loam (fine, montmorillonitic, frigid, Vertic Haplaquoll) indicated that no-till in a 2-year rotation of spring wheat-alternate crop produced yields within 8% of those produced by conventional moldboard-plow tillage (Miller and Dexter, 1982). Following wheat, the yields of barley, flax, corn, sunflower, and sugar beet were greater in no-till than in conventional tillage. Only soybeans and wheat produced less with no-till. All no-till wheat after the alternate crops yielded

Table 21–15. Mean yield of annually cropped spring wheat in eastern North Dakota as affected by full-width fall and spring tillage combinations before planting (adapted from Bauer and Kucera, 1978).†

	Wheat yield for fall tillage treatments					
	Fargo-Bearden complex‡			Heimdal loam-silt loam§		
Preplant spring tillage	None	Moldboard plow	Other¶	None	Moldboard plow	Other
	kg/ha					
None	640	1700	900	660	1790	1000
Moldboard plow	1660	--	1770	1910	--	1930
Field cultivate	--	1910	1600	--	1800	1710

† Planted with a press drill using double-disk opener. Previous crop was spring wheat. Average values from 4 years of comparison.
‡ Fargo silty clay (fine, montmorillonitic, frigid, Vertic Haplaquoll) and Bearden silty clay (fine, silty, frigid, Aeric Calciaquoll).
§ Heimdal loam-silt loam (coarse-loamy, mixed, Udic Haploboroll).
¶ "Other" forms of fall tillage were double disk or noble blade forms which would not bury residue below 10 cm.

about 5% less than with conventional tillage. Again, the performance of no-till was relatively superior after warm and dry spring weather.

Although the progress on CT has been remarkable in the southern part of this TMR, inadequate weed control and lack of systems of adapted crops have hampered progress in the northern and subhumid parts of the region. More evaluations of low-residue crops such as sunflowers and soybeans for no-till in rotation with small grains appear most feasible.

21-4.7   Pacific Northwest

The Pacific Northwest (Fig. 21-1) has warm and dry summers with cool and wet winters (Table 21-7). The Columbia Basin, Columbia Plateau, and Palouse Nez Perce Prairies are characteristic of the eastern portion of the TMR, which has a climate ranging from arid to subhumid. Forty-five percent or less of the precipitation occurs during the growing season, and pan evaporation exceeds precipitation by at least 100%. If the early fall precipitation is discounted, when no dryland crops are actively transpiring, precipitation in the growing season is nearer 25% of the annual. Soil freezing is intermittent and extends less than 15 cm deep. The unusually steep slopes accentuate azimuth effects on the environment and environmental heterogeneity within a field. The low organic matter and clay contents and the volcanic ash origin of the soils account for their low structural characteristic. Yet the climate and other natural factors of the TMR provide an environment with potential for high crop production. Temporal precipitation distribution determines the main crop rotations of wheat-summerfallow, wheat-peas (*Pisum sativum* L.), or wheat-barley-peas (lentils, *Lentilla lens* L.).

Conservation needs in this TMR are water conservation, wind and water erosion control, and a rotation of crops adapted to a growing season, the length of which is determined by water supply from soil storage.

Stubble-mulch tillage was introduced into this TMR at about the same time as in the Northern Great Plains but has persisted only in the drier parts where wheat residues range from 3000 to 6000 kg/ha, weeds are controlled, and wind erosion is a serious hazard. Comparisons of stubble mulch with conventional moldboard-based tillage on a Walla Walla silt loam (coarse-silty, mixed, mesic, Typic Haploxeroll) and a Palouse silt loam (fine-silty, mixed, mesic, Pachic Ultic Haploxeroll) produced the lowest average yield ratio (0.83 to 0.85) of stubble mulch to conventional tillage in the traditional stubble-mulched area (McCalla and Army, 1961). Low yields were attributed to poor weed control and inability to handle wheat residues, which ranged from 8000 to 12 000 kg/ha. A long-term comparison, on a Walla Walla silt loam, between primary tillages of moldboard plowing, sweeping, or discing followed by a uniform set of secondary summerfallow tillage operations is now showing no yield response of wheat to the different primary tillages (P. E. Rasmussen, personal communication, USDA, Pendleton, OR, 1983) because herbicide weed control is successful and machinery will perform in the presence of surface residue. Earlier, when weed control was failing, the primary moldboard-plow tillage gave higher yields, similar to comparisons before 1961 cited by McCalla and Army (1961). Current emphasis is to use a non-inversion primary tillage and reduce the secondary tillage, which will then require drill redesign to seed through unanchored surface residues as concentrated as 4000 kg/ha. A surface residue concentration of 2000 kg/ha after wheat seeding would reduce soil erosion to below the tolerance value on at least 90% of the wheat-summerfallow land in the Columbia Plateau (Allmaras et al., 1980).

Small grains have been seeded by the no-till method into small grain stubble (Cochran et al., 1982; Ramig and Ekin, 1983). Both contact and soil-active herbicides have provided the weed control necessary for water storage responses to surface residues in no-till. Weed control has also permitted the wheat to utilize added stored water. Consequently, no-till and conventionally tilled wheat yielded the same when they followed a small grain. Banded N fertilization, using a new opener design (Wilkins et al., 1982), instead of the conventional broadcasting or shanking with no row orientation, brought no-till annual crop yields nearly up to those from summerfallow (Ramig and Ekin, 1983). This no-tillage system will facilitate the change from a wheat-summer fallow to a wheat-wheat (barley)-summer-fallow rotation whenever the prospect of soil water storage cannot be improved by summerfallow, such as when overwinter precipitation is high, the soil is too shallow to store the overwinter precipitation, or both.

Drastically reduced tillage in a wheat-pea rotation has reduced yields of both crops as much as 10% in a 15-year period where the tillage system treatment has been applied continuously to both crops (R. E. Ramig, personal communication, USDA, Pendleton, OR, 1983). Yet soil water storage in this CT system has increased at least 20%. For both wheat and peas, this drastically reduced tillage system involved a very shallow, preplant, sweep operation followed by planting with a shovel-point opener. In adjacent farm fields of peas-wheat, the rotation of tillage systems has increased wheat and pea yields at least 10%. Wheat after peas is no-till seeded, and peas after wheat are seeded using the conventional tillage-planting system (P. Davis and J. Loiland, personal communication, Athena, OR, 1983). Wheat production in their rotational tillage system ranges from 6000 to 8000 kg/ha; green pea yields range from 3000 to 4500 kg/ha. Preliminary information suggests better control of soil-borne diseases (*Pythium* and *Fusarium* spp.) in the rotation of tillage-planting systems compared to exclusive use of a drastically reduced tillage-planting system for both crops in the rotation.

This analysis indicates good potential for use of CT systems in the Pacific Northwest. Recent improvements in chemical weed control and fertilization indicate that CT systems can be highly productive under the environment of this TMR. Large amounts of small-grain residue are often a problem, but some of these residues can be used in a slot-mulch system (Saxton et al., 1981) to improve erosion control. Slot-mulching is done simultaneously with no-till seeding into a small-grain stubble. Development of a larger number of adapted crops and associated crop rotations would increase flexibility for use of CT systems in this TMR.

21-5 DEVELOPMENT OF MACHINERY FOR CONSERVATION TILLAGE

Adoption of CT frequently follows development of a suitable machine(s). The specific function was usually obvious after the machine was successful, but the amount of trial and error, if any, is not described for the reader. Our experience indicates that multidisciplinary approaches are more productive because function generates the machine design. Function is used here to denote accomplishment of one or more of the environmental requirements in a cultivated field. Ladewig and Garibay (1983) found that a lack of suitable farm machinery was a major reason farmers do not adopt CT.

The steel plow has been improved since its introduction in 1837 (Farm and Industrial Equipment Institute, 1974), but the basic operation including soil inverting, loosening, and pulverizing, remains the same. Other tillage implements, such as disc harrows and field cultivators, likewise have not changed significantly in their mode of operation in the last 100 years. The most dramatic changes in tillage implements have been increased size, operational speed, and combination of tool actions in one machine. Size and speed increased because, until recently, fossil fuel was relatively inexpensive and labor was expensive (Council for Agricultural Science and Technology, 1983).

The average power of farm tractors increased from 27 kW in 1964 to 42 kW in 1979 (USDA, 1980). In 1982, the average power of tractors shipped from the factory was 84 kW (U.S. Dep. of Commerce, 1982). These increases in tractor power permitted more tillage per man hour, and tractor power thus permitted more trips over fields and more tillage. Another result is more timeliness of field operations. The trend toward reduced tillage began in the late 1960s with the realization that overtillage could increase erosion potential by reducing soil surface roughness and crop residue on the surface. The 1973 energy crisis added further encouragement of tillage reduction to reduce costs. Illinois farmers reduced tillage since 1970 to save time and fuel, reduce costs, and reduce erosion (Siemens and Burrows, 1978). However, herbicide use increased.

Adoption of the chisel plow as a primary tillage tool is an example of the trend toward reduced tillage, increased soil surface roughness, and increased surface residue. A chisel plow shatters soil without completely burying or mixing surface materials, but the moldboard plow is designed to partially or completely invert a layer of soil, to bury surface materials, and to pulverize soil. The ratio of delivered chisel plows to moldboard plows increased from 0.43 in 1970 to 0.76 in 1980 (U.S. Dep. of Commerce, 1971 and 1981). Another indication that the moldboard plow has declined in use is that manufacturers shipped 30% fewer plow shares in 1980 than in 1970 (U.S. Dep. of Commerce, 1971 and 1981).

Although numerous tillage tools and planter openers are available for operating through surface residue in CT systems, equipment is still a factor limiting farmer adoption of CT (Office of Technology Assessment, 1982).

Equipment has been developed with too few design criteria on soil effects (Siemens and Burrows, 1978), including distribution of crop residue in the soil profile, soil pulverization, soil compaction, soil thermal and hydraulic characteristics, and plant root response. Not surprisingly, reduced tillage and no tillage have sometimes been a hit-or-miss situation.

Most CT tools have evolved from modifications of existing clean-tillage tools to accommodate and maintain crop residue on the soil surface (Erbach et al., 1983). These modifications include increased ground clearance, coulters to cut residue, increased size and spacing of disc blades, reduced concavity of disc blades, and increased lateral and longitudinal spacing between adjacent tools. Devices have been added to align residue for easier passage of shanks and openers and to clear residue as it begins to collect on the front of shanks and openers. As CT technology progressed, it became evident that tillage and planting equipment must operate through crop residues without clogging. However, only recently have the design and operation of tillage tools been recognized as having direct effects on the soil environment and, subsequently, on plant growth, water management, and erosion control. Recent research has provided knowledge about the effect of specific tillage tools on soil environment in CT systems. These tools include disc harrows (Gill et al., 1982; Vaishnav et al., 1982), chisel plows (Erbach and Cruse, 1981), and seeding equipment (Choudhary and Baker, 1981; Harden et al., 1978; Hyde et al., 1979; Peterson et al., 1983; Schaaf et al., 1981; Wilkins et al., 1982; Wilkins et al., 1983). Future work should include research on methods and techniques for automatic control of individual tillage tools so that tillage and planting equipment can accommodate intrafield variations in soil conditions, crop residue production, and soil erosion potential.

A primary concern in selection of CT equipment is timeliness (Siemens and Burrows, 1978). If a field operation is not completed within an optimum period, there is some penalty. This penalty may be a decrease in crop yield, such as occurs in the Corn Belt if corn is not planted before the middle of May, or it may be a loss of soil (and wheat yield) if wheat is planted too late in the fall in the Pacific Northwest (Cochran et al., 1970). Sometimes, CT systems provide a timeliness advantage over clean-till systems because fewer tillage trips are usually needed before planting. This timeliness advantage of CT systems can be lost if application of herbicides is not timely because CT systems offer fewer opportunities to control weeds with tillage.

Choice and proper use of tillage equipment are critical to the success of CT systems. Pest control, fertilization, soil manipulation, and surface residue management all require equipment. Equipment must be selected so that these functions are integrated into a timely system. Much more must be learned about the tillage requirements for plant growth and for various soil conditions. As research provides knowledge in this area, tillage tools and systems will improve.

21-6 CONCLUSIONS

American agriculture can indeed gain profitability and greater conservation from conservation tillage. However, new technology will be needed to sustain the current adoption rate. Some of the most outstanding technological needs are summarized here.

1. The benefits of CT for control of soil erosion due to wind and water are universally recognized, but the associated tillage management requires focus upon the mechanisms involved. Tillage-related parameters (such as random roughness, erodible aggregate percentage, surface residue cover, and incorporated residue) must be recognized first, and then one can utilize theory to estimate influences of these tillage parameters on the processes involved in soil erosion. This analysis scheme encourages quantification of tillage effects on the soil surface and on surface layers of soil. The effects of random roughness on soil erosion must be researched, especially to understand how soil, rainfall, overland flow, freeze-thaw, and residues affect the decline of tillage-induced roughness. Measurements of tillage effects outside the Corn Belt are badly needed for precise control of soil erosion with tillage management.

2. Water stress was a problem in all seven Tillage Management Regions (TMR). Yet, theoretical relations are lacking between tillage-induced parameters and infiltration and soil water storage. The influences of conservation and conventional tillage on soil water relations must be more adequately clarified to be more useful in management.

3. Conservation tillage can cause dramatic changes in soil properties within the top 10 cm. To recognize these changes and their impact on self-mulching of soils, infiltration, pest management, nutrient availability, and biological activity, we must study soil layers less than 15 or even 7.5 cm thick. Also needed are better measurements of the distribution of buried residue within the whole surface layer of 20 cm to assist with soil erosion and water conservation projections from CT. The ecology of soil fauna and their management in CT systems should receive more research attention, especially as they affect self-mulching of soil and infiltration.

4. Current tillage guides in the Corn Belt indicate an overriding impact of soil drainage on the feasibility of CT. Air and soil temperatures, disease control, nutrient availability, herbicide performance, timeliness, and anoxia are involved. A major research effort is recommended to manage CT on soils with drainage difficulties. Because soil compaction is encouraged by CT and in turn accentuates poor drainage, it should be integral to this management effort.

5. Soil structure and the self-mulching property of soil near the surface ($\leq$ 5 cm deep), are much more critical in conservation and no-till than in conventional tillage. In the tillage guides, soil structure is linked to drill operation, plant population, infiltration, and timeliness, apparently as re-

lated to tractive performance. More use of information on soil consistence is urged to predict tractive performance.

6. Evaluations of crop yields from conservation vs. conventional tillage in seven TMR's indicated that CT and crop rotation are integral parts of each other. Conventional tillage usually facilitates a more intensive and productive crop rotation, and crop rotations facilitate CT systems. Production per unit land area is therefore enhanced by CT in nearly all TMR's except on the poorly drained soils in the Corn Belt. Some authors refer to this management as rotational tillage.

7. Variable and uncertain herbicide performance and the need for less risk of poor weed control was emphasized in nearly all field comparisons of CT and conventional tillage. Many construe this as one of the most critical management barriers to adoption of CT. Because herbicide activity on target and nontarget organisms is affected by herbicide interaction with soil and especially soil microbes, this aspect of CT requires much more attention.

8. Frequently, the comparisons of tillage systems in the seven TMR's indicated that CT was facilitated by some new tillage or planting machine. Suitable machinery systems weigh heavily in farmer acceptance of CT. With the increasing number of functions expected from tillage-planting machinery and the refinement of these functions in CT systems, success will require much more emphasis on multidisciplinary teamwork to develop effective machinery.

REFERENCES

Allmaras, R. R. 1967. Soil water storage as affected by infiltration and evaporation in relation to tillage-induced soil structure. p. 37–43. *In* Tillage for Greater Production, Conf. Proc., Am. Soc. Agric. Engr., St. Joseph, MI. 11–12 December 1967.

----, R. E. Burwell, and R. F. Holt. 1967. Plow-layer porosity and surface roughness from tillage as affected by initial porosity and soil moisture at tillage time. Soil Sci. Soc. Am. Proc. 31:550–556.

----, ----, W. E. Larson, and R. F. Holt. 1966. Total porosity and random roughness of the interrow zone as influenced by tillage. USDA Conserv. Res. Rep. 7. Washington, DC.

----, S. C. Gupta, J. L. Pikul, Jr., and C. E. Johnson. 1980. Soil erosion by water as related to management of tillage and surface residues, terracing, and contouring in eastern Oregon. USDA-AAR-W-10. Washington, DC.

----, E. A. Hallauer, W. W. Nelson, and S. E. Evans. 1977. Surface energy balance and soil thermal property modifications by tillage-induced soil structure. Minn. Agric. Exp. Stn. Bull. 306. St. Paul.

----, J. M. Kraft, J. L. Pikul, Jr., and R. W. Rickman. 1982. Compaction, soil water relations, and ecology of pea-root diseases. Am. Soc. Agron. Abstr. Am. Society of Agronomy, Madison, WI, p. 243.

----, W. W. Nelson, and E. A. Hallauer. 1972. Fall versus spring plowing and related soil heat balance in the western Corn Belt. Minn. Agric. Exp. Stn. Tech. Bull. 283. St. Paul.

Amemiya, M. 1977. Conservation tillage in the western Corn Belt. J. Soil Water Conserv. 32: 29–36.

Bauder, J. W., and G. W. Randall. 1982. Regression models for predicting corn yields from climatic data and management practices. Soil Sci. Soc. Am. J. 46:158–161.

Bauer, Armand, and A. L. Black. 1981. Soil carbon, nitrogen, and bulk density comparisons in two cropland tillage systems after 25 years and in virgin grassland. Soil Sci. Soc. Am. J. 45:1166–1170.

––––, and H. L. Kucera. 1978. Effect of tillage on some soil physicochemical properties and on annually cropped spring wheat yields. N. Dak. Exp. Stn. Bull. 506. Fargo.

Baxter, A. J., G. B. Triplett, Jr., and S. W. Bone. 1971. Evaluation of tillage systems. Ohio Agric. Exp. Stn. Res. Summary No. 55, Aug. Wooster.

Bennett, O. L. 1977. Conservation tillage in the Northeast. J. Soil Water Conserv. 32:9–12.

Black, A. L. 1973. Soil property changes associated with crop residue management in a wheat-fallow rotation. Soil Sci. Soc. Am. Proc. 37:943–946.

––––, and J. F. Power. 1965. Effect of chemical and mechanical fallow methods on moisture storage, wheat yields, and soil erodibility. Soil Sci. Soc. Am. Proc. 29:465–468.

Blevins, R. L. 1982. Does no-till change soil management practices? Soil Sci. News and Views 3:10 (Nov. 1982) Coop. Ext. Serv., Univ. of Ky., Lexington.

––––, D. Cook, S. H. Phillips, and R. E. Phillips. 1971. Influence of no-tillage on soil moisture. Agron. J. 63:593–596.

––––, L. W. Murdock, and G. W. Thomas. 1978. Effect of lime application on no-tillage and conventionally tilled corn. Agron. J. 70:322–326.

––––, M. S. Smith, G. W. Thomas, and W. W. Frye. 1983. Influence of conservation tillage on soil properties. J. Soil Water Conserv. 38:301–305.

––––, G. W. Thomas, and P. L. Cornelius. 1977. Influence of no-tillage and nitrogen fertilizations on certain soil properties after five years of continuous corn. Agron. J. 69:383–386.

––––, ––––, M. S. Smith, W. W. Frye, and P. L. Cornelius. 1982. Changes in soil properties after ten years continuous non-tilled and conventionally tilled corn. Soil and Tillage Res. 3:135–146.

Boels, D., D. B. Davis, and A. E. Johnston (ed.). 1982. Soil degradation. Proc. of a Land Use Seminar, Wageningen, The Netherlands. 13–15 Oct. 1980. A. A. Balkema, Rotterdam, The Netherlands.

Bondy, E., L. Lyles, and W. A. Hayes. 1980. Computing soil erosion by periods using wind-energy distribution. J. Soil Water Conserv. 35:173–176.

Bone, S. W., N. Rask, D. L. Forster, and B. W. Schurle. 1976. Evaluation of tillage systems for corn and soybeans. Ohio Rep. 61(4):60–63. Columbus.

Box, J. E., Jr., S. R. Wilkinson, R. N. Dawson, and J. Kozachyn. 1980. Soil water effects on no-till corn production in strip and completely killed mulches. Agron. J. 72:797–802.

Breece, H. E., H. V. Hansen, and T. A. Hoerner. 1981. Fundamentals of machine operation: planting. Deere and Co., Moline, IL.

Bruce, R. R., D. L. Myhre, and J. O. Sanford. 1968. Water capture in soil surface micro depressions for crop use. Trans. 9th Congr. Soil Sci., Adelaide, Australia, I:325–330.

Buckingham, F. 1976. Fundamentals of machine operation: tillage. Deere and Co., Moline, IL.

Burnside, O. C., G. A. Wicks, and D. R. Carlson. 1980. Control of weeds in an oat (*Avena sativa*)-soybean (*Glycine max*) ecofarming rotation. Weed Sci. 28:46–50.

Burwell, R. E., R. R. Allmaras, and M. Amemiya. 1963. A field measurement of total porosity and surface microrelief of soils. Soil Sci. Soc. Am. Proc. 27:697–700.

––––, and W. E. Larson. 1969. Infiltration as influenced by tillage-induced random roughness and pore space. Soil Sci. Soc. Am. Proc. 33:449–452.

Campbell, R. B., R. E. Sojka, and D. L. Karlen. 1984a. Conservation tillage for soybeans in the U.S. Southeastern Coastal Plains. Soil and Tillage Res. 4:511–529.

––––, ––––, and ––––. 1984b. Conservation tillage for maize in the U.S. Southeastern Coastal Plains. Soil and Tillage Res. 4:531–541.

Carreker, J. R., S. R. Wilkinson, A. P. Barnett, and J. E. Box. 1977. Soil and water management systems for sloping land. USDA, ARS-S-160. Washington, DC.

Chepil, W. S., and N. P. Woodruff. 1959. Estimations of wind erodibility of farm fields. USDA Prod. Res. Rep. 25. Washington, DC.

––––, and ––––. 1963. The physics of wind erosion and its control. Adv. Agron. 15:211–302.

Choudhary, M. A., and C. J. Baker. 1981. Physical effects of direct drilling equipment on undisturbed soil. II. Seed groove formation by a "triple disc" coulter and seedling performance. N.Z. J. Agric. Res. 24:183-187.

Cihacek, L. J., D. L. Mulvaney, R. A. Olson, L. F. Welch, and R. A. Wiese. 1974. Phosphorus placement for corn in chisel and moldboard planting systems. Agron. J. 66:665-668.

Cochran, V. L., L. F. Elliott, and R. I. Papendick. 1977. The production of phytotoxins from surface crop residues. Soil Sci. Soc. Am. J. 41:903-908.

----, ----, and ----. 1982. Effect of crop residue management and tillage on water use efficiency and yield of winter wheat. Agron. J. 74:929-932.

----, R. I. Papendick, and C. D. Fanning. 1970. Early fall crop establishment to reduce winter runoff and erosion on Palouse slopes. J. Soil Water Conserv. 25:231-234.

Cogo, N. P., G. R. Foster, and W. C. Moldenhauer. 1984. Effect of runoff and residue cover on soil erosion. Trans. Am. Soc. Agric. Eng. 28.

----, W. C. Moldenhauer, and G. R. Foster. 1984b. Soil loss reductions from conservation tillage practices. Soil Sci. Soc. Am. J. 48:368-373.

Collins, D. M. 1982. Achieving cost effective conservation. J. Soil Water Conserv. 37:262-263.

Colvin, T. S., J. M. Laflen, and D. C. Erbach. 1981. A review of residue reduction by individual tillage implements. p. 102-110. In J. C. Siemens (ed. chm.) Crop production with conservation in the 80's. ASAE Pub. 7-81. Am. Soc. Agric. Eng., St. Joseph, MI.

Cook, R. J., and K.F. Baker. 1983. The nature and practice of biological control of plant pathogens. Am. Phytopathol. Soc., St. Paul, MN.

----, M. G. Boosalis, and B. Doupnik. 1978. Influence of crop residues on plant diseases. p. 147-163. In Crop residue management systems. Spec. Pub. 31. American Society of Agronomy, Madison, WI.

Cosper, H. R. 1983. Soil suitability for conservation tillage. J. Soil Water Conserv. 38:152-155.

Council for Agricultural Science and Technology. 1983. Agricultural mechanization: Physical and societal effects and implications for policy development. Rep. 96. Council for Agricultural Science and Technology, Ames, IA.

Currence, H. D., and W. G. Lovely. 1970. The analysis of soil surface roughness. Trans. Am. Soc. Agric. Eng. 13:710-714.

Dexter, A. R. 1977. Effect of rainfall on the surface microrelief of tilled soil. J. Terramechanics 14:11-22.

----. 1979. Prediction of soil structures produced by tillage. J. Terramechanics 16:117-127.

Dick, W. A. 1983. Organic carbon, nitrogen, and phosphorus concentrations and pH in soil profiles as affected by tillage intensity. Soil Sci. Soc. Am. J. 47:102-107.

Doran, J. W. 1980. Soil microbial and biochemical changes associated with reduced tillage. Soil Sci. Soc. Am. J. 44:765-771.

Douglas, C. L., Jr., R. R. Allmaras, P. E. Rasmussen, R. E. Ramig, and N. C. Roager, Jr. 1980. Wheat straw composition and placement effects on decomposition in dryland agriculture of the Pacific Northwest. Soil Sci. Soc. Am. J. 44:833-837.

Dyer, J. A., and W. Baier. 1979. Weather-based estimation of field workdays in fall. Can. Agric. Eng. 21:119-122.

Edwards, C. A. 1975. Effects of direct drilling on the soil fauna. Outlook on Agric. 8:243-244.

----, R. R. van der Ploeg, and W. Ehlers. 1979. A numerical study of the effects of non-capillary-sized pores upon infiltration. Soil Sci. Soc. Am. J. 43:851-856.

Edwards, W. M., and W. E. Larson. 1969. Infiltration of water into soils as influenced by surface seal development. Trans. Am. Soc. Agric. Eng. 12:463-468.

Erbach, D. C., and R. M. Cruse. 1981. Chisel plow induced changes in soil conditions. Paper 81-1508. Am. Soc. Agric. Eng., St. Joseph, MI.

----, J. E. Morrison, and D. E. Wilkins. 1983. Equipment modification and innovation for conservation tillage. J. Soil Water Conserv. 38:182-185.

Erikkson, Janne, Inge Hakansson, and Berger Danfors. 1974. The effect of soil compaction on soil structure and crop yields. (In Swedish.) Inst. Agric. Eng., Uppsala. Bull. No. 754. (Eng. trans. by J. K. Aase, ARS, Sidney, MT.)

Fain, Dale. 1981. Moisture management in herbicide fallow for continuous wheat. p. 42–44. *In* Proc. Moisture Manage. Conf., Ardmore, OK. July 1981. Oklahoma State Univ.

Falayi, O., and J. Bouma. 1975. Relationships between the hydraulic conductance of surface crusts and soil management in a Typic Hapludalf. Soil Sci. Soc. Am. Proc. 39:957–963.

Farm and Industrial Equipment Institute. 1974. Men, machines, and land. Farm and Industrial Equipment Institute, Chicago, IL.

Foster, G. R. 1982. Modelling the erosion process. p. 297–380. *In* C. T. Haan, H. P. Johnson, and D. L. Brakensiek (ed.) Hydrological modelling of small watersheds. Am. Soc. Agric. Eng., St. Joseph, MI.

————, C. B. Johnson, and W. C. Moldenhauer. 1982. Critical slope lengths for unanchored cornstalk and wheat straw residue. Trans. Am. Soc. Agric. Eng. 25:935–939, 947.

French, E. W., and Neil Riveland. 1980. Chemical fallow in a spring wheat-fallow rotation. N. Dak. Farm Res. 38(1):12–15.

Galloway, H. M., and D. R. Griffith. 1978. Which tillage method is best for each Corn Belt soil? Crops Soils 30(8):16–18.

————, ————, and J. V. Mannering. 1981. Adaptability of various tillage-planting systems to Indiana soils. Coop. Ext. Serv., Purdue Univ., AY-210, West Lafayette, IN.

Gantzer, C. J., and G. R. Blake. 1978. Physical characteristics of Le Sueur clay loam following no-till and conventional tillage. Agron. J. 70:853–857.

Gardner, W. R., D. Hillel, and Y. Benyamini. 1970. Post irrigation movement of soil water. 2: Simultaneous redistribution and evaporation. Water Resour. Res. 6:1148–1153.

Gaultney, I., G. W. Krutz, G. C. Steinhardt, and J. B. Liljedahl. 1982. Effects of subsoil compaction on corn yields. Trans. Am. Soc. Agric. Eng. 25:563–569, 575.

Gill, W. R., A. C. Bailey, and C. A. Reaves. 1982. Harrow disk curvature—influence on soil penetration. Trans. Am. Soc. Agric. Eng. 25:1173–1179.

Godwin, R. J., and G. Spoor. 1977. Soil factors influencing work days. The Agric. Eng. 32: 87–90.

Greb, B. W., A. L. Black, and D. E. Smika. 1974. Straw buildup in soil with stubble mulch fallow in the semiarid Great Plains. Soil Sci. Soc. Am. Proc. 38:135–136.

————, D. E. Smika, and A. L. Black. 1967. Effect of straw mulch rates on soil water storage during summer fallow in the Great Plains. Soil Sci. Soc. Am. Proc. 31:556–559.

————, ————, and ————. 1970. Water conservation with stubble mulch fallow. J. Soil Water Conserv. 25:59–62.

————, and R. L. Zimdahl. 1980. Ecofallow comes of age in the Central Great Plains. J. Soil Water Conserv. 35:230–233.

Gregory, J. M. 1982. Soil cover prediction with various amounts and types of crop residue. Trans. Am. Soc. Agric. Eng. 25:1333–1337.

Grieve, I. C. 1980. The magnitude and significance of soil structural stability declines under cereal cropping. Catena 7:79–85.

Griffith, D. R. (chair). 1982. A guide to no-till planting after corn or soybeans. Purdue Univ. Coop. Ext. (Tillage) ID-154. Lafayette, IN.

————, J. V. Mannering, H. M. Galloway, S. E. Parsons, and C. B. Richey. 1973. Effect of eight tillage-planting systems on soil temperature, percent stand, plant growth, and yield of corn on five Indiana soils. Agron. J. 65:321–326.

Gupta, S. C., and W. E. Larson. 1982. Modelling soil mechanical behavior during tillage. p. 151–178. *In* D. M. Van Doren and P. W. Unger (ed.) Predicting tillage effects on soil physical properties and processes. Spec. Pub. 44. American Society of Agronomy, Madison, WI.

Haas, H. J., W. O. Willis, and J. J. Bond. 1974. Summerfallow in the western United States. Conserv. Res. Rep. 17. USDA, Washington, DC.

Hance, R. J., ed. 1980. Interactions between herbicides and the soil. Academic Press, New York.

Harden, S. C., J. W. Harden, and L. C. Harden. 1978. No-till plus in-row subsoiling. *In* Proc. 1st Ann. Southeastern No-Till Systems Conf. Georgia Agric. Exp. Stn. Spec. Pub. 5. Experiment, GA.

Hargrove, W. L., J. T. Reid, J. T. Touchton, and R. N. Gallagher. 1982. Influence of tillage practices on the fertility status of an acid soil double-cropped to wheat and soybeans. Agron. J. 74:684–687.

Harrold, L. L., G. B. Triplett, Jr., and W. M. Edwards. 1970. No-tillage corn. Agric. Eng. 51:128–131.

Helling, C. S., P. C. Kearney, and M. Alexander. 1971. Behavior of pesticides in soils. Adv. Agron. 23:147–240.

Hussein, M. H., and J. M. Laflen. 1982. Effects of crop canopy and residue in rill and interrill soil erosion. Trans. Am. Soc. Agric. Eng. 25:1310–1315.

Hyde, G. M., C. E. Johnson, J. B. Simpson, and D. M. Payton. 1979. Grain drill design concepts for Pacific Northwest conservation farming. Paper 79-1035. Am. Soc. Agric. Eng., St. Joseph, MI.

Idso, S. B., R. J. Reginato, R. D. Jackson, B. A. Kimball, and F. S. Nakayama. 1974. The three stages of drying in a field soil. Soil Sci. Soc. Am. Proc. 38:831–837.

Jenkinson, D. S. 1971. Studies on the decomposition of C^{14} labelled organic matter in soil. Soil Sci. 111:64–69.

Johnson, C. B., J. V. Mannering, and W. C. Moldenhauer. 1979. Influence of surface roughness and clod size and stability on soil and water losses. Soil Sci. Soc. Am. J. 43:772–777.

Johnson, W. C., and R. G. Davis. 1972. Research on stubble mulch farming of winter wheat. USDA-ARS Conserv. Res. Rep. 16.

Kaufman, D. D., and P. C. Kearney. 1976. Microbial transformations in soil. p. 29–64. *In* L. J. Audus (ed.) Herbicides. Academic Press, New York.

Kells, J. J., R. L. Blevins, C. E. Rieck, and W. M. Muir. 1980a. Effect of pH, nitrogen, and tillage on weed control and corn (*Zea mays*) yield. Weed Sci. 28:719–722.

––––, ––––, ––––, and ––––. 1980b. Atrazine dissipation as affected by surface pH and tillage. Weed Sci. 28:101–104.

Kemper, W. D., and E. J. Koch. 1966. Aggregate stability of soils from the western United States and Canada. USDA Tech. Bull. 1355.

Ketcheson, J. W., and D. P. Stonehouse. 1983. Conservation tillage in Ontario. J. Soil Water Conserv. 38:253–255.

Kuipers, H. 1957. A reliefmeter for soil cultivation studies. Netherlands J. Agric. Sci. 5:255–262.

Ladewig, Howard, and Ray Garibay. 1983. Reasons why Ohio farmers decide for or against conservation tillage. J. Soil Water Conserv. 33:487–488.

Laflen, J. M., and T. S. Colvin. 1981. Effect of crop residue on soil loss from continuous row cropping. Trans. Am. Soc. Agric. Eng. 24:605–609.

––––, G. R. Foster, and C. A. Onstad. 1985. Simulation of individual-storm soil loss for modeling the impact of soil erosion on crop productivity. p. 285–295. *In* S. A. El-Swaify et al. (ed.) Soil Erosion and Conserv. Proc. Int. Conf. Honolulu, HI. Jan. 1983. Soil Conservation Society, Ankeny, IA.

Langdale, G. W., J. E. Box, Jr., C. O. Plank, and W. G. Fleming. 1981. Nitrogen requirements associated with improved conservation tillage for corn production. Commun. Soil Sci. Plant Anal. 12:1133–1149.

Larson, W. E. 1962. Tillage requirements for corn. J. Soil Water Conserv. 17:3–7.

––––. 1963. Soil parameters for evaluating tillage needs and operations. Soil Sci. Soc. Am. Proc. 28:118–122.

––––, and R. R. Allmaras. 1971. Management factors and natural forces as related to soil compaction. p. 367–428. *In* K. K. Barnes et al. (ed.) Compaction of agricultural soils. Am. Soc. Agric. Eng., St. Joseph, MI.

Linden, D. R. 1979. A model to predict soil water storage as affected by tillage practices. Ph.D. diss. Univ. of Minnesota, St. Paul.

––––. 1982. Predicting tillage effects on evaporation from the soil. p. 117–132. *In* D. M. Van Doren et al. (ed.) Predicting tillage effects on soil physical properties and processes. Spec. Pub. 44. American Society of Agronomy, Madison, WI.

Lindstrom, M. J., W. B. Voorhees, and G. W. Randall. 1981. Long-term tillage effects on interrow runoff and infiltration. Soil Sci. Soc. Am. J. 45:945–948.

Lyles, L., and B. E. Allison. 1976. Wind erosion: The protective role of simulated standing stubble. Trans. Am. Soc. Agric. Eng. 19:61–64.

––––, and ––––. 1981. Equivalent wind-erosion protection from selected crop residues. Trans. Am. Soc. Agric. Eng. 24:405–408.

Mannering, J. V., and C. R. Fenster. 1983. What is conservation tillage? J. Soil Water Conserv. 38:111–143.

––––, and D. R. Griffith. 1981. Value of crop rotation under various tillage systems. Purdue Univ. Coop. Ext. Bull. A Y230. Lafayette, IN.

––––, and L. D. Meyer. 1963. Effects of various rates of surface mulch on infiltration and erosion. Soil Sci. Soc. Am. Proc. 27:84–86.

––––, ––––, and C. B. Johnson. 1966. Infiltration and erosion as affected by minimum tillage for corn. Soil Sci. Soc. Am. Proc. 30:101–105.

McCalla, T. M., and T. J. Army. 1961. Stubble mulch farming. Adv. Agron. 13:125–196.

Meyer, L. D., W. H. Wischmeier, and G. R. Foster. 1970. Mulch rates required for erosion control on steep slopes. Soil Sci. Am. Proc. 34:928–931.

Miller, G. A., T. E. Fenton, and R. M. Cruse. 1982. Full width tillage systems: Soil adaptability. In Farm agricultural resources management, 1982. Proc. Conf. on Conserv. Tillage, Ames, IA. 17–18 March. Iowa State Univ.

Miller, S. D., and A. G. Dexter. 1982. No-till crop production in the Red River Valley. N. Dak. Farm Res. 40(2):3–5.

Mitchell, J. K., J. C. Broch, and E. R. Swanson. 1980. Costs and benefits of terraces for erosion control. J. Soil Water Conserv. 35:233–236.

Moore, I. D. 1981. Effect of surface seal on infiltration. Trans. Am. Soc. Agric. Eng. 24:1546–1552, 1561.

––––, and C. L. Larson. 1979. Estimating micro-relief surface storage from point data. Trans. Am. Soc. Agric. Eng. 22:1073–77.

Moschler, W. W., D. C. Martens, C. I. Rich, and G. M. Shear. 1973. Comparative lime effects on continuous no-tillage and conventionally tilled corn. Agron. J. 65:781–783.

Musick, J. T., A. F. Wiese, and R. R. Allen. 1977. Management of bed-furrow irrigated soil with limited- and no-tillage systems. Trans. Am. Soc. Agric. Eng. 20:666–672.

Nelson, L. R., R. N. Gallaher, R. R. Bruce, and M. R. Holmes. 1977. Production of corn and sorghum grain in double-cropping systems. Agron. J. 69:41–45.

Norum, E. B., B. A. Krantz, and H. J. Haas. 1957. The Northern Great Plains. p. 494–505. In Soil. USDA Yearbk.

Ojeniyi, S. O., and A. R. Dexter. 1979. Soil factors affecting the micro-structures produced by tillage. Trans. Am. Soc. Agric. Eng. 22:339–343.

Office of Technology Assessment. 1982. Impacts of technology on U.S. cropland and rangeland productivity. OTA, U.S. Congress, Washington, DC.

Peterson, C. L., E. A. Dowding, K. N. Hawley, and R. W. Harder. 1983. The chisel-planter minimum tillage system. Trans. Am. Soc. Agric. Eng. 26:378–383, 388.

Phillips, R. E., R. L. Blevins, G. W. Thomas, W. W. Frye, and S. H. Phillips. 1980. No-tillage agriculture. Science 208:1108–1113.

Phillips, S. H. 1978. No-tillage, past and present. p. 1–5. In Proc. 1st Ann. Southeastern No-till Systems Conf., Experiment, GA. 28 Nov. Univ. of Georgia.

Pierce, F. J., W. E. Larson, R. H. Dowdy, and W. A. P. Graham. 1983. Productivity of soils: Assessing long term changes due to erosion. J. Soil Water Conserv. 38:39–44.

Ramig, R. E., and Les Ekin. 1983. No-till annual cropping. p. 23–28. In 1983 Columbia Basin Agric. Res. Spec. Rep. 680. Oregon Agric. Exp. Stn. Corvallis.

––––, and D. E. Smika. 1964. Fallow-wheat-sorghum: An excellent rotation for dryland in central Nebraska. Nebraska Agric. Exp. Stn. Bull. 483. Lincoln.

Rawls, W. J., D. L. Brakensiek, and K. E. Saxton. 1982. Estimation of soil water properties. Trans. Am. Soc. Agric. Eng. 25:1316–1320, 1328.

Reicosky, D. C., D. K. Cassel, R. L. Blevins, W. R. Gill, and G. C. Naderman. 1977. Conservation tillage in the Southeast. J. Soil Water Conserv. 32:13–19.

Richey, C. B., D. R. Griffith, and S. D. Parsons. 1977. Yields and cultural energy requirements for corn and soybeans with various tillage-planting systems. Adv. Agron. 29:141–182.

Robertson, L. S., D. L. Mokma, D. L. Quisenberry, W. R. Meggitt, and C. M. Hansen. 1976. No till corn: 3. Soils. Michigan State Univ. Coop. Ext. Bull. E-906. East Lansing.

Saxton, K. E., D. K. McCool, and R. I. Papendick. 1981. Slot mulch for runoff and erosion control. J. Soil Water Conserv. 36:44–47.

Schaaf, D. E., S. A. Hann, and C. W. Lindwall. 1981. Performance evaluation of furrow openers, cutting coulters, and press wheels for seed drills. p. 76–84. In J. C. Siemens (ed.) Crop production with conservation in the 80's. Am. Soc. Agric. Eng., St. Joseph, MI.

Siddoway, F. H. 1963. Effects of cropping and tillage methods on dry aggregate soil structure. Soil Sci. Soc. Am. Proc. 27:452–454.

Siemens, J. C., and W. C. Burrows. 1978. Machinery selection for residue management systems. p. 231–243. In W. R. Oschwald (ed.) Crop residue management systems. Spec. Pub. 31. American Society of Agronomy, Madison, WI.

Skidmore, E. L., and F. H. Siddoway. 1978. Crop residue requirements to control wind erosion. p. 17–33. In W. R. Oschwald (ed.) Crop residue management systems. Spec. Pub. 31. American Society of Agronomy, Madison, WI.

Slack, C. H., R. L. Blevins, and C. E. Rieck. 1978. Effect of soil pH and tillage on persistence of simazine. Weed Sci. 26:145–148.

Sloneker, L. L., and W. C. Moldenhauer. 1977. Measuring the amounts of crop residue remaining after tillage. J. Soil Water Conserv. 35:231–236.

Smika, D. E. 1970. Summerfallow for dryland winter wheat in the semiarid Great Plains. Agron. J. 62:15–17.

––––, and B. W. Greb. 1975. Nonerodible aggregates and concentration of fats, waxes, and oils in soils as related to wheat straw mulch. Soil Sci. Soc. Am. Proc. 39:104–107.

Soil Conservation Society of America. 1977. Conservation tillage: Problems and potential. Spec. Pub. 20. Soil Conserv. Soc. Am., Ankeny, IA.

Soil Survey Staff. 1975. Soil taxonomy—A basic system of classification for making and interpreting soil surveys. USDA-SCS Agric. Handb. 436. U.S. Government Printing Office, Washington, DC.

Stickler, F. C. 1982. Conservation tillage and the farm equipment industry. In Farm agricultural resources management, 1982. Proc. Conf. on Conserv. Tillage, Ames, IA. 17–18 March. Iowa State Univ.

Swan, J. B., and J.A. True. 1978. Tillage for corn and soybeans. p. 35–60. In Soils, soil management, and fertilizer monographs. Univ. of Minn. Agric. Ext. Serv. Spec. Rpt. 24. St. Paul.

Thomas, G. W., R. L. Blevins, R. E. Phillips, and M. A. McMahon. 1973. Effect of a killed sod mulch on nitrate movement and corn yield. Agron. J. 65:736–739.

Touchton, J. T., and J. W. Johnson. 1982. Soybean tillage and planting method effects on double-cropped wheat and soybeans. Agron. J. 74:57–59.

Triplett, G. B., Jr. 1970. Response of tillage systems as influenced by soil type. Trans. Am. Soc. Agric. Eng. 13:765–767.

––––, and D. M. Van Doren, Jr. 1977. Agriculture without tillage. Sci. Am. 236:28–33.

––––, D. M. Van Doren, Jr., and Samuel W. Bone. 1973. An evaluation of Ohio soils in relation to no-tillage corn production. Ohio Res. Dev. Center. Res. Bull. No. 1068. Wooster.

––––, ––––, and B. L. Schmidt. 1968. Effect of corn (Zea mays L.) stover mulch on no-tillage corn yield and water infiltration. Agron. J. 60:236–239.

Tyler, D. D., and J. R. Overton. 1982a. No-tillage advantages for soybean seed quality during drought stress. Agron. J. 74:344–347.

––––, and ––––. 1982b. Effects of no-tillage and deep seedbed preparation on soybeans. Tenn. Farm Home Sci. No. 121:2–4.

––––, ––––, and A. Y. Chambers. 1983. Soybean tillage effects on soil properties, diseases, cyst nematodes, and yield. J. Soil Water Conserv. 38:374–376.

Unger, P. W. 1968. Soil organic matter and nitrogen changes during 24 years of dryland wheat tillage and cropping practices. Soil Sci. Soc. Am. Proc. 32:427–429.

––––. 1978. Straw mulch effect on soil water storage and sorghum yield. Soil Sci. Soc. Am. J. 42:486–491.

––––, and T. M. McCalla. 1980. Conservation tillage systems. Adv. Agron. 33:1–58.

––––, and J. J. Parker, Jr. 1968. Residue placement effects on decomposition, evaporation, and soil moisture distribution. Agron. J. 60:469–472.

––––, and R. E. Phillips. 1973. Soil water evaporation and storage. p. 42–54. *In* Conservation tillage. Proc. Natl. Conf., Des Moines, IA. 28–30 March. Soil Conserv. Soc. Am., Ankeny, IA.

––––, and A. F. Wiese. 1979. Managing irrigated winter wheat residues for water storage and subsequent dryland grain sorghum production. Soil Sci. Soc. Am. J. 43:582–587.

––––, ––––, and R. R. Allen. 1977. Conservation tillage in the Southern Plains. J. Soil Water Conserv. 32:43–48.

University of Illinois. 1976. Illinois Agronomy Handb., Univ. of Illinois Agric. Exp. Stn., Urbana.

USDA. 1938. Soils and Men, 1938 Yearbook of Agriculture. USDA, Washington, DC.

––––. 1980. Agricultural statistics 1980. USDA, Washington, DC.

––––. 1981. Land resource regions and major land resource areas of the United States. USDA-SCS Agric. Hanbd. 296. U.S. Government Printing Office, Washington, DC.

––––. 1982. RCA, A national program for soil and water conservation: 1982. Final program report and environmental impact statement. USDA, Washington, DC.

U.S. Dep. of Commerce. 1971. Farm machinery and lawn and garden equipment. U.S. Dep. of Commerce, Washington, DC.

––––. 1981. Farm machinery and lawn and garden equipment. U.S. Dep. of Commerce, Washington, DC.

––––. 1982. Tractors (except garden tractors). U.S. Dep. of Commerce, Washington, DC.

Vaishnav, A. S., R. L. Kushwaha, and G. C. Zoerb. 1982. Evaluation of disc coulters as affected by straw and cone index under zero till practices. Paper 82-1517. Am. Soc. Agric. Eng., St. Joseph, MI.

Van Doren, C. A., and R. S. Stauffer. 1943. Effect of crop and surface mulches on runoff, soil losses, and soil aggregation. Soil Sci. Soc. Am. Proc. 8:97–101.

Van Doren, D. M., Jr. 1965. Influence of plowing, discing, cultivation, previous crop, and surface residue on corn yields. Soil Sci. Soc. Am. Proc. 29:595–597.

––––, and R. R. Allmaras. 1978. Effect of residue management practices on the soil physical environment, microclimate, and plant growth. p. 49–83. *In* W. R. Oschwald (ed.) Crop residue management systems. Spec. Pub. 31. American Society of Agronomy, Madison, WI.

––––, and G. J. Ryder. 1962. Factors affecting use of minimum tillage for corn. Agron. J. 54: 447–450.

––––, and G. B. Triplett, Jr. 1973. Mulch and tillage relationships in corn culture. Soil Sci. Soc. Am. Proc. 37:766–769.

––––, ––––, and J. E. Henry. 1976. Influence of long term tillage, crop rotation, and soil type combinations on corn yield. Soil Sci. Soc. Am. J. 40:100–105.

Voorhees, W. B., R. R. Allmaras, and C. E. Johnson. 1981. Alleviating temperature stress. p. 217–266. *In* G. F. Arkin and H. M. Taylor (ed.) Modifying the root environment to reduce crop stress. Monograph 4. Am. Soc. Agric. Eng., St. Joseph, MI.

––––, and M. J. Lindstrom. 1983. Soil compaction constraints on conservation tillage in the northern Corn Belt. J. Soil Water Conserv. 38:307–311.

––––, C. G. Senst, and W. W. Nelson. 1978. Compaction and soil structure modification by wheel traffic in the northern Corn Belt. Soil Sci. Soc. Am. J. 42:344–349.

Whitaker, F. D., H. G. Heinemann, and W. H. Wischmeier. 1973. Chemical weed controls affect runoff, erosion, and corn yields. J. Soil Water Conserv. 28:174–176.

Wicks, G. A., and D. E. Smika. 1973. Chemical fallow in a winter wheat-fallow rotation. Weed Sci. 21:97–102.

Wilkins, D. E., G. A. Muilenburg, R. R. Allmaras, and C. E. Johnson. 1983. Grain-drill opener effects on wheat emergence. Trans. Am. Soc. Agric. Eng. 26:651–655, 600.

----, P. E. Rasmussen, B. L. Klepper, and D. A. Haasch. 1982. Grain drill opener design for fertilizer placement. Paper 82-1516. Am. Soc. Agric. Eng., St. Joseph, MI.

Wilkinson, B. 1975. Soil types and direct drilling—a provisional assessment. Outlook on Agric. 8:233–235.

Williams, J. R., and K. G. Renard. 1984. Assessments of soil erosion and crop productivity with process models (EPIC). p. 67–103. *In* R. F. Follett (ed.) Soil erosion and crop productivity. American Society of Agronomy, Crop Science Society of America, and Soil Science Society of America, Madison, WI.

Wischmeier, W. H. 1975. Estimating the soil loss equation's cover and management factor for undisturbed areas. p. 118–124. *In* Present and prospectie technology for predicting sediment yields and sources. USDA. ARS-S-40. Washington, DC.

----, and D. D. Smith. 1978. Predicting rainfall erosion losses: A guide to conservation planning. Agric. Handb. 537. USDA, Washington, DC.

Woodruff, N. P., C. R. Fenster, W. S. Chepil, and F. H. Siddoway. 1965. Performance of tillage implements in a stubble mulch system. I: Residue conservation. Agron. J. 57:45–49.

----, and F. H. Siddoway. 1965. A wind erosion equation. Soil Sci. Soc. Am. Proc. 29:602–608.

Zingg, A. W., and C. J. Whitfield. 1957. A summary of research experience with stubble-mulch farming in the western states. USDA Tech. Bull. 1166. Washington, DC.

Zuzel, J. F., R. R. Allmaras, and R. Greenwalt. 1982. Runoff and soil erosion on frozen soils in northeastern Oregon. J. Soil Water Conserv. 37:351–354.

22 Simulation of Tillage Residue and Nitrogen Management

J. A. E. Molina
University of Minnesota
St. Paul, Minnesota

M. J. Shaffer, and R. H. Dowdy
Agricultural Research Service
U.S. Department of Agriculture
St. Paul, Minnesota

J. F. Power
Agricultural Research Service
U.S. Department of Agriculture
Lincoln, Nebraska

Tillage disrupts the natural evolution of the soil-crop system. Performed to accomplish time- and space-limited objectives, such as residue and fertilizer incorporation, weed control, seed bed preparation and planting, tillage also induces discontinuities that have a sustained effect on the rates of the numerous chemical, physical, and biological changes that underlie soil fertility and plant growth. The soil-crop system is also affected by managerial choices such as types of crop germplasm and rotation, tillage methods, fertilizer applications, and chemicals used.

First, the management components affecting soil erosion and crop productivity are outlined. Second, we show how computer simulation models can integrate these components. Finally, some futuristic views are given of the possible use of computer-assisted decisions to reduce soil erosion and improve crop productivity and economic returns.

Published in R. F. Follett and B. A. Stewart, ed. 1985. *Soil Erosion and Crop Productivity.*
© ASA-CSSA-SSSA, 677 South Segoe Road, Madison, WI 53711, USA.

22–1 THE COMPONENTS OF SOIL EROSION AND CROP PRODUCTIVITY

22–1.1 Tillage and Residue Management

The chemical, physical, and biochemical properties of the soil-crop system are so interdependent that not one soil property is left unaffected by tillage and residue incorporation. For the sake of simplicity, however, these properties will be treated separately.

22–1.1.1 Physical Properties

We can group the physical components into two categories: those that affect the overall flow of water, nutrients, and heat into or out of the soil root zone (boundary conditions); and those that more directly affect the distribution of these components (and roots) within the system. Infiltration rates and heat fluxes at the soil surface are examples of boundary factors. Soil hydraulic conductivities and thermal properties within the soil profile are examples of factors in the second category. Root growth and uptake of water and nutrients are affected by both categories.

Soil erosion tends to remove or redistribute existing surface layers. The redefined surface materials must then be managed to optimize root growth. Upper boundary conditions, such as infiltration rates and soil surface temperatures, can be managed with appropriate combinations of crop residues and tillage. Gupta et al. (1981) simulated the effect of surface residues on soil temperatures. Seed germination and early plant growth can be managed by manipulating the surface residue cover. The right combination of seedbed temperature and water content is needed to get the young plant established. This can be particularly important when subsoil properties are less than optimal.

Problems associated with subsurface horizons—such as compaction, poor structure, poor internal drainage, and low water holding capacity—are more difficult to manage. Likewise, an impermeable lower boundary becomes much more significant as it approaches the soil surface. Deep chisel plowing and selection of shallow rooting crops sometimes can be used to improve the rooting environment in these cases. The beneficial effect of deep plowing in a compacted soil is shown in Fig. 22–1 (Larson et al., 1983). The improved yield was a direct result of a deeper rooting depth and increased available water.

Supplemental irrigation is another tool for offsetting the effects of a shallow root zone or low water-holding capacity. It may be the only effective means of managing some eroded profiles short of replacing the eroded soil material. An example of supplemental irrigation on an eroded profile is shown in Fig. 22–2.

The effects of changes in soil physical properties on the root system and overall crop growth are complex and not completely understood. The extent to which a root system can compensate for less than optimal condi-

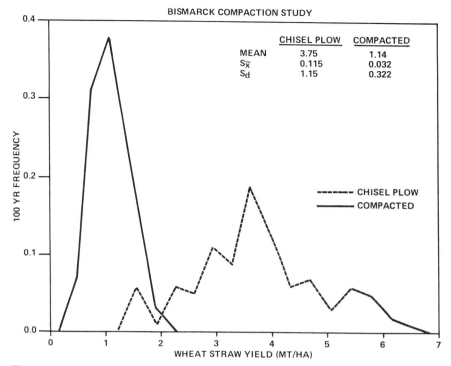

Fig. 22–1. Long-term yield distribution of wheat straw under compacted and chisel plow conditions, Bismark, ND.

tions before yield is affected must be a primary component of any field study or simulation model attempting to describe the processes involved.

Recent reviews of tillage effects and residue management practices on soil physical properties and plant growth were published by the American Society of Agronomy (1978a and 1982).

22–1.1.2 Biological Properties

Residues are a source of energy for the soil microorganisms and provide aromatic nuclei from which stable humus is synthesized. Differences in temperature and moisture conditions between tilled and minimally tilled soils are expected to be reflected in the rates of residue decomposition. More important, however, a shift in the dominant degradative agents may occur as intense tillage practices are replaced by reduced or no-till management. The soil macrofauna and microbes are responsible for residue decomposition and assimilation in the humus fraction (Anderson, 1979). Under forest cover, inside the moist surface litter, the macrofauna plays a primary role by delivering to the microbial population of its intestinal tract a product that has been triturated by mandibles and thus has an increased exposed surface to enzymatic action (Schaller, 1968). Under reduced or no-till situations, surface residue accumulation may create a surface litter where the macrofauna takes added importance over the soil filamentous

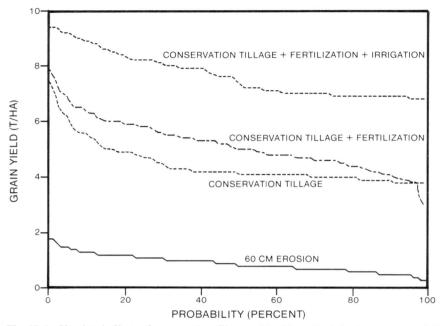

Fig. 22-2. Simulated effects of conservation tillage and fertilizer plus irrigation on an eroded Dakota fine sandy loam.

microorganisms which are usually thought of as dominant for frequently tilled soils. Differences in the type of humus produced by the two types of residue decomposition may affect soil stability and potentially mineralizable nitrogen (Stanford and Smith, 1972), but these effects need documentation.

Rates of decomposition of biological residue are reduced if the microbes do not have enough nitrogen available for the synthesis of proteins corresponding to the assimilated carbon. Residues with a high proportion of resistant or slowly decomposing material, such as cellulose and lignin, are low in nitrogen content. Hunt (1977) established a relationship between the carbon nitrogen ratio (C/N) of plant material and the proportion of resistant fraction. The material can be assumed to be composed of only two fractions: a labile fraction (sugar, starches, proteins, etc.) and a resistant fraction. Each fraction decomposes at a fixed specific rate under optimum environmental conditions. The respective proportion of labile to resistant fractions, which is related to the C/N ratio, affects the rate of residue decomposition. The decomposition rate of residues with high C/N ratios can be increased by applying nitrogen fertilizer. Hunt (1977) and Molina et al. (1983) developed simulation models which account for these effects.

Nitrogen also controls the percentage of carbon used for microbial biosynthesis rather than for carbon dioxide production. If nitrogen is adequate, the efficiency factor of carbon assimilation may be as high as 0.7 (McGill et al., 1981). Recently decomposed materials enriched in biomass, such as manure, have a high mineralization potential and can release nitrogen with sustained rates of mineralization. Thus, in reduced tillage systems, the placement of residues and nitrogen fertilizer can affect residue decomposition and use.

The combination of nitrate, concentrated sources of bioenergy, and high soil water content favors denitrification. Such conditions are enhanced by reduced and no-till management. Some data indicate that denitrification rates increase when the residue is left on the soil surface (Rice and Smith, 1982), although much work remains to define quantitatively the parameters involved. If, for some climatic and soil conditions, loss of nitrogen by denitrification increases with no-till practices, nitrification inhibitors might be used (American Society of Agronomy, 1978b).

22-1.1.3 Chemical Properties

Any reduction in the number of tillage operations reduces soil mixing in the tillage zone. Depending on the method of fertilizer application, reduced soil mixing promotes a concentration of plant nutrients in some zones of the soil, while other areas may become depleted of the same nutrients. In the extreme, steep concentration gradients can develop when N, P, and K are applied broadcast in zero-tillage systems. If the zones of nutrient accumulation do not coincide with zones of high rate development, the potential for reduced nutrient uptake may limit plant growth. Many factors besides the reduced mixing of soil may accentuate this uneven distribution of plant nutrients, such as, insolubility of nutrient source, high adsorption energy and capacity of the soil for the given nutrient, and increased tortuosity of the diffusion path. Randall and Swan (1978) showed that stratification of Bray I extractable P and exchangeable K does occur with reduced tillage systems (Table 22-1). They and Moncrief (1981) showed that diagnostic leaf concentrations of P were not affected by tillage systems and were sufficient for optimum plant growth. Increased surface residue cover in reduced tillage systems may keep the surface soil at a higher water content, thus allowing greater root activity in the zone of P accumulation. However, the potential for plant P deficiencies still exists under situations of lower soil P levels, reduced fertilizer applications, and more droughty growing conditions. If reduced tillage systems result in nutrient deficiencies

Table 22-1. Phosphorus and potassium distribution in a clay loam soil as influenced by three years of tillage.†

Depth	Phosphorus			Potassium		
	No tillage	Fall plow	Fall chisel	No tillage	Fall plow	Fall chisel
m	mg/kg					
0 −0.05	36	17	27	244	156	240
0.05–0.10	18	22	22	122	188	177
0.10–0.15	15	28	16	113	209	134
0.15–0.23	10	20	10	115	190	123
0.23–0.30	4	5	2	107	136	111

† Fertilizer applications consisted of 18 kg/ha of P and 66 kg/ha of K broadcast and 7 kg of P and 17 kg/ha of K starter fertilizer, annually.

over the long-term, fertilizer application likely will shift from broadcast to banding near the seed.

The effects of tillage on the availability of K are not as definitive as for P. As summarized by Stanford et al. (1973) and Moncrief (1981), some studies with reduced tillage show decreased K availability, while others show increased K uptake. Obviously, under some conditions, decreased K availability may limit plant growth, although we may not know the conditions that lead to reduced K uptake. Some of these contradictions may be related to wide differences in clay mineralogy of the various soils studied as well as the exchangeable K status of the soil (Dowdy and Hutcheson, 1963) and the specific climatic regimes under which the measurements were made. The experimental results of the various studies may also be confounded by differences in bulk densities among studies. Bulk density (compaction) has been shown to increase under reduced tillage (Voorhees et al., 1978). Compaction, in turn, has reduced K, Mg, and Ca levels in growing plants (Castillo et al., 1982). Hence, the effects of tillage on the chemistry of soil K is still to be resolved.

22-1.2 Weed Management

One principal reason for tillage in any cropping system is weed control. With the introduction of reduced and no-tillage systems, weed control became a prominant consideration. Hence, the increased availability and use of a broad range of herbicides are essential for economical crop production using reduced tillage systems. This shift in management systems is not without problems, however. Weed species and populations shift readily with the tillage practices used. Richey et al. (1977) summarized the literature by concluding that ". . . the weed problem builds up year after year with continuous no-till culture and rotary strip tillage systems . . ." in corn and soybean production. The same trends toward increased weed populations were noted by Miller and Nalewaja (1979) in spring wheat (Table 22-2). Often, weed populations shift from annual to perennial broadleaf and grass species (Triplett and Lytle, 1972) as reduced tillage systems are introduced, because of lack of mechanical damage to underground reproductive tissue (rhizomes, tubers, bulbs, and stolons).

The increased coverage of the soil surface with crop residues and increased soil water conservation are recognized attributes of reduced tillage

Table 22-2. Weed population in spring wheat as influenced by tillage.

Tillage	Plants	
	Broadleaf	Grass
	plants/m²	
No-till	29	112
Disk	37	100
Chisel	29	106
Plow	17	54

systems, but they also provide excellent environments for weed seed germination, particularly small seeds of annual grasses. In addition, surface residues can physically prevent preemergence herbicides from reaching the soil surface, especially where rainfall is limiting at the time of application. These considerations may require more reliance on postemergence herbicides for weed control. For these and many other reasons, comprehensive soil and crop management models allow researchers to better predict the most effective herbicide program for a given soil, crop, tillage, residue, and climatic situation. Most of the same considerations also apply to other crop pests such as insects and diseases.

22-1.3 Plant Factors

The most important plant factors affecting soil erosion and crop productivity are the quantity and composition of the crop biomass produced and the uses made of this biomass. These are discussed by type of plant—annual crops, perennial legumes, and perennial grasses.

22-1.3.1 Annual Crops

Annual crops are generally produced for seed, root, or total aboveground biomass production. Harvested product is usually physically removed from the field and may or may not be eventually returned to the soil as byproducts of some other enterprise. Crops may be either legumes or nonlegumes.

Annual crops most commonly produced for harvested seed are the cereal grasses and certain annual legumes, such as (soybeans [*Glycine max* L.] and peas [*Pisum sativum* L.]). The quantity of N fixed by legumes varies widely from essentially 0 to 80% of the quantity of N taken up by the crop; however, typically, in crops like soybeans, most of the fixed N is removed in the harvested seed. Also, because such crops are commonly produced on fertile soils in rotation with well-fertilized cereal crops, the relatively high levels of available N in the soil may drastically limit biological N fixation. Thus, for typical situations of soybean production in the Midwest, more N is removed in the harvested seed than is biologically fixed, resulting in a net reduction in soil N.

Legume residues decompose more rapidly than do those of grain crops. Consequently, less residue is available for soil protection, increasing the potential for soil erosion. In addition, the surface of soils from which soybeans have been harvested is characteristically loose, friable, and mellow. Although reasons for this change in soil physical structure have not been adequately identified, detachment and removal of soil particles by falling water occurs much more readily than for soil in grain crops.

The cereal crops—particularly corn [*Zea mays* L.], wheat [*Triticum aestivum* L.], barley [*Hordeum vulgare* L.], and sorghum [*Sorghum bicolor* L.]—may be harvested for either grain or silage. When harvested for grain, ample crop residues are usually left to protect the soil from erosion and to maintain soil levels of organic matter. However, if residues are removed as silage or for other uses, the soil is left exposed and highly erodible, and soil organic matter levels eventually decline (Larson et al., 1972).

The most common root crops produced are sugar beets [*Beta vulgare* L.] and potatoes [*Solanum tuberosum* L.]. Soil erosion, especially wind erosion, is often a major problem for both crops. Historically, a fine, pulverized seedbed is at least partially prepared in the fall for sugar beets, and this seedbed may incur severe wind erosion in winter and spring months. Many of the major potato-producing soils are sandy and, when exposed over winter with little vegetative cover, are highly susceptible to wind erosion. In addition, sugar beets are commonly harvested when the soil is wet, causing additional problems of soil puddling and structure deterioration.

22–1.3.2 Perennial Legumes

Perennial legumes were historically produced in rotation with cereal crops to provide available N to the cereals, but within the past 3 decades in North America, N needs of cereal crops have been largely met through N fertilization. The literature of several decades shows that soil erosion from legume rotations is usually considerably less than from rotations without legumes. Also, much of this literature shows that legumes in rotation reduced the rate at which soil organic matter declined following cultivation. However, in regions of water deficits, grain yields following legumes were often less than those following grain crops because the deep-rooted legumes depleted soil water reserves more thoroughly and to deeper depths than did the grain crops (Haas and Evans, 1957).

Numerous possibilities for intercropping, multiple cropping, cover crops, and other uses of legumes need to be evaluated for U.S. agriculture. Inclusion of legumes in a cropping system always has the potential of not only reducing fertilizer N requirements but also reducing soil loss by erosion.

22–1.3.3 Perennial Grasses

Sod-forming perennial grasses have historically been used to control soil erosion in highly erosive situations and on structures susceptible to erosion. The extensive fibrous root system of perennial grasses holds soil in place and prevents its slippage or detachment. The resulting improvement in soil agregation may persist for several years after the grass is destroyed, thereby reducing erosivity of the soil.

Perennial grasses are the basis for most pasture and grazing systems. An appreciable acreage is still cropped to these grasses in rotation with grain crops in some areas of North America. This presence in the rotation not only greatly reduces soil erosion losses but also helps maintain soil organic matter levels and soil productivity.

The introduction of no-tillage systems make it feasible to convert from a perennial grass to a grain crop without cultivation. Such techniques permit grain crop production on steep slopes and are particularly beneficial to such regions as Appalachia (Bennett et al., 1976). A variant of this procedure is to kill sod in strips with chemicals, leaving live sod in the interrow areas. In rangelands, this technique permits the introduction of legumes or domesticated grass species without cultivation and its attendant problems of wind and water erosion.

22-2 THE INTEGRATOR OF SOIL EROSION AND CROP PRODUCTIVITY COMPONENTS

In 1979, the USDA-ARS initiated a national research program to develop guidelines for the integrated management of tillage practices, crop residues, and nitrogen fertilizers in U.S. agriculture. A number of ARS research locations and cooperating universities have been involved. The development of a computer-simulation model system for nitrogen-tillage-residue management was a prime objective. This family of models, known as the Nitrogen-Tillage-Residue-Management model, was developed by the ARS Soil and Water Management Research Unit in St. Paul, MN, in cooperation with the University of Minnesota, the ARS research unit at Morris, MN, and other cooperating ARS locations and universities. Several systems have been formulated from this cooperative endeavor, including NTRM and NCSWAP.

22-2.1 NTRM

The NTRM model, Fig. 22-3, was developed over about 4 years by utilizing and adapting existing submodels and by developing others. Soil physicists, soil chemists, hydrologists, plant physiologists, soil microbiologists, engineers, and computer scientists contributed to its development.

The primary objective was to develop a comprehensive, mechanistic model of the soil environment and its effects on crop growth. Emphasis was placed on the nitrogen, tillage, and residue aspects of the system. The scope of the model was kept wide enough to allow applications to research, crop and soil management, and teaching.

A cross section of the crop-soil-water continuum and aquifer being simulated with the NTRM model is shown in Fig. 22-4. Note the basic physical, chemical, and biological processes being modeled. The aquifer portions of the model system are a separate set of submodels which require input from the unsaturated-zone simulators. Likewise, the soil temperature submodel is an independent program which generates inputs for the overall model.

The model for irrigation return flow reported by Shaffer et al. (1977) was used as the core of the NTRM model. Their model already contained

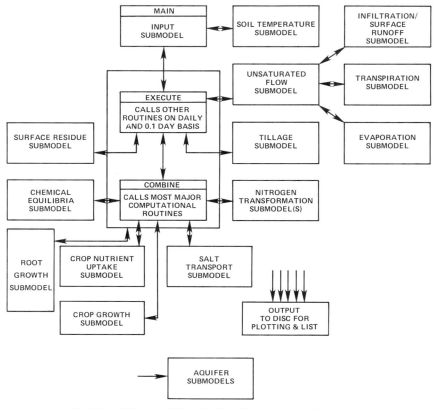

Fig. 22-3. Nitrogen-Tillage-Residue-Management (NTRM) model.

relatively comprehensive submodels for water flow, nutrient transformations, salt reactions, and salt transport.

The Nebraska corn growth model, CORNGRO (Childs et al., 1977),
was incorporated with modifications as a submodel for crop growth. Submodels for additional crops will be added as the model is further developed.
A dynamic root-growth model (Shaffer and Clapp, 1982) was developed
from theory to complement the corn-growth model for top growth and total
dry matter production. The unsaturated flow model of Dutt et al. (1972)
and Shaffer and Gupta (1981) was originally based on works reported by
Hanks and Bower (1962) and was updated to include both the layered case
and upper boundary conditions which were more suitable for rapid water
fluxes and surface evaporation. The evaporation and infiltration submodels
developed by Linden (1979) were incorporated in the overall model.

A second nitrogen transformation submodel (NCSOIL) developed by
Molina et al. (1983) was incorporated in the model. This submodel features
^{15}N tracer capabilities and is more detailed and less empirical than
Shaffer's N transformation model in the areas of crop residue and soil organic matter transformations. Also included in NTRM were the options of
using Shaffer's et al. (1977) transition state submodel for nitrification and

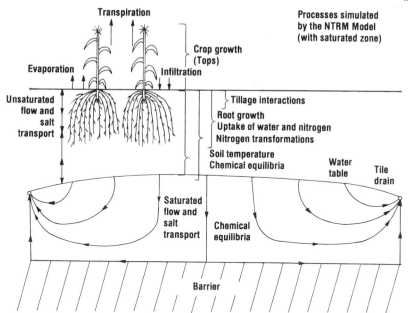

Fig. 22–4. Processes simulated by the NTRM model (with saturated zone).

an expression for urea hydrolysis (Shaffer et al., 1969). Tracer capabilities of ^{15}N were also incorporated into the overall NTRM model.

New models and information transfers were developed to simulate crop residue and tillage interactions. Capability was added to allow decay of residues on the soil surface and interaction of these residues with crop growth, heat flow, and water flow. Similar capabilities were included for residue incorporated into the soil.

The effects of tillage practices on soil physical, chemical, and biological properties can be simulated in the model. The effects of those changes on crop growth can be then modeled using a separate set of relationships.

The soil temperature model reported by Gupta et al. (1981) was incorporated into the NTRM model system. This submodel operates as an independent program with input to the interactive model via transfers of data files.

Capability was retained in the overall NTRM model to generate soil leachate volumes and constituent concentrations suitable for input to aquifer submodels such as the one reported by Ribbens and Shaffer (1976).

Models such as NTRM can be readily applied to simulate the impacts of long-term climate variability on crop yields from eroded profiles. This has already been done for uneroded profiles in the semiarid Great Plains (Larson et al., 1983).

To illustrate the application of the NTRM model to uneroded and eroded conditions in major land resource area (MLRA) 105, a Dakota fine sandy loam soil was elected. This soil is characterized by its shallow depth to coarse material. Model runs were made for the uneroded state, for 30 cm of erosion, and for 60 cm of erosion. Management was assumed to be conventional tillage with 150 kg/ha of fertilizer N applied at the start of the growing season. Weeds, insects, and disease were assumed to be controlled and other nutrients present in sufficient supply.

The impacts of applying a series of management practices to the 60-cm eroded case are shown in Fig. 22–2. Conservation tillage in the form of notill was simulated followed by a combination of no-till and sidedress fertilizer, and then no-till, sidedress, and irrigation. Note the increase in yields obtained from each combination. The yield results obtained with irrigation were higher than with the long-term noneroded soil, indicating the droughty nature of this soil. Note also that the standard deviation increased for the middle treatments and then decreased slightly for the irrigated case.

22–2.2 NCSWAP

Several programs were used to build NCSWAP: NCSOIL (Molina et al., 1983) for nitrogen and carbon transformations in soil; INFIL (Linden, 1979), a multilayered Green and Ampt approach to water microporous infiltration; and REDIS (Linden, 1979), a water redistribution subsystem. The model uses the crop as a biological integrator of the exogeneous (managerial and meterological) variables and soil-status variables. It computes the crop yield from a reference yield which is modified to account for the influence of water, N, and temperature. The computed crop state feeds back on the soil state. The reference crop is defined as controllable inputs by the kinetics of plant mass growth and N percentage for a set of air temperatures, a specific field, and the assumption of neither water nor N limitations. Other nutrient limitations inherent to the field can be built into the reference kinetics. Losses from pests and weed competition are not considered.

The objectives of NCSWAP are therefore to simulate water, C, N, and ^{15}N dynamics in the soil profile and to translate these status variables into changes in the state of the reference crop through the concepts of water, N, and temperature stress, expressed as reduction-factor functions. A guide for the execution of NCSWAP with a description of the input-output format, including a demonstration case, is available for those who would like to participate in the validation process (Molina and Richards, 1983).

Two types of input variables, environmental and process, are considered. Environmental input variables are divided into five categories: time-frame, crop, soil, meteorology, and management. Process variables include specific rates of transformations, reduction-factor functions, etc. Model tuning is used as a research tool to obtain information about those process variables for which experimental data are either not available or have been obtained for environmental conditions outside the range specified by the environmental input data.

An efficient tuning requires the ability to obtain rapidly a global diagnostic of NCSWAP dynamic performance for each set of input conditions.

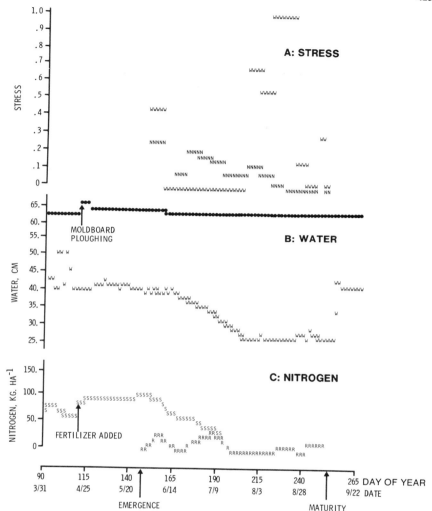

Fig. 22-5. Simulation by NCSWAP. a: Weekly average stress nitrogen (N) and water (W). b: Water content, actual (W) and at field capacity (•), in the soil profile. c: Soluble inorganic nitrogen in the profile (S) and the root zone (R).

For this purpose, the kinetics of selected state variables are displayed on a set of 11 graphs. Following are examples of such graphs to illustrate their usefulness in the analysis of one soil-crop-management situation, in this case, corn grown under nitrogen and water stress in a deep, well-drained, tilled soil.

At emergence, on day 150, a minor nitrogen stress (Fig. 22-5a) appears as a result of leaching of soluble inorganic nitrogen from the topsoil by abundant rainfalls (376 mm up to day 104). The nitrogen percentage in the young plant is drastically reduced (Fig. 22-6), then restored to optimum values when the roots catch up with the upper nitrogen front (Fig. 22-5c).

From day 165 to 195, the plant nitrogen uptake decreases the soil reserves. Soluble inorganic nitrogen is present in the soil profile but is rapidly depleted from the root zone (day 160 to 170). This process is compensated by a net nitrogen mineralization (Fig. 22-7). During the same period, evapotranspiration is not balanced by precipitation (58 mm from day 105 to 211). Gradually, the soil water reserves are depleted (Fig. 22-5b). Water stress in some soil layers of the root zone also limits the plant availability of soluble inorganic nitrogen (Fig. 22-5c), and on day 200, water stress levels are reached for the whole profile. On day 225, the crop has reached physiological maturity before the fall rains can recharge the soil profile and release the

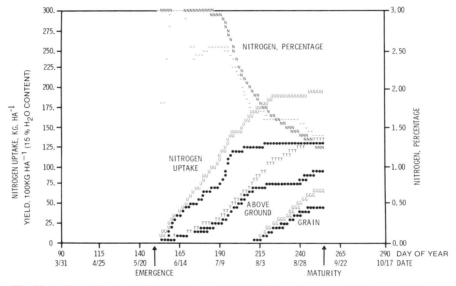

Fig. 22-6. Simulation by NCSWAP. Crop yield at 15.5% water content. Reference yields are shown by T for above ground yields and G for grain yields; actual yields are shown by •. Reference for nitrogen top plant uptake (kg/ha) is shown by U and actual top plant uptake by •. Reference for top nitrogen percentage on a dry basis is shown by N and actual by —.

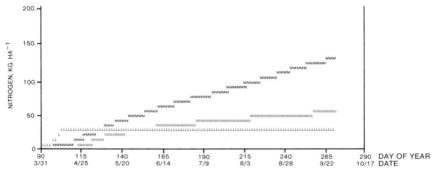

Fig. 22-7. Simulation by NCSWAP. Cumulative mineralization = MI, immobilization = I, and leaching = L.

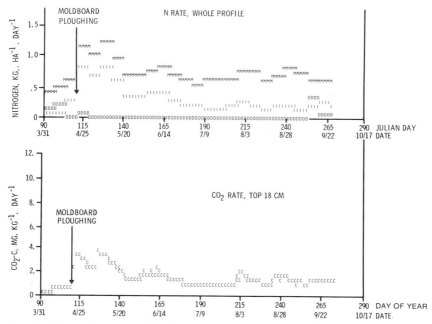

Fig. 22-8. Simulation by NCSWAP. Weekly average rates of mineralization (M), immobilization (I), and denitrification (D), in the whole profile. Weekly average rates of CO_2 production in the top 18 cm of the soil profile.

water stress. The crop is stressed by water when it should start, on day 215, to dilute its nitrogen content by growth without nitrogen uptake. Therefore, the crop ends up with a nitrogen percentage higher than the one expected under optimum conditions (Fig. 22-6). At physiological maturity, the crop has undergone a water deficit of 80 mm and a cumulative nitrogen uptake deficit of 39 kg/ha. Had the water limitations been offset by irrigation, 65 kg/ha of nitrogen fertilizer would have been needed to reach maximum potential yield. The consequences of tillage performed on day 110 can be seen as increases in field capacity (Fig. 22-5b), rates of CO_2 evolved, mineralization, and immobilization (Fig. 22-8). Inorganic nitrogen, 15 kg/ha of both NO_3-N and NH_4-N, were added on day 109 (Fig. 22-5c).

22-3 CONCLUSION: COMPUTER-ASSISTED MANAGERIAL DECISIONS

The preceding discussion provides examples of how our present knowledge of factors affecting soil erosion and crop productivity can be integrated through computer simulation modeling. As knowledge increases, computer simulations such as these can also be updated, and appropriate changes can be made to integrate this new knowledge. This approach provides us with an excellent tool for developing insights into how various crop-production factors interact. This should aid scientists in identifying the best management practices and critical areas for more research.

The information derived from these models can now be used as guides in making management decisions. More accurate estimates of optimum tillage practices, fertilizer rates and placement, irrigation practices, weed-control techniques, and other management factors will be available through these models as they are developed and perfected.

Ultimately, the information available through continued research and model development can be integrated into the performances of various pieces of farm equipment. Through the use of onboard sensors and microprocessors linked to servomotors, planting, fertilizing, spraying, tilling, and harvesting equipment can respond to changing field conditions. Thus, each acre would be managed near its optimum, rather than as an average for the whole field. The hardware exists to accomplish this, but we usually lack the research base to properly program the microprocessor. However, if we knew the optimum depth of planting and the optimum amount of pressure needed for the press wheels for the range of soil types, water contents, and bulk densities present in a field, we could program a soil map into the microprocessor, plus the calibrations for depth and pressure adjustments as a function of bulk density and water content. Then, with proper onboard sensors for bulk density and water content, adjustments would automatically be made so that all seed was placed in the optimum possible environment.

Before this scenario can become a reality, however, we need to expand greatly our basic knowledge of many of the interactions among soil properties. In the example described, for instance, we need to know for each soil type the relationships between applied pressure and porosity, between porosity and water and air movement, etc. Computer simulation techniques will help us identify the more critical experiments that we need to provide the required database.

REFERENCES

American Society of Agronomy. 1978a. Crop residue management systems. Spec. Pub. 31. Madison, WI.

----. 1978b. Nitrification inhibitors—potentials and limitations. Spec. Pub. 38. Madison, WI.

----. 1982. Predicting tillage effects on soil physical properties and processes. Spec. Pub. 44. Madison, WI.

Anderson, J. M. 1979. Animal/microbial interactions in soil biological processes. p. 311, 312. *In* E. G. Grossbard (ed.) Straw decay and its effect on disposal and utilization. John Wiley and Sons, New York.

Bennet, O. L., E. L. Mathias, and C. B. Sperow. 1976. Double cropping for hay and no-tillage corn production as affected by sod species with rates of atrazine and nitrogen. Agron. J. 68:250–254.

Castillo, S. R., R. H. Dowdy, J. M. Bradford, and W. E. Larson. 1982. Effects of applied mechanical stress on plant growth and nutrient uptake. Agron. J. 74:526–530.

Childs, S. W., J. R. Gilley, and W. E. Splinter. 1977. A simplified model of corn growth under moisture stress. Trans. Am. Soc. Agric. Engr. 20(5):858–865.

Dowdy, R. H., and T. B. Hutcheson, Jr. 1963. Effect of exchangeable potassium level and drying upon availability of potassium to plants. Soil Sci. Soc. Am. Proc. 27:521–522.

Dutt, G. R., M. J. Shaffer, and W. J. Moore. 1972. Computer simulation model of dynamic bio-physiochemical processes in soils. Tech. Bull. 196. Agric. Exp. Stn., Univ. of Arizona, Tucson.

Gupta, S. C., J. K. Radke, and W. E. Larson. 1981. Predicting temperature of bare and residue covered soil with and without a corn crop. Soil Sci. Soc. Am. J. 45:405–412.

Haas, H. J., and C. E. Evans. 1957. Nitrogen and carbon changes in Great Plains soils as influenced by cropping and soil treatment. USDA Tech. Bull. 1164. Washington, DC.

Hanks, R. J., and S. A. Bower. 1962. Numerical solution of the moisture flow equation for infiltration into layered soils. Soil Sci. Soc. Am. Proc. 26:530–534.

Hunt, H. W. 1977. A simulation model for decomposition in grasslands. Ecology 58:469–484.

Larson, W. E., J. B. Swan, and M. J. Shaffer. 1983. Soil management for semi-arid regions. *In* W. K. Stone (ed.) Plant production and management under drought conditions. Elsevier Publishing Co., Amsterdam, The Netherlands.

----, C. E. Clapp, W. H. Pierre, and Y. B. Morachan. 1972. Effect of increasing amounts of organic residues on continuous corn: Organic carbon, nitrogen, phosphorus, and sulfur. Agron. J. 64:204–208.

Linden, D. R. 1979. A model to predict soil water storage as affected by tillage practices. Ph.D. diss. Univ. of Minnesota, St. Paul.

McGill, W. B., H. W. Hunt, R. G. Woodmansee, and J. O. Reuss. 1981. Phoenix, a model of the dynamics of carbon and nitrogen in grassland soils. *In* R. E. Clark and T. Rosswell (ed.) Terrestrial nitrogen cycles: Processes, ecosystem strategies, and management inputs. Ecol. Bull. (Stockholm) 33:49–115.

Miller, S. D., and J. D. Nalewaja. 1979. Weeds in wheat with tillage and herbicides. Proc. North Cent. Weed Control Conf. 34:39.

Molina, J. A. E., C. E. Clapp, M. J. Shaffer, F. W. Chichester, and W. E. Larson. 1983. NCSOIL, a model of nitrogen and carbon transformations in soil: Description, calibration, and behavior. Soil Sci. Soc. Am. J. 47:85–91.

----, and K. Richards. 1984. Simulation model of the nitrogen and carbon cycle in the soil-water-plant system, NCSWAP; Guide for the preparation of input data files and execution of NCSWAP. Soil Series 116. Dep. of Soil Science, Univ. of Minnesota, St. Paul.

Moncrief, J. F. 1981. The effect of tillage on soil physical properties and the availability of nitrogen, phosphorus and potassium to corn (*Zea mays* L.). Ph.D. diss. Univ. of Wisconsin, Madison. (Diss. Abstr. 82-03184).

Randall, G. W., and J. B. Swan. 1978. Conservation tillage study. p. 134–140. *In* Univ. of Minnesota, St. Paul, Dep. Soil Sci., Soil Series 103.

Ribbens, R. W., and M. J. Shaffer. 1976. Irrigation return flow modeling for the Souris Loup. Am. Soc. Agric. Engr., St. Joseph, MI.

Rice, C. W., and M. S. Smith. 1982. Denitrification in no-till and plowed soils. Soil Sci. Soc. Am. J. 46:1168–1173.

Richey, C. B., D. R. Griffith, and S. D. Parsons. 1977. Yields of cultural energy requirements for corn and soybeans with various tillage-planting systems. Adv. Agron. 29:141–182.

Schaller, F. 1968. Soil animals. The University of Michigan Press, Ann Arbor.

Shaffer, M. J., and C. E. Clapp. 1982. Root growth submodel. Chapter X. *In* M. J. Shaffer and W. E. Larson (ed.) Nitrogen-tillage-residue management (NTRM) model. Technical documentation. USDA-ARS and Univ. of Minnesota, St. Paul.

----, and S. C. Gupta. 1981. Hydrosalinity models and field validation. *In* I. K. Iskandar (ed.) Simulating land treatment of wastewater. John Wiley and Sons, New York.

----, G. R. Dutt, and W. J. Moore. 1969. Predicting changes in nitrogenous compounds in soil-water systems. p. 15–28. *In* Collected papers regarding nitrates in agricultural wastewater. Water Pollution Control Res., Series 13030 Ely 12/69. U.S. Government Printing Office, Washington, DC.

----, R. W. Ribbens, and C. W. Huntley. 1977. Prediction of mineral quality of irrigation return flow, Vol. V, Detailed return flow salinity and nutrient simulation model. USEPA-6001, 2-77-179e. National Technical Information Service, Springfield, VA.

Stanford, G., O. L. Bennett, and J. F. Power. 1973. Conservation tillage practices and nutrient availability. *In* Conservation tillage. Soil Conserv. Soc. of Am., Ankeny, IA.

––––, and S. J. Smith. 1972. Nitrogen mineralization potentials of soils. Soil Sci. Soc. Am. Proc. 36:465–472.

Triplett, G. B., and G. D. Lytle. 1972. Control and ecology of weeds in continuous corn growth without tillage. Weed Sci. 20:453–457.

Voorhees, W. B., C. G. Senst, and W. W. Nelson. 1978. Compaction and soil structure modification by wheel traffic in the northern Corn Belt. Soil Sci. Soc. Am. J. 42:344–349.

23 Structures and Methods for Controlling Water Erosion

J. M. Laflen
Agricultural Research Service
U.S. Department of Agriculture
Ames, Iowa

R. E. Highfill
Soil Conservation Service
U.S. Department of Agriculture
Washington, DC

M. Amemiya
Iowa State University
Ames, Iowa

C. K. Mutchler
Agricultural Research Service
U.S. Department of Agriculture
Oxford, Mississippi

Soil erosion, in most cases, is not controlled by a single practice but by a system, composed of a number of components. Each component performs one or several functions. The functions a system must perform are (i) control rill and interrill erosion, (ii) deposit eroded material, and (iii) convey runoff water in a nonerodible manner.

This paper describes the use and benefits of soil conservation practices as components of soil erosion control systems. These soil conservation measures include conservation tillage, sod-based rotations, contour farming, stripcropping, terracing, grassed waterways, underground outlets, water and sediment control basins, and vegetative cover. We are as quantitative as possible, presenting the available database upon which evaluations of practices have been made and hypothesizing where required about the effectiveness of combinations of practices for controlling water erosion.

The basic mechanics of erosion are presented elsewhere in this volume. Because conservation tillage has been adequately reviewed by others, we only discuss briefly its role in erosion control.

Published in R. F. Follett and B. A. Stewart, ed. 1985. *Soil Erosion and Crop Productivity.*
© ASA-CSSA-SSSA, 677 South Segoe Road, Madison, WI 53711, USA.

23-1   CONSERVATION TILLAGE

Conservation tillage has been defined as "any tillage sequence that reduces loss of soil or water relative to conventional tillage" (Soil Conservation Society of America, 1982, p. 33). Although the adequacy of this definition has been discussed at length, there has been little argument that conservation tillage has great potential for reducing rill and interrill erosion.
Conservation tillage is a major practice in reducing erosion on nearly all
lands. Conservation effectiveness varies greatly, depending on the tillage
system and amount of crop residue available for erosion control. Erosion
reductions can range from none to more than 90%. McGregor and Mutchler
(1982) in Mississippi, Laflen and Colvin (1981) in Iowa, and Dickey et al.
(1983) in Nebraska (as well as other studies) demonstrate that conservation
tillage methods can be used to control soil erosion for much of the USA.

Conservation tillage is the major method in the USA for control of rill
and interrill erosion. Crosson (1981) has conservatively estimated that 50 to
60% of U.S. cropland will be farmed with conservation tillage by 2010.
Many writers are seemingly not concerned about crop production with conservation (Behn, 1982; Young, 1982; Hayes, 1982; Hughes, 1980; and Phillips and Young, 1973). Economic studies as early as 1974 (Nicol et al.)
showed that conservation tillage was an efficient means for controlling erosion. Erbach (1982), while showing that corn and soybean yields in a corn-
soybean rotation were little affected by the tillage system, showed that
yields for continuous corn were reduced as tillage was reduced. This finding
is common for the cool, wet soils in the northern Corn Belt. Tillage guides
for the best selection of tillage systems for specific soils have been prepared
for many states.

While conservation tillage has been shown to reduce greatly rill and interrill erosion, its effect on runoff volumes has been mixed. Conservation
tillage has reduced runoff volumes (Laflen et al., 1978), occasionally by a
tremendous amount (Edwards, 1982), while in other situations the reduction, if any, has been minimal (Laflen and Colvin, 1981). Hence, conservation systems that include conservation tillage must include components for
delivering surface runoff to channels in a nonerosive manner.

23-2   SOD-BASED ROTATIONS

Man has not devised a better system for controlling rill and interrill
erosion than keeping land continuously in meadow. Crop rotations that include meadow are called sod-based rotations. As early as the mid-1700s,
early American agriculturalists advocated the use of sod in rotation for
building the soil (McDonald, 1941).

Sod-based rotations reduce erosion by temporarily decreasing the soil's
erodibility after the sod is tilled and by reducing the frequency of row crops,
which decreases the period when the soil surface is bare and cultivated. The

Table 23-1. Relative soil erosion for sod-based rotations and continuous row crops in Iowa with conventional tillage.

Rotation (fall plowing)	Relative soil loss	
Continuous corn	Beaconsfield-measured	1.00
Corn-oats-meadow	Beaconsfield-measured	0.11
Corn-soybeans	Beaconsfield-measured	1.15
Continuous corn	Castana-measured	1.00
Corn-oats-meadow-meadow	Castana-measured	0.18
Continuous corn	USLE estimate	1.00
Corn-oats-meadow	USLE estimate	0.23
Corn-oats-meadow-meadow	USLE estimate	0.18
Corn-soybeans	USLE estimate	1.09

expected and measured relative soil losses for two sod-based rotations are shown in Table 23–1 for two Iowa locations (J. M. Laflen, unpublished data, 1963–1969). A sod-based rotation, including both a row crop and a small grain, each for one-third to one-fourth of the years, would be expected to have about 20% as much erosion as a continuous corn crop, and less than 20% as much erosion as a corn-soybean rotation. Estimates based on the universal soil loss equation (USLE) (Wischmeier and Smith, 1978) compare reasonably well with those measured at the two Iowa locations.

Within a sod-based rotation, soil erosion in meadow is very low. This is shown by coefficients in the USLE (Wischmeier and Smith, 1978). In a high-producing grass and legume meadow, soil erosion is about 20% of the erosion from corn under no-tillage cultivation. Even the most erodible meadow is only slightly more erodible than no-till following corn. After the meadow is tilled, soil erosion is less than that following a cultivated crop. This reduction is 40 to 75% the first year after meadow and 5 to 50% the second year.

Benefits of a sod-based rotation are not restricted to reduction of soil erosion. A well-managed, sod-based rotation can maintain, over an extended period, a desirable level of soil organic matter (Smith, 1942). Also, when erosion control costs are included as a crop production cost, sod-based rotations may be more profitable under some conditions than continuous row cropping (Ervin and Washburn, 1981).

23-3 CONTOUR FARMING

Thomas Jefferson advocated "horizontal plowing" in an 1813 letter to C. W. Peale (Jefferson, 1813):

Our country is hilly and we have been in the habit of ploughing in straight rows whether up and down hill, in oblique lines, or however they lead; and our soil was all rapidly running into the rivers. We now plough horizontally, following the curvatures of the hills and hollows, on the dead level, however crooked the lines may be. Every furrow then acts as a reservoir to receive and retain the waters, all of which go to the benefit of the growing plant, instead of running off into the streams. In a farm horizontally and deeply ploughed, scarcely an ounce of soil is carried off from it.

More recent conservationists, while convinced of the value of contouring, do not ascribe as much benefit in terms of erosion control. Data on soil loss from contoured land, compared with up-and-down hill farming, are shown in Table 23–2. Generally, the judgment is that, for storms with small amounts of runoff, contouring is extremely effective in controlling soil erosion. Where runoff rates are high, however, contouring is less effective.

On long slopes, contoured rows may be topped, concentrating runoff and causing considerable rilling or gullying, and soil erosion may actually be increased because of contouring. Evidently beyond some critical length, contouring loses effectiveness. Contouring apparently has a small effect on soil erosion on gently sloping areas, where erosion is very low. Runoff velocities and erosion due to rilling increase as slopes become steeper. Critical slope length decreases as slope increases and at some point contouring becomes relatively ineffective. Commonly accepted limits from Wischmeier and Smith (1978) are given in Table 23–3.

Contouring is an attractive economic alternative for land management since it does not add significantly to farmers' costs. Ervin and Washburn (1981) concluded that the differences in net returns between up-and-down hill cultivation and contouring were small (sometimes favoring contouring, other times up-and-down hill) and were not likely to be a significant factor in a farmer's choice between the two alternatives. Walker and Timmons (1980), in a study of policies for controlling soil loss and sedimentation in

Table 23–2. Comparisons of soil loss from contoured and up-and-down hill tillage.

				Soil loss	
Slope	Slope length	Cropping	Location	Up-and-down hill	Contour
%	m			t ha^{-1} yr^{-1}	t ha^{-1} yr^{-1}
13†	22.1	Continuous corn	Zanesville,‡ OH	130.0	36.3
13	22.1	Continuous corn	Zanesville, OH	210.0	225.9
12	22.1	Corn in corn-oats rotation	Castana,§ IA	56.5	22.6
7	54.9	Continuous corn	Guthrie,¶ OK	85.6	37.2
7	82.3	Corn-wheat-meadow	Bethany,# MO	15.7	7.6
4	22.1	Corn-oats-cotton	Temple,†† TX	22.9	8.7
3.5	51.2	Continuous cotton	Temple, TX	35.2	13.2
3	168.9	Cotton-corn-cotton-oats	Temple, TX	8.7	8.7
3	128.0	Continuous corn	McCredie,‡‡ MO	19.9	14.1
3	128.0	Continuous corn	McCredie, MO	--	10.8
3	27.4	Continuous corn	McCredie, MO	3.4	--
2	54.9	Corn in corn-oats	Urbana,§§ IL	9.0	5.6
2	54.9	Corn in corn-soybean	Urbana, IL	7.6	4.3
2	54.9	Soybean in corn-soybean	Urbana, IL	6.1	1.3
2	54.9	Oats in corn-oats	Urbana, IL	1.8	1.1
2–7¶¶		Corn-oats-meadow	Lafayette,## IN	4.9	1.1

† Ridged rows.
‡ Borst et al. (1945).
§ Moldenhauer and Wischmeier (1960).
¶ Daniel et al. (1943).
# Smith et al. (1945).

†† Hill et al. (1944).
‡‡ Jamison et al. (1968).
§§ Van Doren et al. (1950).
¶¶ Watersheds from 0.8 to 1.5 ha.
Bedell et al. (1946).

Table 23-3. P values and slope-length limits for contouring (Wischmeier and Smith, 1978).

Slope	P value	Maximum length
%		m
1–2	0.6	122
3–5	0.5	91
6–8	0.5	61
9–12	0.6	37
13–16	0.7	24
17–20	0.8	18
21–25	0.9	15

western Iowa on an area with a high erosion hazard, found that a subsidy to encourage contouring would likely not reduce soil erosion very much because most of the land that would be contoured with a subsidy would be contoured anyway. Crop yields and costs were higher with contouring, but the net result was that contouring was cost effective.

While contouring may be an attractive economic alternative for land management, it frequently cannot alone control soil erosion. Additionally, large farm machinery cannot follow contours exactly, leading to portions of fields that are not contoured and row slopes that may be excessive. The database for evaluating the effects of deviations from the contour is limited.

23–4 CONTOUR STRIPCROPPING

Contour stripcropping combines the good features of contouring and sod-based rotations. Stripcropping is the alternation of equal-width strips of different crops, with at least every other strip in a close-growing crop or sod. The crops are rotated each year on each strip. Contour stripcropping is very similar economically to a sod-based rotation, adjusted slightly to account for contouring. Wischmeier and Smith (1978) recommended maximum slope lengths and maximum strip widths for contour stripcropping.

An additional benefit of contour stripcropping not included for sod-based rotations or contouring is that soil eroded from row-crop strips is generally captured in strips of meadows or close-grown crops, keeping soil relatively near its point of origin and on the field where it can contribute to future crop production. Runoff water must be conveyed from stripcropped areas so that land degradation is prevented.

Although untested, conservation tillage can be used instead of conventional tillage in stripcropping. In some situations, such strips might be used to reduce the frequency of sod or close-grown crops in the stripcropping.

23–5 TERRACING

A terrace is "an artificial measure designed primarily for control of runoff in high-rainfall areas and for conservation of water in low-rainfall areas. Control of erosion is the ultimate objective in the more humid areas and a very important objective in the dry regions'' (Bennett, 1939, p. 443). Terraces can be classified by alignment, cross section, grade, and outlet. Terraces may or may not be parallel, may or may not be farmed over, may or may not have steep front or backslopes, may be level or on a grade, and may have surface or underground outlets, both, or neither. Additionally, the interval between terraces may be partially or wholly level (or nearly so) for better use of available moisture in crop production.

Terraces reduce erosion by decreasing the slope length and by preventing or reducing damage caused by surface runoff. Conservation tillage can maintain the soil resource by reducing rill and interrill erosion between terraces and can reduce maintenance costs and extend terrace life. A major impact of terraces is to reduce delivery of sediment from a field.

Terracing to control soil erosion is not a new practice, having been used for centuries in other countries. Nearly 70 years ago, Ramser (1917) described most of the terraces used today, even including terraces with underground outlets. Of course, much less emphasis was placed on making terraces parallel, but it was recognized that farmers objected to sharp bends in terraces and therefore that terraces should be built higher and wider in these areas.

Spomer et al. (1973), in studying the cost effectiveness of level terraces in western Iowa, reported slightly higher corn yields from unterraced fields. The terraces were very efficient in terms of erosion control, with sediment yield about 4% that of unterraced fields. Spomer et al. speculated that unterraced fields would eventually become so gullied that they would be difficult to farm.

Rosenberry et al. (1980), in a southern Iowa study, evaluated the effects of conservation practices on costs of soil erosion when soil loss limits were met. They found that more row crops were grown when terracing was included than when other means were used to achieve soil loss limits. However, alternatives that included terraces were not the most economically feasible. Economic analyses by the Center for Agricultural and Rural Development (Nicol et al., 1974; Meister et al., 1976; Wade and Heady, 1976) show that terraces are required on a broad scale when stringent soil loss limits are applied nationally, if high levels of production are to be maintained. Recent analyses have shown that terracing is more expensive per ton of soil erosion reduction than most other alternatives for soil erosion control (USDA, 1981).

Foster and Highfill (1983) developed equations for computing the conservation practice factor, P, for the USLE for various kinds of terraces. Their analyses allowed for considerable benefit due to deposited soil above the terraces, which they assume will be important in future crop production. Additionally, they included a factor for estimating sediment delivery from the terrace. Estimates of the percentage of soil eroded between terraces that should benefit future crop production and the percentages of soil eroded between terraces that are retained on the field are given in Table 23-4. The fraction retained on the field that contributes to future crop production is a judgment decision. As shown in Table 23-4, even for narrow terrace inter-

Table 23-4. Fraction of eroded soil above terraces that is retained on the field and that benefits future crop production.

Terrace interval	P_f†	Closed outlets	Terrace channel grade—percent					
			0	0.2	0.4	0.6	0.8	>0.8
m		Fraction of eroded soil retained on field						
--		0.95	0.90	0.83	0.71	0.51	0.17	0
		Fraction of eroded soil benefiting future crop production‡						
<33	0.5	0.48	0.45	0.42	0.36	0.26	0.09	0
33–42	0.4	0.38	0.36	0.33	0.28	0.20	0.07	0
43–54	0.3	0.29	0.27	0.25	0.21	0.15	0.05	0
55–68	0.2	0.19	0.18	0.17	0.14	0.10	0.03	0
69–90	0.1	0.10	0.09	0.08	0.07	0.05	0.02	0
>90	0	0	0	0	0	0	0	0

† P_f is the fraction of eroded soil retained on the field that contributes to future crop production.

‡ The fraction contributing to future crop production is computed as the product of the fraction retained on the field and P_f.

vals, less than half the soil eroded between terraces is expected to benefit future crop production because of the localized nature of deposition.

To improve the efficiency of farm machine operation, terraces are usually constructed parallel to other terraces, sometimes even parallel to field boundaries, with considerable potential for soil erosion between terraces. In such situations, more effective conservation tillage and sod-based crop rotations may be required to reduce erosion to acceptable levels. Mitchell and Beer (1965) have shown considerable benefit to farm machine operation from parallel terracing.

23-6 GRASSED WATERWAYS

Runoff water is usually conveyed to watershed outlets by open channels. If unprotected, these channels are a major contributor to watershed sediment yields (Vanoni, 1975) and lead to dissection of fields. Because of increased slope steepness due to channel degradation, rill and interrill erosion on the field may be greatly increased (Johnson and Shahvar, 1982). A major means of decreasing the threat posed by erosion in open channels is to confine such flow to nonerodible, grassed waterways. The design of such channels is well established, with design information available in guides and texts (Schwab et al., 1981). Not so readily available is information relating to construction and maintenance costs and to sediment trapping.

The amount of sediment trapping in grassed waterways has largely been ignored. For watersheds in the Blacklands of Texas, Williams and Berndt (1972) estimated that grassed waterways were trapping about 70% of the soil entering the waterways. Because of the considerable soil erosion above grassed waterways and because of deposition within waterways, establishment and maintenance of waterways is difficult and costly. Additionally, many grasses in waterways are susceptible to herbicide damage.

Although considerable cropland is required for waterways, many fields would be unproductive without them. Bondurant and Laflen (1978) studied conservation on two fields in Iowa and estimated for surface drained terraces that land in waterways was 5 and 7% of the field area. Such land may not produce higher-income row-crops but can produce substantial amounts of hay, if properly maintained. Grassed waterways are highly variable in initial cost, ranging from as low as $10/ha of watershed area to much larger amounts where tree removal and earthmoving costs are incurred.

23-7 UNDERGROUND OUTLETS

Underground outlets are sometimes an attractive alternative to grassed waterways for conveying runoff water to watershed outlets. They perform much like urban storm drains and require careful engineering, construction, and maintenance to ensure adequate operation at a minimum cost over a long period.

Underground outlets are used to drain surface runoff ponded on a watershed. Pondage can occur above terraces, above water and sediment control basins and diversions, and in field depressions. Because runoff is ponded, peak runoff rates are greatly reduced, and considerable deposition occurs in the pondage area. Because ponding is temporary, little land is removed from production.

Underground outlets must be carefully engineered so that discharge rates are adequate yet not excessive and so that the underground outlet is protected from damage due to excessive water pressure and high velocities. Also, when several ponded areas drain via a single outlet, good engineering is required to prevent overtopping of lower ponded areas because of the discharge into lower ponded areas of runoff water from ponded areas of higher elevations.

Maintenance is required to keep underground outlets and inlets to underground outlets free of blockages. Failure to do so results in unfarmable ponded areas, damage to underground outlets, and topping of structures.

Underground outlets permit much latitude in field design where structures are involved. They are often necessary to attain the straight, parallel rows required by large farm machinery.

23-8 WATER AND SEDIMENT CONTROL BASINS

Water and sediment control basins are formed by structures across major waterways where excess runoff is temporarily ponded, usually over standing crops, until the runoff discharges through underground outlets. They perform very similarly to structures studied by Laflen et al. (1972), constructed by Jacobson (1967), described by Samstad (1964), and recommended for certain situations by Ramser (1917). The basins are primarily designed to control runoff to prevent gullying, to induce deposition of eroded sediment, and to improve the efficiency of farm equipment.

Water and sediment control basins usually have little effect on rill and interrill erosion, a major distinguishing difference with terraces. They are generally located parallel to field boundaries so that point rows are avoided. The effect of the basins on sediment yield from a field can be estimated with the impoundment element of CREAMS (Chemicals, Runoff, and Erosion from Agricultural Management Systems) (Knisel, 1980) given by Foster et al. (1981). Usually, sediment trapping in excess of 80% of the sediment delivered to the basins would be expected.

Maintenance costs of water and sediment control basins vary widely, depending on the watershed. Where conservation above a basin is not practiced, deposition might be so great as to fill the storage volume in a short time, perhaps less than a decade. Where good erosion control is practiced, for example, with rotations and conservation tillage, the life of a basin, without maintenance, might be extended by a factor of ten or more. Maintenance costs might also be reduced a commensurate amount.

Because they do very little, if anything, to reduce erosion on the watershed above the basins, most of the benefits of water and sediment control basins are in reducing gully growth and off-site sediment damage.

23–9 VEGETATIVE COVER

Many areas within fields cannot be efficiently cropped or, if cropped, are extremely susceptible to soil erosion. Such areas include irregularly shaped areas, service roads, turn areas at row ends, unproductive areas with physical or chemical problems, and other extremely erodible areas. For most such areas, vegetation is an important means of controlling soil erosion, permitting access, and increasing farming efficiency.

Vegetated areas are also effective sediment filters, usable in both agricultural and nonagricultural areas. Kao et al. (1975) showed that grass filter strips could be very effective in reducing soil loss from construction sites.

Turn areas at row ends have become an increasing problem as machinery has increased in size and as land has become more valuable. Turn areas for 12-row-wide equipment, at 0.75-m row width, and for row lengths of about 800 m (0.5 mile) would occupy about 2.3% of the field. Percentages increase as row lengths decrease. Turn areas are subject to considerable tillage, have poor crop growth, and are frequently up-and-down hill, particularly where contouring is practiced. Extra tillage is usually required to eliminate compaction due to harvesting, grain hauling, and tillage. The use of grass strips in such areas helps reduce such problems and provides a good transport area. These grass strips are also deposition areas for eroded material.

Vegetation in problem areas eliminates them as a source of soil erosion. Grass establishment may be difficult for soils with severe physical or chemical problems, but some grass species can usually be established on such areas. Care must be taken that vegetation does not harbor insects and diseases that threaten crop production.

In addition to providing vegetative cover for special areas of fields, interseeding of fields with cover crops, grass, or small grains is an excellent practice for reducing soil erosion after harvest of the earlier crop and before or during establishment of a succeeding crop. Many variations are possible. Depending on the farming enterprise and several other factors, such as climate, prior and succeeding crop, such a practice may or may not be economically viable.

23–10 SUMMARY

Soil erosion can be controlled by using a system of readily usable soil conservation practices. The components must reduce rill and interrill erosion to an acceptable level, then must transport runoff water from fields to watershed outlets, and must deposit eroded soil before it enters the transport system. Special treatment for unusual areas must also be evaluated before erosion control on a field can be considered satisfactory.

Some alternatives maintain the soil resource better than others in specific instances. In every case, however, a satisfactory system of maintaining the soil resources should be designed so that erosion is reduced to an acceptable level, runoff water is delivered to an outlet in a nonerosive manner, and structures that induce sediment deposition are located where needed.

REFERENCES

Bedell, G. D., H. Kohnke, and R. B. Hickok. 1946. The effects of two farming systems on erosion from cropland. Soil Sci. Soc. Am. Proc. 10:522–526.

Behn, E. E. 1982. More profit with less tillage. Wallace-Homestead, Des Moines, IA.

Bennett, H. H. 1939. Soil conservation. McGraw-Hill, New York.

Bondurant, D. T., and J. M. Laflen. 1978. Design and operation of gradient terrace systems. Paper 78-2520. Am. Soc. Agric. Engr., St. Joseph, MI.

Borst, H. L., A. G. McCall, and F. G. Bell. 1945. Investigations in erosion control and the reclamation of eroded land at the Northwest Appalachian Conservation Experiment Station, Zanesville, Ohio, 1934–42. USDA, Tech. Bull. 888.

Crosson, P. 1981. Conservation tillage and conventional tillage: A comparative assessment. Soil Conserv. Soc. Am., Ankeny, IA.

Daniel, A., H. M. Elwell, and M. B. Cox. 1943. Investigations in erosion control and reclamation of eroded land at the Red Plains Conservation Experiment Station, Guthrie, Oklahoma, 1930–1940. USDA Tech. Bull. 837.

Dickey, E. C., C. R. Fenster, J. M. Laflen, and R. H. Mickelson. 1983. Effects of tillage on soil erosion in a wheat-fallow rotation. Trans. Am. Soc. Agric. Engr. 26:814–820.

Edwards, W. M. 1982. Predicting tillage effects on infiltration. p. 105–115. In Predicting tillage effects on soil physical properties and processes. American Society of Agronomy, Madison, WI.

Erbach, D. C. 1982. Tillage for continuous corn and corn-soybean rotation. Trans. Am. Soc. Agric. Engr. 25:906–918.

Ervin, D. E., and R. A. Washburn. 1981. Profitability of soil conservation practices in Missouri. J. Soil Water Conserv. 36:107–111.

Foster, G. R., and R. E. Highfill. 1983. Effect of terraces on soil loss: USLE P factor values for terraces. J. Soil Water Conserv. 38:48–51.

––––, L. J. Lane, J. D. Nowlin, J. M. Laflen, and R. A. Young. 1981. Estimating erosion and sediment yield on field-sized areas. Trans. Am. Soc. Agric. Engr. 24:1253–1262.

Hayes, W. A. 1982. Minimum tillage farming. No-till farmer, Inc., Brookfield, WI.

Hill, H. O., W. J. Peevy, A. G. McCall, and F. G. Bell. 1944. Investigations in erosion control and reclamation of eroded land at the Blackland Conservation Experiment Station, Temple, Texas, 1931–41. USDA Tech. Bull. 859.

Hughes, H. A. 1980. Conservation farming. Deere and Company, Moline, IL.

Jacobson, P. 1967. Keeping soil and water on the farm. J. Soil Water Conserv. 22:54–57.

Jamison, V. C., D. D. Smith, and J. F. Thornton. 1968. Soil and water research on a claypan soil. USDA Tech. Bull. 1379. Washington, DC.

Jefferson, T. 1813. Letter to C. W. Peale, Apr. 17, 1813. p. 509. In Thomas Jefferson's garden book, annotated by E. M. Betts. American Philosophical Society, Philadelphia, PA (1944).

Johnson, H. P., and Z. Shahvar. 1982. Computer simulation of soil erosion effects on topography. Unpublished project report. Dep. of Agric. Eng., Iowa State Univ., Ames.

Kao, D. T. Y., B. J. Barfield, and A. E. Lyons. 1975. On-site sediment filtration using grass strips. p. 73–82. Proc. 1975 Natl. Symp. on Urban Hydr. and Sediment Control, Pub. UKY BU 109, 28–31 July 1975, Lexington, KY. Univ. of Kentucky.

Knisel, W. G. (ed.) 1980. CREAMS: A field-scale model for chemicals, runoff, and erosion from agricultural management systems. USDA Conserv. Res. Rep. 26. Washington, DC.

Laflen, J. M., and T. S. Colvin. 1981. Effect of crop residue on soil loss from continuous row cropping. Trans. Am. Soc. Agric. Engr. 24:1227–1229.

––––, H. P. Johnson, and R. C. Reeve. 1972. Soil loss from tile-outlet terraces. J. Soil Water Conserv. 27:74–77.

––––, J. L. Baker, R. O. Hartwig, W. F. Buchele, and H. P. Johnson. 1978. Soil and water loss from conservation tillage systems. Trans. Am. Soc. Agric. Engr. 21:881–885.

McDonald, A. 1941. Early American soil conservationists. USDA Misc. Pub. 449. Washington, DC.

McGregor, K. C., and C. K. Mutchler. 1982. C factors for no-till and reduced-till corn. Paper 82-2024. Am. Soc. Agric. Engr., St. Joseph, MI.

Meister, A. D., E. O. Heady, K. J. Nicol, and R. W. Strohbehn. 1976. U.S. agricultural production in relation to alternative water, environmental, and export policies. CARD Report 65. Center for Agricultural and Rural Development, Iowa State University, Ames.

Mitchell, J. K., and C. E. Beer. 1965. Effect of land slope and terrace systems on machine efficiencies. Trans. Am. Soc. Agric. Engr. 8:235–237.

Moldenhauer, W. C., and W. H. Wischmeier. 1960. Soil and water losses and infiltration rates on Ida silt loam as influenced by cropping systems, tillage practices, and rainfall characteristics. Soil Sci. Soc. Am. Proc. 24:409–413.

Nicol, K. J., E. O. Heady, and H. C. Madsen. 1974. Models of soil loss, land and water use, spatial agricultural structure, and the environment. CARD Report 49 T. Center for Agricultural and Rural Development, Iowa State University, Ames.

Phillips, S. H., and H. M. Young, Jr. 1973. No-till farming. Reiman Associates, Milwaukee, WI.

Ramser, C. E. 1917. Prevention of the erosion of farmlands by terracing. USDA Bull. 512.

Rosenberry, P., R. Knutson, and L. Harmon. 1980. Predicting the effects of soil depletion from erosion. J. Soil Water Conserv. 35:131–134.

Samstad, L. E. 1964. An introduction to the hydrosol intransitive land engineering method of soil and water control. Paper 64-727. Am. Soc. Agric. Engr., St. Joseph, MI.

Schwab, G. O., R. K. Frevert, T. W. Edminster, and K. K. Barnes. 1981. Soil and water conservation engineering. John Wiley and Sons, New York.

Smith, D. D., D. M. Whitt, A. W. Zingg, A. G. McCall, and F. G. Bell. 1945. Investigations in erosion control and reclamation of eroded Shelby and related soils at the Conservation Experiment Station, Bethany, Missouri, 1930–42. USDA Tech. Bull. 883.

Smith, G. E. 1942. Sanborn field: Fifty years of field experiments with crop rotations, manure, and fertilizing. Missouri Agric. Exp. Stn. Bull. 458.

Soil Conservation Society of America. 1982. Resource conservation glossary. Soil Conserv. Soc. of Am., Ankeny, IA.

Spomer, R. G., W. D. Shrader, P. E. Rosenberry, and E. L. Miller. 1973. Level terraces with stabilized backslopes on loessial cropland in the Missouri Valley: A cost-effectiveness study. J. Soil Water Conserv. 28:127–131.

USDA. 1981. Soil, water and related resources in the United States: Analysis of resource trends. 1980 RCA Appraisal, Part II. USDA. Washington, DC.

Van Doren, C. A., R. S. Stauffer, and E. H. Kidder. 1950. Effect of contour farming on soil loss and runoff. Soil Sci. Soc. Am. Proc. 413–417.

Vanoni, V. A. (ed.) 1975. Sedimentation engineering. Am. Soc. Civil Engr., New York.

Wade, J. C., and E. O. Heady. 1976. A national model of sediment and water quality: Various impacts on American agriculture. CARD Report 67. Center for Agricultural and Rural Development, Iowa State University, Ames.

Walker, D. J., and J. F. Timmons. 1980. Costs of alternate policies for controlling agricultural soil loss and associated stream sedimentation. J. Soil Water Conserv. 35:177–183.

Williams, J. R., and H. D. Berndt. 1972. Sediment yield computed with universal equation. J. Am. Soc. Civil Engr. (Hyd. Div.). HY12:2087–2098.

Wischmeier, W. H., and D. D. Smith. 1978. Predicting rainfall erosion losses. USDA Agric. Handb. 537.

Young, H. M., Jr. 1982. No-tillage farming. No-Till Farmer, Inc., Brookfield, WI.

24 Methods for Controlling Wind Erosion

D. W. Fryrear
Agricultural Research Service
U.S. Department of Agriculture
Big Spring, Texas

E. L. Skidmore
Agricultural Research Service
U.S. Department of Agriculture
Manhattan, Kansas

Wind erosion is a serious problem in portions of the USA and becomes more widespread and severe during droughts. Wind erosion can become a major problem whenever the soil is loose, dry, finely divided, bare or nearly bare, and the wind velocity exceeds the threshold velocity for the soil. Although wind erosion may occur in humid and subhumid climates, it is more prevalent in semiarid to arid areas and is extensive in the Great Plains.

In this report, we identify various practices used to reduce wind erosion and we describe their advantages and limits. Because of the tremendous variation in soils, climate, and crops across the USA, no single erosion control technique will be applicable to all areas. By combining two or more control techniques, however, wind erosion can be reduced to tolerable levels in most areas.

24-1 CONTROL METHODS

24-1.1 Surface Residues

The basic method of reducing wind erosion is to keep the soil protected with surface residues. While applicable to all areas, surface residues are more widely accepted in cropping areas where they do not cause planting or harvesting problems. Some residue is left in the field after harvest of most crops, but even high-residue crops may not produce sufficient residues to protect the soil when the erosion hazard is severe (Table 24–1). Of the 24 835 600 ha of cropland in the Great Plains, 42.2% will not be protected

Published in R. F. Follett and B. A. Stewart, ed. 1985. *Soil Erosion and Crop Productivity.* © ASA-CSSA-SSSA, 677 South Segoe Road, Madison, WI 53711, USA.

Table 24-1. Harvested area, and residues available on wide fields of barley, oats, corn, grain sorghum, and wheat[†] in the major land resource areas (MLRAs) of the Great Plains[‡] (Skidmore et al., 1979).

MLRA	Barley Harvested area	Barley Residues available	Oats Harvested area	Oats Residues available	Corn Harvested area	Corn Residues available	Sorghum Harvested area	Sorghum Residues available	Wheat Harvested area	Wheat Residues available	Total area	% not protected
	kha	t/ha	kha	t/ha	kha	t/ha	kha	t/ha	kha	t/ha	kha	
52	245.1	−0.6	25.2	−0.5	0.1	−1.2	0	−4.2	803.9	−0.4	1074.3	100
53	148.5	0	223.5	0.1	43.7	−1.9	4.1	−2.1	1478.1	−0.8	1897.9	80
54	69.8	0.4	126.8	0.4	2.5	−1.5	0	−2.3	510.1	−0.6	709.2	72
55	577.3	0.4	455.8	0.6	234.7	−1.3	17.9	−1.2	2025.5	−0.3	3311.2	69
56	464.5	1.3	239.3	1.2	82.4	0.4	0	−3.0	1150.9	0.9	1937.1	0
57	38.6	2.2	90.8	2.2	32.5	1.9	0	−1.4	65.9	1.7	227.8	0
58	64.3	0.6	24.8	0.6	2.7	1.0	0	−3.5	182.8	0.7	274.6	0
59	54.9	0.3	26.2	0.3	0.5	0.2	0	−3.9	241.5	−0.1	323.1	75
60	5.5	−0.2	6.4	0.3	4.0	1.4	0.3	−2.0	25.2	0.4	41.4	14
61	2.6	−0.3	6.4	0	0.9	−0.5	0.2	−2.1	29.7	1.2	39.8	9
62	0.2	0	0.6	0.2	0.1	−0.4	0	−2.2	0.8	0.5	1.7	6
63	9.9	0	38.6	0	5.3	−1.0	18.7	−1.8	156.1	0.5	228.6	10
64	3.9	0.1	8.3	0.1	7.0	1.3	0.5	−2.0	49.0	0.8	68.7	1
65	2.2	0.2	10.4	−0.2	99.2	2.1	2.0	−1.4	17.5	0.6	129.5	8
66	9.2	0.2	36.1	0.7	32.8	−1.2	14.9	−1.3	30.8	1.0	123.8	39
67	27.2	0	13.0	−0.4	113.7	1.9	46.0	−2.4	747.5	−0.3	947.4	85
68	4.6	0.9	0	−2.8	28.6	2.6	0.4	−1.7	74.6	−0.5	108.2	69
69	4.5	0.5	0.7	−0.3	14.5	1.9	42.7	−1.6	147.3	−0.8	209.7	91
70	0.9	−0.7	0	−3.4	4.6	0	15.2	−3.3	30.2	−1.0	50.9	100
71	1.2	0.5	5.9	0.6	318.0	2.8	23.7	−0.7	58.4	1.9	407.2	6
72	5.9	−0.4	4.5	−0.7	464.6	2.0	205.4	−1.2	1728.9	0	2409.3	9
73	1.9	0.3	7.2	0.3	144.5	2.6	285.5	−1.0	869.6	1.0	1308.7	22
74	0.6	1.0	5.5	0.6	14.5	1.3	112.6	−0.1	302.1	1.9	435.3	26
75	2.4	1.3	22.8	1.3	568.1	3.5	602.4	0.3	715.8	2.2	1911.5	0
76	4.4	1.6	11.5	1.0	29.4	0.9	190.9	0.4	200.4	2.2	436.6	0
77	12.1	0.4	4.4	−0.7	280.5	2.8	1287.3	−0.7	1144.5	−0.9	2728.8	89
78	29.9	−0.4	66.4	−0.3	1.2	2.0	297.9	−1.5	1196.3	0	1591.7	25
79	1.9	0.3	1.2	−0.3	18.8	2.8	89.9	−1.1	439.2	0.8	551.0	17
80	36.7	1.0	34.1	0.8	4.2	2.3	77.2	−0.2	1198.4	1.9	1350.6	6

† All wheat (spring, winter, durum).   ‡ T (Tolerable soil loss) = 11.2 t ha⁻¹ yr⁻¹ (5.0 t acre⁻¹ yr⁻¹); K′ (soil ridge roughness factor) = 1.0.

from wind erosion with residue from the five "residue crops" listed. In major land resource areas 52 (north-central Montana) and 70 (eastern New Mexico), none of these crops will produce sufficient residue to protect the soils. Areas that do not produce enough residue to protect the soil surface are concentrated along the western boundary of the Great Plains, which coincides with areas of limited rainfall and high winds.

24–1.1.1 Crop Residues

Dr. J. D. Bilbro (unpublished data, 1984) revealed that winter wheat (*Triticum aestivum* L.) at Big Spring, TX, will produce no residue 30% of the time and estimated erosion will be 39 t ha^{-1} yr^{-1} 50% of the time. The quantities of various crop residues needed to protect soils from wind erosion have been determined (Chepil, 1944; Chepil et al., 1963; Siddoway et al., 1965; Skidmore and Siddoway, 1978) and compared to an equivalent amount of flat small grain (Lyles and Allison, 1980 and 1981) (Fig. 24–1).

Standing residues are more effective than flattened residues (Chepil et al., 1963), and rows of crop residue perpendicular to wind direction control

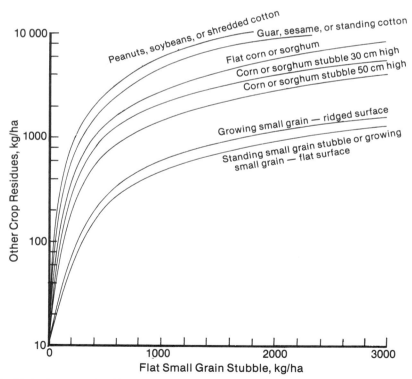

Fig. 24–1. Amount of residue of various crops needed to equal a given amount of flat small grain stubble (USDA-SCS, 1973, Chart #3).

wind erosion more effectively than parallel rows (Englehorn et al., 1952; Skidmore et al., 1966; Lyles et al., 1973). With Lyles and Allison's (1976) model of standing crop residues, if dimensions of residues are known, the estimated erosion can be calculated for any soil. The relationships are for smooth, leaf-free residues and do not consider possible branching.

24–1.1.2 Stubble-Mulch Tillage Practices

The goals of stubble-mulch tillage are to reduce the number of tillage operations and maintain residues on the soil surface for conserving water and controlling erosion. The advent of stubble-mulch, sweep-tillage machines in the late 1930s made possible the control of weeds without destroying the protection provided by the stubble. Woodruff et al. (1965 and 1972) have shown that large sweeps reduce residue levels of small grains about 10% and disc implements about 50% (Table 24–2). The moldboard plow usually buries all residues. The amount of residue buried with tillage implements depends on soil conditions, operating speed, proper clearance (0.6 m), and flexibility of adjustment and implement frame. In recent years field size, tractor horsepower, implement size, and operating speed have all increased, which generally reduces the amount of residues on the soil surface.

As surface residues are exposed to weathering and deterioration, their weight decreases. This does not necessarily mean, however, that the protection provided will decrease in proportion to the weight change. The relationship between physical properties of the residue and erosion was modeled by Lyles and Allison (1976). Erosion control should remain constant if the orientation of the residue or its physical dimensions do not change with time. Using stalk densities of 1.57 and 1.37 kg/m^2 and stalk diameters of 2.78 and 17.7 mm for wheat and sorghum (*Sorghum vulgare*), respectively, Lyles and Allison (1976) showed that the weight required to

Table 24–2. Tillage machine and percentage of surface residue lost with each operation (Woodruff et al., 1972).

Tillage machine	Residue lost
	%
Stirring or mixing machines:	
One-way disk (0.60- to 0.66-m disks)	50
One-way disk (0.46- to 0.56-m disks)	40
Tandem or offset disks	50
Power disk	60
Field cultivator (0.41- to 0.46-m sweeps)	20
Chisel plow (50-mm chisels 0.3 m apart)	25
Mulch treader (spade-tooth)	25
Mulch treader (spike-tooth)	30
Sidewinder rotary tiller (0.30-m tilled on 1-m center)	30
Subsurface machines:	
Blades (0.91 m or wider)	10
Sweeps (0.60 to 0.91 m)	15
Rodweeders (plain rod)	10
Rodweeders (with semichisels or shovels)	15

completely cover the soil surface compared well with values reported by Fryrear and Koshi (1971) (Fig. 24–2). We believe that the percentage of the soil surface covered is easier to estimate in the field than is the weight of residues on the soil surface and that the former will be related to wind erosion. Erbach (1982) used the percentage of soil surface covered by residues to evaluate residue reduction with various tillage systems. The percentage of soil cover can be measured by the meter-stick method (Hartwig and Laflen, 1978).

24–1.1.3 Permanent Vegetation

Permanent vegetation is the ultimate means of protecting soils from wind erosion. Properly managed grasses are the most reliable method of reducing erosion on deep sandy soils. But, if the grass is overgrazed, erosion can be very severe, particularly during prolonged droughts (Lyles, 1980). One advantage of grasses is that the plant crowns and root systems provide some cover even as erosion becomes severe.

24–1.1.4 Limits of Application

The basic method of protecting the soil with crop residues is universally applicable, but farmer acceptance depends on short-term economics. Farmers growing high residue crops may be more receptive of advances in stubble mulching than those that must switch from a high-value crop to a

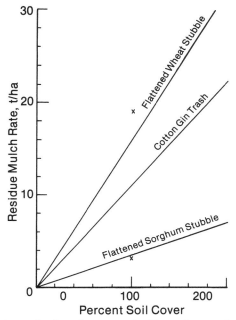

Fig. 24–2. Percentage of soil surface covered with sorghum, cotton gin trash, and wheat mulches (Fryrear and Koshi, 1971, Fig. 2). X indicates values calculated from Lyles and Allison (1976).

residue crop. Residues do reduce evaporation losses from soils, thus conserving soil water. On the minus side, residues intercept herbicides, and their use may require a modification of weed control practices. Residues also break down with time, and little residue may remain on the soil surface during prolonged droughts. To control wind erosion as surface residues deteriorate, some other practice, such as barriers or tillage, must be used until more residues are produced.

Gross income in semiarid areas is usually greater from cultivated crops than from grassland. Many fields replanted to grass during or immediately following a drought are plowed again as favorable rainfall and higher crop prices return. But, improvements in grass-establishment techniques and production in semiarid regions and improved cattle prices could reverse this economic situation.

24–1.2 Reducing Field Width

Wind erosion increases with the length of the eroding surface until some maximum is reached. The relation of distance to maximum erosion flux and soil type is shown in Table 24–3 (Chepil, 1957). The values in the table are for a wind blowing perpendicular to the strip at a velocity of 17.8 m/s at 15 m above the ground. The width of field strips required to control erosion decreases as wind velocities increase or the wind direction approaches parallel to the strip. Reducing field width is most effective when all erosive winds are from the same direction (Skidmore and Woodruff, 1968). Although the distinction between crop barriers and crop strips has not been defined, in this report nonerodible strips less than 3 rows wide (usually about 1 m) will be called crop barriers and will include tree shelterbelts; strips wider than 3 rows will be called crop strips.

Table 24–3. Average distance for soil flux to reach maximum and width of field strips required to control erosion from a 17-m/s wind at 15 m above the ground blowing perpendicular to strip and 0.3-m high stubble on windward side of eroding area (Chepil, 1957, Table 1).

Soil	Distance†	Field strip‡
	m	
Sand	30	6
Loamy sand	51	8
Silty clay loam	219	25
Granulated clay	280	30
Clay loam	419	45
Sandy loam	719	76
Silty clay	811	86
Loam	1006	105
Silt loam	1289	131

† Average distance for soil erosion to reach a maximum.
‡ Average width of field strip required to keep average soil flux below 0.01 kg m^{-1} width s^{-1}.

24-1.2.1   Crop Strips

Strip cropping, using alternate strips of small grains, corn (*Zea mays* L.), or sorghum with a fallow strip, has been used for years on erodible soils in the Northern and Central Great Plains to reduce wind erosion (Chepil, 1957). It has been very successful in these areas because of the prevailing direction of erosive winds. The widths of the fallow and crop strips are usually equal and are determined by even units of machinery widths. For all soil textures except a loam or silt loam, the strips must be less than 100 m wide to completely protect the soil from wind erosion (Table 24-3).

24-1.2.2   Shelterbelts and Crop Barriers

Trees or tall vegetation have been used to shelter soils in Russia, China, central Texas, south-central Kansas, and eastern Montana for many years, and basic principles of shelterbelt designs have been documented (Black and Siddoway, 1971; Fryrear, 1963; Denisov, 1960; Sheng and Kang, 1961; and Great Plains Agriculture Council, 1976). The higher the wind velocity, the smaller the protected zone, but Hagen (1976, p. 31) reported that "maximum wind and erosion reduction extends over a larger leeward area when windbreak porosity is near 40% as compared with a less-porous windbreak." Erosion can be reduced appreciably to the lee of barriers with moderate reductions in wind velocities (Fig. 24-3). Because of the funneling

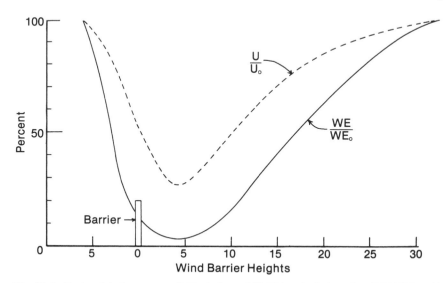

Fig. 24-3. Ratio of shelter to open field wind speed (U/U_o) and wind erosion (WE/WE_o) with all windspeeds above threshold velocity and normal to a 40% porous windbreak. Wind speeds measured at 0.12H above soil surface (Hagen, 1976, Fig. 2).

effect created by a gap in the barrier, 2-row barriers are usually recommended to minimize gaps (Hagen et al., 1972).

Historically, perennial barriers have been grown where they can trap drifting snow to help provide sufficient water for their growth. In hot and arid climates, and in the absence of a water table, the roots of the barrier must move a lateral distance about 2.5 to 3 times the height of the barrier (Greb and Black, 1961). Unless the barrier produces a harvestable crop, it removes an appreciable area of cropland from production. In more arid areas, annual crop barriers or grass strips are used because of less competition for soil moisture, but the annual crops must be established each year. Because of different growth habits, trees or crops will protect different distances leeward, but the usual distance is considered to be 10 times the height of the barrier (Table 24-4).

24-1.2.3 Field Orientation

Reducing the width of a field or installing shelterbelts or crop strips will not be effective unless the field is oriented perpendicular to the prevailing erosive wind direction. Skidmore and Woodruff (1968) have prepared tables showing the prevailing direction of erosive winds during different months of the year for the USA. With a high proponderance value, field width can be reduced with strips or shelterbelts to effectively reduce wind erosion. With low preponderance values, orientation against the wind is less effective.

24-1.2.4 Interplanting

In some vegetable-producing areas of the Midwest, rows of a protective crop are planted in conjunction with a crop sensitive to wind damage. While

Table 24-4. Leeward distance protected by various shelterbelt trees or annual crop barriers (Woodruff et al., 1972).

Windbreak	Factors for determining protected distances†
Trees and shrubs‡	
2-row (mulberry)	18.2
5-row (plum, cedar, mulberry, elm, olive)	15.0
1-row (Osage orange)	12.0
3-row [cedar (2), shrub]	11.0
1-row (Siberian elm)	9.5
Annual crops	
Kochia	12.0
Sudangrass	7.5
Grain sorghum	6.0
Forage sorghum	4.0
Broomcorn	1.0

† To find the distance protected, multiply barrier height by the appropriate number in the right-hand column.
‡ Mulberry (*Morus alba* f. *tartarica*), plum (*Prunus americana*), cedar (*Duniperus virginiana*), elm (*Ulmus pumila*), olive (*Elaegnus angustifolia*), osage orange (*Macolura pomifera*).

the practice can be very effective, it requires a high level of management. This practice is usually limited to areas where water stress on the cash crop is not a problem.

24–1.2.4 Limits

Shelterbelts, crop strips, or crop barriers are very effective in reducing erosion in areas with a dominant prevailing wind direction during the wind erosion period. Most trees or shrubs require several years before they attain their design height, and the establishment of trees in semiarid regions is difficult. Because trees must live on available rainfall during prolonged droughts, mortality within the shelterbelt can be a problem. The sheltered area provides homes for wildlife and may improve the microclimate for adjacent crops, but it can also harbor nonbeneficial insects. In warm, semiarid areas the perennial barrier must extend its root system laterally to survive and thus competes with the cash crop for soil water and nutrients.

24–1.3 Soil Roughness, Clods, and Stabilizers

Next to residues, surface roughness and clods are the most widely used methods of reducing wind erosion. Although clods are temporary, they can be re-formed in cohesive soils. They are most effective when used in combination with residues and field orientation but require careful management to optimize benefits. The interaction between soil texture, tillage method, cloddiness, and crop yields must be recognized. No tillage method can effectively reduce wind erosion of a deep sand that contains no silt or clay.

24–1.3.1 Tillage and Clods

Clods are the result of proper tillage performed at the most appropriate time and soil moisture content. They are desirable for wind erosion control, but large clods are not desirable for a good seedbed. The lister and moldboard plow produce the highest number and most stable clods (Lyles and Woodruff, 1962). These are the dominant tillage implements used to reduce wind erosion in the Southern Plains on coarse textured soils with no residue. Fryrear (1980) found a positive relation between tillage methods that leave a cloddy soil surface and cotton yields on an Amarillo fine sandy loam soil (fine-loamy, mixed, thermic, Aridic Paleustalfs). The increased cloddiness from listing or moldboard plowing was evident for 12 months. Clods are usually stable in the absence of rainfall or freezing and thawing. Intense rainstorms will melt down clods, particularly on coarse-textured soils, but, if soils are tilled soon after a rain, more clods can be formed (Fryrear, 1980). The key is timeliness of the tillage operation. Although tilling a wet soil increases the hazards of soil compaction, the potential wind erosion hazard on coarse-textured sandy soils is great enough to justify the practice.

24–1.3.2 Soil Roughness

In addition to leaving clods on the soil surface, listing effectively roughens the soil by creating ridges and furrows. Soil ridges alone can reduce erosion 50 to 90% (Armbrust et al., 1964; D. W. Fryrear, unpublished data, 1984). For soil ridges to be most effective, they must have erosion-resistant soil clods on the surface. If a cloddy ridge surface results from listing, wind erosion will be controlled until the clods are broken down by additional tillage, weathering, or erosion. Tillage to reduce wind erosion will be more effective where residue crops or crops with extensive root systems are grown that will increase or maintain soil organic matter. Decomposing surface residues improve cloddiness and stability of the clods (Chepil, 1955b). Tillage also may be used to roughen the soil and reduce the hazard of wind erosion when the wind is parallel to shelterbelts or crop strips, or while shelterbelts are being established.

24–1.3.3 Soil Stabilizers

Chemicals for stabilizing soil surfaces against wind erosion have been evaluated (Armbrust and Dickerson, 1971; Armbrust and Lyles, 1975; Chepil, 1955a; Chepil and Woodruff, 1963; Chepil et al., 1963; Lyles et al., 1969; Lyles et al., 1974a). Several products successfully controlled wind erosion for a short time, but many were more expensive than equally effective wheat straw anchored with a rolling disk packer (Chepil et al., 1963). Armbrust and Lyles (1975) found five polymers and one resin-in-water emulsion that reduced erosion for two months, did not adversely affect plant emergence or growth, and were easily applied without special equipment. They added, however, that before soil stabilizers can be used on agricultural lands, methods must be developed to apply large volumes rapidly. Also, reliable, preemergent, weed-control chemicals for coarse-textured soils must be developed, as well as films that are resistant to raindrop impact yet allow water and plant penetration and are environmentally safe.

24–1.3.4 Emergency Tillage

When soil surface residues are depleted and a wind erosion hazard exists, emergency tillage is often the last resort (Woodruff et al., 1957). The use and type of emergency tillage varies with locality and climatic condition. While listing or chiseling in midwinter in the Central or Northern Great Plains may be considered emergency tillage, listing is the dominant control method in the Southern Great Plains and is not normally considered emergency tillage. If surface clods on listed sandy soils are broken down by rainfall, a sand fighter or rotary hoe is used to disturb the soil and leave new clods on the surface. The sand fighter and rotary hoe could be considered emergency tillage implements since they are used to control wind erosion, but they are not effective if the soil has been blowing and the surface few millimeters of soil is dry.

If winter wheat does not protect soil from blowing, the field must be roughened to reduce wind erosion. A chisel is used because it destroys less wheat and takes less horsepower than a lister. A spacing of 0.60 to 0.80 m between narrow-point chisels is most effective when operated just deep enough to bring clods to the surface and at a speed of at least 1.8 m/s. Chiseling on 0.76-m spacing did not reduce winter wheat yields in Kansas (Lyles and Tatarko, 1982). All emergency tillage operations should be done perpendicular to the wind direction (Woodruff et al., 1972).

24-1.3.5 Limits

Clods are compact, coherent masses of soil formed by tilling the soil. To effectively reduce wind erosion, most of the soil surface must be covered with nonerodible clods. This is possible for most soils if they are properly tilled before wind erosion begins. Generally, the finer the soil texture the greater the number and stability of clods formed. Coarse-textured soils must be tilled after each rain to bury loose sand grains and bring more clods to the surface. Because most crops are seeded in the surface 0.05 m of soil, the farmer must compromise to have the minimum clods to control wind erosion and still have a satisfactory seedbed.

24-2 INTEGRATED CONTROL METHODS

Wind erosion is a problem on a wide variety of soils, climatic regions, and crop conditions, and maximum control may require the use of several measures. While we cannot describe all possible combinations, we discuss the basic ones.

24-2.1 Residue and Tillage

Adequate surface residues are the major control method, but in semiarid regions residue crops may not produce sufficient cover to protect the soil. In semiarid areas and during drought periods in all areas, tillage may be necessary to reduce wind erosion. An Amarillo fine sandy loam has a potential soil loss in excess of 100 t ha^{-1} yr^{-1}, but tilling this soil and applying gin trash to the surface reduces soil loss below the assumed tolerable loss of 11 t ha^{-1} yr^{-1} (Fryrear and Koshi, 1971).

Winter wheat is often too small in midwinter to protect erodible soils from wind erosion. If the wheat fields start to erode, farmers use emergency tillage to roughen the soil. This combination of growing wheat and tillage can be very effective if tillage is done before erosion becomes excessive.

24-2.2 Field Width, Residues, and Tillage

When permanent barriers, such as shelterbelts or perennial plants are being established, adequate surface residues and/or tillage must be used to control wind erosion. As the barrier attains its design height, the quantity of surface residue needed to protect the soil can be reduced and lower residue-producing crops grown. An alternative is to use a more intensive tillage program to reduce wind erosion during barrier establishment. Because barriers are usually oriented perpendicular to the prevailing wind, soil ridges can control wind erosion very effectively.

24-3 OPTIMUM CONTROL PERIODS

24-3.1 Soil Damage

The objective of practices to control wind erosion is to reduce erosion below the soil loss tolerance established for a particular soil. Chepil (1960) described the visible effects of wind erosion for various levels of annual soil loss (Table 24-5). It is difficult or even impossible to visually detect wind erosion losses of less than 41 t ha^{-1} yr^{-1}. If we assume that wind erosion damages soils when surface soil is removed from the land, then we are concerned with the removal of soil fines, primarily silt, clay, and very fine sands. These fractions of soil are responsible for holding water and nutrients within the root zone and for maintaining good soil tilth. The resultant short-term impact on soil productivity depends on soil depth, but the long-term impact is basically an accelerating deterioration in productivity. Since more erosion occurs within a few months each year, efforts to control wind erosion should be concentrated during these months to be most effective. The specific time varies, but at Big Spring, TX, 75% of the dust storms occur during January through May, and 50% occur in February, March, and April (Fryrear, 1981).

24-3.2 Plant Damage

Field and wind-tunnel tests have established the relative tolerance of various crops to blowing sand (Table 24-6) (Fryrear et al., 1975). Some plants, such as onions (*Allium cepa*), may not be killed, but their growth is delayed several weeks and yields are significantly reduced. For most crops, tolerable soil loss is less than the assumed tolerable loss that will ensure continued productivity. In the Southern Great Plains, farmers recognize the need to control wind erosion during the critical planting and crop establishment period. As crops emerge, the soil surface is roughened after each rain. As crops mature they become more tolerant to wind erosion damage, and the wind erosion hazard decreases.

While blowing sand before harvest would have little effect on the yield or sorghum or wheat, it would reduce the marketability of leaf crops such as tobacco (*Nicotiana tabacum*), cabbage (*Brassiea oleracea*), or lettuce (*Lactuca sativa*) (Armbrust, 1979; Downes et al., 1977). Though shelter-

Table 24-5. Relationships among quantity of wind erosion and visible effects of erosion (Chepil, 1960, Table 3).

Degree of erosion	Description of erosion	Annual soil loss†
		t/ha
None to insignificant	No distinct visible effects of soil movement.	>41
Slight	Soil movement not sufficient to kill winter wheat in boot stage.	41–136
Moderate	Removal and associated accumulations to about 25 mm depth sufficient to kill wheat in boot stage.	136–454
High	About 25–50 mm removal and associated accumulations.	454–906
Very high	50–75 mm removal with small dune formations.	906–1360
Exceedingly high	More than 75 mm removal with appreciable piling into drifts or dunes.	>1360

† Occurring in the vicinity of Garden City, KS, during 1954 through 1956.

Table 24-6. Crop survival as influenced by duration of exposure to a 15 m/s wind with sand flux of 0.05 kg m^{-1} width s^{-1} on plants 9 or 10 days old (Fryrear and Downes, 1975, Table 3).

Crop	Survival rates at three exposure times (min)		
	5	10	20
	%		
Pepper	75	8	0
Onion	100	100	100
Cabbage	100	87	56
Southern pea	100	94	72
Carrot	91	10	4
Cucumber	100	100	46
Cotton	100	85	15
Sunflower	91	88	72
Avg.	95	72	46

belts, crop barriers, or crop strips help reduce erosion damage, many farmers roughen the soil as soon as possible following a rain while the crop is small.

24-4 SUMMARY

Properly managed crop residues can effectively reduce wind erosion wherever they are normally grown. In areas with insufficient residues to protect the soil, proper tillage can greatly reduce the erodibility of cultivatable sandy soils. Timing of the tillage operation is extremely important to produce the combination of surface clods and roughness needed to control wind erosion. When winter cover crops do not protect the soil surface, tillage is used as an emergency wind erosion control practice. Crop strips, crop barriers, or shelterbelts reduce field width and erosion from perpendicular winds. Principles of conservation tillage can supplement the barrier's influence, and tillage is essential as the barriers are being established.

REFERENCES

Armbrust, D. V. 1979. Wind- and sandblast-damage to tobacco plants at various growth stages. Tobacco Sci. XXIII:117–119.

----, W. S. Chepil, and F. H. Siddoway. 1964. Effects of ridges on erosion of soil by wind. Soil Sci. Soc. Am. Proc. 28(4):557–560.

----, and J. D. Dickerson. 1971. Temporary wind erosion control: Cost and effectiveness of 34 commercial materials. J. Soil Water Conserv. 26:154–157.

----, and Leon Lyles. 1975. Soil stabilizers to control wind erosion. Soil Cond. 7:77–82.

Black, A. L., and F. H. Siddoway. 1971. Tall wheatgrass barriers for soil erosion control and water conservation. J. Soil Water Conserv. 26(3):107–111.

Chepil, W. S. 1944. Utilization of crop residues for wind erosion control. Sci. Agric. 24:307–319.

----. 1955a. Effects of asphalt on some phases of soil structure and erodibility by wind. Soil Sci. Soc. Am. Proc. 19:125–128.

----. 1955b. Factors that influence clod structure and erodibility of soil by wind: V. Organic matter at various stages of decomposition. Soil Sci. 80:413–421.

----. 1957. Width of field strips to control wind erosion. Kansas Agric. Exp. Stn., Tech. Bull. 92. p. 16 Dec.

----. 1960. Conversion of relative field erodibility to annual soil loss by wind. Soil Sci. Soc. Am. Proc. 24(2):143–145.

----, and N. P. Woodruff. 1963. The physics of wind erosion and its control. Adv. Agron. 15:211–302.

----, ----, F. H. Siddoway, D. W. Fryrear, and D. V. Armbrust. 1963. Vegetative and non-vegetative materials to control wind and water erosion. Soil Sci. Soc. Am. Proc. 27:86–89.

Denisov, I. 1960. Screens for crops and on fallow land on the steppes. Sel'skoe Khoziaistove Sibiri. 5(May):18–20.

Downes, J. D., D. W. Fryrear, R. L. Wilson, and C. M. Sabota. 1977. Influence of wind erosion on growing plants. Trans. Am. Soc. Agric. Engr. 20(5):885–889.

Englehorn, C. L., A. W. Zingg, and N. P. Woodruff. 1952. The effects of plant residue cover and clod structure on soil losses by wind. Soil Sci. Soc. Am. Proc. 16:29–33.

Erbach, D. C. 1982. Tillage for continuous corn and corn-soybeans rotation. Trans. Am. Soc. Agric. Engr. 25(4):906–911, 918.

Fryrear, D. W. 1963. Annual crops as wind barriers. Trans. Am. Soc. Agric. Engr. 6(4):340–342, 352.

----. 1980. Tillage influences monthly wind erodibility of dryland sandy soils. p. 153–163. *In* Crop Production with Conserv. in the 80s. Conf. Proc., Chicago, IL. 1–2 Dec. Am. Soc. Agric. Engr., Pub. 7-81.

----. 1981. Dust storms in the Southern Great Plains. Trans. Am. Soc. Agric. Engr. 24(4):991–994.

----, D. V. Armbrust, and J. D. Downes. 1975. Plant response to wind erosion damage. Proc. Soil Conserv. Soc. Am., 36th Ann. Meeting, San Antonio, TX. 10–13 Aug.

----, and J. D. Downes. 1975. Consider the plant in planning wind erosion control systems. Trans. Am. Soc. Agric. Engr. 18(6):1070–1072, 1075.

----, and P. T. Koshi. 1971. Conservation of sandy soils with a surface mulch. Trans. Am. Soc. Agric. Engr. 14(3):492–495, 499.

Great Plains Agriculture Council. 1976. Shelterbelts on the Great Plains. Proc. of the Symp., Denver, CO. 20–21 Apr. Great Plains Agric. Council Pub. 78.

Greb, B. W., and A. L. Black. 1961. Effects of windbreak plantings on adjacent crops. J. Soil Water Conserv. 16(5):223–227.

Hagen, L. J. 1976. Windbreak design for optimum wind erosion control. p. 31–36. *In* Proc. of the Symp. on Shelterbelts on the Great Plains, Denver, CO. 20–21 April. Great Plains Agric. Council Pub. 78. Lincoln, NE.

––––, E. L. Skidmore, and J. D. Dickerson. 1972. Designing narrow strip barrier systems to control wind erosion. J. Soil Water Conserv. 27(3):269–272.

Hartwig, R. O., and J. M. Laflen. 1978. A meter stick method for measuring crop residue cover. J. Soil Water Conserv. 33:90–91.

Lyles, Leon. 1980. The U.S. wind erosion problem. p. 16–24. *In* Crop Production with Conservation in the 80s. Conf. Proc., Chicago, IL., 1–2 Dec. 1980. Am. Soc. Agric. Engr., Pub. 7-81.

––––, and B. E. Allison. 1976. Wind erosion: The protective role of simulated standing stubble. Trans. Am. Soc. Agric. Engr. 19(1):61–64.

––––, and ––––. 1980. Range grasses and their small grain equivalents for wind erosion control. J. Range Manage. 33:143–146.

––––, and ––––. 1981. Eqivalent wind-erosion protection from selected crop residues. Trans. Am. Soc. Agric. Engr. 24(2):405–408.

––––, D. V. Armbrust, J. D. Dickerson, and N. P. Woodruff. 1969. Spray-on adhesives for temporary wind erosion control. J. Soil Water Conserv. 24:190–193.

––––, N. F. Schmeidler, and N. P. Woodruff. 1973. Stubble requirements in field strips to trap windblown soil. Res. Pub. 164. Kansas Agric. Exp. Stn.

––––, R. L. Schrandt, and N. F. Schmeidler. 1974a. Commercial soil stabilizers for temporary wind-erosion control. Trans. Am. Soc. Agric. Engr. 17(6):1015–1019.

––––, ––––, and ––––. 1974b. How aerodynamic roughness elements control sand movement. Trans. Am. Soc. Agric. Engr. 17(1):134–139.

––––, and John Tatarko. 1982. Emergency tillage to control wind erosion: Influences on winter wheat yields. J. Soil Water Conserv. 37(5):344–347.

––––, and N. P. Woodruff. 1962. How moisture and tillage affect soil cloddiness for wind erosion control. Agric. Engr. 43(3):150–153, 159.

Sheng, T.-C., and H. Kang. 1961. Windbreaks in Taiwan. Chinese-American joint comm. on rural reconstruction. Forestry Series 7, Taipei, Taiwan.

Siddoway, F. S., W. S. Chepil, and D. V. Armbrust. 1965. Effect of kind, amount, and placement of residue on wind erosion control. Trans. Am. Soc. Agric. Engr. 8(3):327–331.

Skidmore, E. L., M. Kumar, and W. E. Larson. 1979. Crop residue management for wind erosion control in the Great Plains. J. Soil Water Conserv. 34(2):90–94.

––––, and F. H. Siddoway. 1978. Crop residue requirements to control wind erosion. p. 17–33. *In* W. R. Oschwald (ed.) Crop residue management systems, Spec. Pub. 31. American Society of Agronomy, Madison, WI.

––––, N. L. Nossaman, and N. P. Woodruff. 1966. Wind erosion as influenced by row spacing, row direction, and grain sorghum population. Soil Sci. Soc. Am. Proc. 30:505–509.

––––, and N. P. Woodruff. 1968. Wind erosion forces in the United States and their use in predicting soil loss. USDA Agric. Handb. 346.

Woodruff, N. P., W. S. Chepil, and R. D. Lynch. 1957. Emergency chiseling to control wind erosion. Tech. Bull. 90. Kansas Agric. Exp. Stn.

––––, C. R. Fenster, W. S. Chepil, and F. H. Siddoway. 1965. Performance of tillage implements in a stubble mulch system. I. Residue Conservation. Agron. J. 57:45–49.

––––, L. Lyles, F. H. Siddoway, and D. W. Fryrear. 1972. How to control wind erosion. p. 22. *In* USDA Agric. Info. Bull. 354.

25 Transferring Soil Conservation Technology to Farmers

R. W. Jolly
Iowa State University
Ames, Iowa

B. Eleveld
Oregon State University
Corvallis, Oregon

J. M. McGrann
Texas A&M University
College Station, Texas

D. D. Raitt
Economic Research Service
U.S. Department of Agriculture
Columbia, Missouri

This paper deals with the process and problems associated with the transfer of soil conservation technology to farmers. The importance of this activity is unquestioned—new knowledge or technology is of little value to society until it is applied. However, concentrating solely on the transfer of new technology ignores a more fundamental issue, the development and maintenance of appropriate technology. Applying research findings at the farm level is the final stage in a process that begins with the identification of the research problem. Transferring a technology that significantly improves the welfare of the recipient is considerably easier than transferring a technology that does not. In many situations technology transfer can be enhanced more by better research and development than by improving education and extension efforts.

We begin with a review of the technology transfer process in agriculture. This section draws primarily on research conducted over the past 25 years by sociologists and economists. Next, we describe soil conservation

Published in R. F. Follett and B. A. Stewart, ed. 1985. *Soil Erosion and Crop Productivity.*
© ASA-CSSA-SSSA, 677 South Segoe Road, Madison, WI 53711, USA.

technology and compare some of its characteristics with other agricultural technologies. Some perceived or measured barriers to adoption of soil conservation technology are listed. We then describe the technology development process in the USA from an institutional perspective and as an information system.

Finally, we review methods for transferring technology and describe a few conservation research/extension programs that attempt to improve the transfer of conservation findings to farmers.

25–1   HOW FARM PEOPLE ACCEPT NEW IDEAS

The title for this section is taken from a venerable Iowa State University extension bulletin first published almost 30 years ago (Cooperative Extension Service, 1955). The title is noteworthy in its finality. Its authors apparently believed they understood the process, at least well enough to share their findings with the public. Social scientists have studied the process of transferring new technology for a long time and have developed a number of concepts to describe it, such as diffusion, adoption, embodiment, or technical change. Most of the vast literature has been written (usually independently) by sociologists and economists. Rogers (1983) reports counting over 3000 publications on the sociology of diffusion alone. Economists have been equally prolific on technical change, technology choice, allocation of research resources, and related topics (Ruttan, 1982; Feder et al., 1982). We summarize only a few key ideas that emerge from this extensive research effort. Because sociologists and economists have rarely collaborated on technology transfer research, we review their accomplishments along discipline lines.

25–1.1   Sociological Models

A major research topic pursued by rural sociologists was the lag time between the discovery of a new farming practice or technology and its general use by farmers. This process is referred to as diffusion. An empirical model of this process was developed by a number of sociologists during the 1940s and 1950s. Cooperative Extension Service (1955) provides one of the clearest descriptions of this early model and its explanation of the diffusion process, and Rogers (1983) presents a modern treatment of adoption and diffusion. We review the model's components and then consider its implications for influencing the adoption of new technology.

25–1.1.1   The Adoption Process

Adoption is a social psychological process of individual decision making. Because the adoption of new technology takes time, the adoption model is also time dependent. The potential adoptor of new technology is assumed to move through several stages in the acceptance process.

25-1.1.1.1 Awareness. The adoption process begins with an individual discovering a new technology or learning of its existence. Awareness may occur through mass media, demonstrations, or interaction with neighbors.

25-1.1.1.2 Interest. During this stage, an individual begins to gather background information on new technology. In effect, the second stage represents the beginning of a self-directed learning process. Mass media is still an important information source, but search activities tend to become more focused.

25-1.1.1.3 Evaluation. In this stage the farmer begins to evaluate the appropriateness of the new technology specifically for his or her own situation. This may involve simply thinking about how the new technology might affect the farming operation or family, or a more formal evaluation might be conducted using budgeting methods or other analytical techniques. Because the farmer's information needs have now become very specific, mass media, large extension meetings, and similar outlets tend to be replaced by reliable sources of specific information.

25-1.1.1.4 Trial. After evaluating the range of likely outcomes from adoption, the farmer may try the new technology on a small scale. This stage serves both to verify information obtained in the previous stages and to develop new techniques and skills required by the new technology. In the trial stage, one-on-one contact between the adoptor and change agents, such as extension agents or machinery dealers, is often required. General information sources are typically not employed. The success or likelihood of a trial stage is also a function of the technology. A small or divisable technology such as a new corn variety or herbicide is amenable to trial. A lumpy technology such as a confinement swine unit or a large combine is not. Information and learning costs associated with the trial stage need to be considered.

25-1.1.1.5 Adoption. This last stage in the adoption process occurs, ideally, when the new technology completely supplants the old. This implies an on-going process in which the new technology is maintained or supported by the farmer. If the performance of the innovation is not satisfactory, the new technology is rejected. In any event, the adoption process in this simple model is considered complete at this stage.

25-1.1.2 Technological Factors in the Adoption Process

The characteristics of the new technology obviously influence the adoption process. Some factors identified by sociological researchers include costs associated with learning requirements or new capital outlays. New seed varieties may require few changes in equipment or practices, but the introduction of a new enterprise may require a drastic alteration of capital flow, labor, and managerial skills.

25-1.1.3   Social and Personal Factors in the Diffusion Process

Early sociological diffusion models looked to community and individual values for an explanation of why the commencement and duration of the adoption process vary among individuals. Personal characteristics were considered to be important factors explaining the diffusion of innovations.

25-1.1.3.1   Innovators. This group is the first to adopt a new technology. They are experimenters. Generally, they possess the required wealth and education to permit the risk of experimentation. Often, innovators will establish direct contacts with researchers or other technology developers. Early researchers found that other members of the community did not view innovators as reliable sources of information.

25-1.1.3.2   Early adopters. This category includes the first significant, identifiable group within a community that adopts a new technology. They tend to be younger, better educated, and more active in community affairs than average. Because of their visibility and leadership in the community, the early adoptors were identified as an important target group for change agents such as extension and soil conservation workers.

25-1.1.3.3   Early majority. When this group begins to adopt a new technology, the rate of adoption tends to increase rapidly. The early majority was identified as older than average, better educated, active in community organizations, and informal leaders.

25-1.1.3.4   Late majority. This group was described as below average in education, somewhat older, and less active in community organizations.

25-1.1.3.5   Non-adoptors or laggards. This group was identified as those members of a community that either do not adopt or are very slow to adopt a new technology. As expected, this group comprised the older, least educated members of a community. They tended to be isolated from community activities and did not assume leadership roles.

25-1.1.4   Implications for Technology Transfer

Diffusion is viewed essentially as an information system. Information on a new technology is acquired, verified, modified, and transmitted to definable groups within a community. These groups are distinguished, in part, by age, wealth, status, and education. In addition, this process of information transfer is influenced by characteristics of the technology itself.

Given this model, diffusion of new technology can be encouraged by investing in or improving the information system that drives the process. This might include investments in extension programs, demonstrations, or community leadership training designed to identify innovators or early adoptors and make them more effective in their community.

The model also gives some insight into targeting information for specific groups depending on their stage in the adoption process. Extension programs and tillage demonstrations may be appropriate for early adoptors and early-majority individuals who are in the interest and evaluation stages. However, when these individuals move into the trial stage, technical assistance and direct support by field advisors may become more important.

25–1.2 Economic Models

Economists, too, study the process of technology transfer. In many situations their efforts are combined under the slightly broader rubric of technical or institutional change. Technical change, in its simplest sense, looks first at how research and development activities alter the basic relationships among inputs and outputs (a production function approach). Once the altered production relationships have been measured, the economist identifies and measures the economic incentives to adopt the new technology. In this section, we review some of the major results of this effort.

25-1.2.1 The Role of Economic Incentives

The economic model of technical change enhances the information-based model of the sociologists. Economists, as one might anticipate, were interested in the impacts of economic variables on the transfer of new technology. One of the first measurements they identified was the profitability of the new technology relative to existing practices. Later economists added risk impacts, investment requirements, and the basic resource endowment. Sociologists had previously identified economic efficiency and returns as factors in the diffusion process but had made few attempts to quantify their impact.

We can imagine a potential adoptor preparing a budget of projected costs and returns generated by the new technology. These estimates are then compared with the costs and returns from the technology currently in use. If the new technology performs better economically than the old, it is adopted. Of course, risk, learning costs, and changes in the farm's capital stock are usually taken into account as well.

The economic model considers certain incentives to adopt, but it does not adequately consider the dynamic process by which individuals obtain the information needed to complete the economic analysis, evaluate the new technology, or incorporate it into an ongoing farming operation. This aspect provides a link to the descriptive sociological models.

One of the first, and certainly most important, economic analyses of technology transfer was Griliches' (1957) study of diffusion of hybrid corn (*Zea mays* L.). Griliches used time series data from states and crop districts on relative area planted to hybrids as a measure of adoption. At this level of aggregation, Griliches hypothesized that sociological factors would tend to be offsetting, leaving the economic forces to explain adoption. Griliches simplified the diffusion process by dividing it into two components: the supply of new technology from public and private organizations, and the demand for new technology by farmers.

Griliches believed that economic incentives would explain differences among geographic regions with widely differing adoption histories. He related differences in the beginning of the adoption process to profit incentives, development costs, and distribution costs faced by seed corn suppliers. He related differences in the speed of adoption and its long-run equilibrium position to the difference in farmers' profitability between open-pollinated and hybrid varieties. Measures of profitability to new technology suppliers and adoptors appeared to explain most of the variation in adoption parameters. Griliches' conclusions were controversial. They deemphasized the importance of personal and community factors and focused on the role of economic incentives both for suppliers and adoptors of new technology.

In the 25 years since Griliches' article, the economic dimensions of technical change have been subjected to intense research. Additional economic aspects were added to Griliches' basic model—for example, credit availability, risk, farm size, tenure arrangements, and market performance. Attempts were made to merge some sociological (or human capital) factors with the economic model, such as, education levels, participation in extension programs, and demographic characteristics. Despite these refinements, Griliches' original hypothesis seems to remain valid.

A review by Perrin and Winkelmann (1976) of factors influencing adoption of high yielding wheat (*Triticum aestivum* L.) and maize varieties in several developing countries provides some insight. First, economic incentives (usually measured by some profit or income criterion) are a dominant variable in explaining adoption. The addition of personal or institutional variables (both sociological and economic) account for some, but not very much, of the variability in adoption rates. In fact, Perrin and Winkelmann conclude that many public programs intended to subsidize input costs, assist with education programs, or mollify risk impacts appear to be important only where the economic advantages of the new technology are weak.

The key consideration in understanding adoption, Perrin and Winkelmann point out, is to understand what lies behind the economic performance of a new technology. Their conclusion is that economic performance is inextricably linked to the biological performance of the new technology in a local environment.

> The impression from these studies is that the most pervasive explanation of why some farmers do not adopt new varieties and fertilizer while others do is that the expected increase in yield for some farmers is small or nil, while for others it is significant, due to differences (sometimes subtle) in soils, climate, water availability or other biological factors (Perrin and Winkelmann, 1976, p. 893).

25-1.2.2 Implications for Technology Transfer

The implications of the economic adoption model for transfer of new technology are fairly direct. To encourage adoption of new technology, major attention must be given to profitability or to the direct income benefits to the potential adoptor. Although a host of demographic, personal, and institutional factors can affect profitability, economic performance will be strongly influenced by technological characteristics. Further, the biological, chemical, or mechanical performance of a new technology may be extremely localized or site-specific. This characteristic is also important for economic factors such as the impact of an individual's financial structure and risk-bearing ability on the adoption process.

25-2 SOIL CONSERVATION TECHNOLOGY: CHARACTERISTICS AND POTENTIAL FOR ADOPTION

25-2.1 Some Definitions

In this section we consider adoption of soil conservation technology from economic and sociological perspectives. We define soil conservation technology broadly as any agricultural practice that reduces soil erosion. This definition includes structures such as terraces or grassed waterways, tillage practices, and crop sequences, as well as supporting technology such as machinery, crop varieties, and pesticides. We assume that erosion rates are, in general, too high. Further, we assume that the transfer of soil conservation technology is inadequate and partly explains the excessive rates of soil loss. It is beyond the scope of this paper to evaluate these assumptions. Clearly, their validity is crucial to the development and transfer of conservation technology.

A direct superior technology is one in which the benefits accrue directly to the adoptor and the net benefits are greater than the existing technology over the decisions maker's planning horizon. A technology is directly inferior if the benefits accrue to the adoptor but do not exceed those of the existing technology. The direct inferior technology would not, in all likelihood, be widely adopted until changes in relative prices or other factors sufficiently increase the flow of net benefits.

A technology is considered indirect if the benefits do not accrue to the adoptor, even though, in a societal context, the benefits may exceed the costs. The indirectness of benefits may be the result of market structure, institutions, or other external factors.

This classification of technologies is not mutually exclusive, particularly when a given innovation produces multiple benefits and costs. We can, however, use the taxonomy to predict the likelihood of adoption. A direct superior technology will be adopted. A direct inferior technology will not be adopted but could be were prices or other economic factors to change. An indirect technology will not be adopted unless the economic agent gains satisfaction from other individuals' benefiting from his or her activities. To be used, an indirect technology requires institutional change or public intervention to redirect the benefit flows to the adoptor.

The inputs for soil conservation technology, as for any other agricultural innovation, include capital, labor, fertilizer, fuel, and management. The output, however, is reduced loss in soil productivity and reduced sediment damage. Certainly, reducing soil productivity losses may directly benefit the potential adoptor of soil conservation technology through higher yields, lower costs, higher land values, wealthier heirs, and the like, but the major beneficiaries may be the general public and/or present and future users of water resources.

In the absence of governmental intervention, the costs and risks of adopting soil conservation technology are borne directly by farmers. The output or benefits from the technology may accrue to individuals outside the adoptor's immediate family or business relations. Furthermore, the market or price system may not directly reward individuals for adopting soil conservation technology. A bushel of corn is a bushel of corn, whether it is produced using soil saving or soil exploiting technology. Information on the value of soil in the future may not be transferred very effectively through traditional commodity markets. Similarly, economic incentives or information on the value of water quality to water resource users may not be transmitted effectively to adoptors of soil conservation technology.

The potential failure of the market to provide and transmit economic incentives that link the output of soil conservation technology to the inputs creates a major stumbling block to adoption. In other words, even if it were possible to determine the vlaue to society of conservation investment, this value might not be directly transferred to or serve as an incentive for farmers to adopt or maintain conservation technology.

As we look at soil conservation technology, the benefits of maintained soil productivity and improved water quality are either direct and potentially beneficial (that is, inferior to an existing technique) or indirect. Technologies with these characteristics are unlikely to be adopted without a change in prices or compensating public action. When direct economic benefits are absent, the economic and sociological models of adoption appear to break down.

25-2.2 A Sociological Perspective

Rural sociologists recently debated the appropriateness of the adoption-diffusion model for conservation technology. Several unique characteristics of conservation technology were identified that could require modification of the classical adoption-diffusion model.

Rogers (1982) suggested that conservation is a preventive technology, similar to family planning, automobile seat belts, smaller fuel-efficient cars, or solar energy. These preventive technologies require action today to avoid an undesirable outcome in the future. In economic terms the preventive technologies can be viewed as investments. That the benefit is the avoidance of a cost makes little difference economically, but these investments are direct and potentially beneficial because the benefits can flow to the adoptor. The motive to adopt is left intact.

Heffernan and Green (1982) also stress the importance of noneconomic factors (perhaps nonpecuniary factors would be more accurate) as a component in a conservation adoption model. They point out that the utility of work, or the status and comfort associated with an innovation, were often more important in explaining adoption than were economic considerations. Job satisfaction or comfort are, in fact, economic considerations because they can be included as a determinant of an individual's well-being similar to income or wealth.

The much-heralded conservation ethic can also be viewed as a noneconomic factor. The conservation ethic refers to a set of beliefs or values that establishes the ultimacy of the preservation or stewardship of the earth's resources—or at least some of them. Furthermore, this stewardship requirement transcends individual desires for wealth, status, or other short-run benefits. However, the conservation ethic can also be considered a dimension of individual utility, and it thereby becomes an economic (but nonpecuniary) factor. It may influence the adoption decision and may interact with or modify the importance of other factors.

Noneconomic factors are not unique to conservation technology. Presumably, they will always influence adoption to some degree. They merit attention by researchers, but, in themselves, do not require a modification of the adoption-diffusion model. The key factors are their importance in influencing adoption and their suitability as policy instruments.

A third possible shortcoming of the adoption model is related to problem perception (Nowak, 1982; Heffernan and Green, 1982). Soil erosion is difficult to measure even with instruments such as the universal soil loss equation (USLE). Although farmers may espouse the social desirability of soil conservation, they may not recognize their own erosion situation. Soil erosion is often perceived as somebody else's problem.

The costs and benefits of soil conservation are equally arcane. Many farmers and researchers have difficulty estimating crop production costs for a single growing season. Estimating a lifetime flow of costs and benefits from conservation, and the salvage value of land in an environment of uncertainty and technological change, is indeed difficult.

These problems are, however, not unique to conservation technology. Estimation (or perception) problems are inherent in any decision. Farmers may have estimated all too well the individual benefits to conservation which society is pricing through our market system.

Finally, Nowak (1982) emphasized the importance of adaptation or reinvention. Adaptation recognizes the fact that adoption itself is a dynamic process. Even after an individual has acquired new knowledge or the capital, equipment, seeds, or chemicals involved in the conservation technology, the societal goals of increased conservation may not be forthcoming. The technology may require extensive modification to meet local conditions, and the learning process may extend well beyond the original purchase of the needed equipment or supplies. The adaptation issue is also common to many other technologies. Pesticides, for example is a direct, superior technology. However, to use pesticides successfully, many location-specific modifications were required and many were developed by farmers. The farmer's purchase of a herbicide didn't guarantee weed control. The management or behavior of the farmer must adapt to make the new technology effective. Similarly, a farmer's investment in a conservation tillage planter or terraces does not necessarily mean that soil erosion problems have disappeared.

Looking at these examples of self-criticism, it appears that most of the perceived shortcomings of the sociological model of adoption and diffusion and its concomitant information flows are not particularly serious or unique. The missing factor is the strong incentive to obtain information, learn, experiment, invest, bear risk, and adopt conservation technology.

25-2.3 An Economic Perspective

The suitability of the economic model of technical change for conservation technology has not been seriously questioned. The simple decision model presented earlier can be viewed as a plausible positive model of short-term economic behavior. Decision makers evaluate costs and benefits to themselves, although a bequest motive may also be included. Most of the comparisons will be made subject to error and uncertainty. Noneconomic or nonpecuniary factors may be important as determinants of individual satisfaction. As with the sociological perspective, the economic model of technology adoption seems quite valid when applied to conservation.

25-2.4 Implications

Is conservation technology unique? Certain aspects are unique, but our understanding of adoption, diffusion, and technological change is nevertheless applicable to conservation innovations or research. Farmers considering adopting conservation technology face a choice similar to many other technologies. Both economic and noneconomic characteristics of the technology are involved, and evaluation of individual costs and benefits is subject to errors in estimation or perception. The dispersion and relative magnitude of the estimated net benefits from conservation technology influence the onset and rate of the adoption-diffusion process. So too does the efficiency of the community information system and related factors.

Conservation technology is unique because the benefits from reduced soil loss or improved water quality are indirect, or if direct, not sufficiently strong to displace current practices. In making this assertion, we address only the dimensions of soil productivity, sediment damage, and water quality.

We began this section with the assumption that erosion rates are too high and therefore that conservation technology transfer needs to be improved. We identified indirect or weak incentives as a major obstacle to adoption of this technology. Therefore, to study the transfer of soil conservation technology, we must also identify a set of policy instruments (artificial incentives) that will drive the adoption process.

The most common instruments for increasing adoption of conservation technology are education, regulation, taxation, and subsidization or cost sharing. Farmers can be educated on existing soil loss levels, their likely impacts, and some control methods. Regulation involves requiring farmers to use a specific set of production practices or make investments that allow them to meet prescribed erosion levels. Taxes on soil loss or erosion-enhancing inputs or outputs can create incentives to control erosion. Alternatively, tax credits can be granted to reduce the after-tax cost of investment in conservation technology. Subsidies on soil saved (a negative tax) or cost sharing can be used to reduce investment or leasing costs. These instruments are well known and frequently discussed.

However, there are also a number of less conventional policy instruments. The first involves investing public funds to develop high-income cropping technology that reduces soil loss. This is essentially an attempt to side-step the externalities present in soil and water resource markets. The extensive public investment in conservation tillage is an example of this approach. Conservation technology is generally promoted as a means to maintain or improve profitability and still (or incidentally) save soil. This technology does not create direct rewards for controlling soil losses or water degradation. Rather, it opens direct economic benefits to the producer that will lead to adoption independently of soil productivity or water quality benefits.

Another option requires cross-compliance for participants in other government programs. To receive corn price supports, for example, a farmer would be required to meet certain conservation standards. Again, the risk protection and direct income benefits from commodity programs are used to create incentives to conserve soil. The underlying market failure is not corrected.

The Role of Technology

Since soil and water conservation is characterized by weak private economic incentives, further adoption of conservation technology will be strongly influenced by public policy. The next question that we need to consider is the role of research or technological advance in improving conservation of the soil resource.

Improved technology cannot alter the underlying market failure or externalities that result in soil and water exploitation. Science can increase the ability of conservation technology to compete in the short run with the conventional cultural methods without significantly altering the mix of products. For example, improvements in conservation tillage such as better planters, cold-tolerant varieties, or less costly weed control practices indirectly increase conservation.

Some conservation technologies may never be profitable to an individual. Terraces, for example, will likely always require some cost sharing. Improved technology in this case might reduce construction or maintenance costs of the structures, thereby making more efficient use of public subsidies. Some practices that are potentially profitable, such as meadow-based rotations, may be made profitable by technological improvements. Research in this case represents an attempt to shift agricultural production in favor of less erosive enterprises.

Finally, research can improve conservation management techniques. For example, improved measurement and analysis of conservation technology's benefits, costs, and impacts on future productivity and water quality could facilitate adoption as well as the application of public policies. Specific examples include improvements in relating residue cover to erosion or estimating the impact of a production system on soil characteristics over a long period.

25-3 TECHNOLOGY DEVELOPMENT AND TRANSFER IN AN INSTITUTIONAL SETTING

Technology is developed and transferred within complex institutional framework with economic (Ruttan, 1982), political (Hadwiger, 1982), and sociological (Busch, 1980) dimensions. Figure 25-1 represents information flows through a college, department of agriculture, or other agency with research capabilities. The research unit defines, designs, and conducts research. Extension repackages or synthesizes results into a form appropriate for the identified clientele of the organization. The technology may be adapted or modified by innovators interacting with adoptors.

Figure 25-1 is a model of simplicity. The flow of information is direct and unambiguous. Clients' needs are passed to extension which reinterprets them for researchers. Research results pass through extension to the clients.

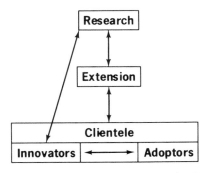

Fig. 25-1. A simplistic view of the process of technology development and transfer.

We have even allowed for direct interaction between innovators and researchers.

Despite the appeal of the diagram, we doubt that the process ever worked so neatly, even during the early periods of the state experiment stations and extension services. Figure 25-2 is a more realistic model of the institutional framework for research. The arrows indicate information (and in some cases resource) flows between various government, business, farm, and research agencies and their perceived clients. The process of problem identification, resource acquisition, experimentation, and transfer now appears enormously complex. Although the research/dissemination infrastructure is highly developed in the USA, important problems occur. Research conducted by state or national scientists, funded by action agencies, and published in scholarly journals may have little relationship to clients' actual needs. Clients' needs may be lost in the system as they are passed between private and public agencies. This problem becomes even more important when the potential research issues identified by clients (were they asked) are very site-specific, differ significantly across regions, and may not be important to many people.

One additional complication needs to be added. Figure 25-2 best represents the information flows within a given discipline, such as agricultural engineering, agronomy, or economics. To a degree, universities, governments, private industry, and even farm organizations or commodity groups tend to segregate along discipline lines. If we consider research on an integrated production systsem, we might have a separate Figure 25-2 for each discipline. The only common element among disciplines might be the clients —the potential adoptors.

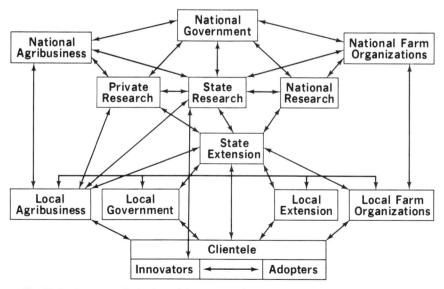

Fig. 25-2. A more realistic view of the process of technology development and transfer.

25-4 IMPROVING CONSERVATION TECHNOLOGY TRANSFER

We began with the assertion that technology that doesn't significantly and directly benefit the adoptor is difficult to transfer. For a variety of reasons, conservation technology that can only reduce productivity losses and improve water quality fails to meet this basic criterion. Therefore, conservation researchers must attempt to make the technology more immediately beneficial to the adoptor. They might accomplish this by changing the relative competitiveness of specific enterprises, such as forage-based rotations and livestock; by making conservation-policy instruments more effective, for example, lower-cost terraces; by improving evaluation methods for farm management and technology, or by altering existing institutions and markets to reward the individual producer for investing in conservation. We also asserted that the development of a complex private and public research/transfer infrastructure may make it difficult for individual scientists to obtain information on conservation problems or client needs and to monitor the actual performance of new technologies following their initial development.

A review of the extensive research on technological change, adoption, and diffusion supprots these assertions. The sociological literature provides a model of how the diffusion process occurs within a community. It can guide the development of more effective educational and technical-assistance programs. The economic literature identifies the major incentives to adopt a new technology and provides criteria by which new conservation technologies can be designed and evaluated. We also identified several characteristics of conservation technology that make it unique. The benefits of reducing soil productivity loss, reducing off-sight sediment damage, and preserving water quality are difficult for the adoptor to perceive and capture. Consequently, government or policy intervention becomes inextricably linked to the adoption process. Given this knowledge base, we now consider ways to improve conservation technology transfer.

25-4.1 Transfer Methods

Three basic methods can be used to transfer conservation technology: (i) education, (ii) technical assistance, and (iii) embodiment.

Education includes formal attempts to transfer information or concepts to large and often remote groups of individuals. Examples include the preparation of technical bulletins and reports, field days, demonstration plots, correspondence courses, and classroom lectures.

Technical assistance includes all one-on-one counseling and planning activities. The development of farm plans by SCS personnel and individual field checks by extension workers or farm input suppliers are examples. Technical assistance is intensive, location specific, and frequently includes coaching, on-going feedback and response between the change agent and the adoptor.

Embodiment refers to the incorporation of technological developments into existing inputs or practices. One example might be the incorporation of cold-stress resistance into a corn hybrid. We also include the development of planters that perform well under both conventional and high-residue conditions. To a degree, embodiment is a passive method in that new technology may be transfered without requiring major decisions or changes by the adoptor.

These broad categories describe transfer mechanisms. The transfer process, particularly as it relates to the technology development process, can be improved in three major ways: (i) improving technology quality and appropriateness, (ii) improving the performance of conservation policy instruments, and (iii) directly improving the transfer methods.

Improving technology quality and appropriateness begins with better identification and anticipation of problems faced by technology users. Experimental designs are required that will test technologies using criteria meaningful to adoptors. Frequently, this requires a systems approach in which technical performance is evaluated using economic criteria.

A second area that will improve conservation transfer is more effective policy instruments, such as efficient cost-share or tax programs. For example, Iowa is currently experimenting with a revolving fund program to provide low- or no-interest loans as a substitute to direct cost sharing.

Policy design also requires an improved research base for evaluating the effects of various programs or instruments at the farm and national level. Policies that are not targeted toward actual problems are no more helpful than inadequately designed technology. Consequently, problem identification and appropriate design are also critical steps in policy development. For example, changes in machinery investment and learning costs are frequently mentioned barriers to adoption of conservation tillage. A program that allows farmers to rent a conservation planter and learn how to operate it satisfactorily might be a more effective policy than a per acre subsidy.

Finally, conservation technology transfer can be improved by improving the transfer system itself. In other words, we can attempt to improve educational programs, technical assistance, or embodiment techniques. Improving educational programs to make them more appropriate or supportive of adult learning patterns is one example. Adults tend to be eclectic learners who frequently rely on consultants when they are actually adopting a new practice. Educational programs must reflect their learning patterns.

Transfer methods, as with technology and policy instruments, need to be appropriate. If profitability and weed control are measured obstacles to conservation adoption, the educational and technical assistance programs must address these issues. Economic analysis of conservation systems frequently consider only two equilibrium positions—the beginning and the end. Little information is available to guide adoptors along adjustment paths. A researcher may tell a farmer that a given conservation system is profitable when the farmer's question is "What is the first step I should take?"

25-4.2   Examples of Innovative Programs

In this section we will briefly describe a few conservation transfer programs that attempt to incorporate technology and policy development with improvements in the transfer system.

25-4.2.1   Interdisciplinary Research and Extension in Range Management

Conservation research activity in Texas involves a close working relationship between the SCS, the Texas Agricultural Experiment Station, the agricultural research division of Texas A&M University, and such stations as the USDA-ARS Southwestern Great Plains Research Center at Bushland. Soil and water conservation problems in Texas are extremely varied. They range from brush control and management on 28 million acres of rangelands (Whitson and Scifres, 1981) to water erosion in central Texas and wind erosion in the Texas High Plains, an area that accounts for 35% of land damaged by wind in the USA (Fryrear and Bogusch, 1979).

As is typical of conservation efforts, much of the brush-management technology developed for rangelands has not been evaluated in the economic and financial framework of a total ranch (Whitson, 1982; Harris et al., 1979). The income tax implications most important for the investment decisions of the individual producer have not been incorporated in the evaluation of alternatives (Whitson, 1982; Shiflet and Pendleton, 1982). This research and extension deficiency is not only a barrier to producer adoption of technological innovations, but it also inhibits the direction of research to the most cost-effective projects. The present cost-price squeeze and high interest rates make investments in conservation practices with long-term economic payoffs very risky. Technical research must be complemented with financial analysis to help the individual producer assess the potential for adopting range-improvement techniques.

Scifres (1981) used a systems concept of brush management rather than the single-treatment approach to brush control. The project involves scientists and extension workers in range management, animal science, and economics. He identifies combinations of brush control alternatives, such as prescribed burning, herbicides, and livestock-management needs for rangeland. This method places research in a managerial frame that should make technology more transferable to individual producers.

Research findings resulting from this project focus on prescribed burning as an alternative to the increasingly costly chemical and mechanical brush control practices. The Texas Agricultural Experiment Station determined the optimum time for burning, pre- and post-grazing management practices, and burning procedures. The results have been published at the producer level (Hamilton et al., 1981; Welch, 1982) and in scientific journals (Whitson and Scifres, 1981). This information is also brought to the producer's attention through workshops and burning demonstrations sponsored by the Extension and SCS.

25–4.2.2 Development of Farm-Level Computer Models

Frequently, conservation research tends to be location specific and therefore of limited usefulness in other areas. A major research effort is now underway in Illinois that may be useful over a very wide, perhaps even nationwide, area. The objective is to develop a user-oriented simulation model to compute the long-run costs and benefits of a large number of potential conservation-management strategies. The first version of this model, known by the acronym SOILEC, was described by Eleveld et al. (1983). A related model was developed by Raitt (1981). The USDA-SCS entered into a cooperative agreement with the University of Illinois to improve this model and make it operational. The SOILEC model requires technical data specific to the soils or fields being analyzed, along with economic information such as prices of outputs and costs of production. Many of these data are being collected by SCS district conservationists and extension advisors as part of their current involvement in farmers' conservation planning.

The SOILEC model calculates annual soil erosion (as estimated by USLE) for a combination of crop rotations, tillage, and conservation practices. The estimate of topsoil loss is then used to adjust soil productivity or crop yields and to increase the costs of production through a long-run planning period of up to 50 years. Net incomes are calculated for each management alternative and discounted to give an annuity-equivalent, net-income figure which includes the discounted residual value of the land at the end of the planning period.

The user of the program must specify a base management system—usually the system currently used on the land under analysis—against which other alternatives are compared. The model can rank the alternatives according to increasing costs of production on a per acre basis. This feature may be most appropriate for farm planning. The user may also specify a ranking according to increasing soil loss or annual cost per ton of soil loss reduction. The latter two options are useful for policy makers' analyses in targeting assistance payments.

Because of the importance of economic performance in technology transfer, computer models that integrate the technical characteristics into a managerial framework are extremely important. Furthermore, models of this sort can assist with the design and implementation of policy.

25–4.2.3 Research on Technology Transfer Systems

Several projects have been undertaken in Iowa to develop more effective methods for transfer of conservation technology. These studies focus on the social, economic, and institutional dimensions of transfer. One involves a field trial implementation of the Iowa Soil 2000 Program (Miller et al., 1982), which provides a framework for educational and technical assistance in conservation. Although procedures for developing conservation plans under Iowa Soil 2000 are well established, the operating rules, their efficiency, and impacts are not well understood. The purpose of the research project, therefore, is to make the program operational and thereby improve the transfer process.

A related project (Nowak, 1983) considers the design of educational programs to "promote" conservation. This project, too, considers the economic and institutional or agency setting in which education and technical assistance occurs.

Fink et al. (1982) reviewed the conservation planning process in four Iowa soil conservation districts and examined the potential use of computers and telecommunications technology in local conservation planning. In addition, they reviewed the suitability of information generated by conservation planning for the economic analysis of the plan. Results from their study provide some guidance for increasing the value of conservation planning for conservation technology transfer.

25-4.2.4   Solutions to Environmental and Economic Problems (STEEP)

This project was initiated in 1972 by scientists working in Washington, Oregon, and Idaho (Oldenstadt et al., 1982). Coordination at the Federal level resulted in an integrated, multiagency research project that was eventually funded in 1976. STEEP represents an interdisciplinary, long-term commitment to develop and transfer soil conservation technology in the Pacific Northwest. Research is directed toward five components of conservation technology: tillage and plant management, plant design, pest management, erosion and runoff prediction, and economics and socioeconomics of erosion control. This project encourages better cooperation among existing conservation-related agencies as well as representatives of producer groups. Extension efforts primarily involve preparation of educational materials such as pamphlets and slide/tape sets. Efforts are currently underway to increase the use of demonstration plots.

25-4.2.5   A Lesson from the Third World

Although this brief overview of innovative conservation research and transfer programs cannot be complete, a few characteristics do emerge. The successful programs are interdisciplinary. They attempt to synthesize research into a form that is meaningful to a farm manager, and they integrate policy and technological issues. They acknowledge the importance but not the dominance of transfer techniques and strategies.

On the other hand, most of these programs appear fragmented and lack an integrated flow of information between client and researcher. Given the complexity of the infrastructure of U.S. agricultural research, this is understandable. We see many efforts to improve the components of technology transfer but few examples of integrated development and transfer systems.

The problems confronting the development and transfer of conservation technology are similar to problems faced by the international development centers such as the International Maize and Wheat Improvement Center (CIMMYT). Economic incentives in developing countries to change practices may be weak or difficult to measure. Variability in climate, soils, and topography make universal truths extremely difficult to discover, and specific truths are extremely costly. The problems of developing appropriate or site-specific technology in the Third World have led to the creation of farming systems research (FSR) or on-farm research (OFR) (Byerlee and Collinson, 1980). The FSR/OFR techniques probably share a common heritage with the farm management programs developed in the U.S. shortly after World War II (Johnson, 1982). Farming systems research has several characteristics. It is applied and directed toward relatively rapid turnaround problems. The research team is interdisciplinary, including biological, physical, and social scientists. To a large extent the topics are defined by farmers, and farmers frequently conduct the field trials and experimentation. This direct contact between scientists and farmer tends to merge research and extension functions. On-farm research incorporates appropriate institutional factors, such as land tenure, market structure, and government programs, into the design and analysis. Finally FSR/OFR supports basic experiment station research by identifying gaps in existing scientific knowledge.

Figure 25-3 summarizes some of these basic functions. Understandably, the OFR/FSR structure in the United States must be incorporated into the complex infrastructure shown in Figure 25-2, but, because the linkages in OFR/FSR between researcher, extension worker, and farmer client are much more direct, the information may flow more freely.

The process of OFR/FSR as described by CIMMYT (Byerlee and Collinson, 1980) begins with identification of a target region or group of farmers. Next, an informal or unstructured survey of farmers can be conducted to identify broad areas that lead to inefficiencies in food production, low farm incomes, or other factors such as excessive erosion or sediment pollution. This survey data can be augmented with secondary data from census or similar sources. A formal or verification survey can follow to check results or inferences from the exploratory survey in a statistically measurable framework. Information from the surveys can be used to identify potential research problems, describe production techniques, or develop a demographic or resource database.

The third major step is to use primary and secondary data to organize an experimental strategy. This might include prescreening technologies, that

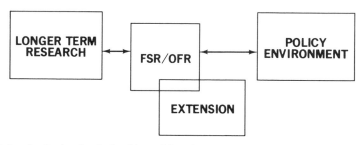

Fig. 25-3. Institutional relationships of farming systems research and on-farm research (FSR/OFR) (adapted from Byerlee et al., 1982).

is, looking for research areas that will yield the highest short-term returns. Research problems that require carefully controlled conditions or special equipment might be relegated to an experiment station. Other research, particularly those problems that involve modifying or adapting existing production or marketing practices, could be performed on cooperating farms. Analysis of the data might involve augmenting traditional statistical methods with more management-oriented, decision science techniques (Anderson, 1974).

Information and techniques generated from an OFR program should be relevant to farmers within the target region. The final stages in this process involve extending information through farmer educational programs, written materials, extension workers, and conservation district personnel. An on-going program of evaluation and adaptive research may continue on the original research problem. This would include follow-up and coaching activities by researchers and extension field staff.

This brief description of FSR/OFR is intended to describe a method that formally incorporates many of the features described in the innovative research/extension programs. Recent reviews by Byerlee et al. (1982), Hildebrand (1982) and Norman (1980) provide good overviews of these techniques.

Although the FSR/OFR model does provide some insight into how the transfer of soil conservation technology can be made more effective, it is certainly not without problems. With restricted resources, targeting the intensive site-specific research required by FSR/OFR is difficult economically and politically. Reward and incentive systems for researchers work against FSR/OFR efforts. Publishing in national journals remains the *sine qua non* among university researchers (Lacy and Busch, 1982). The FSR/OFR projects are not likely to result in this type of publication or product. The early farm management programs that predate FSR/OFR and the current work in Third World countries are both characterized by a relatively simple research/transfer infrastructure. In contrast, the U.S. system is highly complex, and development of a conservation FSR/OFR program would be difficult in the USA, particularly in view of our large competitive private sector. However, projects such as STEEP and some of the integrated pest management programs indicate that coordinated research and extension programs are feasible in the USA.

The transfer of conservation technology must be integrated with an appropriate research program, which in turn must be linked to expressed and anticipated client needs. Perhaps the FSR/OFR process can serve as a model for conservation research projects.

ACKNOWLEDGMENT

The authors wish to thank John Miranowski, Peter Korsching and Min Amemiya for reviewing earlier drafts of this paper.

REFERENCES

Anderson, J. R. 1974. Risk efficiency in the interpretation of agricultural production research. Rev. Mark. Agric. Econ. 3(42):131–184.

Busch, Lawrence. 1980. Structure and negotiation in the agricultural sciences. Rural Sociology 45(1):26–47.

Byerlee, Derek, and Michael Collinson. 1980. Planning technologies appropriate to farmers—concepts and procedures. CIMMYT (International Maize and Wheat Improvement Center), Mexico City, Mexico.

————, Larry Harrington, and D. L. Winkelmann. 1982. Farming systems research: Issues in research strategy and technology design. Am. J. Agric. Econ. 64(5):897–904.

Cooperative Extension Service. 1955. How farm people accept new ideas. North Cent. Regional Pub. 15. Agricultural Extension Service, Iowa State University, Ames.

Eleveld, B., G. V. Johnson, and R. G. Dumsday. 1983. SOILEC: Simulating the economics of soil conservation. J. Soil Water Conserv. 38(5):387–389.

Feder, Gershon, R. E. Just, and David Zilberman. 1982. Adoption of agricultural innovation in developing countries. World Bank Staff Working Paper 542. World Bank, Washington, DC.

Fink, R. J., R. W. Jolly, and G. A. Miller. 1982. A management information system for conservation planning. Report to the Iowa Dep. of Soil Conserv. Cooperative Extension Service, Iowa State University, Ames.

Fryrear, D. W., and H. C. Bogusch. 1979. Tillage for wnd erosion control. p. 18–25. In B. L. Harris, and A. E. Colburn (ed.) Conservation tillage in Texas. Texas Agricultural Extension Service, College Station.

Griliches, Zvi. 1957. Hybrid corn: An exploration in the economics of technological change. Econometrica. 25(4):501–522.

Hadwiger, D. F. 1982. The politics of agricultural research. University of Nebraska Press, Lincoln.

Hamilton, W. T., C. J. Scifres, D. N. Ueckert (ed.) 1982. Summaries of brush management and range improvement research, 1980–81. Texas Agric. Exp. Stn. Pub. CPR 3968-4041B.

Harris, B. L., E. Burnett, and C. L. Williams. 1979. Potentials for conservation tillage systems in Texas. p. 1–17. In B. L. Harris and A. E. Colburn (ed.) Conservation tillage in Texas. Texas Agricultural Extension Service, College Station.

Hefferman, W. D., and Gary Green. 1982. Applicability of the adoption-diffusion model to resource conservation: The pro position. Paper presented at the Rural Sociological Soc. annual meeting, San Francisco, CA, 3 Sept. 1982.

Hildebrand, P. E. 1982. Farming systems research: Issues in research strategy and technology design: discussion. Am. J. Agric. Econ. 64(5):905–916.

Johnson, G. L. 1982. Small farms in a changing world. p. 7–28. In Proc., 1981 Farming Systems Research Symposium. Kansas State University, Manhattan.

Lacy, W. B., and Lawrence Busch. 1982. Guardians of science: Journal and journal editors in the agricultural sciences. Rural Sociol. 47(3):429–448.

Miller, G. A., Minoru Amemiya, R. W. Jolly, S. W. Melvin, and P. J. Nowak. 1982. Soil erosion and the Iowa soil 2000 program. PM-1056. Cooperative Extension Service, Iowa State University, Ames.

Norman, D. W. 1980. The farming systems approach: Relevancy for the small farmer. MSU Rural Devt Paper 5. Department of Agricultural Economics, Michigan State University, East Lansing.

Nowak, P. J. 1982. Adoption and diffusion of soil and water conservation practices. Proc. RCA Symp. on Future agricultural technology and resource conservation. USDA, Washington, DC.

————. 1983. The selling of soil conservation: A test of the voluntary approach. Sociol. Rep. Iowa State Univeristy, Ames.

Oldenstadt, D. L., R. E. Allen, G. W. Bruehl, D. A. Dillman, E. L. Michaelson, R. I. Papendick, and D. J. Rydrych. 1982. Solutions to environmental and economic problems (STEEP). Science 217:904–990.

Perrin, Richard, and Don Winkelmann. 1976. Impediments to technical progress on small versus large farms. Am. J. Agric. Econ. 58(5):888–894.

Raitt, D. D. 1981. A computerized system for estimating and displaying short-run costs of soil conservation practices. USDA/ERS Tech. Bull. 1959. Washington, DC.

Rogers, E. M. 1982. Comments on the applicability of the diffusion model to conservation innovations. Paper presented at the Rural Sociological Soc. annual meeting, San Francisco, CA, 3 Sept. 1982.

––––. 1983. Diffusion of innovations. 3rd ed. The Free Press, New York.

Ruttan, Vernon W. 1982. Agricultural research policy. Univ. of Minnesota Press, Minneapolis, MN.

Scifres, C. J. 1981. Applications of prescribed burning in brush control: The systems concept of brush management. Paper presented at Animal Agric. Conf., Texas A&M Univ., College Station, TX, 6–7 April 1981.

Shiflet, T. N., and D. T. Pendleton. 1982. Concept to application-technology transfer: The agency and the private landowners. p. 94–103. In D. D. Briske and M. M. Kothmann (ed.) Proc., A nat. conf. on grazing manage. tech. Dep. of Range Science, Texas A&M Univ., College Station.

Welch, T. G. (ed.) 1982. Prescribed range burning in central Texas. Texas Agricultural Extension Service. College Station.

Whitson, R. E. 1982. Economic risk in grazing management. p. 134–135. In D. D. Briske and M. M. Kothmann (ed.) Proc., A nat. conf. on grazing manage. tech. Dep. of Range Science, Texas A&M University, College Station, TX.

––––, and C. J. Scifres. 1981. Economic comparison of honey mesquite control methods with special reference to the Texas rolling plains. J. Range Manage. 34:412–415.

26 A Framework for Analyzing the Productivity Costs of Soil Erosion in the United States

Pierre Crosson
Resources for the Future
Washington, DC

Paul Dyke
Economic Research Service
U.S. Department of Agriculture
Temple, Texas

John Miranowski
Economic Research Service
U.S. Department of Agriculture
Washington, DC

David Walker
University of Idaho
Moscow, Idaho

Erosion imposes costs both on-site (losses in soil productivity) and off-site (sedimentation of reservoirs, damages to water quality, etc.). In this paper we treat only on-site costs, not because we believe off-site costs are less important, but because much less is known about them than about productivity losses. This lack of information is a major obstacle to analysis of off-site costs, particularly analysis aimed, as ours is, at forming a judgment about the importance of the costs in a national perspective.

With respect to on-site costs, we discuss estimates both of past erosion-induced productivity losses and of what these losses may be in the future. The discussion deals only with erosion from cropland, which is the largest single source of erosion at present and may increase significantly in the future.

Published in R. F. Follett and B. A. Stewart, ed. 1985. *Soil Erosion and Crop Productivity.*
© ASA-CSSA-SSSA, 677 South Segoe Road, Madison, WI 53711, USA.

We would like to estimate the costs of productivity loss for the nation, but data limitations prevent quantitative estimates of these costs. We present instead an analytical framework for making these estimates when data become available. Despite the absence of data, we can make empirically based statements about the effects of erosion on costs relative to the effects of other factors.

The paper is organized in three main sections. The first deals with definitions and concepts of costs of erosion-induced productivity losses. The second presents and discusses estimates of productivity losses from the end of World War II to 1980 as well as estimates of future productivity loss. Two of these forward-looking estimates assume continuation of 1977 rates of cropland erosion over periods of 50 and 100 years. In addition, we consider the implications for future productivity loss of erosion rates substantially higher than in 1977. The third part of the paper discusses implications of the analysis for thinking about soil conservation policy.

26–1   COSTS OF PRODUCTIVITY LOSS: DEFINITIONS AND CONCEPTS

We define productivity as crop output per hectare of cropland per unit of time. In the empirical portions of the paper this means national average annual yields of main crops: wheat (*Triticum aestivum* L.), corn (*Zea mays* L.), and soybeans [*Glycine max* (L.) Merrill]. All studies of the effects of erosion on productivity use this definition because the concept of productivity relates physical quantities of inputs and outputs, and this definition fits that concept.

Estimates of erosion-induced productivity losses according to this definition have many uses, but by themselves they provide no measure of the costs of the losses. In ordinary usage, and as used here, cost is an economic concept, and estimates of productivity loss lack an economic dimension. Without it, we have no way of valuing the losses and hence cannot estimate their social cost. We suggest that the cost of erosion-induced losses of productivity should be measured by their effect over time on unit costs of crop production. We discuss two aspects of this way of valuing productivity losses: the cost of the losses in some year in the future, and the total cost of the losses over a number of years.

26–1.1   Costs in Some Future Year

The costs of crop production at any time reflect the interplay of factors affecting the supply of and demand for crop output. Changes in cost over time reflect shifts in these supply and demand factors. Demand shifts primarily because of growth or decline in population and per capita income in both domestic and foreign markets and because of changes in consumer preferences. The principal supply shifters are changes in prices of production inputs, technological advance, and erosion-induced losses of productivity.

The relationships among these supply and demand factors at two dates and their effects on crop prices and production costs are illustrated in Fig. 26–1 and Fig. 26–2. In Fig. 26–1 the lines called D_0 and D_{100} are demand curves "now" and 100 years later. The S_0 and S_{100} lines are the corresponding supply curves. Q_0 and Q_{100} are the quantities of production "now" and 100 years later, and P_0 and P_{100} are the corresponding prices. The illustration assumes that erosion reduces crop productivity by 10% over the 100 years, resulting in a leftward shift of approximately 10% in the supply curve and a consequent increase in production costs.[1] The cost increase occurs for one or some combination of two reasons. Farmers may respond to the productivity loss by using more land-substituting inputs. The additional inputs probably would include more fertilizer per acre, more pesticides for improved insect and weed control, investment in better storage to reduce postharvest losses, as well as other measures, all aimed at substituting the productivity of non-land inputs for the lost productivity of the soil. These additional inputs may offset some or all of the erosion-induced productivity loss, but their cost is a cost of erosion. If the prices of the non-land inputs are too high relative to their productivity, the farmer will simply accept the erosion-induced loss. In this case his total costs remain unchanged, but his yields decline, so unit costs rise.

In Fig. 26–1 erosion-induced productivity loss is the only supply shifter, input prices and technology being assumed constant. Costs may increase also because farmers choose to invest in erosion-control practices to avoid or reduce the threatened loss of productivity rather than to compensate for it. In this case the upward or leftward shift in the supply curve occurs not because soil productivity declines but because farmers incur costs to prevent it. The higher costs are attributable to erosion, but it is potential rather than actual erosion.

Conceptually, from a production standpoint, increases in costs to reduce erosion are no different from higher input costs to compensate for it. Both enter costs of production, and in principle both are reflected in the higher supply curve, S_{100}. In practice, however, not all the costs of controlling erosion would appear in S_{100} because a part of them are paid by the federal government through various cost-sharing and technical assistance programs. Farmers pay only a portion of these erosion-control costs, so they are only partially passed on in crop prices. How important this understatement of erosion costs may be is an empirical matter which cannot be pursued here. Our guess is that the understatement is small. It is noteworthy, however, that the Federal conservation effort has funded about $20

---

[1] The assumption of a 10% reduction in productivity is based on the work of William Larson and associates (1983) with corn yields in the cornbelt as discussed below. The productivity decline for other crops and other regions could be greater. Preliminary results obtained by Walker and Young (1981) and others in the Palouse of the Northwest indicate a 20% decline in wheat yields from 50 years of erosion and a 50% decline in 100 years. The empirically derived yield-erosion relationship for wheat on Palouse soils is expressed by $y = 36.44 + 47.01 (1 - e^{-.098640})$ t, where y = wheat yields in bu/ac, and t = number of years of erosion. Initial yield is 72.7 bu, topsoil depth is 37.5 cm, and erosion is 54 MT ha^{-1} yr^{-1}.

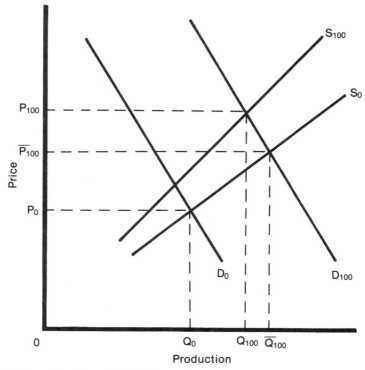

Fig. 26-1. Illustrative effects of shifts in demand and supply on production costs and prices—Case I.

billion in capital improvements and technical assistance over the last 45 years.

Figure 26-1 shows that because of erosion-induced losses of productivity, *unit* production costs in 100 years are higher than they otherwise would be. But what is the effect of erosion on *total* costs in that terminal year? The increase in total cost depends upon both the increase in unit costs and on the volume of production. In Fig. 26-1 the increase in total costs is represented by the area between the two supply curves bounded on the left by their points of intersection with the vertical axis and on the right by curve D_{100}.

Figure 26-1 assumes that erosion-induced productivity loss is the only shifter of the supply curve, leaving out shifts due to changes in input prices and technological advance. Figure 26-2 depicts a situation which takes these supply factors into account, in addition to the effect of erosion on supply. As in Fig. 26-1, the lines in Fig. 26-2 called D_0 and D_{100} are, respectively, the demand curves for crops at present and 100 years later, and S_0 and S_{100} are the corresponding supply curves. The line S_{100} includes the effects of technology and erosion. Q_{100} is production 100 years from now. The line $\bar{S}_{100}$ is the supply curve 100 years from now if erosion had no effect on productivity, and $\bar{Q}_{100}$ is the corresponding output. $\bar{S}_{100}$ lies to the right of S_0,

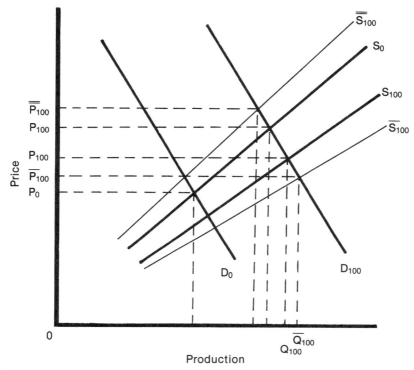

Fig. 26-2. Illustrative effects of shifts in demand and supply on production costs and prices—Case II.

reflecting the combined effect on supply of input prices and technological advance over the 100 years. S_{100} lies above $\bar{S}_{100}$, reflecting a 10% decline in soil productivity because of erosion.

Note that P_{100}, the price 100 years from now, is higher than P_0, the present price. Part of this increase is because of erosion. But, even if erosion had no effect on productivity, the price in 100 years would be higher than at present by the difference between P_0 and $\bar{P}_{100}$. This means that quite apart from the effect of erosion on productivity, prices rise because the combined effect of technological advance and input prices in increasing supply is insufficient to accommodate the growth of demand at prices now.

Just as Fig. 26-1 understates the real increase in costs due to erosion because the supply curve does not include the costs of erosion control paid for by public funds such as federal government cost-sharing, Fig. 26-2 likewise understates such increases. In that figure, however, an additional cost of erosion may not be reflected in curves S_{100} and $\bar{S}_{100}$. This occurs if, in response to potential erosion-induced productivity losses, public funds are invested in development of new land-saving technology to offset the losses. These development costs are not reflected in the prices farmers pay for the new technology, so they are not represented in either S_{100} or $\bar{S}_{100}$. Nevertheless, they are costs of erosion in the sense that they would not be incurred were it not for the threat of erosion to productivity.

We have no estimates of how important this hidden cost of erosion may have been in the past, although we suspect it was small. We know of no evidence indicating that investments by the federal government or other public agencies to develop new agricultural technologies were spurred by the threat of erosion to productivity. But this is an empirical issue that does not concern us here. Our point is the principle that in estimating costs of erosion, these public investment costs must be taken into account.

Population and per capita income growth in the USA and the rest of the world almost surely will increase demand for U.S. crop output far more than 10% over the next 100 years. Technological advance in crop production, even if at the relatively sluggish pace of the 1970s, will tend to shift the supply curve to the right substantially more than 10%. This is likely even with some increase in real input prices. Consequently, if erosion reduces soil productivity by only 10% over the next 100 years, the effect on production costs likely will be small relative to the effects of demand growth, input prices, and technology. This is illustrated in Fig. 26-2 where the assumed shifts in demand and supply are conservative for a 100-year period. Yet these shifts have a substantially greater impact on costs and prices than does erosion. If input prices and technology have no effects on supply, erosion being the only supply shifter, then over the 100 years price rises from P_0 to $\bar{\bar{P}}_{100}$. Of this increase, demand growth is responsible for the rise from P_0 to P_{100}, about 80% of the total, and erosion is responsible for the rest.

Because of input prices and technological advances, price after 100 years is 20 to 25% less than it otherwise would be (P_{100} divided by $\bar{\bar{P}}_{100}$ minus 1), and because of erosion price is 10 to 15% higher (P_{100} divided by $\bar{P}_{100}$ minus 1). Thus, in this illustration the effect of input prices and technology on prices is substantially greater than the effect of erosion. These conclusions about relative magnitudes are based on plausible assumptions about shifts in supply and demand and are suggestive of one possible scenario. Other scenarios are plausible, and in any case, these results are not definitive for the Corn Belt and should not be generalized for other crops and other regions.

26-1.2   Costs Over a Number of Years

Knowledge of the effects of erosion on production costs in some future year and of the magnitude of these effects relative to those imposed by other supply and demand factors is useful for judging the importance of erosion as it affects productivity. However, the productivity loss typically is not confined to the terminal year of the period considered but accumulates over much if not all of the period. Consequently, knowledge of the effect on costs in the terminal year is incomplete in two ways: (i) it understates the full cost of erosion, and (ii) it reveals nothing about when it would pay to undertake measures to control erosion or to offset its effects on productivity.

To redress these deficiencies, one must calculate the present value of the productivity losses.[2] As indicated in the discussion of Fig. 26-1, the value of the loss in any year is measured by the area between the two supply curves bounded on the left by their points of intersection with the vertical axis and on the right by demand curve D_{100}. The sum of these discounted values for each year in the period considered gives the present value of the losses over the entire period.

This value depends crucially on three elements in the calculation: (i) the annual values of the losses; (ii) the interest rate used to discount the annual values; and (iii) the time distribution of the losses. The annual values of the losses depend on the effect on production costs of the productivity losses and on all the other factors tending to shift demand and supply curves over time. Conceptualizing these relationships is easy enough, although actually calculating the annual losses would be quite another matter.

The interest rate used to discount the annual values has a powerful effect on their present value, and there is a considerable literature on what determines the appropriate rate. Without discussing this literature, we note only the obvious fact that, for a given set of annual values of productivity loss, their present value will be higher the lower the rate of discount. Accordingly, people may agree completely about the long-term impact of erosion on productivity and even about the annual values of the losses, yet disagree vigorously about the importance of the losses because they adopt different rates of discount.

Knowledge of the time distribution of the losses is crucial to the calculation of their present values, yet little is known empirically about this distribution. Figure 26-3 is a stylized representation of two plausible possibilities. Curves L and C both depict a situation in which erosion reduces productivity 10% over 100 years. The time distribution of the decline, however, is quite different in the two cases, with important consequences for the present values of the losses. In the first 75 to 80 years of the period, the annual productivity losses are greater with the L distribution, and in the last 20 to 25 years they are greater with the C distribution. Calculated in year 0, the present value of the L distribution losses will be greater than that for the C distribution, and this will continue to be true for a number of years. However, at some time (we have not calculated when, but it would be before years 75 to 80) the larger losses toward the end of the period of the C distribution will cause their present value to exceed that of the L distribution.

The difference the time distribution of losses makes for their present value is not just a matter of academic interest. Both for individual farmers and for society as a whole, decisions about whether and when to intervene

---

[2] In principle, the present value of the productivity losses would be calculated into perpetuity since the losses are assumed to be permanent. In the illustration used here, however, we assume calculations over 100 years to remain consistent with our treatment so far. Of course, with any reasonable rate of discount the present value of the losses beyond 100 years would be very small.

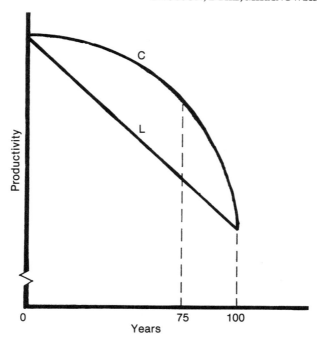

Fig. 26–3. Illustrative time distribution of productivity losses from soil erosion.

to control erosion, or to offset its productivity effects, are strongly influenced by, if not dependent on, a comparison of the present value of the productivity losses with the present value of the costs of corrective measures. To the extent that economics controls these decisions, intervention is justified whenever the present value of the losses exceeds the present value of the cost of the corrective measures. If the losses are distributed evenly over time, as with curve L, their present value is at a maximum at time zero and declines with each subsequent year. With a distribution like curve C, however, the present value of the losses increases for a time because the losses toward the end of the period are so much higher than at the beginning. If the costs of the corrective measures are the same for both distributions, the appropriate time to take these measures obviously may not be the same.

Clearly, this method of deciding when to intervene to control erosion has implications for both individual farmers and those with responsibilities for soil conservation policy. This approach provides a substitute for traditional T values as a standard for judging when erosion is excessive. By the T standard, intervention is called for whenever erosion promises to exceed T for a prolonged period. By the present-value standard, this may or may not be the right time to intervene. We return to this issue in section 26–3.2.

26-2   ESTIMATES OF PRODUCTIVITY LOSS: PAST AND FUTURE

26-2.1   Losses from 1950 to 1980

In research undertaken at Resources for the Future (RFF) over the last several years, a regression model was developed to estimate the effects of erosion and a small number of other factors on the growth of corn, soybeans and wheat yields in major producing areas for those crops. The analysis focuses on 1950 through 1980. It is based on county-level data and seeks to explain intercounty differences in yield growth by intercounty differences in erosion and the other variables. The dependent variable is the trend of yield growth by county rather than yields in a single year by county because the latter reflect intercounty differences in soil quality and climate in addition to differences in erosion and other factors. Soil quality and climate differences among counties should be relatively stable over a 30-year period, however, so their effect on differences in yield growth should be negligible.

Intercounty differences in rates of change of technology and managerial skills of farmers clearly may be of major importance in explaining differences in yield growth. The RFF research is still under way, and in time technology and management may be fully incorporated in the analysis. In the model presented here, however, technology is only weakly represented, and management not at all. Nevertheless, the results showing the effect of erosion on yield growth are both interesting and significant.

26-2.1.1   Discussion of Variables

The dependent variable is the annual trend of county yields of corn, soybeans, and wheat from 1950 to 1980. The trend for each county was found by fitting a simple least squares equation to the annual yield data. The data were from the crop reporting services in the various states included in the analysis. The names of the states are given in Table 26-1.

The independent variables are as follows:

USLE is the annual erosion in tons per acre of land in each crop in 1977, taken from the 1977 National Resources Inventory (NRI).

Y_{52} is the average yield in each county from 1950 to 1954.

ACD is a dummy variable. For counties in which land in the crop between 1950 and 1980 was never less than 2025 ha, ACD = 1. For all other counties, ACD = 0. Data came from state crop reporting services.

IRD is a dummy variable. For all counties in which 2% or more the the land in the crop was irrigated in 1977, IRD = 1. For all other counties, IRD = 0. Data came from NRI.

Table 26–1. Information from regressions of trends in yields of corn, soybeans, and wheat as a function of erosion and other variables, 1950–80.

| Regression | Equation significance | | | | USLE coefficient significance | | | | | |
| | None | Probability | | | Expected sign | | | Probability | | |
		1%	5%	10%	Right	Wrong	None	1%	5%	10%
Corn (616 counties)†										
1. All USLE		x			x			x		
Corn (341 counties)‡										
3. All USLE		x			x		x			
4. USLE ≤ 5§		x				x				
5. USLE > 5		x			x					x
6. 5 ≤ USLE ≤ 10	x				x		x			
7. 10 ≤ USLE ≤ 20		x			x				x	
8. USLE > 20		x			x		x			
Soybeans (299 counties)‡										
9. All USLE		x			x			x		
10. USLE ≤ 5	x					x				
11. USLE > 5		x			x					x
12. 5 ≤ USLE ≤ 10	x				x		x			
13. 10 ≤ USLE ≤ 20				x	x		x			
14. USLE > 20		x			x		x			
Wheat (191 counties)¶										
15. All USLE		x			x		x			
17. USLE ≤ 5		x				x				
18. USLE > 5				x	x		x			

† Counties in Indiana, Illinois, Iowa, Missouri, western Tennessee, Nebraska, South Dakota, and North Dakota.
‡ Counties in Indiana, Illinois, Iowa, and western Tennessee.
§ USLE numbers are short tons per acre. Equivalent values in metric tons per hectare are 5 = 11.2; 5–10 = 11.2–22.4; and 10–20 = 22.4–44.8.
¶ Counties in Kansas, Nebraska, South Dakota, North Dakota, and in the Palouse region of Idaho and Washington.

The effect of erosion on the growth of crop yields probably is cumulative, so in the regressions one would like to use total erosion over the entire period instead of erosion in a single year. However, county-level erosion data are available only for 1977. The use of those data as a proxy for total erosion is permissible if two conditions were met: (i) erosion differences among counties were stable over time, and (ii) intercounty differences in the effect of erosion on the trend of crop yields were proportional to differences in the amount of erosion in 1977.

With respect to the first condition, the R, K, and S factors in the USLE should be relatively stable over a 30-year period, as should L (slope length) except where terraces or other drastic land-forming operations were undertaken. The distribution among counties of the C and P factors, however, might be more changeable. To check this possibility, we ran several regressions in which we substituted RKLS for USLE. The results were virtually identical. We conclude that differences among counties with respect to the C and P factors were relatively stable in 1950 to 1980.

The second condition about erosion probably is not as well met as the first. The effect of erosion on crop yields depends not only on the amount of soil lost but also on the nature of the subsoil and on rooting-zone depth. If counties vary widely with respect to these soil characteristics, then inter-county differences in erosion in 1977 would not fully capture the effect of total erosion in the period on differences in crop yield trends.

Our hypothesis is that, other things the same, erosion reduces not only the *level* of crop yields at any given time but also the *growth* of crop yields over time. Accordingly, we expect counties with higher rates of erosion to have slower rates of yield increase. More specifically, in all the regressions, USLE is expected to be negatively related to the trend of crop yields.

Average yield in 1950 to 1954 was included in an attempt to represent at least some of the differences among counties in rates of technological ad-vance. Yields generally began to rise in the 1930s and 1940s. If they were about the same to start with in all counties, then counties where they were higher in 1950 to 1954 would have been on a faster rising trend than counties where they were lower. In this case, intercounty differences in the trend of crop yields from 1950 to 1980 would be positively related to differ-ences in average yields in 1950 to 1954. However, if yields differed in the 1930s and 1940s and the effect of technological change was to narrow the differences, then intercounty differences in trend would be negatively re-lated to yield differences in 1950 to 1954. Accordingly, we have no hypothesis about the sign of Y_{52} in the regressions.

Yields may have grown more rapidly in counties with a significant amount of land in the crop throughout the period than in counties where the crop was not consistently grown. The dummy variable ACD distinguishes between these two sets of counties and is expected to be positively related to yield trends.

In the Plains States and Missouri, significant amounts of land in corn and wheat are irrigated, and much of the expansion in irrigation took place between 1950 and 1980. This would give an upward tilt to yields in the counties where it occurred. The irrigation dummy, IRD, is intended to pick this up and is expected to be positively related to yield trends.

26-2.1.2   *Presentation of Results*

The complete results of the regression analysis are reported in Crosson (1983). The summary in Table 26-1 presents key findings for 16 of 18 re-gressions. (The other two regressions substituted RKLS for USLE in the re-gressions for corn, but the substitution made no difference, so the two re-gressions are not reported.) Since greater amounts of erosion are expected to have more severe effects on yield growth, we made several groupings of counties according to amount of erosion.

The findings are as follows:
1. Eleven of the 16 regressions were significant at the 1% level of probabili-ty and 2 at the 10% level. Three were not significant.
2. Thirteen of the 16 regressions had the right (negative) sign for the USLE coefficient. One each for corn, soybeans, and wheat had a positive sign for USLE, in each case the regression in which USLE $\leq$ 11.2 mt/ha.

3. The USLE coefficient was significant at the 1% level in the regressions for 616 corn counties and for 299 soybean counties. In the regression for 341 corn counties where USLE was between 22.4 mt and 44.8 mt, the USLE coefficient was significant at the 5% level. Among the 341 corn counties and 299 soybean counties where USLE was greater than 11.2 mt, the coefficient was significant at the 10% level. It was not significant in any of the three regressions for 191 wheat counties.

These results strongly suggest that from 1950 to 1980 erosion significantly retarded the increase of corn and soybean yields in important areas growing those crops. The amount of the retardation, however, seems to have been small, as suggested by Table 26-2. The table presents results for 11 regressions, but we discuss only those for all the 616 corn counties (regression 1) and all the 299 soybean counties (regression 9). Only in these regressions was the USLE coefficient significant at the 1% level.

Table 26-2 shows that, for each crop, erosion reduced the 1950–1980 trend of yields by 4% from what it otherwise would have been. This was found by calculating the erosion-induced reduction in yield when all variables were taken at their means and by expressing the result as a percentage of the mean yield trend. For example, the mean value of erosion for the 616 corn counties was 19 mt/ha and the USLE coefficient was −0.254. With all variables at their means, therefore, erosion reduced the growth of corn yields by 4.83 kilos ha^{-1} year^{-1}. This is 4% of the mean corn yield trend of 123.58 kilos ha^{-1} year^{-1} (1.97 bu/acre).

This calculation gives the effect of erosion on the *growth* of yields. The effect on the *level* of yields in 1980 was less than on the growth of yields

Table 26-2. Estimated effects of erosion on the trend of crop yields, 1950–80.

	Regression number	Reduction in yield trend due to erosion (percentage of mean yield trend)
Corn (616 counties)		
All USLE	1	4**
Corn (341 counties)		
All USLE	3	1
USLE > 5†	5	3‡
10 ≤ USLE ≤ 20	7	18*
USLE > 20	8	3
Soybeans (299 counties)		
All USLE	9	4**
USLE > 5	11	4‡
10 ≤ USLE ≤ 20	13	22
USLE > 20	14	2
Wheat (191 counties)		
All USLE	15	1
USLE > 5	18	--§

 * Regression coefficient for USLE significant at 5% probability.
** Regression coefficient for USLE significant at 1% probability.
 † USLE values are short tons per acre. For equivalent values in metric tons per hectare see Table 26–1.
 ‡ Regression coefficient for USLE significant at 10% probability.
 § Less than 1%.

since yield growth did not start from zero in 1950. The regression results allow us to calculate that because of erosion the trend value of corn yields in 1980 was about 2.5% less than it otherwise would have been.

26-2.1.3   Effect on Costs

The results described above refer to the effect of erosion on yield growth—productivity as we have defined it. But what of the effect on production costs, the real test? The effect must have been to increase costs over the period 1950 to 1980, but not by very much, judging from the small effect on yield growth. In any case, real (inflation-adjusted) costs of producing corn evidently declined over the period and costs for soybeans did not increase. We conclude this because real prices received by farmers for corn declined by 36% from 1950-1954 to 1975-1979, and prices received for soybeans were unchanged (USDA, 1972 and 1980). To be sure, movements in prices are not a perfect indicator of movements in production costs because prices in any year are affected by other factors as well. The decline in corn prices in particular probably overstates the decline in costs because government price support programs for corn were stronger in 1950-1954 than in 1975-1979. This would not likely account for all the decline in corn prices, however, and there were no price supports for soybeans in either period.

Movements in real prices of corn and soybeans measure changes in production costs for the crops, but they understate the rightward shift in their supply curves. This is because prices reflect the interaction of shifts in both supply and demand, and demand was increasing in the period referred to. Thus, the rightward shift in the supply curve of corn was sufficient to more than offset the effect on costs of both erosion *and* the growth of demand. The shift in the supply curve for soybeans was approximately equal to the combined effect on costs of erosion and rising demand.

The prices for corn and soybeans do not reflect the costs of public investment in cost sharing programs to reduce erosion or in development of technology to offset its effects on productivity. However, these hidden costs must have been small to negligible. As noted earlier, Federal expenditures on soil conservation programs totalled about $20 billion from the mid-1930s to the end of the 1970s. Expressed in constant prices of 1982, this comes to about $60 billion. Taken by itself, this is a substantial amount, but spread over the entire period it was less than 1% of total annual farm expenses. Investments *induced by erosion* to develop newland-saving technology probably were even less. Apart from recent small investments in research on conservation tillage, we are unaware of any evidence that any public investment in new technology in this period was induced by concern about erosion.

Corn and soybean prices did not rise, despite the decrease in soil productivity and increases in prices of land and other inputs (particularly labor), because of advances in technology and management, input substitution, and economies of scale. The effect of these advances was to increase the combined productivity of all resources used in agriculture by 58%

from 1950–1954 to 1975–1979 (USDA, 1982). This increase was enough, or more than enough, to offset the effects on production costs of erosion-induced productivity loss and higher input prices. Although there is no evidence that the improvements in technology and management were significantly spured by concern about erosion, the improvements demonstrated the possibilities for substituting these factors for land if that became an attractive response to erosion in the future. We return to this theme in section 26–3.3.

26–2.2 Future Losses

26–2.2.1 With 1977 Erosion

Based on data from the 1977 NRI (and other sources), two models have been built and used to project erosion-induced losses of crop yield for major producing areas and for the country as a whole. One model, called the Yield-Soil Loss Simulator (Y-SLS), was developed as part of the 1980 Resource Conservation Assessment (RCA), and its results were published in the final RCA report (USDA, 1981, Part II). The other model was developed by William Larson and associates at the University of Minnesota (Larson et al., 1983; Pierce et al., 1983).

When work was begun on the 1980 RCA, previous research on erosion-yield relationships was recognized as inadequate for projecting the effects of erosion on national average yields over the long term. The Y-SLS was deveoped to fill this gap (see Crosson, 1983, ch. 5, for more details). The basic component of the Y-SLS is a set of 210 equations for yields and soil loss, one for each of 10 crops in each of 21 water-resource regions and subregions. In the equations crop yield is made a function of depth of topsoil and of two subsoil horizons, average slope of the land, and soil texture, and of several other variables.

The Y-SLS was used to estimate the loss of yields over 50 years if cropland erosion continued at 1977 rates. The projections are apart from the growth of yields expected from advances in technology and management. The results showed that, under the assumed conditions, continuation of 1977 erosion would reduce yields in 2030 by 8% from what they otherwise would be on the 117 million ha of cropland considered in the model.

Larson and associates (1983) used a two-step methodology for projecting erosion-induced losses of productivity over the long term. The first step was development of a crop rooting model which relates yield to three soil characteristics: bulk density, available water capacity, and pH. Nutrient content of the soil is not specified because it is assumed that fertilizer will be applied to compensate for losses of naturally occurring nutrients. The combined effect on yield of the four soil characteristics varies by soil layer, becoming generally less favorable at lower layers. Larson and associates distinguish three layers to a depth of 1 m. The yield from 1 ha of soil 1 m deep is the weighted average of the contribution to yield of each of the three layers where the weights are the relative rooting concentration in each layer. Yield is expressed as an index varying from 0 to 1. In an uneroded or slightly eroded soil, rooting concentration is greatest in the relatively favorable top layer. On such soils the yield index is 1 or close to 1. As erosion occurs, the weight of the topsoil layer declines relative to the generally less favorable lower layers, so the erosion index falls.

In the second step, data on erosion on specific soils, taken from the 1977 NRI, were fed into the model to calculate percentage declines in yield of those soils on the assumption that 1977 rates of erosion continue for specified periods of time. Larson et al. (1983) did this to project losses of corn yields over 50 and 100 years in two major land resource areas (MLRAs), one (MLRA 107) along the Missouri River in western Iowa, and the other (MLRA 134) along the Mississippi River in Arkansas, Tennessee, Louisiana, and Mississippi. MLRA 107 had 3.2 million ha of cropland which eroded at an average rate of 38 t/ha in 1977. In MLRA 134, 2.8 million ha eroded at an average rate of 30.9 t/ha. The model was used also by Pierce et al. (1983) to project the long-term effect of 1977 erosion on corn yields in MLRA 105 (southeastern Minnesota). This region contained 370 000 ha of cropland, which in 1977 were eroded at an average rate of 25 t/ha.

The effect of 1977 erosion over 50 and 100 years on yields in these three MLRAs is as follows:

MLRA	Percentage yield loss	
	50 years	100 years
	%	
105	3	5
107	2	3
134	3	5

These percentages are weighted averages of losses on soils grouped by degree of slope. In general, losses were higher on more sloping land. The maximum loss was 56% on 8500 ha of land in MLRA 105 with slopes of 20 to 45%. In MLRA 134, the maximum loss in 100 years was 23%, and in MLRA 107 it was 5%. In both these MLRAs, the maximum loss was on land with a 6 to 12% slope.

Larson et al. (1983, p. 464) interpret their results as indicating that continuation of 1977 rates of erosion for 100 years would reduce national average crop yields by not less than 5 to 10% from what they otherwise would be. This estimate clearly is highly speculative but not implausible since 45% of U.S. cropland has a slope of 2% or less and 70% has a slope of 6% or less (USDA, 1981, Part I).

The projections by Larson et al. indicate a smaller impact of 1977 erosion on crop yields than those derived from the Y-SLS. (Five to ten percent over 100 years compared with 8% over 50 years.) However, in view of the quite different methodologies underlying the two sets of projections, their similarity is more striking than their difference. Moreover, both appear consistent with the results of the RFF regression analysis, which showed corn and soybean yields reduced 2 to 3% by erosion over 30 years. The three sets of estimates seem to point to the conclusion that from the end of World War II to the end of the 1970s the effect of erosion on national average yields of main crops was small, and that, if 1977 erosion continues for 100 years, the yield effect will continue to be small.

26–2.2.2   With Higher Rates of Erosion

There are plausible scenarios in which erosion increases substantially from the 1977 level. Should this happen, the productivity loss no doubt would be greater than projected by the Y-SLS and by Larson et al. One such scenario was developed by Crosson and Brubaker (1982), in which cropland erosion rises about 80% from 1977 to 2010. The reason is an increase of some 25 million ha of land in corn, soybeans, and wheat, much of this land being more susceptible to erosion than land already in crops. In this scenario, farmers' demand for cropland increases because the growth of demand for crops outpaces the growth of yields. Yield growth is relatively slow because rising input prices and sluggish development of land-saving technologies increases the economic attractiveness of technologies that use relatively much land.

This scenario is also obviously highly speculative. It assumes steady expansion of foreign demand for U.S. grains and soybeans, although not as rapidly as in the 1970s. The scenario also assumes no major breakthroughs in development of land-saving technologies. If demand growth is slower and/or technological advance more rapid than assumed, then the demand for land, and erosion will not increase as much and perhaps not at all.

Nonetheless, the possibility of a significant increase in cropland erosion over the next several decades is real. Should this happen, yields will probably decline more on the additional land than on land in production in 1977. As noted, rates of erosion on the additional land likely would be higher than on land in production in 1977, and this alone suggests more yield decline. If the additional land also has thinner topsoils and less favorable subsoils than 1977 land (which is plausible since the better land likely would be drawn into production first), then the yield decline on the additional land probably would be greater than the higher rate of erosion alone would suggest.

We conclude that should erosion rise by 2010, as it does in this scenario, *and be maintained at the higher level*, then the decline in yield over 100 years would be higher than projected by Larson et al. The amount of the increase is completely uncertain, but a doubling is not implausible. Instead of the 5 to 10% decline projected by Larson et al., it could be 10 to 20%.

26–2.2.3   Effect on Future Production Costs

In the discussion of Fig. 26–2, we assumed an erosion-induced productivity decline of 10% over 100 years. If the productivity loss is greater, as it would be in the high-erosion scenario, we would expect production costs to rise proportionately more.

From the perspective of social interest, how important would these erosion-induced yield declines and resulting cost increases be? Obviously, if the yield decrease is 5% over 100 years (the low end of the Larson et al. projection) it will be viewed as less important than if it is 20% (the high end of the high-erosion projection). But, in either case, the behavior of *total* production costs is likely to condition judgments of the importance of the erosion-induced cost increases. If the combined effect of input prices and technological advance is such that rising demand for food and fiber over the next 100 years can be met at constant or declining real cost, despite the erosion-induced cost increase, then erosion likely will not be viewed as a major problem. However, if total costs rise, say because of relatively sluggish technological advance, then erosion likely will be viewed more seriously, even if its effect on costs is the same in both cases.

The argument here is that, to judge the severity of erosion-induced increases in production costs, the increases must be considered in the context of total cost increases and of the contribution of other factors to the increases. This way of viewing the severity of the erosion problem has important implications for thinking about soil conservation policies.

26-3 IMPLICATIONS FOR THINKING ABOUT POLICY

An issue for soil conservation policy arises when erosion is perceived to present a threat to the social interest in the land. Two questions arise: What is the social interest in the land? and How do we know when erosion threatens that interest? A wide and deep consensus exists in the conservation community that both the present and all future generations have an interest in maintaining the productivity of the soil, and that it is the duty of each generation to manage the land to protect this interest. The interest is threatened wherever erosion is such that, if it persists, the productivity of the soil eventually will decline and production costs will rise, imposing an unfair burden on future generations.

These are the concepts of interest and threat which underlie T values as a standard for the conduct of soil conservation policy. The perspective taken in this paper is consistent with the T value standard in that both taken behavior of future production costs as the proper measure of performance. We are quite prepared to accept the ethical imperative underlying the T value standard—that each generation should so manage its resources that long-run costs of production do not rise. However, the perspective taken here leads to marked differences from the T value standard in deciding *how* and *when* to intervene to counter the threat of rising production costs.

26-3.1 How to Intervene

The difference about how to intervene arises because, by the cost criterion, as we shall call it, attention is focused not only on erosion but on all factors affecting production costs. Consequently, when analysis indicates that long-run production costs threaten to rise, the first policy question is, Which factors are responsible for the threat? And the second question is, What mix of policies will be most cost-effective in meeting the threat? Soil conservation measures may well be a part of this policy mix, but, when they are, they may be of secondary importance relative to other measures, for example, those that foster faster development of land-saving technologies.

By contrast, the T value standard requires that conservation measures be taken wherever erosion threatens to reduce productivity and raise costs. The bearing of other factors on costs is not considered, nor are other measures as alternatives to soil conservation for meeting the threat of higher costs.

The cost criterion, therefore, requires consideration of a broader set of cost factors and a broader mix of policy alternatives than the T value criterion. We emphasize that the various alternatives are not mutually exclusive. The objective is to find the mix of soil conservation and other policies that is most cost-effective in meeting the cost criterion. To do this, both the costs and the benefits over time of the various alternatives have to be considered. The socially optimal mix of policies, of course, will depend on numerous factors affecting future supply-demand relations for food and fiber. But the mix almost surely would include some combination of soil conservation and technology development.

In comparing the conservation and technology alternatives, account must be taken of the possibility that sustained erosion may lower the payoff to new technology. This possibility has been discussed by Walker and Young (1982). It is suggested by the fact that on many soils sustained erosion reduces water-holding capacity, soil tilth, and the crop rooting zone. Deterioration in these characteristics makes crop yields from these soils less responsive to new technology than if erosion were less. Walker and Young have produced some empirical evidence from the Palouse suggesting that this effect of erosion on yield of new technology is real in that region. How important it may be in other regions is unclear. Wherever it is important, however, it clearly would reduce the payoff of the technology alternative relative to that to conservation.

Under the cost criterion, the environmental costs and payoffs of the various alternatives also must be considered. Conservation almost surely would yield an environmental payoff in improved water quality, and it should impose no environmental costs. The environmental account, therefore, should unequivocally increase the payoff to conservation. The environmental effects of the technology alternative appear more mixed. By increasing crop yields, new technology permits production to be concentrated on less land and less erosive land. It therefore reduces sediment delivered to water bodies, a clear increase in environmental payoff. However, if the new technology is heavily dependent on chemicals, such as inorganic fertilizers and pesticides, it likely will introduce more of these materials into the environment, both directly as effluents from farms and indirectly as wastes from manufacturing operations. Any resulting environmental damage would reduce the payoff to the new technology alternative. The alternative's net effect on the environment is quite uncertain. Clearly, it would depend on the specific characteristics of the technology and on the ecological and institutional conditions in which it was used.

26-3.2 When to Intervene

Under the T value standard, the time to institute conservation measures is "now" on all soils where erosion exceeds T. In practice, of course, that does not happen on all soils, largely because many farmers do not agree that for them "now" is the time to act. Conservationists vary in their interpretation of this. Some attribute it to a difference between the farmer and society about the time horizon over which maintenance of soil productivity counts. This is simply a difference in perspective between the private and social interest in the land. Sometimes, however, the farmer's failure to act is attributed to his own short-sightedness or ignorance of his own long-term interest, or to one or more institutional constraints, for example, unavailability of credit or short-term leasing arrangements.

Under the cost criterion, the time to intervene, whether by conservation or some other alternative to prevent production costs from rising, is when the marginal present value of the cost of the preventive measures is less than the marginal present value of the production losses averted. With respect to conservation measures, the situation is the one illustrated by Fig. 26-3. The time to adopt conservation measures may or may not be "now," even on soils where erosion is a present threat to productivity. The present value of the costs of erosion control must be compared with the present value of the productivity loss. The rate of discount and the time distribution of the loss and of control costs have a major impact on calculations of the present value. The time to adopt conservation, if that is the most cost-effective alternative, clearly is variable. Just as clearly that time may be—but may not necessarily be—"now."[3]

The cost-criterion definition of T provides a more complex guide to soil conservation policy than the traditional definition. Despite that disadvantage, the cost criterion has some marked advantages, especially for conservationists in the field dealing with farmers about conservation measures.

26-3.3 Additional Reflections on the Cost Standard

Soil conservation policy should be viewed in the context of a mix of policies, the objective of which is to manage agricultural resources so that long-term costs of producing food and fiber do not rise. Pursuit of such policies requires the application of economic criteria, both in deciding *which* mix of policies is most effective and *when* to apply them.

---

[3] An erosion damage function (Walker, 1982), which compares the long-run cost of eroded soil productivity with the short-run cost of conservation adoption, is one method for determining when conservation pays.

Central to this view is the argument that over a wide range of amounts of land used, non-land inputs may efficiently substitute at the margin for soil in production of agricultural commodities. The limit to substitution is not known, although as a practical matter it clearly should be regarded as finite. That is, some amount of land will be required for efficient agricultural production in the indefinite future. The amount, however, is unknown and unknowable without taking account of substitution possibilities with non-land inputs.

Few conservationists would seriously dispute the argument that, with the existing agricultural land base in the USA, non-land inputs may efficiently substitute for soil over a range at the margin of land use. Many might even accept the argument that non-land inputs embodied in new technology would permit production at constant costs over the next 100 years, even if erosion reduces soil productivity by 20%, as in the high-erosion scenario. But 100 years, although long in the perspective of individuals, is short in the time span of the human race. What will the substitution possibilities look like in 500 or 1000 years? Conservationists point out that many of the non-land inputs currently embodied in new technologies are derived from exhaustible resources. At some future time, not even the knowledge and technical skill accumulated between now and then will be sufficient to extract these resources except at high and rising cost. At that time, runs the argument, we will badly need the soil productivity we have lost to erosion, but it will not be available. The age of substitution of non-land inputs for land will have come to an end, and we will find the capacity of the land severely diminished relative to the demands upon it.

There is sense in this scenario. Many resources on which modern civilization depends are exhaustible, fossil fuels, which are particularly critical to current agricultural technologies, being among the most important. At any positive rate of use, these resources eventually will be exhausted. When this happens—indeed, well before it happens—the technologies in all sectors, that rely on these resources will become prohibitively expensive. Current levels of income will be unsustainable, and civilization as we know it put at hazard.

But is this dire outcome inevitable? Clearly, it behooves us to think seriously about management of exhaustible mineral and energy resources, not only because of their importance to agriculture. Such thought provides interesting insights to management of another exhaustible resource, the land. First, no one argues that because fossil fuels are exhaustible we should reduce their use now to zero to preserve the stock intact for future generations. If each generation applied this rule, no generation would benefit from the resources. Conservation to this end would be a *reductio ad absurdum*, which is why no one advocates it. This is not to say that conservation is not a legitimate policy issue with respect to these resources. The issue concerns the *rate* of mining the resource, not the fact of mining itself.

Second, when thinking about management of exhaustible mineral and energy resources, we pay special attention to development of substitutes for them which will permit us to phase them out as they become increasingly scarce and expensive. This was a major theme in the discussion of responses to the escalation of petroleum prices in the 1970s. Short-term alternatives included switching at the margin of use from oil to coal, development of oil shale and biomass technologies, and increased reliance on nuclear energy. For the longer term, attention was, and is, being given to development of fusion power and to the ultimate renewable resource, solar energy.

In thinking about management of exhaustible mineral and energy resources, therefore, both conservation and development of substitutes emerge as key policy elements. The main policy problem is to find the balance between the two elements that will be most cost effective in meeting important social objectives such as rising, or at least not falling, per capita income. Finding this balance is a difficult and unending task. It involves continuous monitoring and analysis of the effect of long-term trends in demand, input prices, and technology on the costs of producing these resources. All this is attended by much uncertainty since it deals with future events. The greater the uncertainty about development of economical substitute technologies, the greater the weight likely to be given to conservation in the policy mix. Yet at all times both conservation and alternative technologies are included in the mix. Tension continues between conservationists, who would like to slow the rate of resource use, and those inclined to rely more on development of new technology. But all agree that both approaches belong in the mix of policies.

Adoption of the cost criterion in thinking about soil conservation policy simply applies the same approach to management of the land as now is applied to management of other exhaustible resources. If development of technological substitutes, with all the uncertainty that entails, is an integral part of policies to manage these other resources, why should it not be also for policies to manage the land? As with these other resources, the issue with respect to land is not *either* conservation *or* new technology, but finding the balance between the two alternatives which is most cost effective in meeting the cost criterion.

Acceptance of this way of thinking about policies to manage the land encounters much resistance from soil conservationists. The idea that substitution of technology for land might be a viable long-term component of the policy mix is not easily accepted. Adherence is strong to the standard of avoiding *any* loss of soil productivity and to T values as the maximum soil loss consistent with this. Yet the conservationists' position is ambiguous regarding the usual definition of T values. These values are based on the rate of topsoil formation, not the rate of new soil formation, a much slower process. Achievement of T values on all soils would represent the complete success of present soil conservation policies. Yet we would be mining the soil, and eventually its productivity would fall as rooting zone limits were encountered. Some conservationists have used this to argue that T values ought to be redefined to reflect rates of new soil formation (McCormack et al., 1982). All recognize, however, that reducing erosion to achieve this definition of T would exact an enormously high cost, both in lost current output and in measures to control erosion. Thus, it appears that without putting it explicitly in present value terms, all soil conservationists recognize that the present value of the costs of not mining the soil are far higher than the present value of the productivity losses which mining implies.

We believe that soil conservationists would get a readier response from farmers than they do now if they explicitly substituted the present value criterion for the T value criterion for deciding when and where erosion is excessive. A common lament running through the soil conservation literature is that many farmers refuse to accept recommendations of conservationists, even where erosion is well in excess of T. As noted above, this often is attributed to lack of insight by farmers or to various institutional constraints. These no doubt are operative in some cases. But the main reason most farmers ignore the conservationists' recommendations probably is that farmers apply an expected present-value criterion in deciding whether to invest in erosion control, and the conservationist applies the T standard. This is not to say that farmers go through the complex and detailed operations needed to calculate present values. It is enough that they think roughly about how much erosion likely will cost them over the long-term and compare that cost with the cost of adopting erosion control now. To the farmer, these economic dimensions are crucial. To the conservationist applying the traditional T standard, they are irrelevant. Small wonder that the farmer frequently fails to follow the conservationists' advice. If they would adopt the cost criterion for judging the need for erosion control, conservationists would directly address the key economic concern of the farmer: his profit and loss. The advice and influence of the conservationist could not fail to carry more weight with farmers than it does now.[4]

ACKNOWLEDGMENT

We are grateful to David Ervin and James Nielsen for their comments.

---

[4] We deal only with the costs of lost soil productivity. The costs of off-farm erosion damages present an entirely different set of policy issues.

REFERENCES

Crosson, P., and S. Brubaker. 1982. Resource and environmental effects of U.S. agriculture. Resources for the Future, Washington, DC.

----. 1983. Productivity effects of cropland erosion in the United States (with A. T. Stout). Resources for the Future, Washington, DC.

Larson, W. E., F. J. Pierce, and R. H. Dowdy. 1983. The threat of soil erosion to long-term crop production. Science 219:458–465.

McCormack, D. E., K. K. Young, and L. W. Kimberlin. 1982. Current criteria for determining soil loss tolerance. p. 95–111. In B. L. Schmidt, R. R. Allmaras, J. V. Mannering, and R. I. Papendick (ed.) Determinants of soil loss tolerance. Spec. Pub. 45. American Society of Agronomy and Soil Science Society of America, Madison, WI.

Pierce, F. J., W. E. Larson, R. H. Dowdy, and W. Graham. 1983. Productivity of soils—assessing long-term changes due to erosion. J. Soil Water Conserv. 38:39–44.

USDA. 1972/1980. Agricultural statistics. U.S. Government Printing Office, Washington, DC.

----. 1981. Soil, water, and related resources in the United States: Analysis of Resource Trends. 1980 RCA Appraisal, Parts I and II. U.S. Government Printing Office, Washington, DC.

----. 1982. Economic indicators of the farm sector: Production and efficiency statistics, 1980. Economic Research Service, Stat. Bull. 679. Washington, DC.

Walker, D. J. 1982. A damage function to evaluate erosion control economics. Am. J. Agric. Econ. 4(4):690–698.

----, and D. L. Young. 1981. Soil conservation and agricultural productivity: Does erosion pay? Paper presented at annual meeting of the Western Agric. Econ. Assoc., Lincoln, NE, 20 July 1981.

----, and ----. 1982. Technical progress in yields—no substitute for soil conservation. Coop. Ext. Serv., Agric. Exp. Stn., Univ. of Idaho, Moscow.

27 Direction and Politics of Soil Conservation Policy in the United States

Sandra S. Batie

Virginia Polytechnic Institute and State University
Blacksburg, Virginia

Soil erosion policy, like other public policies, is a product of increment-al politics. That is, our system does not operate in an apolitical manner with "technical experts providing objective, factual information on demand, to the decision maker" who then decides on the best course of action (Randall, 1982, p. 39). Rather, it is characterized by individuals or groups of in-dividuals with various endowments of resources, various access to in-formation, and, therefore, various amounts of political power. Change, which is mainly incremental, occurs when convincing arguments for change coalesce individual dissatisfactions into political majorities that then support proposals for new directions (Randall, 1982, p. 44).

The recent experience with the Soil and Water Resources Conservation Act (RCA) process is illustrative of this contention. In part, RCA was born in response to criticisms that past soil conservation efforts were not always well directed and were having too little impact. One purpose of RCA was to provide a more scientific basis for budget and program decisions, and thereby to reduce what has been a long continuing conflict between the legislative and executive branches of government (Allee, 1982).

The RCA has raised the visibility of soil conservation on the public agenda. It has provided considerable improvement in databases with which to appraise existing soil and water resources. Because appraisal data were to be used for policy evaluation and policy choice, however, RCA was (and is) a political process. "The RCA sought to integrate evaluation into planning and budgeting, but in doing so it posed . . . [a] danger, that the preferred conclusions influence the analysis more than the analysis influences the con-

Published in R. F. Follett and B. A. Stewart, ed. 1985. *Soil Erosion and Crop Productivity.*
© ASA-CSSA-SSSA, 677 South Segoe Road, Madison, WI 53711, USA.

clusions'' (Leman, 1982a). Predictably, politics did influence soil conservation policy choices, and hence influenced which evaluations would be conducted. For example, Leman commented with reference to the initial policy alternatives considered in early drafts of the RCA program:

> The seven alternatives formally raised in RCA did leave out some major options in dealing with soil conservation problems. In particular, the strategies of reducing or taxing exports and of rearranging land uses were not explored. Some SCS officials would have liked to explore these possibilities, but the political climate of the department under both Democrats and Republicans said no, and Congress in any case would never have tolerated it. . . . The seven existing alternatives created significant political conflict themselves. For example, the mention of cross-compliance provoked congressional resistance that nearly derailed the whole RCA process (Leman, 1982b, p. 78).[1]

Incremental politics are subject to various institutional constraints that can change with time. However, most changes are small adjustments from the status quo, and hence a careful examination of the status quo politics and existing institutions can provide insights into the probable redirections of soil conservation policy. This paper examines the influences of political actors and the political environment on proposed redirections of soil conservation policies.

27-1 THE DIRECTION OF SOIL CONSERVATION POLICY

Primarily because of the RCA process, more is known about the condition of our rural natural resources than ever before. There is enough data to fashion improved soil conservation strategies. Furthermore, the public is better informed than ever before. Recent polls reflect broad public awareness of soil erosion and a willingness to support conservation efforts (Louis Harris and Associates, 1980). Coupled with the fact that several environmental organizations have decided to make the protection of croplands a priority issue, the political environment appears to be favorable for fashioning improved conservation policy.

Also, the first "final" soil conservation program of this phase of the RCA process was recently released (USDA, 1982). This program has several components:

Priority emphasis or reducing excessive soil erosion on U.S. croplands,

Targeting of up to 25% of Federal technical and cost-share conservation dollars to critical resource-problem areas over the next 5 years,

Encouragement of the use of conservation tillage,

Establishment of a system of Federal matching grants to soil and water conservation districts for use in planning and implementing soil conservation practices,

---

[1] However, an idea which is political folly at one time can be conventional wisdom at another, and an idea which is political folly for one agency can be supported by another. Thus, while RCA's analysis did not examine the linkages between export policies and erosion, an 1980 Congressional Research Service report did examine these relationships (Leman, 1982b, p. 79).

Encouragement of pilot projects to test new conservation incentives, and
Establishment of a requirement that farmers seeking Federal loans have
at least applied to USDA for a conservation plan for their farms
(USDA, 1982).

These program components are themselves the result of numerous
political compromises, and the compromises will continue. To predict where
these compromises will lead, one must begin by examining the actors in the
soil conservation policy process: in most general terms, the Reagan Administration and its watchdog agency, the Office of Management and Budget
(OMB); career personnel in the agencies; Congress; special interest groups;
farmers and ranchers; and the general public.

President Reagan delivered the final program of the RCA to Congress
on 22 December 1982, with stated concern that our resources needed to be
better managed if the USA is to meet foreign demand for grain. Reagan
specifically mentioned, however, that he might not be able to support
program funding at the USDA minimum recommended level of $735
million. He cautioned that the demands for funds for these purposes must
be weighed against other national goals and interests (Anon., 1983, p. 237).
Since then, the proposed budget for SCS was released with funding levels at
$475 million, a significant reduction from the $578 million available in
1982.

The Reagan Administration has little flexibility for generosity to conservation budgets even if this were its natural proclivity. It clearly is committed to reducing Federal funding in a variety of areas, while increasing
defense spending. However, because deficit financing is deemed inflation-
ary, because a strong demand not to increase taxes is perceived, and because
reducing social security payments is considered politically impossible,
discretionary funds—those funds OMB can truly control—are going to be
carefully rationed (Shannon, 1982). Thus, soil conservation funding has
been and still is an obvious target for fiscal restraint and retrenchment.

Because of the budgetary constraints, an important actor in determining the future direction of soil conservation policy is OMB. The Administration's stance of reduced funding is implemented by the OMB. When the
RCA final program draft was being prepared, USDA proposed various
spending alternatives, ranging from $735 million a year to $1 billion a year
(Risser, 1982b). During the internal clearance procedures for the final
program, OMB argued for the lowest figure. Indeed, one frustrated SCS
official was rumored to have complained that "Zero . . . [funding for con-
servation] would have been fine with them (OMB)" (Risser, 1982b).

One observer concludes that reference to OMB's influence: "there is
no lack of cynicism within USDA about the way the RCA analysis has been
bent, spindled and multilated to fit the budget mill" (Cook, 1982b, p. 214).
The then chief of SCS, Norman Berg, further reflected: . . . "It's become a
political document beyond what any of us imagined" (Cook, 1982b, p. 214).

The Reagan Administration realizes that such a posture on funding soil
conservation is not costless. This realization is considered by several observers to be the reason that the final FCA document was literally locked up

in a secret location until its public release on December 22, despite its printing three months earlier and its scheduled release in November. Reportedly, the Administration feared that releasing the final document before the Congressional elections might be a political liability to Republican candidates. This was supposedly because the final document recommended low funding levels, excluded some popular tax incentives for soil conservation which had been vetoed by OMB, and omitted farmland retention issues. The official reason for the lock-up, however, was that OMB wanted to study the proposal longer (Risser, 1982a).

Countervailing forces to OMB's resistance to expanding the budget for soil conservation come from Federal agencies and Congress. Although fewer members of Congress are now concerned with rural issues than in the past, and probably fewer now understand the nature of agribusiness, those that remain are generally in positions of influence. Senator Roger Jepsen (Rep., Iowa), for instance, chairs the Senate Subcommittee on Soil and Water Conservation and is a vocal critic of the Reagan Administration's soil conservation priorities. Mississippi (Dem.) Congressman Jamie Whitten is chairman of the House Appropriations Committee and is an effective defender of ASCS. In many cases, these individuals and others like them have managed to protect soil conservation expenditures, frequently by tying such expenditures to other programs, such as farm income support programs (Hadwiger, 1982, p. 275). Also SCS and ASCS have built a constituency of program participants and locally based government employees (including SCS employees, ASCS committees, and office workers and district personnel.) To obtain the popular support of farmers, both agencies have historically spread program benefits widely. Furthermore, thousands of farmers have been involved in service on local ASCS committees and Soil and Water Conservation District Boards. In the process, the agencies have built a political consensus that has sustained them for many years (Leman, 1982b, p. 54).

Adding to the complexities of the politics of conservation policy are the special interest groups, many of whom are well organized for political effectiveness. These include the National Wildlife Federation, The Sierra Club, the Audubon Society, American Farmland Trust, the American Land Forum, the National Association of Conservation Districts, and the agricultural commodity groups such as National Association of Wheat Growers and American Soybean Association as well as the American Farm Bureau Federation, The American Agriculture Movement, and the National Farmers Organization. The groups have divergent goals, and, although most have historically supported conservation programs, they have done so for different reasons.

Finally, the general public is an actor of sorts. Public opinion polls have shown that citizens do support protection of a quality environment. Yet "there are few 'identifiable beneficiaries' of soil conservation. Instead, benefits are 'diffuse', most of them probably flowing to future generations" (Hadwiger, 1982, p. 275). Thus, while public opinion appears to support conservation, major grass-roots coalitions are unlikely to form on the issue.

To discern the direction of soil conservation policies, one must be cognizant of the ability of all these groups to influence the policy agenda. Very, very few of the individuals and groups involved are interested only in the protection of the nation's soil and water. Rather, most are interested in the protection of the natural resources of this nation, subject to certain constraints such as achievement of their own group's unique goals.

27-2 POLITICAL DIMENSIONS TO ACHIEVING COST-EFFECTIVE SOIL CONSERVATION STRATEGIES

Because of the nature of the political environment, there are numerous dimensions to achieving cost-effective strategies for conserving soil and water in the nation. Analysts can examine various strategies for reducing soil conservation, and they can highlight the tradeoffs of efficiency and equity involved in policy choices. Whether the analyses ultimately influence policy design and selection, however, will depend on whether the various political actors can coalesce with political strength to champion a particular policy design.

27-2.1 Conflict of Policy Goals

In many cases, national goals other than those of conservation are politically paramount. For example, Ken Cook partially ascribes the current depression in the agricultural economy to a rational political gamble taken by the Reagan Administration. If the Reagan Administration or Congress had established effective production controls on 1982 crops, today's farm incomes might have been higher, but so would consumer food prices. This was considered an unacceptable position for an election year. "Better to let the farmers hang, and hope for relief through increased exports or a poor domestic crop" (Cook, 1983, p. 26). The gamble was that this strategy would cost less ultimately then the 1983-1984 set-asides, paid diversions, and the PIK[2] program. Cook estimates that by "erring on the side of plenty" with our commodity policies, however, we "may have lost in one year the equivalent of the soil saved by the ACP [Agricultural Conservation Program] in three to five years, at a cost of $238 million to $427 million" because of the additional croplands which were in production (Cook, 1983, p. 26). This conflict of goals—soil conservation, balance of payments, farm income, consumer prices, and budget deficits—is not unique.

27-2.2 Conservation Tillage

Because of the dilemmas of conflicting policy goals, the adoption of conservation tillage has appeal as a conservation strategy, and it is receiving

---

[2] PIK is a term representing Payment in Kind. Farmers in the PIK program were paid in crops for land they voluntarily removed from production (set aside) in 1983.

considerable emphasis in the present restructuring of the USDA soil conservation program. It is profitable on most farms, and it is effective in reducing water-causing erosion. It allows simultaneous production and soil conservation. However, even this policy choice has problems. Conservation tillage usually requires increased use of herbicides and insecticides to be effective. Not only does this require a skilled farm manager[3], but it can also mean reduced water quality and increased problems of pesticide-resistant insects and weeds (Hinkle, 1983). Conservation tillage is therefore not viewed as a panacea by many, and particularly not by those favoring organic agriculture or by those who worry that overall environmental quality may suffer if conservation tillage is adopted as the *only* answer. While these critics of widespread adoption of conservation tillage might wage more caution in pursuing such a policy and might desire more research to improve alternative agricultural enterprises (for example, organic) and to develop integrated pest-management programs, they are presently not strong enough to influence current USDA directions.

27-2.3   Targeting: By Whom?

Since almost 70% of the soil erosion in the nation that exceeds 11.21 t ha^{-1} year^{-1} is concentrated on less than 8.6% of the total acreage, targeting of funds and practices to those areas experiencing some of the worst problems makes economic sense (Ogg and Miller, 1981). Despite this concentration of erosion problems, past cost-sharing programs have dispersed assistance widely. Less than 19% of the soil conservation practices installed have been placed on the most erosive lands, and over half the cost-shared practices have been placed on lands with erosion rates of less than 11.21 t ha^{-1} year^{-1} (USDA-ASCS, 1981). Indeed, until recently the distribution of conservation payments via the ACP was essentially unaltered from the distributions that prevailed in 1936, even though the locations of agricultural production and hence erosion have shifted considerably (Miranowski, 1979).

Policies could be made more cost effective if this pattern were changed and public programs were directed at farmers whose lands account for the lion's share of erosion problems. But there are several obstacles, both political and technical, to achieving such targeting. Who decides where the funds are to be targeted is extremely important.

Targeting from the national level would not accord with the New Federalism concept associated with the Reagan Administration. New Federalism decentralizes political power and places authority for allocating financial resources at non-federal levels. Presumably, targeting could be conducted at nonfederal levels with block grants or matching grants (as is

---

[3] One soil conservationist described some problems with conservation tillage succinctly: "No-till will never by used by a majority of farmers until some insect and weed control and fertilizer placement problems are solved and until certain fallacies about no-till are overcome" (Walter, 1982, p. 215).

now proposed) from Federal to local and state entities. But Congress is not enthusiastic about giving up a very powerful tool, control of funding. Thus, it is not surprising that the "firm opposition from Congressman Jamie Whitten (D-Miss.) . . . to any form of district-administered financial assistance (cost-sharing) that could rival the Agricultural Stabilization and Conservation Service, . . . makes it unlikely that the program will be more generously funded in the future" (Cook, 1982b, p. 213). This opposition remains even if the grants are made categorical as opposed to matching.

Also, many of the nation's 2950 Soil and Water Conservation Districts (SWCD's) contain very little cropland with significant erosion problems as estimated by the 1977 National Resources Inventory (NRI). Because of this regional diversity, targeting conducted nationally would affect different croplands than if each state received Federal funds and directed them at their worst erosion problems. Similarly, the effect would be different if each SWCD received funds and SWCD personnel selected the targets within their boundaries. Quite understandably, in districts with less severe erosion, SWCD personnel and their farmer clientele feel threatened by a national or state targeting policy and argue that they have made considerable improvements already and should not be penalized for having less erosive lands or for being better soil stewards within their district boundaries.

Thus, even the modest proposal to target 25% of the conservation budget by 1986 is criticized. But, without a reform of the present broad diffusion of funds, erosion might increase even if the conservation budget were larger (Leman and Miranowski, 1982).

27-2.4 Targeting: To What

Another political dimension to the targeting strategy is the establishment of criteria for selecting critical areas. If maintaining soil productivity is the paramount objective, then it may make sense to target areas with fertile but shallow topsoils, regardless of erosion rates.[4] If water quality goals are primary, than targeting might be focused on areas with high erosion rates, regardless of topsoil depth.

The RCA final program states that maintenance of soil productivity will be the highest priority for conservation expenditures, which is the result of political compromise. Presumably, targeting for soil productivity will focus on soils with erosion rates that exceed soil-loss tolerances, or T values. T values are defined as the maximum annual soil losses that can be sustained without adversely affecting the productivity of the land. For soils in the USA, the USDA assigned T-values that usually range from 2.24 to 11.21 t/ha, depending on the properties of the soil.

---

[4] Targeting dollars to areas of greater erosion potential, however, should not mean that every highly eroding area is protected or reclaimed. Some areas are so severely eroded that thousands of dollars per acre could be spent with little improvement in the land's productivity. The relative benefits to be gained should be balanced against the costs if truly cost-effective strategies are to be achieved.

The validity of these numbers in representing maximum, sustainable soil losses is doubtful, however. On many soils, current T values have been set too high to ensure the long-term productivity of the land, and on others soil loss tolerances are too low.[5]

But the relevant issue is not really soil re-formation but protection of long-term productivity. For T values to represent maintenance of long-term productivity, they need to reflect the impact of technology on crop yields. For T values also to represent economic conditions, they have to reflect the costs and benefits of soil maintenance as well. For, as Crosson (1983) notes, what "if the minimum cost of achieving the T-values exceeds the cost of the productivity loss . . . ?" (p. V-3). Because these technological and economic influences are uncertain, however, their incorporation into T values would be difficult and somewhat ambiguous.

Furthermore, incorporating these concerns might mean lower T values for numerous fields and higher ones for others, facts that have political ramifications. Larry Vance, chief of Ohio's Division of Soil and Water Districts and formerly director of Iowa's Department of Soil Conservation, summarized these dimensions of the issue:

If we changed T-values, a new farmer probably wouldn't know the difference. Some guy, who is supposed to know, will still be telling him 'there are some acres that will lose productivity over time.' But to an older farmer, changing T-values may suggest to him that he has invested money unwisely. We have something called the 4-ton club out in Iowa. They have taken care of their land for many years. Spent a lost of time and money. And they feel good about it. All of a sudden, they get the impression it really wasn't important after all. I'd say then you've got a real credibility problem on your hands (Cook, 1982a, p. 92).

Of course, criteria other than T values may be used to define critical areas, but if the criteria are not well specified, then the distribution of targeted funds may be no different than the present distribution (Leman, 1982a, p. 9). While the choice of targeting criteria remains unresolved, its resolution is crucial to the ultimate impact of the conservation program and therefore will be subjected to political pressures.

27–2.5 Water-Quality Goal

One of the strongest arguments for public funding of conservation practices can be made for reducing off-farm damages of soil erosion. Yet these issues have been given less priority in the RCA program. This may be an appropriate USDA stance, given that protection of the environment has historically been the purview of the Environmental Protection Agency (EPA). However, EPA is also being subjected to cuts in funding for programs. Funding for the continuation of Rural Clean Water Program (Section 208 funding of the 1972 Federal Water Pollution Control Act) is doubtful. Thus, despite their importance, water-quality issues will not receive the emphasis in the near future that they did in the past.

---

[5] Bartelli (1980) estimated that some Illinois soils, with modern farming techniques, would regenerate at not less than 26.9 t ha^{-1} year^{-1}.

27-3   THE LONG VIEW

In the short term, the directions of Federal soil conservation policy are for the most part structured by the proposed RCA final program. Because the final program lacks specifics on how the program will be implemented and which budgets will be cut, many political battles are yet to come as the process moves away from broad generalities. The boundaries of these battles are mostly determined, although they will be difficult fights since they involve reallocating a shrinking budget. Nevertheless, if one accepts the contention that the RCA process was initiated, in part, to obtain more conservation per dollar spent (Libby, 1982) some progress toward this objective will apparently be made.

What of the longer term? If the various publics involved with soil conservation continue to demand that the issues be clarified and that critical questions be asked, more fundamental changes will be made in the soil conservation programs of our nation. Some possible long-term changes include a more complete recognition that the various agricultural and other national goals intertwine and impact one another. Problems of agricultural surpluses relate to foreign trade policies, which relate to soil conservation problems, which relate to the structure of agriculture, which relates to other governments' food and resource policies, etc. Fewer symposiums will focus only on domestic commercial agricultural policy to the exclusion of natural resource issues, trade issues, foreign development issues, or macro-economic policy issues. I also suspect that water-quality issues will, in the long run, assume prominence once again on the national public agenda.

More concern may develop the distributional impacts of our policies, far transcending the concern now focusing on the targeting of conservation dollars. Conservation is fundamentally a concern for intergenerational distribution:

> It is somewhat incongruous and surprising, therefore, that we have not taken keener interest in the just distribution within generations of both agricultural resources, and the goods those resources produce. Pursuing such an interest raises some rather difficult questions . . . . As Historican High Stretton writes, 'conservation is not worth having if it merely shifts hardships from rich to poor, or from later to now.' The size of the resource pile we leave to our descendents is of little moral consequence if we have not shown them the need to use those resources more justly than we have (Cook, 1983, p. 27–28).

If these distributional concerns garner more attention in the policies of the future, so too will questions of equity. Whether the right to private ownership is equivalent to the right to let one's land erode may be re-interpreted. More attempts will probably be made to add conservation to other programs, but rationally, so that fewer programs are in direct conflict or are internally inconsistent. I hope that some innovative attempts will be made at compensable methods of selecting which lands are to be cultivated, so that some of the most erodible lands will not be plowed except in times of extreme shortages. Because such innovative strategies may be best pursued at the local level, some fundamental rearranging of the roles of the various levels of government in soil conservation policy may even occur. RCA has already encouraged state and local conservation units to broaden their own conservation agendas (Libby, 1982, p. 125).

Many political and financial constraints hamper the adoption of new conservation programs: vested interests in old programs, limited budgets, limited personnel, traditional views of property rights, and conflicts with other policy objectives. Also continued resistance to redistribution of soil conservation funds risks discrediting the entire public commitment to soil conservation. Yet in the new political climate, many groups have coalesced on the wisdom of protecting the soil resource in a cost-effective manner. The challenge remains to translate strong societal desires to avoid scarcity and to maintain a quality environment into those institutional changes that will motivate farmers to conserve our nation's soil when and where it is appropriate to do so.

ACKNOWLEDGMENT

Numerous people provided critical contributions to the improvement of this paper. While their agreement with this paper's statements and conclusions should not be assumed, credit is due Norman Berg, Frederick Buttel, Kenneth Cook, David Ervin, John Fedkiw, Mack Gray, Maureen Hinkle, Randall Kramer, Christopher Leman, Wayne Rassmussen, W. Neill Schaller, Leonard Shabman, and Daniel Taylor.

REFERENCES

Allee, D. J. 1982. Implementation of RCA: A problem accomodating economics in soil and water conservation. p. 93–108. *In* Halcrow, Heady, and Cotner (ed.) Soil conservation policies, institutions, and incentives. Soil Conserv. Soc. Am., Ankeny, IA.

Anonymous. 1983. Reagan transmits RCA program to Congress. J. Soil Water Conserv. 38: 33.

Bartelli, L. 1980. Soil development deterioration and regeneration. Paper presented at the Soil Transformation and Productivity Workshop, Washington, DC. 16–17 Oct. Nat. Res. Council.

Cook, K. 1982a. Commentary soil loss: A question of values. J. Soil Water Conserv. 37: 89–92.

––––. 1982b. RCA: No federalism is new federalism. J. Soil Water Conserv. 37:213–214.

––––. 1983. Commentary: Surplus madness. J. Soil Water Conserv. 38:25–28.

Crosson, P. (with A. T. Stout) 1983. Productivity effects of cropland erosion in the United States. Resources for the Future, Washington, DC.

Hadwiger, D. F. 1982. Commentary: Political support for soil conservation. J. Soil Water Conserv. 37:225–241.

Hinkle, M. K. 1983. Problems with conservation tillage. Soil Water Conserv. 38:201–206.

Leman, C. K. 1982a. Evaluating the evaluators: RCA in retrospect. Renewable Resour. J. 1:19–23. I (2 & 3) (Fall 1982 and Winter 1983).

Leman, C. K. 1982b. Political dilemmas in evaluating and budgeting soil conservation programs: The RCA process. p. 47–88. *In* Halcrow, Heady, and Cotner (ed.) Soil Conservation Policies, Institutions, and Incentives, Soil Conserv. Soc. Am., Ankeny, IA.

Leman, C. K., and J. A. Miranowski. 1982. How farm soil is sliding into the pork barrel. Christian Science Monitor, Nov. 3, p. 15.

Libby, L. W. 1982. Interaction of RCA with state and local conservation programs. p. 112–128. *In* Halcrow, Heady, and Cotner (ed.) Soil Conservation Policies, Institutions, and Incentives. Soil Conserv. Soc. Am., Ankeny, IA.

Louis Harris and Associates. 1980. Poll on rural environment resources conducted for the soil Conservation Service. USDA, Washington, DC.

Miranowski, J. A. 1979. More Federal government involvement in soil conservation. p. 66–71. *In* Increasing understanding of public problems and policies—1982. Farm Found., Oak Brook, IL.

Ogg, C., and A. Miller. 1981. Minimizing erosion on cultivated land: Concentration of erosion problems and the effectiveness of conservation practices. p. 6. *In* Policy Research Notes. USDA, Washington, DC.

Randall, A. 1982. The Federal role in natural resource management. p. 34–44. *In* Increasing understanding of public problems and policies—1982. Farm Found., Oak Brook, IL.

Risser, J. 1982a. Soil conservation plan locked up by USDA; 'sensitive' election issue. Des Moines Register, Oct. 29, p. 5.

————. 1982b. Reagan cites soil woes, but proposes funds cut. Des Moines Register, 22 Dec., p. 1.

Shannon, J. 1982. New federalism: The search for new balances. p. 177–127. *In* Increasing understanding of public problems and policies—1982. Farm Found., Oak Brook, IL.

USDA-ASCS. 1981. National summary evaluation of the agricultural conservation program—Phase I. Washington, DC.

USDA. 1982. A national program for soil and water conservation: 1982 final report and environmental impact statement. U.S. Government Printing Office, Washington, DC.

Walter, J. 1982. Hard rain's gonna fall! On selling conservation in Woodbury, Iowa. J. Soil Water Conserv. 37:214–217.

28 Concerns and Policy Directions of the U.S. Department of Agriculture

Peter C. Myers
U.S. Department of Agriculture
Washington, DC

The farmer is going to have the last word on applying many of the erosion control techniques discussed in this volume. If Federal soil conservation programs remain voluntary—as I certainly hope they will—the farmer will have to make the decision about whether or not to become a conservation cooperator. If we are going to get erosion under control, the farmer, and nobody else, must adopt farming systems that combine adequate crop yields with adequate resource protection. Research stations and Federal agencies can supply new data and new techniques, and even financial and technical assistance, but the final action is up to the farmer.

Make no mistake: the American farmer knows that soil erosion is an unmitigated evil and that it will eventually eat into crop yields. Although we can't predict when the productivity of specific acres of cropland will nosedive, farmers know with certainty that, unless we change our ways, we will do serious, irreparable damage to our topsoil and will decrease for years to come its ability to produce high crop yields. That threat is waiting for us somewhere in the future, and it is threat enough for me to consider soil erosion as the most serious resource problem in this country today. Most farmers I know agree with me.

But personal convictions are not enough when discussing budget and appropriate figures with Congress and the Office of Management and Budget. We need hard data about the impact of erosion, facts and figures backed up by scientific experiment and field experience. That is why SCS enthusiastically supports research on soil erosion and crop productivity. It

Published in R. F. Follett and B. A. Stewart, ed. 1985. *Soil Erosion and Crop Productivity.*
© ASA-CSSA-SSSA, 677 South Segoe Road, Madison, WI 53711, USA.

is our highest priority research need and includes economic as well as physical and biological aspects.

This emphasis is consistent with the aims of the USDA's national conservation program—developed under the 1977 Soil and Water Resources Conservation Act—and with the recently published ARS program plan. We are pleased that so many state experiment station directors and the Economic Research Service are also giving high priority to similar research. When the research is complete, SCS and the Extension Service will help get the word to farmers, ranchers, lawmakers, and other interested people through information and educational and technical assistance programs.

Today, the SCS, is able to help other farmers improve and expand their conservation efforts. SCS administers several conservation programs of USDA, including conservation technical assistance to farmers and ranchers. This important part of our work was funded during 1984 at $275 million, about 44% of the total 1984 appropriation for SCS of $622 million.

For 1985, the total proposed budget for SCS is somewhat lower—$603 million—but the amount requested for conservation technical assistance is even higher—$279 million. This increase indicates the importance that this Administration attaches to the SCS program of professional assistance to producers. It is our primary activity for helping reduce soil erosion and conserve water. I know that many in Congress share the interest of the Administration in keeping conservation technical assistance funded to deal effectively with our most pressing resource problems.

The Department's policy directions on soil and water conservation have never been more clearly stated or more open to public scrutiny than they are today. Publication of *A National Program for Soil and Water Conservation* (USDA, 1982) leaves no doubt about USDA conservation priorities and planned activities. Mandated by the Soil and Water Resources Conservation Act of 1977 (RCA), the *Program* is the result of 5 years of healthy discussion and debate by USDA and White House agencies, along with the advice and comments of thousands of farmers and other interested Americans.

In addition to the *Program*, we published a 2-volume *Appraisal* of soil and water conditions and trends, based in large part on data collected during the 1977 National Resources Inventory (USDA, 1981a and 1981b). Taken together, the *Appraisal* and *Program* represent the most searching and comprehensive examination of Federal soil and water programs ever undertaken.

President Reagan sent the RCA *Program* to Congress on 21 December 1982, accompanied by a Presidential policy statement on soil and water conservation (Office of the Federal Register, 1982). In that statement, the President warned that "soil erosion appears to be increasing again," adding that "about one-third of America's cropland is currently experiencing soil erosion . . . at rates which threaten the long-term productivity of the land." He also pointed out that about 89% of the excessive erosion occurs on about 10% of U.S. cropland, and that agricultural water is being used, and wasted, in greater amounts than ever.

Reviewing the history of Federal soil and water programs, President Reagan noted that some 27 conservation programs, involving conservation research, education, technical assistance, cost-sharing, and loans, are administered by eight agencies of USDA. "Some of these programs, while popular with farmers and ranchers," he said, "do not clearly address the Nation's most critical soil and water resource problems. Further, after nearly half a century of Federal conservation assistance programs, a substantial number of farmers have not applied needed conservation measures. Too much soil continues to erode at rates that threaten productivity and impair water quality. Too much water is not efficiently managed, resulting in a threat of water shortage. Too much land is subject to excessive flood damages." I know of no more succinct statement of the concerns of this Administration over soil and water resource problems and past attempts to deal with those problems.

Throughout the RCA discussions, it became evident that the missions of the Department's conservation agencies had become muddled and diffused over the years by too many different goals. We were trying to be all things to all people. Clearly, national priorities were needed to guide us through 1983 to 1987, so that we could focus our funds, people, and energies. The first RCA priority is to reduce excessive soil erosion on crop, range, pasture, and forest lands. The second priority is to conserve water used in agriculture and to reduce flood damage in upstream areas.

Because conservation needs vary from place to place, the program also will provide assistance for local and state priorities. These priorities can include the two national priorities and any of these concerns of national significance:

Improvement of range, pasture, and forest land;

Improvement of water quality;

Conservation and development of natural resources in urban areas and rural communities;

Improvement of fish and wildlife habitat; and

Management of organic wastes.

In line with these priorities, the RCA program calls for new or redirected Department activities to encourage farmers and ranchers to apply more soil and water conservation to their land. The first of these activities is targeting more funds and people to areas with critical resource problems, while maintaining a base level of assistance in other areas. The two agencies most involved in targeting are SCS and ASCS, which administers the Agricultural Conservation Program, a cost-sharing program for farmers and ranchers.

By the end of 1982, our two agencies had completed more than a year of trials with targeting, and the results reported by the first 15 states with targeted counties were very encouraging. Erosion-targeted areas lost about 3.5 million fewer metric tons of soil than in previous years. In the West,

irrigation farmers in targeted areas saved 160 million m³ of water on cropland alone. By zeroing in on the most serious resource problems, we accomplish more conservation work for every tax dollar invested in the program.

Typical of targeted areas are the rapidly eroding wheatlands of the
Palouse, the croplands of western Tennessee, the erodible loess soils of
western Iowa and northwestern Missouri, and irrigation-farming areas of
Colorado, Utah, Idaho, and Oregon. So far, the cooperation of farmers
and ranchers with SCS and ASCS in these critical areas has been excellent.

In fiscal year 1984, targeting was expanded to 44 states and Puerto
Rico. ASCS targeted $19 million in financial assistance, and SCS targeted
$27.5 million. Both agencies are maintaining that level of funding in fiscal
year 1985. Some people were surprised that so many states identified areas
with serious soil and water problems, but these are the findings of the National Resources Inventory, borne out again by the 1982 Inventory. In some
states, the critical areas include only a few counties; in others, like Iowa, the
critical areas are much larger. When we select an area with critical erosion,
our sole criterion is not total tons of soil lost. We are just as concerned
about moderate erosion in areas with thin soils that cannot afford much soil
loss. We are focusing on places where productivity is threatened or already
declining, whether the topsoil is a foot thick or an inch. I liked what Secretary of Agriculture John Block said about targeting: "It makes good
sense," he said, "to spend a little more where the problem is the worst. It's
kind of a logical thing to do."

As time goes by, many of the critics of targeting will support this concept. Right now, some people fear that targeting will shortchange their conservation districts. Actually, Administration enthusiasm over targeting is
one reason that the budget continues to call for substantial sums for conservation technical assistance. It is definitely a plus, not a minus sign.

Targeting is nothing new to USDA. One example of targeting assistance is the Great Plains Conservation Program, in which all the funds and
assistance go to counties in just 10 states. Another example is the Resource
Conservation and Development program, in which accelerated assistance
goes to certain counties with economic difficulties that can be partly
remedied through better resource use. Targeting is effective but hardly
revolutionary.

Another key feature of the RCA program is the increased emphasis on
farming systems that begin with conservation tillage. Since the current
financial plight of many farmers does not leave room for the installation of
expensive conservation practices, we are not going to sell farmers on the
basis of resource protection alone. At least some of the conservation
alternatives we recommend must be cost-effective, and they must include
reliable data on costs and benefits.

To help us obtain this kind of information, we have contracted with the
University of Illinois to develop a computer program to show farmers the
relative cost or savings, and the amount of soil saved, through the application of various conservation practices, singly and in combination. The program is called SOILEC. When competed in 1985, we can furnish farmers
schematic diagrams on the costs and benefits of alternative practices. For

most soils, SOILEC printouts will show savings in dollars *and* soil for the farmer who substitutes conservation tillage for conventional tillage.

Talking to thousands of farmers, I see first-hand that mulch tillage, ridge tillage, no-till, and all the other forms of conservation tillage are increasing fast. The Conservation Tillage Information Center surveys (1982 and 1983) show that, in 1982, 24% of total acres planted were planted under some form of conservation tillage and that, in 1983, 31% were planted under conservation tillage—a 7% increase.

The evidence is also persuasive that once farmers try conservation tillage, they stick with it. A study of farmer attitudes in 15 states conducted by Pioneer Hi-Bred International (1982) found that 96% of farmers using conservation tillage are either moderately satisfied or highly satisfied with results. Two-thirds of the farmers cited reduced soil erosion as a reason for satisfaction.

Not all the problems connected with conservation tillage have been solved, however, and farmers need the help of public and private researchers to help them overcome the remaining roadblocks. The Soil Conservation Service came up with 11 priority needs for the scientific community for 1984. The list was sent to all Federal and state research stations and to many private facilities.

The second priority need calls for research to deal with several problems slowing the adoption of conservation tillage. Farmers answering the Pioneer survey listed inadequate weed control as their leading reservation about conservation tillage. In particular, farmers need practical, safe, and inexpensive methods to control a number of deep-rooted grasses and certain broadleafed weeds.

Other roadblocks to conservation tillage will be overcome by farmers themselves, often with the help and encouragement of SCS soil conservationists, Extension people, and industry representatives. The most important element in making a success of conservation tillage is the desire of the farmer to make it work. Problems always arise in switching to any new system, but the determined farmer will solve his problems and make conservation tillage fit his operation.

Not all serious erosion will be stopped by conservation tillage, of course. Conservation tillage alone is not a panacea, so we need to keep looking for cost-effective farming systems that perform well with conservation tillage. Moreover, on some land being cropped today, only satisfactory answer to erosion control is to switch the land out of crops and into grass or trees—permanently. The Payment-in-Kind program provided many farmers with an opportunity to take marginal, hard-to-protect acres out of crops and put them into a soil-conserving use. Also contributing to cropland retirement are two more pillars of the RCA program: increased use of long-term agreements with farmers and the use of pilot projects to test new approaches for dealing with resource problems.

After discussions with other USDA agencies, we developed criteria and guidelines for our pilot projects. One project now underway encourages producers farming Class IV, VI, or VII land to agree to put the acres under some form of permanent cover for at least 10 years. The farmers or ranchers receive cash incentive payments as part of long-term agreements, as well as cost-share payments for the required conservation practices. We've started this project in three widely separated states—Alabama, Idaho, and South Dakota—and will evaluate the results carefully before deciding whether or not to extend the plan.

The land-conversion approach is only one kind of pilot project under consideration. There will be others, all tried on a limited basis to see how they work before being launched on a larger scale.

The RCA program also calls for cross-compliance among USDA programs by changing the eligibility requirements for some loans from the Farmers Home Administration. An applicant for a farm-ownership loan or soil and water loan will be required to have, or request, a conservation plan for the land on which the loan is to be used, and to make a commitment to correct serious erosion problems.

That is as far as the RCA program goes with cross-compliance, but its extended use as a conservation tool may not prove as objectionable to farmers as many believe. In 1982, several Colorado groups of farmers and ranchers initiated a selective cross-compliance proposal in the form of Senator William Armstrong's "sodbuster" bill. The bill would make those farmers who plow up fragile soil have, and use, a conservation plan on that land, or forgo their price supports and other USDA benefits.

Another key feature of the RCA program calls for strengthening the conservation role of existing local organizations and agencies. Programs developed at local and state levels will form the basis for activities carried out through the national program.

Where state and local people identify resource problems as severe, USDA proposed to match state or local funds by awarding grants to conservation districts. The maximum Federal share would be 75%. The SCS budget proposal for 1983 included $10 million for these matching grants, but the proposal was not funded by the Congress. For 1985, no funds were requested, but the Ad Hoc Committee on the Budget of the House Agriculture Committee recommended $10 million for matching grants. We continue to think matching grants to districts should be considered.

Meanwhile, state and local governments continue to demonstrate that they can take effective action on their own against erosion. For example, in Weld County, CO, in spring 1982, foreign investors and local farmers were plowing some of the area's most fragile grasslands to plant wheat. Residents feared another Dust Bowl. By May, the Weld County commissioners passed an ordinance stipulating that anyone wishing to plow rangeland that had remained unbroken during the last 5 years must submit to the county a legal description and map of the proposed site and must obtain a permit before plowing. A conservation plan for the land also must be approved by local conservation agencies. The penalty for plowing without a permit in Weld County can be $300 for each day of noncompliance and 90 days in jail.

While nobody has served time yet for plowing his land, the indiscriminate plowing of fragile land in that county has stopped. To date, 36 farmers have applied for permits, and 35 have been granted. In addition, several farmers, however, talked with SCS people about plowing for wheat and were told that, since their land is Class VI, they probably would not be granted a permit. The farmers backed off, and the county ordinance achieved its aim. Five more counties in Colorado have since passed similar ordinances.

This situation provides an excellent example of local action leading to the appropriate regulating solution.

These are the salient features of the RCA program sent to Congress by the President. Most of the stories and editorials about the program focused on proposed spending levels, which were criticized as too low by several writers and editors. But regardless of Administration, critics of Federal soil conservation programs always equate dollars with a serious commitment to reducing soil erosion. Their formula for determining the value of any program is, "How much?"

We already spend a great deal of money for soil conservation programs in USDA—about $1 billion last year. That's a lot of money. But as Secretary Block said,

There's no way we're going to solve all the conservation problems by buying terraces on all the land that could use terraces, or building structures everywhere that we could build structures, because there isn't that much money in the Federal government or in the States.

"The real solution to erosion is going to be provided by the farmer—on his land. He's going to do it once he becomes fully convinced that conservation tillage and other improved tillage techniques are in his best interest. It will be in his interest because it keeps his land in place for his children. Or because if he wants to sell the land, it's going to sell for more. Or because he can make more money by using conservation tillage."

The Secretary went on to say that no matter what we do, some people will say you should spend more money. "Frankly," he said, "it's not the money as much as it is the attitude and approach and the commitment. That's what really counts." And in SCS, we can't add a thing to that statement.

REFERENCES

Conservation Tillage Information Center. 1982. National Survey of Conservation Tillage Practices. National Association of Conservation Districts, Washington, DC.

Conservation Tillage Information Center. 1983. National Survey of Conservation Tillage Practices. National Association of Conservation Districts, Washington, DC.

Pioneer Hi-Bred International, Inc. 1982. Soil Conservation Attitudes and Practices: The Present and the Future. Pioneer Hi-Bred International.

Office of the Federal Register, National Archives Records Service, General Services Administration, Public Papers of the Presidents of the United States. Ronald Reagan. 1982. Book II, p. 1628.

USDA. 1981a. Soil, Water, and Related Resources in the United States: Status, Condition, and Trends. RCA Appraisal Part I. Washington, DC.

----. 1981b. Soil, Water, and Related Resources in the United States: Analysis of Resource Trends. RCA Appraisal Part II. Washington, DC.

----. 1982. A National Program for Soil and Water Conservation: 1982 Final Program Report and Environmental Impact Statement. Washington, DC.

----. 1984. Soil and water conservation: Research and education progress and needs. SCS, Washington, DC.

List of Common and Scientific Names

Common Name	Scientific Name
Alfalfa	*Medicago sativa* L.
Barley	*Hordeum vulgare* L.
Beans	*Phaseolus vulgaris* L.
Bluegrass	*Poa pratensis* L.
Cabbage	*Brassica oleracea* L.
Corn	*Zea mays* L.
Cotton	*Gossypium hirsutum* L.
Cottonwood	*Populus* spp.
Fesque	*Festuca arundinacea* Schreb.
Flax	*Linum usitatissimum* L.
Grape	*Vitis vinifera* L.
Indigo	*Indigofera anil* L.
Lentils	*Lentilla lens* L.
Lettuce	*Lactuca sativa* L.
Mesquite	*Prosopis* spp.
Oats	*Avena sativa* L.
Onions	*Allium cepa* L.
Orchardgrass	*Dactylis glomerata* L.
Peanuts	*Arachis hypogaea* L.
Peas	*Pisum sativum* L.
Potato	*Solanum tuberosum* L.
Rose	*Rosa* spp.
Rye	*Secale cereale* L.
Sacaton, alkali	*Sporobolus airoides* Torr.
Sacaton, big	*Sporobolus wrightii* Munro.
Sorghum	*Sorghum bicolor* L.
Soybean	*Glycine max* L.
Sugarbeet	*Beta vulgaris* L.
Sugarcane	*Saccharum officinarum* L.
Sunflower	*Helianthus annuus* L.
Sweetclover	*Melilotus* spp.
Timothy	*Phleum pratense* L.
Tobacco	*Nicotiana tabacum* L.
Trefoil, birdsfoot	*Lotus corniculatus* L.
Vetch, hairy	*Vicia villosa* Roth.
Wheat	*Triticum aestivum* L.
Willow	*Salex* spp.

Common names of plants are used in this book. The editors compiled this list to identify them by scientific names.

Subject Index